MODERN SIZE-EXCLUSION LIQUID CHROMATOGRAPHY

MODERN SIZE-EXCLUSION LIQUID CHROMATOGRAPHY

Practice of Gel Permeation and Gel Filtration Chromatography

SECOND EDITION

André M. Striegel
Wallace W. Yau
Joseph J. Kirkland
Donald D. Bly

WILEY
A JOHN WILEY & SONS, INC., PUBLICATION

Published by John Wiley & Sons, Inc., Hoboken, New Jersey.
Published simultaneously in Canada.

Library of Congress Cataloging-in-Publication Data:

Modern size-exclusion liquid chromatography / André M. Striegel . . . [et al.].—
2nd ed.
p. cm.
Includes index.
ISBN 978-0-471-20172-4 (cloth)
1. Gel permeation chromatography. I. Striegel, André M., 1967-
QD272.C444Y38 2009
543′.8–dc22 2008036261

Printed in the United States of America

10 9 8 7 6 5 4 3 2 1

CONTENTS

FOREWORD

From the very beginning, synthetic polymers were so immensely useful that their development and commercialization followed almost immediately after their invention. The same was true for size-exclusion chromatography (SEC or gel-permeation chromatography, GPC) as a method for polymer characterization. SEC yielded eminently useful information (complete molar mass distributions) much more easily and more rapidly than did previous methods. In addition, the simultaneous development of high-pressure liquid chromatography for "small" (low-molar-mass) molecules meant that SEC soon became highly precise (i.e., repeatable), robust, and automatic. SEC was — and is — embraced by industry, and the greatest experts have learned the trade there through extensive personal experience or apprenticeship. In industry, publishing the tricks of one's trade is generally discouraged, and those who do publish are often frowned upon. If we combine this with the gigantic effort it takes to write a book, the very existence of the monumental first edition of *Modern Size-Exclusion Liquid Chromatography* by Wallace Yau, Jack Kirkland, and Donald Bly may be considered a near miracle.

I am looking through my copy for the umpteenth time. I had to retrieve it from the lab. It usually finds its way onto the desk of one of the Ph.D. students — a good sign. It is decorated with a number of yellow Post-it notes marking important passages — another good sign. It is remarkable how much this 30-year old book is being used. It is also understandable and even commendable that this is the case. Reading through the book is still a humbling experience. It makes me realize how many things I don't know. It is, as the subtitle reads, a guide to *The Practice of Gel-Permeation and Gel-Filtration Chromatography*. It is also much more. It is an excellent introduction to the principles of size-exclusion chromatography and of a great number of related subjects. It reflects vast knowledge, but more importantly, it displays a thorough understanding. It is a great book.

André Striegel has accepted the daunting task of rewriting the book. I hardly think it is possible to improve the quality of the text, as this would imply producing something greater than great. Maintaining the quality of the text is already a challenging ambition. Fortunately, he has been getting the best possible help through the active involvement of the original authors.

There is, however, one aspect in which the first edition of *Modern Size-Exclusion Liquid Chromatography* can be significantly improved. We do not need something greater than great, but we do need something more up to date than what was "modern" 30 years ago. Putting the word *modern* in the title entails the danger of a text not living up to expectations; it also provides encouragement for renewing the material. It has taken quite some time for someone to realize the latter implication, but here we are. The new edition describes twenty-first-century SEC. A large number of new developments are described and new chapters are added.

The most important question that remains is whether SEC is as important now as it was 30 years ago. Surely, measuring property distributions of polymers has become much more important, because there are many more different polymer formulations for many more applications. Moreover, both the formulations and the applications are increasingly sophisticated. We need very good tools to measure distributions. We need other liquid-chromatographic techniques to characterize other types of distributions, such as those describing the chemical composition or number and type of functional groups. In principle, we may use mass spectrometry to measure molar mass distributions and to obtain additional chemical information. However, for all but the narrowest distributions with the most homogeneous ionization profiles, SEC is still the preferred technique. In most cases this may easily remain true for the next 30 years.

We need SEC more than ever in research laboratories where polymers and materials are being investigated and applied; in material science, life science, food science, and many other fields. And perhaps most important, SEC remains an invaluable tool in industry. Chromatographers, polymer scientists, and many others should benefit from entering the era of truly *Modern Size-Exclusion Liquid Chromatography*.

Amsterdam
June 2008

PETER J. SCHOENMAKERS

PREFACE

Much has changed in size-exclusion chromatography (SEC) since publication of the first edition of this book in 1979. As a result, this second edition is an almost complete rewrite of the first, to take into account the many changes that have occurred in SEC since then. While the fundamentals of the method were well understood at the time, advances in both column and detector technology have been transformative. A half-century after its inception, the principal use of SEC remains determining the molar mass averages and distributions of natural and synthetic polymers. While this is still generally accomplished through the application of calibration curves, the popularization of robust, easy-to-use light-scattering photometers now allows users to measure these properties in absolute, calibrant-independent fashion. Similarly, the combination of multiple detection methods allows for obtaining a truly impressive variety of polymer properties. Indeed, the use of multidetector SEC has ushered in a new era of polymer analysis. A variety of chemical and physical properties of macromolecules can now be determined as a continuous function of molar mass, with many other parameters obtained from the same set of analyses.

The applicability of SEC has also extended into both smaller and more complex realms. Column advances, dictated by sample performance as well as legal requirements, have advanced the area of oligomeric SEC. Characterization and quantitation of polymers is now possible: in many cases, down to a single, monomeric repeat unit. Meanwhile, the complexity of real-world polymers and the need to understand their characteristics in order to optimize processing and end-use properties has served to further the development of polymer two-dimensional liquid chromatography (2D-LC). Because of its premier status in characterizing the molar mass distribution, SEC is virtually always one of the dimensions of separation.

In light of all of the above, we have tried to bring this book up to date on developments in multidetector, oligomeric, and two-dimensional analysis, among others. We place special emphasis on the wealth of information that can be obtained from a multidetector SEC experiment. As with the first edition, we have tried to keep this as much a "how to" book as a "why?" book. Because our main audience is the practitioner of SEC, we try to guide this scientist in designing experiments, carrying them out, and interpreting the results. No aspect of the technique is treated as a "black box," and we have tried to share with the reader as much of our (often hard-earned) practical experience as possible.

Those familiar with the first edition will see that detection techniques and structure–property relations are treated much more heavily in this second edition, as noted by the inclusion of individual chapters dealing with physical detectors (Chapter 9), chemical detectors (Chapter 10), and polymer architecture and dilute solution thermodynamics (Chapter 11). We also devote new, individual chapters to aqueous SEC (Chapter 12), to oligomeric SEC (Chapter 13), and to the role the technique plays in 2D-LC (Chapter 14). Techniques that are becoming more widespread, such as high-speed SEC, as well as niche methods such as inverse and recycle SEC, are treated in Chapter 15. Connections with rheology are explored in the final chapter (Chapter 16). This is the only chapter in the book that presupposes some familiarity by the reader with the subject matter.

The fundamental chapters dealing with retention (Chapter 2), band broadening (Chapter 3), and resolution (Chapter 4) have been updated where appropriate. The same is true of the chapters dealing with calibration methods and column technology (Chapters 8 and 6, respectively). Less emphasis is placed in this edition on column-packing techniques, for example, due to the fact that most current users employ commercially available columns. Also, the chapter on data handling in the first edition has been eliminated, due to the fact that the overwhelming majority of practitioners employ commercially available software packages for data acquisition and handling.

The original chapters on operating variables and laboratory techniques have been combined into the current chapter on experimental variables and techniques (Chapter 7). This combined chapter has also been updated with respect to a more refined understanding of analytical procedures, often due to advances in hardware. Here, the user is likely to find a good deal of practical information regarding experimental design (from selecting columns to selecting a solvent), sample preparation, execution of experiments, instrument care, and troubleshooting. For parameters that can have an adverse effect on results, we try to explain how these effects are brought about and what can be done to avoid or minimize them.

We would like to express our thanks to family, friends, and associates who have provided encouragement and support in bringing about the second edition of this book. We are particularly grateful to Professors John G. Dorsey and Peter J. Schoenmakers for their critical review of several chapters and for their insightful comments and suggestions. Any errors that remain are entirely our own fault!

Tallahassee, Florida
November 2008

A. M. Striegel
W. W. Yau
J. J. Kirkland
D. D. Bly

1

BACKGROUND

1.1 INTRODUCTION

This book is about modern size-exclusion chromatography (SEC). Size-exclusion chromatography is a liquid column chromatographic technique that sorts molecules according to their size in solution. The sample solution is introduced onto the column, which is filled with a rigid-structure, porous-particle column packing, and is carried by solvent (mobile phase) through the column. The size sorting takes place by repeated exchange of the solute molecules between the bulk solvent of the mobile phase and the stagnant liquid phase within the pores of the packing. The pore size of the packing particles determines the molecular size range within which separation occurs.

Throughout the book we use the term *size-exclusion chromatography*, which is meant to include the techniques originally (and sometimes still) referred to as gel permeation chromatography (GPC) and gel filtration chromatography (GFC). The term *GPC* was traditionally used when referring to analyses employing organic solvents and mobile phases for the separation. When aqueous solvents and mobile phases were used, the term *GFC* was used. Nowadays, gels are not always used as column packing materials. Also, one might employ aqueous solvents for separation one week and organic solvents the next, while the separation mechanism remains the same. Hence, the more general, all-inclusive term *size-exclusion chromatography* is preferred.

Modern Size-Exclusion Liquid Chromatography: Practice of Gel Permeation and Gel Filtration Chromatography, Second Edition
By André M. Striegel, Wallace W. Yau, Joseph J. Kirkland, and Donald D. Bly

1.2 HISTORY

Size-exclusion chromatography has its roots in conventional liquid chromatography (LC). Ettre's interesting paper, "The Development of Chromatography" [1], describes how David Talbot Day demonstrated in 1897 that crude oil fractions could be separated through pulverized fuller's earth. Unfortunately, Day did not properly interpret the phenomenon that was occurring and, because of this, the original founding of chromatography is often ascribed to Michael S. Tswett. In 1903–1906, Tswett clearly described the chromatographic separation of colored vegetable pigments in petroleum ether on calcium carbonate and recognized the method as a general process. From Tswett's early beginning, a large number of workers have continued to develop liquid chromatography into its present high-performance capabilities. Today, high-performance liquid chromatography is used widely in various forms within many scientific disciplines [2].

The origin of gel filtration chromatography is generally attributed to J. Porath and P. Flodin [3]. In 1959, these workers of the Institute of Biochemistry of the University of Uppsala (Porath) and of the Pharmacia Research Laboratories (Flodin), in Sweden, demonstrated that columns packed with cross-linked polydextran gels, swollen in aqueous media, could be used to size-separate various water-soluble macromolecules. The gels for this technique were made commercially available and have been used extensively for biomolecule separations in low-pressure systems. The technique has been reviewed by Porath [4] and, more recently, by Flodin [5].

In 1964, J. C. Moore of the Dow Chemical Company disclosed the use of cross-linked polystyrene "gels" for separating synthetic polymers soluble in organic solvents [6] and, with this event, conventional gel permeation chromatography (GPC) was born. It was recognized immediately that with proper calibration, gel permeation chromatography was capable of providing molar mass (M) and molar mass distribution (MMD) information for synthetic polymers. Because this information was difficult to obtain by other methods, gel permeation chromatography came rapidly into extensive use. The inception of GPC was reviewed some years later by Moore himself [7], while the background and applications of conventional early gel permeation chromatography have been reviewed by Bly [8].

The column packing materials used by Porath and Flodin for gel filtration and by Moore for gel permeation were particles of lightly cross-linked, porous, semirigid, organic-polymer networks. As such, they could be packed into columns and used with various mobile phases only at relatively low flow rates and pressures, less than 17 bar or 250 psi. At high pressures and flow rates, these packings collapse, and separations cannot be made. Because of these limitations, both conventional gel filtration chromatography and gel permeation chromatography are relatively slow techniques.

Modern, high-performance size-exclusion chromatography is a result of the development of small, more rigid porous particles for column packings. The first small particles introduced commercially for SEC were μ-Styragel (a trade name for microparticle cross-linked polystyrene gel) by Waters Associates, Milford, Massachusetts. Packed into efficient columns, these semirigid 10-μm particles withstand relatively high pressure (e.g., 2000 to 3000 psi) and provide performance approximately 10 times better than that of the macroparticle cross-linked polystyrene (e.g., 70 to 150 μm Styragel) widely used previously. Subsequent to the introduction of μ-Styragel, completely rigid inorganic-based particle packings were developed (Chapter 6). Unger et al. [9,10] and Kirkland [11,12] have described porous silica particles, and Sato et al. [13] have discussed porous alumina for SEC.

1.3 UTILITY OF SEC

For water-soluble macromolecules of biochemical origin, separation by size-exclusion chromatography is normally desired for one or more of the following reasons:

1. To prepare molecular fractions for characterization or further use
2. To serve as a method for desalting or buffer exchange (i.e., to act as a substitute for dialysis)
3. To estimate molar mass using calibration standards or an absolute method (e.g., light scattering)
4. To estimate molecular association constants:
   a. Complexes of small molecules with macromolecules
   b. Macromolecular aggregation

Many examples of these uses are presented throughout this book, especially in Chapter 12.

The utility of aqueous size-exclusion chromatography is illustrated in Figure 1.1, where the separation of a number of protein molecules is made in a matter of minutes. Traditionally, this analysis takes several hours to perform. A calibration relating the molar mass of carbohydrate-free globular proteins in water to their retention volume is shown in Figure 1.2. This calibration plot, which was obtained in a few hours, would have taken much longer to obtain by large-particle-based conventional gel filtration techniques. Reference 14 provides a good review of the size-exclusion chromatography separation of proteins in both denaturing and nondenaturing solvents.

It is well known that many macromolecules, both natural and synthetic, are polydisperse with respect to molar mass. This is the case for biopolymers such as cellulose and the starch fractions amylose and amylopectin [17] and for all synthetic polymers, which can range from being narrowly to broadly polydisperse. As seen in Figure 1.3, in addition to an MMD, macromolecules can possess distributions in a variety of chemical and physical properties, including branching (long- and short-chain), chemical heterogeneity, and polyelectrolytic charge. A generic example of how the distribution of several of these properties as a function of M may overlay the MMD of a polymer is shown in Figure 1.4.

The applications of polymers are often determined by the distributions of the chemical and physical properties present. The breadth of the MMD, for example,

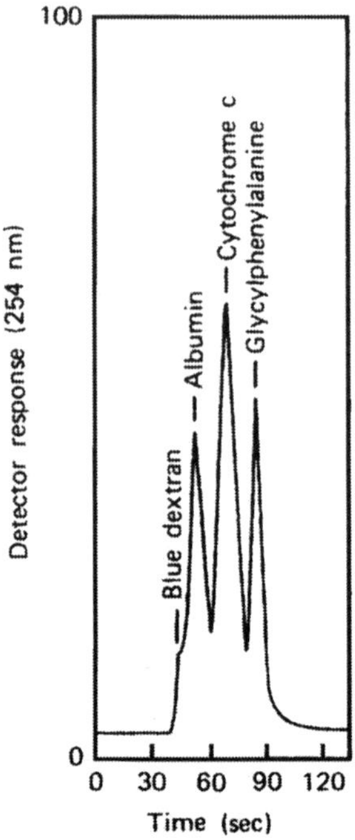

Figure 1.1 Chromatogram for size-exclusion chromatography of proteins. Column, 30 × 0.41 cm stainless steel packed with 5 to 10-μm Glycophase G/CPG, 100-Å pore diameter; temperature, 25°C; velocity, 0.7 cm/s at 2700 psi; mobile phase, 0.1 *M* KH_2PO_4 (pH 6). (Reprinted with permission from Ref. 15.)

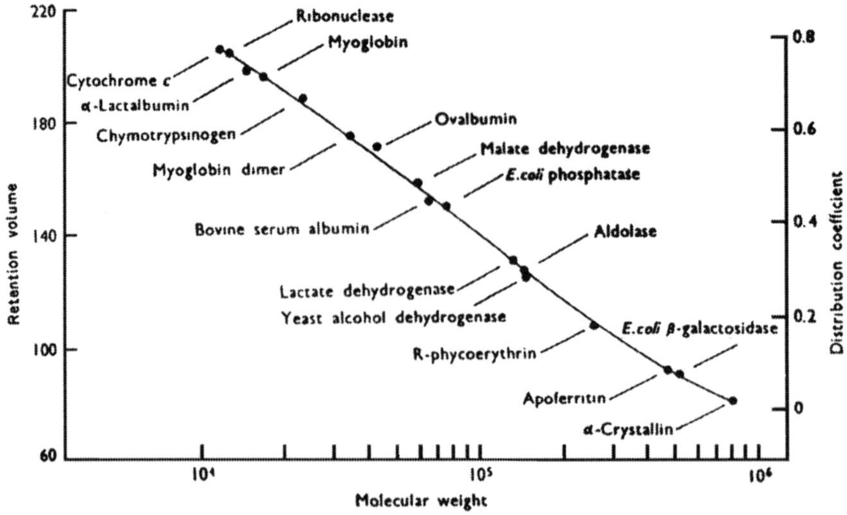

Figure 1.2 Relationship between molar mass and retention volume for certain proteins in water. (Reprinted with permission from Ref. 16.)

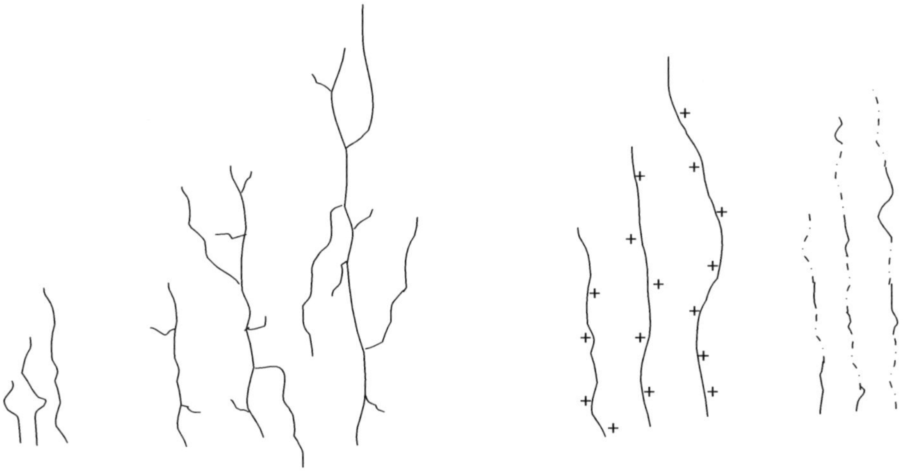

Figure 1.3 Examples of macromolecular distributions. From left: molar mass, long- and short-chain branching, polyelectrolytic charge, chemical heterogeneity.

can affect the elongation and tensile strength of the macromolecule and adhesive properties of the final product; long-chain branching has a profound impact on such rheological properties as the viscosity of melts and solutions and the shear strength of formed products; chemical heterogeneity can affect the toughness, brittleness, and biodegradability of plastics. Table 1.1 lists the types of macromolecular property

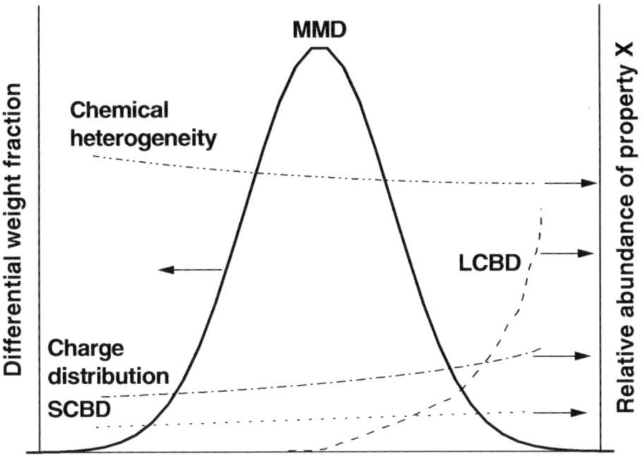

Figure 1.4 Distribution of chemical and physical properties. Property *X* refers to LCB, SCB, charge, and % co-monomer. MMD, molar mass distribution; LCBD, distribution of long-chain branches as a function of *M*; SCBD, distribution of short-chain branches as a function of *M*; charge distribution, distribution of polyelectrolytic charge as a function of *M*; chemical heterogeneity, distribution of the percentage of one component of a copolymer as a function of copolymer *M*.

Table 1.1 Macromolecular distributions: their measurement and end-use effects[a]

Macromolecular Property	Representative End-Use Properties Affected	Separation Method Used for Determination[b]
Molar mass	Elongation, tensile strength, adhesion	SEC, FFF, HDC, TGIC, CEC, SFC, MALDI-MS, rheology
Long-chain branching	Shear strength, tack, peel, crystallinity	SEC-MALS, SEC-VISC, rheology enzymology
Short-chain branching	Haze, stress-crack resistance, crystallinity	SEC-IR, SEC-NMR, TREF,[c] CRYSTAF,[c] enzymology
Cross-linking	Gelation, vulcanization, surface roughness	SEC-MALS, SEC-VIS, rheology
Architecture	Flow modification, diffusion, encapsulation	SEC-MALS-QELS-VISC
Tacticity	Crystallinity, anisotropy, solubility	SEC-NMR, TGIC, LCCC,
Chemical composition	Morphology, miscibility, solubility	GPEC, TGIC
Chemical heterogeneity	Toughness, brittleness, biodegradability	SEC-spectroscopy/ spectrometry, LCCC, PFC
Chemical composition vs. molar mass	Mechanical properties, blending, plasticization	2D-LC (e.g., SEC-GPEC)
Block sequence	Dielectric properties, reactivity, miscibility	SEC-spectroscopy, 2D-LC (e.g., PFC-SEC)
Base-pair sequence	Genetic code, heredity, sequencing, mutations	Automated DNA sequencing, MALDI-MS
Polyelectrolytic charge	Flocculation, transport, binding of metals	SEC-conductivity
Particle size	Packing, drag, friction, mixing	FFF, HDC, PSDA, sieving

Source: Ref. 20.

[a]Many techniques require a concentration-sensitive detector (e.g., a differential refractometer), not included here for simplicity.

[b]SEC, size-exclusion chromatography; FFF, field-flow fractionation; HDC, hydrodynamic chromatography; TGIC, temperature-gradient interaction chromatography; CEC, capillary electrokinetic chromatography; SFC, supercritical fluid chromatography; MALDI-MS, matrix-assisted laser desorption/ ionization mass spectrometry; MALS, multiangle light scattering; VISC, viscometry; IR, infrared spectroscopy; NMR, nuclear magnetic resonance spectroscopy; TREF, temperature-rising elution fractionation; CRYSTAF, crystallization fractionation; QELS, quasielastic (dynamic) light scattering; LCCC, liquid chromatography at the critical condition; GPEC, gradient polymer elution chromatography; PFC, phase fluctuation chromatography; 2D-LC, two-dimensional liquid chromatography; PSDA, particle-size distribution analyzer.

[c]For crystalline polymers only.

distributions that can exist or coexist in polymers, how these properties affect both processing and end use, and the types of separation methods used for measuring these distributions. As can be seen, SEC is the most widely represented technique in the table, especially when combined with a number of analytical techniques that can serve as detection methods: light scattering, viscometry, mass spectrometry, conductivity, spectroscopic methods, and so on [18,19]. The power of multidetector SEC will be a recurrent theme in this book.

Several nonseparation techniques are also listed in the last column of Table 1.1. These include enzymology, matrix-assisted laser desorption/ionization mass spectrometry (MALDI-MS), rheology, and sequencing. All these provide information which can, in select cases, closely complement that obtained by the separation methods. For example, the polysaccharide pullulan can possess an MMD, determined most accurately and conveniently using SEC with both a concentration-sensitive detector (e.g., a differential refractometer) and a static light-scattering detector [21]. Pullulan is composed of maltotriose units joined to each other via α-(1 $\rightarrow$ 6) linkages, but pullulan also possesses about 6.6% maltotetraose units. Whether these maltotetraose units were distributed uniformly and linearly along the pullulan backbone, were located at the chain ends, or were arranged along the backbone such as to form short-chain one- to three-glucose unit branches was not known originally. The matter was resolved using enzymatic analysis, which showed that the maltotetraose units were distributed along the pullulan backbone and were linked terminally (i.e., without resulting in short-chain branching) [22].

1.4 MOLAR MASS AVERAGES AND MOLAR MASS DISTRIBUTION

Size-exclusion chromatography normally is used as an analytical procedure for separating molecules by their difference in size and to obtain molar mass averages (M_n, M_w, M_z) or information on the molar mass distribution (MMD) of polymers. At times, however, it is also used for preparing various molar mass fractions for further use (Chapter 15). The raw-data SEC curve is a molecular size-distribution curve. If a concentration-sensitive detector is used, the SEC curve is really a size distribution curve in weight concentration. With calibration (Chapter 8) or static light-scattering detection (Chapter 9), the raw data are converted to a molar mass distribution curve and the respective molar mass averages can be calculated. Because determining molar mass averages and distributions remains the principal use of SEC, we present here a short overview for polymers of the meaning of molar mass distribution and molar mass averages (M_n, M_w, and M_z).

Various reaction mechanisms are employed for the synthesis of high polymers. Examples are the addition reaction to form polyethylene from ethylene, and the condensation polymerization of hexanedioic acid and hexamethylenediamine to form the polyamide (nylon). During the course of a polymerization reaction, a large quantity of polymer chains are initiated, grow, and then are terminated (i.e., stop growing). The number and length (or weight) of the polymeric chains formed during the reaction vary with the reaction mechanism and the reaction conditions employed. At times, the distribution of these chains is accurately predictable from statistical considerations; at other times (nonequilibrium processes), a priori predictions are not accurate. In either case SEC can be used to determine experimentally the distributions and the molar mass averages of the polymer formed.

One convenient way of measuring the "average" chain length in a polymer sample provides a quantity known as M_n, the number-average molar mass. M_n is historically significant because for many years it has been a characterizing value obtained directly in the laboratory by colligative property methods. M_n also has been correlated with a number of polymer physical properties (Table 1.2) and is defined as the mass of the sample in grams $\sum W_i$, or $\sum N_i M_i$, divided by the total number of chains present, N, which is $\sum N_i$. Here W_i and N_i are the weight and number of molecules of molar mass M_i, respectively, and i is an incrementing index over all molar mass present. Thus,

$$M_n = \frac{\sum N_i M_i}{\sum N_i} = \frac{\sum W_i}{\sum (W_i/M_i)} \tag{1.1.a}$$

and from SEC,

$$M_n = \frac{\sum_{i=1}^{N} h_i}{\sum_{i=1}^{N} (h_i/M_i)} \tag{1.1.b}$$

where h_i is the SEC curve height at the ith volume increment and M_i is the molar mass of the species eluted at the ith retention volume. The equation assumes that h_i is proportional to solute concentration and M_i is sampled in equal volume increments.

Another molar mass average that can be correlated with physical properties is the weight-average molar mass, M_w, which is determined in the laboratory from static light scattering (Section 9.3) and ultracentrifugation measurements as well as from SEC. It is defined as

$$M_w = \frac{\sum N_i M_i^2}{\sum N_i M_i} = \frac{\sum W_i M_i}{\sum W_i} \tag{1.2.a}$$

and from SEC,

$$M_w = \frac{\sum_{i=1}^{N} (h_i M_i)}{\sum_{i=1}^{N} h_i} \tag{1.2.b}$$

Some observations about the relative properties of M_n and M_w have been made [15]. The value of M_w is always larger than M_n, except that the values are identical for a monodisperse system. The ratio M_w/M_n, termed the *molar mass polydispersity* or, more simply, the *polydispersity*, is a measure of the breadth of the polymer molar mass distribution. M_w/M_n, is equal to unity for monodisperse systems, has a value of 2 for a Flory most probable distribution, and is exceedingly large for a

Table 1.2 Examples of effect of molar mass or molar mass distribution on various polymer properties

A. General Correlations[a]

	Tensile Strength	Elongation	Yield Strength	Toughness	Brittleness	Hardness	Abrasion Resistance	Softening Temperature	Melt Viscosity	Adhesion	Chemical Resistance	Solubility
Increase the molar mass	+	+	+	+	+	+	+	+	+	−	+	−
Narrow the molar mass distribution	+	−	−	+	−	−	+	+	+	−	+	0

B. Specific Correlations

Polymer	Property	Correlation
Poly(11-hydroxyundecanoic acid),[b] a polyester	Fiber and film strength, polymer solubility	Strength increases with increasing M_n while solubility decreases with increasing M_n
Polyesters from ω-hydroxydecanoic acid[c]	Fiber strength	Increases with increase in M_n
Nylon 6,6[d]	Fiber tenacity	Increases with M_n
Styrene–butadiene rubber[e]	Die swell	Increases with increase in MMD
Poly(methyl methacrylate)[f]	Sensitivity as an electron resist	Increases with higher M_n and increases with narrower MMD
Polyalkylacrylates[g]	Solution viscosity and shear stability index	Decrease with a decrease in M_w caused by shearing
Polyolefins[g]		
Polystyrenes[g]		
Polyethylene (PE)[h]	Strength, toughness	Increase with increasing M_n
	Melt fluidity, film friction	Decrease with increasing M_n
PE[h]	Strength, toughness	Increase with decreasing M_w/M_n
	Fluidity (ease of processing)	Decrease with decreasing M_w/M_n
Epoxy resins[i]	"Acceptance quality" of circuit boards	Overall SEC curve (MMD) profile
Cellulose triacetate[j]	Density (d) and shrinkage (s) of films	d increases with MMD, s decreases with MMD

Source: (A) Reprinted in part from E. A. Collins, J. Bareš, and F. W. Billmeyer, Jr., *Experiments in Polymer Science*, Wiley, New York, 1973, p. 312, with permission.

[a] Profile of performance property dependence on molecule–structure parameters for typical parameters. Key: +, property goes up; −, property goes down; 0, little change.

[b] V. V. Korshak and S. V. Vinogradovia, *Polyesters*, translated from the Russian by B. J. Hazzard, Pergamon Press, New York, 1965, p. 310.

[c] W. H. Carothers and F. J. van Natta, *J. Am. Chem. Soc.*, **55**, 4715 (1933).

[d] J. Zimmerman, *Text. Manuf.*, **101**, 19 (1974).

[e] W. Mills and F. Giurco, *Rubber Chem. Technol.*, **49**, 291 (1976).

[f] J. H. Lai and L. Shepherd, *J. Appl. Polym. Sci.*, **20**, 2367 (1976).

[g] D. E. Hillman, H. M. Lindley, J. I. Paul, and D. Pickles *Br. Polym. J.*, **7**, 397 (1975).

[h] F. W. Billmeyer, Jr., *Textbook of Polymer Science*, Wiley, New York, 1972, p. 382.

[i] *Ind. Res.*, Jan. 1977, p. C1.

[j] N. P. Zakurdaeva and T. A. Ivanova, *Plast. Massy*, **9**, 68, 1976; *Chem. Abstr.*, **85**: 193430b.

cross-linked polymer. High-molar-mass species particularly influence the value of M_w, whereas the value obtained for M_n is influenced more by species at the lower end of the molar mass distribution. If *equal weights* of molecules with $M = 10{,}000$ and $M = 1{,}000{,}000$ are mixed, $M_w = 55{,}000$ and $M_n = 18{,}200$; if *equal numbers* of each kind of molecule are mixed, $M_w = 92{,}000$ and $M_n = 55{,}000$ [23].

The molar mass distribution (MMD) can be expressed graphically in integral form as the cumulative weight fraction or cumulative number fraction versus molar mass (M) (or X, the number of repeat units in the chain). The MMD may also be in the differential form as the weight fraction or number fraction versus M (or X). As used here, M is a generic term for the molar mass, which is obtained by multiplying the repeat unit M by the number of repeat units X. The true MMD can be deduced from the SEC curve only via careful application of calibration curves or by the use of static light-scattering detection. Figure 1.5 shows the differential MMD of a sample of brominated polystyrene, PSBr, as determined by SEC with both differential refractive index and static multiangle light-scattering detection (both detection methods are described in Chapter 9) [24–26]. Marked on the curve are the number-, weight-, and z-averages of the molar mass (M_z is described below). It is worth noting the broad molar mass range covered by this sample's MMD, extending from 2×10^4 to 5×10^6 g/mol.

By proper selection of columns and other experimental conditions, the molar mass range accessible by SEC can be very large. Figure 1.6 shows a calibration curve based on narrow polydispersity linear polystyrene (PS) standards. The molar

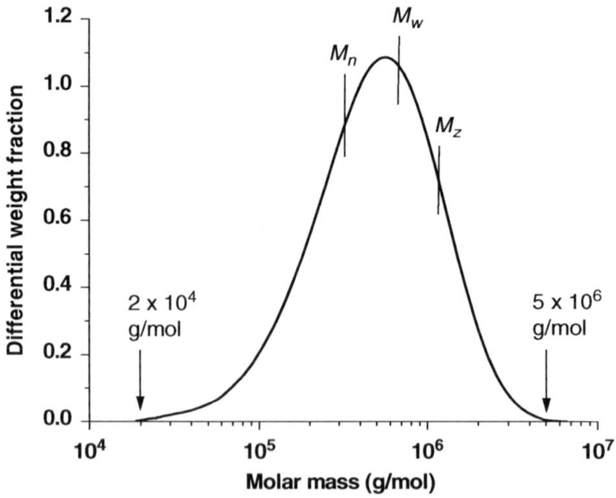

Figure 1.5 Molar mass averages and distribution of brominated polystyrene, PSBr. MMD and molar mass averages determined by SEC with differential refractive index and static multiangle light-scattering detection. Solvent, DMAc/0.5% LiCl; temperature, 35°C; flow rate, 1 mL/min; columns, three PSS GRALlinear 10-μm columns and one PSS GRAL10000 10-μm column, preceded by a guard column. $M_n = 3.26 \times 10^5$ g/mol, $M_w = 6.74 \times 10^5$ g/mol, $M_z = 1.17 \times 10^6$ g/mol, $M_w/M_n = 2.07$. (Adapted from Ref. 26.)

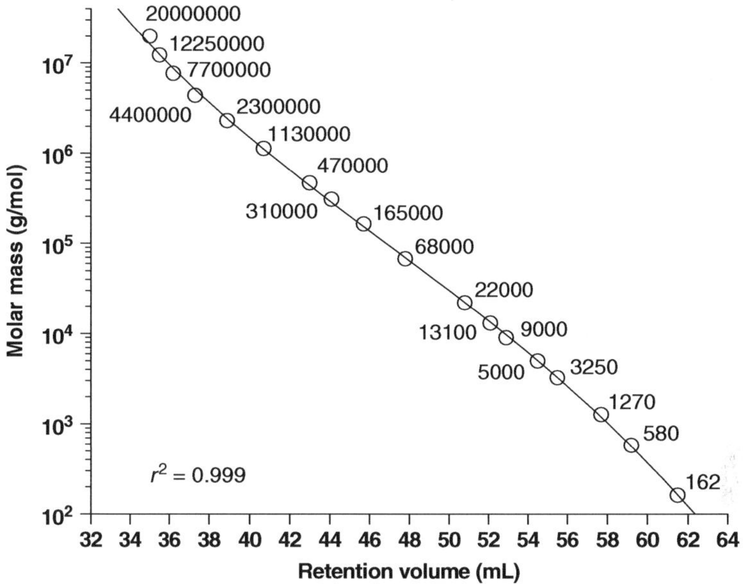

Figure 1.6 Separation range of SEC: elution of linear polystyrene standards. Circles denote average elution time of triplicate injections of each narrow polydispersity PS standard, with error bars substantially smaller than data markers and therefore not shown. Numbers next to markers denote the peak-average molar mass, M_p, of each standard in g/mol. Solid line is a third-order fit to the data, with $r^2 = 0.999$. Solvent, 1,2,4-trichlorobenzene (with 1.5 mg/mL Santonox); temperature, 135°C; columns, PLgel Mixed A; flow rate, 0.1 mL/min; detector, DRI. (Reprinted with permission from Ref. 27.)

mass range covered by this curve spans over five orders of magnitude, from 162 to 2×10^7 g/mol!

Historically, before SEC became available, the MMD curves were very difficult to obtain. Examples of some of the various M and MMD parameters are shown in Figures 1.7 to 1.9, which represent theoretical plots for condensation polymers (e.g., nylon) and other distribution functions. In the figures, the extent of reaction p is defined as the mole fraction (of all functional groups available for polymerization both in monomer and in growing polymer chains) that has reacted at various times. The great utility of M_n, M_w, and the MMD is shown in Table 1.2, where various correlations with physical properties for synthetic polymers are compiled. Calculations of M_n, M_w, M_z, and MMD are performed routinely by most commercial SEC software.

It is not always necessary to calculate the molar mass averages or MMD to obtain useful information about a sample from the SEC curve. Simple inspection of chromatograms often reveals important information. For example, Figure 1.10 shows raw-data chromatograms of two batches of supposedly the same epoxy resin. Inspection indicates immediately, however, that batch 1443 is missing a significant amount of material on the low-molar-mass side of the main peak. This absence of certain material could account for differences in sample properties. There also might be

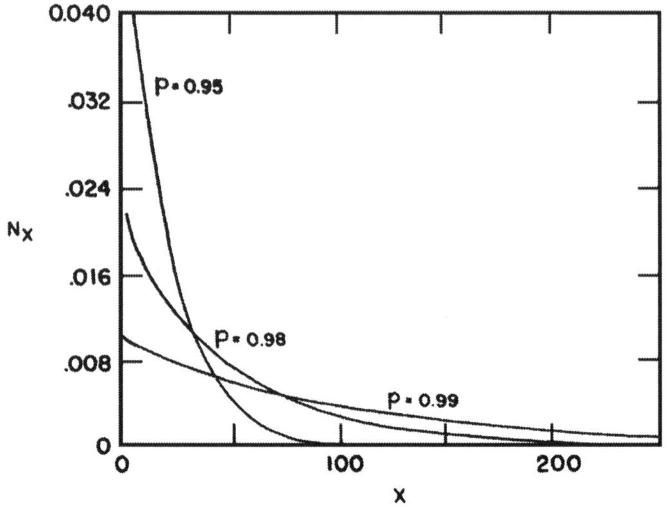

Figure 1.7 Mole fraction distribution of chain molecules in linear condensation polymers for several extents of reaction p. (Reprinted with permission from Ref. 28.)

differences in M_n or M_w between these lots, but the values obtained would not indicate where differences occur in the overall MMD.

As mentioned above, values of M_w/M_n have often been used traditionally to express the breadth of the molar mass distribution. Figure 1.11 shows, however, that three different distribution curves can provide identical values of M_n, M_w, and M_z.

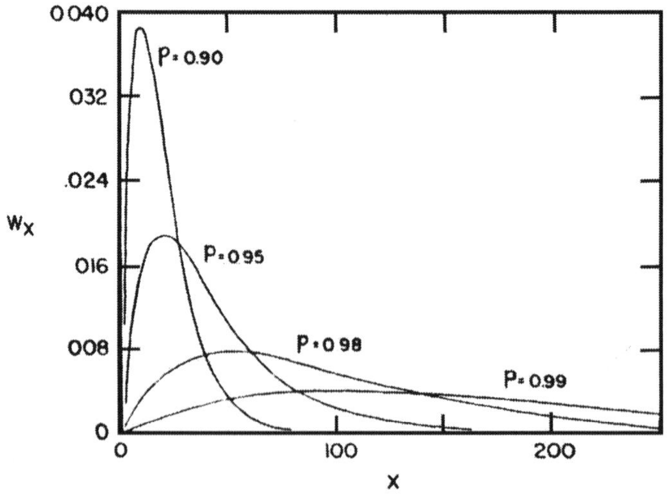

Figure 1.8 Weight fraction distributions of chain molecules in linear condensation polymers for several extents of reaction p. (Reprinted with permission from Ref. 28.)

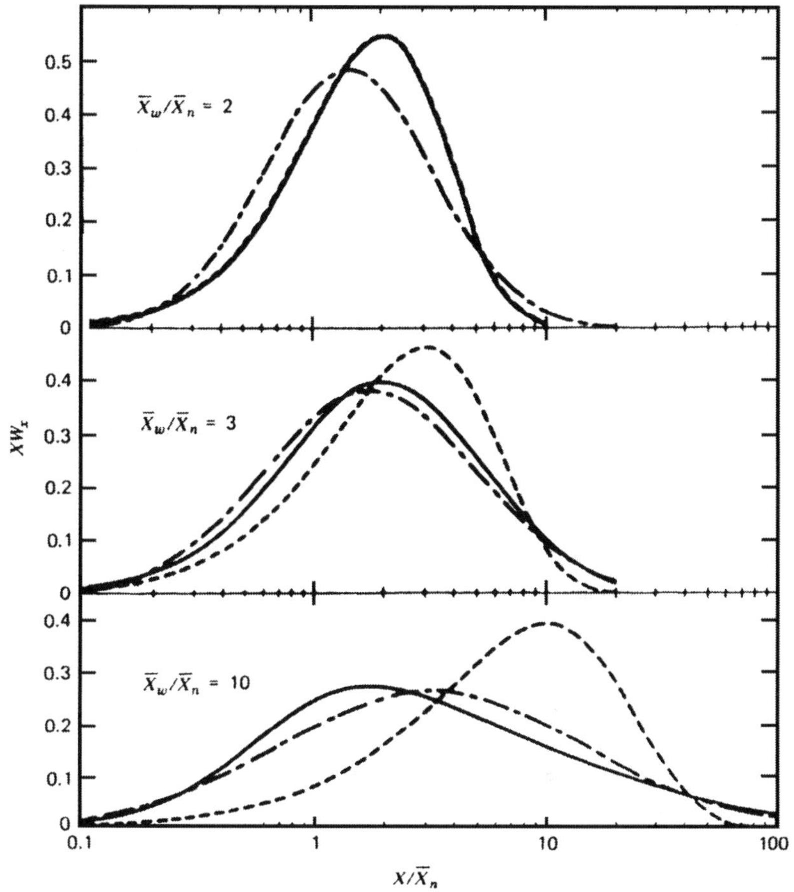

Figure 1.9 Theoretical size-exclusion chromatograms for three values of X_w/X_n according to various distribution function formulations. Dashed-dotted curves, logarithmic normal; dashed curves, Schulz–Zimm; solid curves, modified Stockmayer; X_w, weight-average chain length; $\bar{x}_n$, number-average chain length. (Reprinted with permission from Ref. 29.)

The parameter M_z is related to a higher moment of the distribution defined by

$$M_z = \frac{\sum N_i M_i^3}{\sum N_i M_i^2} \tag{1.3}$$

At times, M_z is correlated to polymer processing properties, in particular to properties such as flex life and stiffness that are governed by the longest chains in the MMD. If molar mass values were obtained for these three distributions by light scattering, osmometry, or centrifugation, all the polymers would have identical M_n or M_w or M_z values and identical polydispersity M_w/M_n. Yet, clearly, the distributions are not alike, and physical properties of materials fabricated from these polymers

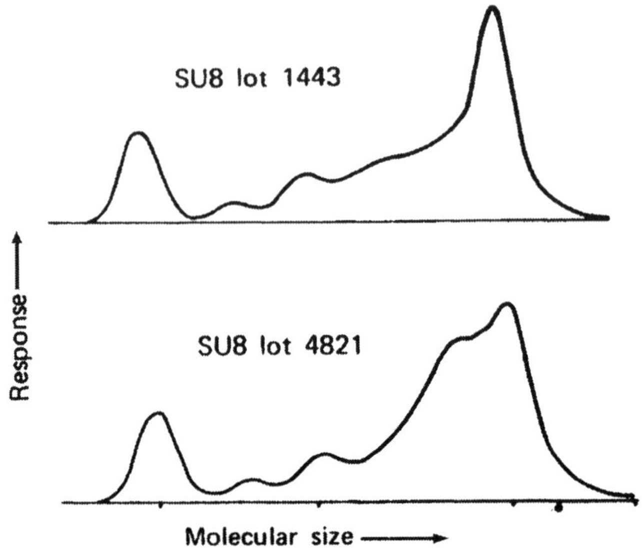

Figure 1.10 Comparison of two lots of SU-8 resin by SEC showing batch variations. (Reprinted with permission from Ref. 30.)

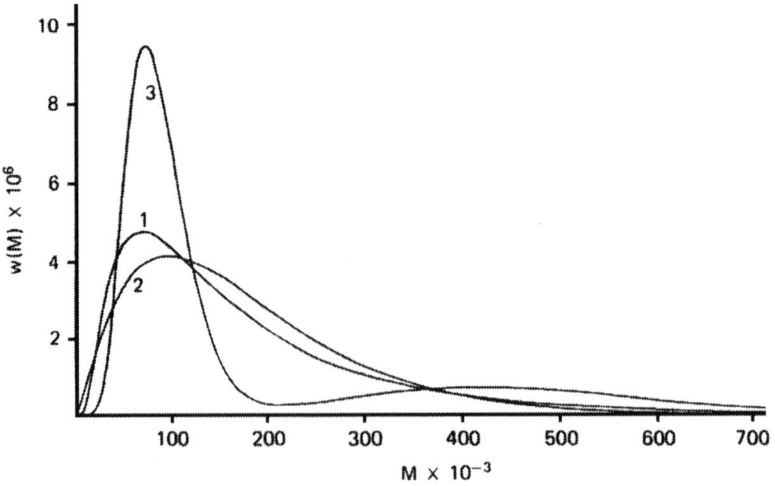

Figure 1.11 Three differential weight distribution curves corresponding to identical values of M_n, M_w, and M_z. Curve 1 is a logarithmic normal function; curves 2 and 3 are sums of two exponential functions. (Reprinted with permission from Ref. 31.)

could be different. This information illustrates the utility of the entire MMD profile as provided by SEC.

Two other molar mass averages are used in this book and will be encountered in the literature and in daily use. These are the peak-average molar mass, M_p, and the viscosity-average molar mass, M_υ or M_η. The peak-average molar mass is simply the molar mass of the slice eluting at the peak apex in an SEC chromatogram. It is used primarily in assigning molar masses when constructing peak-position calibration curves based on narrow MMD standards (see Section 8.2).

The viscosity-average molar mass is defined as

$$M_\upsilon = M_\eta = \left[\frac{\sum_{i=1}^{N} h_i (M_i)^a}{\sum h_i}\right]^{1/a} \tag{1.4}$$

The term a corresponds to the exponent in the Mark–Houwink equation (Equation 8.2). Molding properties and polymer extrudability have often been found to correlate with M_υ. The viscosity–average molar mass is unlike M_n, M_w, and M_z. The latter three averages are "absolute" in the sense that, properly measured, their values are independent of the solvent–temperature conditions of analysis. M_υ, however, will depend on experimental conditions; the latter, as we will see in Chapters 8 and 9, enter the equation through the a term.

1.5 STRUCTURE OF THE BOOK

The next three chapters (Chapters 2 to 4) serve to introduce the reader to the fundamental chromatographic aspects of size-exclusion chromatography: retention, band broadening, and resolution. The treatment of these topics is rather detailed in the hopes of establishing a strong foundation on which to design and optimize separations. In Chapter 5 we describe the various components of an analytical SEC system, concentrating on the hardware that precedes the column. The latter is the focus of Chapter 6, where we describe the types of columns and column packing materials available and how packing materials are synthesized and columns packed. Chapter 7 provides a lengthy discussion of experimental variables, an extremely practical discussion about most of the considerations that an actual SEC practitioner must take into account to obtain reliable, reproducible data in a safe manner, all the while ensuring that the equipment is taken care of.

In the chapter on calibration techniques, Chapter 8, we differentiate between the various types of calibration effected using narrow polydispersity standards, giving the relative advantages and disadvantages of each. We also discuss calibration methods based on broad MMD standards, the accuracy and linear ranges of the various calibrations, and recent developments regarding band-broadening corrections for certain types of calibration methods.

Chapters 9 and 10 deal with physical and chemical detection methods, respectively. The discussion in Chapter 9 revolves mostly around the methods themselves.

In Chapter 10 we also describe the type of information obtained from the chemical detection methods, as these methods are likely to be more familiar to the reader than the physical methods from Chapter 9. Because of this, we devote Chapter 11 to the architectural and thermodynamic information obtainable when a multiplicity of physical detection methods is used. Indeed, the use of multiple detection methods in SEC has transformed the technique over the last two decades [18,19,32].

Because of the types of analytes that are water-soluble (e.g., proteins and peptides) and the types of effects that can be encountered when using water as a solvent and chromatographic mobile phase, we have dedicated one chapter (Chapter 12) to aqueous SEC. As discussed, not only is aqueous SEC used for proteins and peptides but also for analyzing a variety of polysaccharides and synthetic polymers, including dendrimers and polyelectrolytes.

Like the use of multiple detection methods, another area where SEC has experienced tremendous growth in the last decade is in the analysis of oligomers [32]. This is due to the great advances in column technology for oligomeric analysis, driven in many ways by regulatory requirements. Oligomeric SEC is the subject of Chapter 13.

Two current areas of growth for SEC are two-dimensional (2D) chromatography and high-speed analysis [32]. Understanding the physicochemical composition of complex polymers is not always straightforward, but is vital to optimizing the processing and end use of materials. This "deformulation" of a material is best done using more than one separation dimension. The capability of SEC to separate analytes based on size (which can then be related to molar mass) affords it a preeminent role in 2D-LC macromolecular analysis, the subject of Chapter 14.

High-speed SEC analysis, vital for high-throughput screening, for combinatorial research, and to meet the increasing quality assurance and quality control (QA/QC) demands of industrial production, is treated in Chapter 15. We also discuss a number of other "special techniques," niche methods such as recycle, inverse, vacancy, and differential SEC, as well as more widespread applications such as preparative SEC and size-exclusion electrochromatography.

In the final chapter, Chapter 16, we look at high-temperature SEC (used primarily, although not exclusively, in the study of polyolefins) and at connections between SEC and rheology. This chapter distinguishes itself from the others in that some familiarity by the reader with rheological methods and terminology is assumed. The particular connections we explore are the rheological determination of the MMD of polymers, which is a primary application of SEC; how to obtain rheological properties of polymers from SEC measurements; and how SEC and rheology combine in the study of dilute oligomer solutions. New theories, based on a generalized M-averaging concept, are developed to help to close the gap between SEC and rheology measurements.

REFERENCES

1. L. S. Ettre, *Anal. Chem.*, **43**, 20A (1971).
2. L. R. Snyder and J. J. Kirkland, *Introduction to Modern Liquid Chromatography*, 2nd ed., Wiley, New York, 1979.

3. J. Porath and P. Flodin, *Nature*, **183**, 1657 (1959).
4. J. Porath, *Lab. Pract.*, **16**, 838 (1967).
5. P. Flodin, *Polym. Eng. Sci.*, **38**, 1220 (1998).
6. J. C. Moore, *J. Polym. Sci. A*, **2**, 835 (1964).
7. J. C. Moore, *J. Polym. Sci. C*, **21**, 1 (1968).
8. D. D. Bly, in *Physical Methods in Macromolecular Chemistry*, Vol. **2**, B. Carroll ed., Marcel Dekker, New York, 1972, Chap. 1.
9. J. Probst, K. Unger, and H. J. Cantow, *Agnew. Makromol. Chem.*, **35**, 177 (1974).
10. K. Unger, R. Kern, M. C. Ninou, and K. F. Krebs, *J. Chromatogr.*, **99**, 435 (1974).
11. J. J. Kirkland, *J. Chromatogr. Sci.*, **10**, 593 (1972).
12. J. J. Kirkland, *J. Chromatogr.*, **125**, 231 (1976).
13. S. Sato, Y. Otaka, N. Baba, and H. I. Iwasaki, *Bunseki Kagaku*, **22**, 673 (1973).
14. R. C. Montelaro, in *Aqueous Size-Exclusion Chromatography*, P. L. Dubin, ed., Elsevier, Amsterdam, 1988, Chap. 10.
15. S. H. Chang, K. M. Gooding, and F. E. Regnier, *J. Chromatogr.*, **125**, 103 (1976).
16. P. Andrews, *Br. Med. Bull.*, **22**, 109 (1966).
17. A. M. Striegel and J. D. Timpa, *Carbohydr. Res.*, **267**, 271 (1995).
18. A. M. Striegel, ed., *Multiple Detection in Size-Exclusion Chromatography*, ACS Symp. Ser. 893, American Chemical Society, Washington, DC, 2005.
19. A. M. Striegel, *Anal. Chem.*, **77**, 104A (2005).
20. A. M. Striegel, in Ref. 18, Chap. 1.
21. A. M. Striegel and J. D. Timpa, in *Strategies in Size Exclusion Chromatography*, ACS Symp. Ser. 635, M. Potschka and P. L. Dubin, eds., American Chemical Society, Washington, DC, 1996, Chapter 20.
22. B. J. Catley and W. J. Whelan, *Arch. Biochem. Biophys.*, **143**, 138 (1971).
23. F. W. Billmeyer, Jr., *Textbook of Polymer Science*, 3rd ed., Wiley, New York, 1984.
24. A. M. Striegel, *Anal. Chem.*, **74**, 3013 (2002).
25. A. M. Striegel, *Polym. Int.*, **52**, 1863 (2003).
26. A. M. Striegel, in Ref. 18, Chap. 4.
27. A. M. Striegel, unpublished results.
28. P. J. Flory, *Chem. Rev.*, **39**, 137 (1946).
29. H. L. Berger and A. R. Shultz, *J. Polym. Sci. A*, **3**, 3643 (1965).
30. T. D. Zucconi and J. S. Humphrey, *Polym. Eng. Sci.*, **16**, 11 (1976).
31. R. Koningsveld, *Adv. Polym. Sci.*, **7**, 1 (1970).
32. A. M. Striegel, *Anal. Bioanal. Chem.*, **390**, 303 (2008).

2

RETENTION

2.1 INTRODUCTION

In column chromatography, sample components migrate through the column at different velocities and elute separately from the column at different times. As a solute moves along with the carrier fluid (mobile phase), it is at times held back momentarily either by the surface of the column packing, by a contained stagnant phase of the column packing (stationary phase), or by both. Since solutes move only when they are in the mobile phase, the distribution of solute molecules between the mobile and the stationary phases determines the average solute migration velocity. Molecules that favor the stationary phase migrate more slowly and elute from the column later.

All forms of chromatography are therefore simply differential migration separation processes where sample components are selectively retained to different degrees by a stationary phase (an exception to this is hydrodynamic chromatography, discussed in Section 2.6.2). The mobile phase in the chromatographic process normally is a gas (gas chromatography), a liquid [liquid chromatography (LC)], or a supercritical fluid (supercritical fluid chromatography). The LC stationary phase can be a solid surface, as for liquid–solid chromatography, or a stagnant liquid, as for liquid–liquid chromatography (LLC). According to the mechanism of solute retention, LC methods can be classified into four categories: ion-exchange, adsorption, liquid-partition, and size-exclusion chromatography (SEC). As we will see, the retention mechanism in SEC is virtually unique in that solute distribution between phases is established by entropy instead of enthalpy differences.

Modern Size-Exclusion Liquid Chromatography: Practice of Gel Permeation and Gel Filtration Chromatography, Second Edition
By André M. Striegel, Wallace W. Yau, Joseph J. Kirkland, and Donald D. Bly

Since both the basic separation mechanism and the information obtainable from SEC are quite different from those of other LC methods, technologies have been developed specially for SEC in determining polymer molar mass (M). Many early advances in SEC were made by biochemists and polymer chemists. As a result, nomenclature and conventions derived for SEC are often not consistent with those for other LC methods. Since the general LC instrumentation and column techniques have become an integral part of SEC, it is useful that practitioners be acquainted with the equivalences and the differences in SEC and general LC terminology. The general nomenclature and conventions for LC peak retention are reviewed in the following section, and special SEC peak retention terminology is discussed in Section 2.3.

2.2 SOLUTE RETENTION IN LC

There are four ways of reporting conventional LC peak retention: retention time, t_R; retention volume, V_R; retention factor, k'; and solute distribution coefficient, K_{LC}. The term t_R can be measured most directly by experiment, but it is the least definitive parameter for identifying sample components. On the other hand, the term K_{LC} is the most difficult parameter to measure, but it is the most fundamental quantity for describing peak retention.

The simple experimental value of t_R, measured by the time required for a peak to elute from the column following sample injection (see the bottom of Figure 2.1), is useful only for comparing peaks that have appeared in the same chromatogram. The value of t_R is sensitive to changes in experimental conditions, such as flow rate and the specific columns used; therefore, it is not very specific for defining sample components. The retention volume V_R is a more fundamental quantity in that it accounts for flow-rate differences. To calculate V_R, the mobile-phase volume flow rate, F, must be known as well as the t_R values, since $V_R = Ft_R$. While peak retention reported as V_R is not subject to flow rate change, it can still vary with differences in column size and instrumental dead volume. Such variations are inherently compensated for with the more basic retention parameter, k'. Physically, k' represents the ratio of the *weight of solute* in the stationary phase to that in the mobile phase. Thus, the weight fraction of solute remaining in the mobile phase is $1/(1+k')$. For an unretained peak, $t_R = t_0$, $k' = 0$, and the value for the solute weight fraction in the mobile phase equals unity, meaning that the solute resides only in the mobile phase. Since solutes migrate only when in the mobile phase, the retention time should be inversely proportional to this weight fraction:

$$1 : \frac{1}{1+k'} = t_R : t_0$$

or

$$k' = \frac{t_R - t_0}{t_0} \qquad (2.1)$$

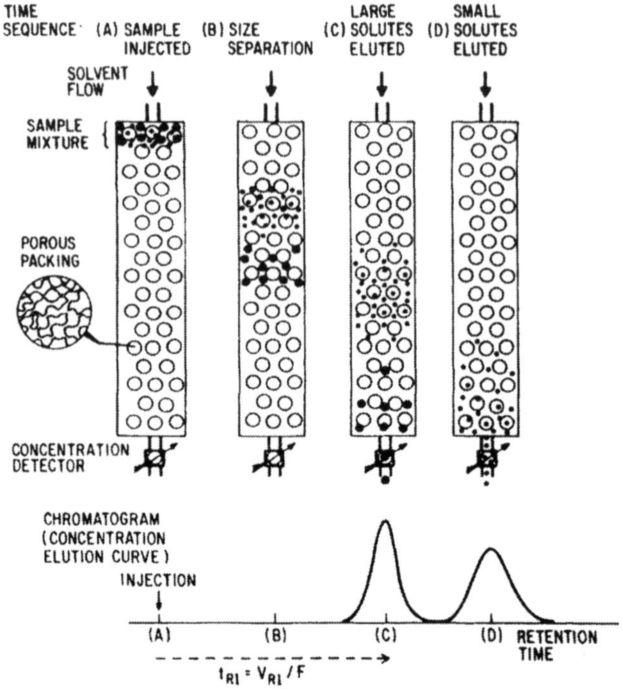

Figure 2.1 Development and detection of size separation by SEC.

and

$$k' = \frac{V_R - V_M}{V_M} \tag{2.2}$$

assuming constant flow rate, where $V_M = Ft_0$ for the retention volume of the unretained solute. [There are several ways to determine t_0 and V_M. One simple method is to inject a monomer having similar structure and size as the mobile-phase solvent (e.g., pentane when hexane is the mobile phase) and to detect the unretained peak with a differential refractometer.] Although widely used for comparing conventional LC data, values of k' still do not compensate for differences in the stationary-phase concentration caused by the difference in the surface area and porosity of the column packing. Peak retention, or value of k', increases with increasing stationary-phase loading.

To account for differences in stationary-phase loading, the parameter K_{LC} should be used to define retention. In fact, K_{LC} is the only parameter that can uniquely define the retention characteristics of different organic compounds in conventional LC experiments with a specified column packing, mobile phase, and temperature. Physically, K_{LC} is the ratio of *solute concentration* in the stationary phase to that in the mobile phase. For a given mobile phase and column packing, values of K_{LC} uniquely reflect the basic thermodynamic balance of solute between the phases. Assuming that the equivalent liquid volume for a stationary phase is V_s (the actual liquid volume for LLC, or the volume equivalent to the surface-effect retention in adsorption, or the weight of absorbent in ion exchange), K_{LC} is related to k' by

$$k' = \frac{K_{\mathrm{LC}}\, V_s}{V_M} \tag{2.3}$$

Inserting this relationship into Equation 2.2, one can show that

$$V_R = V_M + K_{\mathrm{LC}} V_s \tag{2.4}$$

Equation 2.4 represents the equilibrium theory of conventional LC peak retention. It explains why the experimental value of V_R is determined solely by the thermodynamic balance of solute distribution between phases. The validity of the equilibrium LC retention theory is supported by experimental observations. However, since V_s of conventional LC is difficult to determine accurately by experiment, values of K_{LC} are difficult to measure and are not commonly used in practice. The volume elements in Equation 2.4 are illustrated in Figure 2.2a for partition LLC as an example of where the mobile-phase volume V_M is subdivided into two parts: the moving mobile-phase volume, V_o, and the stagnant mobile phase, V_i.

Direct use of terminology traditional to LC (henceforth called *LC terminology*) in SEC applications can sometimes be awkward. In SEC (Figure 2.2b), the size separation occurs only within the mobile-phase volume, V_M, where different-size solutes distribute differently between V_o and V_i, that is, between the solvent moving outside the packing and the stagnant solvent inside the pores of the packing. The distribution favors V_o more for larger solutes. According to LC terminology and Equation 2.2, the SEC chromatogram would have to be interpreted with awkward negative values of k', since $V_R \leq V_M$ in SEC as solute elutes before the solvent peak (i.e., $t_R < t_0$ and $k' < 0$, according to Equation 2.1). This is why the distribution coefficient K_{SEC} is used in SEC as the peak retention index instead of k' as in conventional LC (Section 2.2). (For the same reason, the separation factor α, defined in conventional LC as the ratio of k' for two solutes, is not used in SEC.)

According to Equation 2.2, failure to distinguish the moving and the stagnant parts of the mobile phase does not affect the estimation of k'. However, it does cause an error in the calculation of solvent velocity, v:

$$v = \frac{L}{t_0} = L\frac{F}{V_M} = L\frac{F}{V_o + V_i} \tag{2.5}$$

where L is the column length. For porous packings that contain a stagnant mobile phase, calculation of mobile-phase velocity according to Equation 2.5 will underestimate the true solvent velocity in the column. The calculated value in this case is actually the volume-averaged velocity of the moving and the stagnant mobile phases

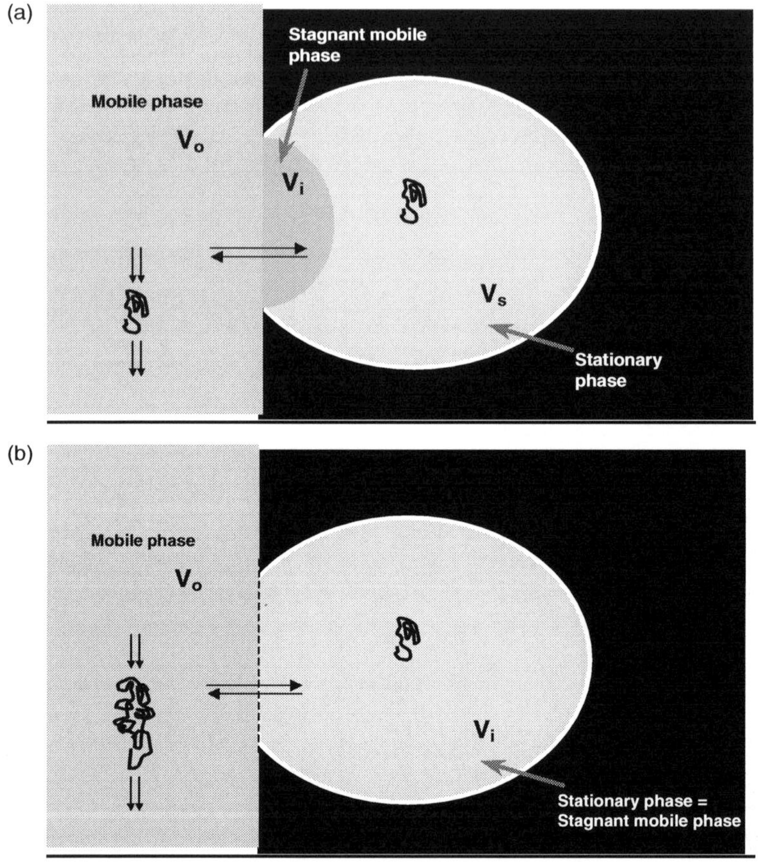

Figure 2.2 Liquid chromatographic retention mechanism: (a) partition; (b) exclusion.

(Figure 2.2). (The true solvent linear velocity is given by Equation 2.9, discussed in the next section.)

2.3 SOLUTE RETENTION IN SEC

In LC methods other than SEC, sample components are retained by the column packing and elute after the unretained solvent peak. However, in SEC, solutes are partially excluded from the column packing and elute ahead of the solvent peak. As a solute band moves along with the solvent down the column and around the packing particles, the solute molecules repeatedly permeate or diffuse in and out of the pores of the packing. The driving force for this process is the concentration gradient between the phases. The development and detection of a size separation in SEC are illustrated in Figure 2.1. Here it is shown that larger solute molecules elute faster than the smaller molecules because they have less penetration into the pores of the packing. Solutes of two distinct sizes can be resolved into two peaks, as shown in the chromatogram. All SEC peaks detected at the end of the column are of finite width, as illustrated in the figure. Even for solutes of only one size, the elution peak is still necessarily broader than that expected from the finite injected sample volume because of the mixing effects in the column, detector, and connecting tubing. Peak broadening processes affect the performance of SEC analyses, and Chapters 3 and 4 are devoted to these particular subjects.

Whereas in the SEC separation of oligomers or naturally occurring macromolecules, several distinct peaks may be obtained, in synthetic polymer analyses the SEC chromatogram or elution curve is usually just a broad, continuous elution pattern. To extract polymer MMD information from an SEC chromatogram, the exact M versus V_R calibration relationship of the SEC column is required. Pertinent calibration methods and data-handling techniques for SEC-MMD calculations are discussed in Chapter 8.

In discussing LC retention, the volumes of the mobile phase inside and outside the pores of column packing are grouped into one volume term, V_M, the retention volume of the solvent peak (Equation 2.4). Since all peaks in the other LC methods elute after V_M, it is not as important to distinguish between the stagnant versus the moving parts of the mobile-phase volume, V_M. Subdivision of V_M is necessary to explain SEC (where the term *mobile phase* simply means, in SEC, the carrier solvent), because the stagnant part of the "mobile" phase residing in the pores is, in effect, the "stationary" phase for SEC separation (Figure 2.2b). To avoid confusion with the stationary-phase volume V_s in the other LC methods, the stagnant solvent in the porous packing structure in SEC is designated as V_i, the internal pore volume. The remaining liquid volume in an SEC system is designated as the void volume V_o, which is mainly the interstitial liquid volume between the packing particles. By definition, then,

$$V_M = V_o + V_i \tag{2.6}$$

Size separation in SEC is the result of differential solute distribution between the solvent spaces outside and inside the pores of the column packing. This solute distribution can be described by the SEC distribution coefficient K_{SEC}, which represents the ratio of the average solute concentration in the pores to that outside the pores. Because of the size-exclusion effect, not all the pore volume V_i is accessible to large solutes. Solute concentration inside the pore decreases with increasing solute size. In effect, then, the total accessible liquid volume for different-size solutes is not $V_o + V_i$ but $V_o + K_{\mathrm{SEC}} V_i$. Substitution of this accessible liquid volume for V_M in Equation 2.4 leads to the general retention equation

$$V_R = V_o + K_{\mathrm{SEC}} V_i + K_{\mathrm{LC}} V_S \tag{2.7}$$

In SEC practice it is important that the last term in Equation 2.7 be minimized by using inert column packings to avoid interference of surface effects on SEC solute

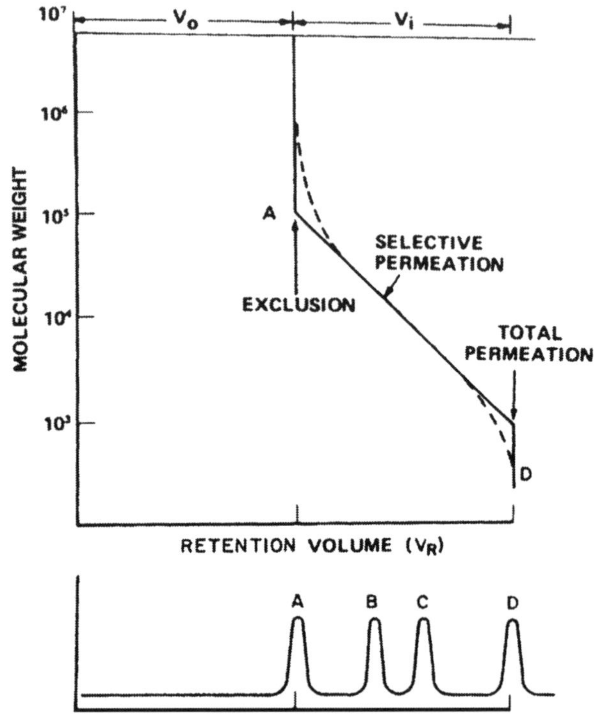

Figure 2.3 SEC calibration and separation range. (Reprinted with permission from Ref. 1.)

retention. With negligible surface effects, SEC retention can be approximated as

$$V_R = V_o + K_{\mathrm{SEC}} V_i \tag{2.8}$$

The functional dependence of Equation 2.8 on solute M constitutes the SEC calibration relationship, as illustrated by Figure 2.3 and discussed below. To cover wide molar mass ranges in SEC separations, the SEC calibration curve is plotted conventionally with molar mass in the logarithmic scale of base 10. Peak retention in SEC should be recorded in V_R (not t_R) units to minimize the distortion of the elution curve shape due to possible flow-rate variations. (For high-speed SEC analyses it is important to compensate adequately for flow-rate variation to assure the accuracy in molar mass of the SEC results; see Section 7.2.) The detailed features of the SEC elution curve are important because they are used in direct interpretation of polymer sample MMD.

In Figure 2.3, a high-molar-mass solute, designated as solute A, elutes at the void or exclusion volume, V_o, of the SEC column. This solute migrates down the column only through the interstitial spaces between the packing particles. The velocity of this solute, which can be calculated as column length divided by V_o/F, provides a true measure of the solvent linear velocity:

$$v(\text{true}) = L\frac{F}{V_o} \tag{2.9}$$

Since $V_o < V_M$, the true solvent velocity calculated by Equation 2.9 is necessarily larger than the average solute velocity according to Equation 2.5.

As the molar masses of the polymer solutes decrease (peaks B and C in Figure 2.3), the fraction of the pore volume accessible to the solutes increases, causing peaks to elute later. For solute D, which is small enough to access all the pore volumes, elution occurs at the total permeation limit. The fact that the retention volume for peak D is interpreted as "total permeation" in SEC but as "unretained peak" in conventional LC reflects an interesting contrast of conventions and viewpoints. When comparing data from different SEC experiments or in discussing SEC theory, values of K_{SEC} calculated according to Equation 2.8 are preferable to values of V_R for describing SEC peak retention. The values of K_{SEC} compensate for column-size variations. For any SEC column, regardless of size, $K_{\text{SEC}} = 0$ at exclusion and $K_{\text{SEC}} = 1$ at total permeation. The dashed line in Figure 2.3 illustrates the gradual approach of the usual experimental SEC calibration curve to the column exclusion and permeation limits. The solid straight line, the linear approximation to the calibration curve, is commonly used in SEC to facilitate MMD calculations (Chapter 8).

It should be noted that peak A may be comprised of a number of solutes, all of size greater than the largest pore volume. These solutes will then elute together at the exclusion volume of the column. Similarly, peak D may be comprised of a number of solutes. In this case, while the solutes may be of different sizes, they are all essentially infinitely smaller than the smallest pore volume. Consequently, these small solutes will elute together at the total permeation volume of the column.

The information in Figure 2.3 also suggests that SEC intrinsically is a low-resolution technique. Unlike other LC methods, which can be developed to resolve up to hundreds of component peaks representing many column volumes and extended retention times, SEC separations are constrained to occur within the limits of the packing pore volume. Thus, only a few peaks can be fully resolved in SEC. Limited SEC peak capacity is a practical constraint to the SEC analyses of small molecules (Chapters 13). However, the relatively low resolution of SEC does not prevent one from using the technique to obtain important polymer molar mass information, in addition to information regarding polymer architecture and dilute solution thermodynamics. The individual molar mass components of a sample need not be well resolved for determining the MMD features of the whole polymer. The concept of SEC resolution and molar mass accuracy is a subject of discussion in Chapter 4. Absolute molar mass detectors are discussed in Chapters 9 and 10. Applying SEC analysis to determine the architecture of polymers and/or the thermodynamics of dilute polymer solutions is covered in Chapter 11.

The large difference in peak capacity between SEC and other forms of LC can also be explained in terms of basic retention parameters. The value of K_{SEC} is constrained to be between 0 and 1, which means that solute distribution favors the unrestricted space outside the pore. On the other hand, values of K_{LC} are unlimited, which means that solute distribution favors the stationary phase, as is the case for most LC peaks. This difference in peak capacity between SEC and the other LC methods is indicative that different thermodynamic balances are involved in controlling solute distribution. As described next, SEC is uniquely different from the other LC methods in that it is a chromatographic process controlled by entropy, not enthalpy.

2.4 SEC RETENTION MECHANISM

As solute molecules migrate through the chromatographic column, they transfer back and forth between the moving and stationary solvent phases, constantly redistributing themselves between the phases to satisfy the thermodynamic equilibrium. Under normal chromatographic conditions, solute distribution approximating thermodynamic equilibrium is achieved. (This is true even for the large, slowly diffusing solutes in SEC, as proved by the flow-rate studies and the static mixing experiments described later in this section.) Thermodynamic equilibrium of solute distribution is defined as the condition in which the chemical potential of each solute component is the same in the two phases [2]. For dilute solutions at equilibrium, solute distribution can be related to the standard free-energy difference (ΔG°) between the phases at constant temperature and pressure:

$$\Delta G^\circ = -RT \ln K \tag{2.10}$$

with

$$\Delta G^\circ = \Delta H^\circ - T\Delta S^\circ \tag{2.11}$$

where K is the solute distribution coefficient, R the gas constant, T the absolute temperature, ln the natural logarithm (base e), and ΔH° and ΔS° standard enthalpy and entropy differences between the phases, respectively.

Solute partitioning in the other forms of LC occurs largely because of solute–stationary phase interactions. Whether absorption or adsorption is involved, the transfer of solutes between phases is associated with intermolecular forces and substantial enthalpy changes. The entropy change in the other LC methods is generally small and can usually be ignored. Therefore, by combining Equations 2.10 and 2.11 and neglecting the ΔS° term, one can derive K_{LC} as

$$K_{LC} \simeq e^{-\Delta H/RT} \tag{2.12}$$

The value for ΔH° is usually negative (corresponding to an exothermic sorption for an attractive solute–stationary phase interaction), resulting in K_{LC} values being larger than unity, according to Equation 2.12, and LC peaks eluting later than the solvent peak. On the other hand, solute distribution in SEC is governed mainly by the entropy change between phases [3,4]. Again by combining Equations 2.10 and

Table 2.1 Temperature independence of K_{SEC}[a]

	K_{SEC}	
Oligosaccharide	25°C	37°C
Maltose	0.683	0.675
Maltoheptaose	0.430	0.415
Cellobiose	0.657	0.648
Cellopentaose	0.431	0.419
Isomaltose	0.626	0.613
Isomaltoheptaose	0.356	0.341
Laminaribiose	0.657	0.643
Laminariheptaose	0.369	0.354
α-Cyclodextrin	(0.890)[b]	(0.758)[b]
γ-Cyclodextrin	(0.718)[b]	(0.647)[b]

Source: Data from Ref. 7.

[a]Solvent: H_2O, pH 7.4.

[b]K_{SEC} values for α- and γ-cyclodextrin data are reported here for comparison purposes only. Elution of these cyclodextrins did not occur by a strict size-exclusion mechanism but, instead, reflects a substantial enthalpic contribution to the separation.

2.11 but with $\Delta H^\circ \simeq 0$, K_{SEC} is derived as

$$K_{SEC} \simeq e^{\Delta S^\circ/R} \tag{2.13}$$

Because solute mobility becomes more limited inside the pores of the column packing, solute permeation in SEC is associated with a decrease in entropy, or a negative value of ΔS° (discussed further in Section 2.4). This effect causes K_{SEC} values to be less than unity, according to Equation 2.13, and solutes to elute before the solvent peak.

Equation 2.12 indicates that a direct temperature dependence exists for peak retention with the other LC methods, and the temperature independence of SEC peak retention is predicted by Equation 2.13. This theory is well substantiated by experimental observations, as shown in Table 2.1 and Figure 7.10. In Table 2.1, changes in K_{SEC} of several oligosaccharides of only a few (four or less) parts per hundred were observed in aqueous solvent over a range of 12°C [7]. In organic solvents (DMAc and DMAc/LiCl), the change in K_{SEC} was on the order of less than 10 parts per hundred over a range of 25°C [5,6,8]. These changes are substantially smaller than the changes in distribution coefficient expected when the temperature of enthalpically dominated separations, such as reverse-phase LC, is varied over a similar range. For the cyclic oligosaccharides α- and γ-cyclodextrin, however, the contribution from ΔH° to K at aqueous conditions cannot be ignored. As shown in Table 2.1, the same 12°C variation caused changes of 15% and 10% in K_{SEC} of α- and γ-cyclodextrin, respectively. Additional evidence of nonideal SEC behavior for the cyclodextrins in aqueous solvent was provided by molecular dynamics computer modeling [7].

In Figure 7.10, a large temperature change from 25°C to 150°C had only a small effect on the SEC retention characteristics for polystyrenes and polyisobutenes of different molar masses. Temperature changes have only a small indirect effect on SEC retention, as they affect the size of the polymer solute molecules, which in turn affects the ΔS° value. In good solvents, the size of the polymer molecules changes very little with temperature. This is in agreement with the small shifts of SEC retention observed at different temperatures. The effect of temperature on the calibration curves is small only in the context of verifying that the SEC separation is an entropy-controlled process. Even these small curve shifts will significantly affect the accuracy of the molar mass results in calibration-curve-based polymer MMD analyses. Also, temperature does have a significant influence on SEC peak broadening as with all LC peaks (Section 3.4). Therefore, large temperature fluctuations in SEC experiments should be avoided.

The validity of explaining SEC retention only in terms of thermodynamic considerations requires that the solute distribution in the SEC experiment be close to thermodynamic equilibrium. The fact that this occurs in SEC columns is supported by two sets of studies, one showing that SEC retention is independent of flow rate [10,11] and the other providing measurement of equilibrium solute distribution through a series of simple static mixing experiments [12,13]. As seen in Figure 2.4, flow rate has little effect on SEC retention. This is true even for large 37- to 42-μm SEC packings, as shown in Figure 2.5. For skewed SEC peaks at very high flow rates, the values of V_R at the center of mass, not the maximum, of each peak should be used in studying the effects of flow rate on SEC retention. Experimental results indicate that the kinetic or mass transfer process does not influence the retention mechanism in the SEC separation [14,15], a fact recognized by John Moore in his original publication on gel permeation chromatography [9].

Thus, SEC separation is controlled by the differential *extent* of permeation rather than by the differential *rate* of permeation. Further proof for this contention is provided by static polymer and porous packing mixing experiments [12,13]. Here, a polymer solution of a known volume and initial concentration, C_i, is mixed with a known amount of dry, porous packing material. The mixture is allowed enough time for complete solute permeation. The concentration, C_o, of the final solution is then measured and compared with C_i. The change in solution concentration is a direct measure of the equilibrium solute distribution. If solute distribution in SEC separations reaches thermodynamic equilibrium, the experimental values of K_{SEC} for solutes of different molar masses should vary linearly with the corresponding values of $1 - C_i/C_o$ obtained in the independent mixing experiment. The data shown in Figure 2.6 fully support this proposition and the equilibrium theory.

The results of the temperature, flow-rate, and static mixing experiments clearly show that SEC *retention is an equilibrium, entropy-controlled, size-exclusion process*. This mechanistic model indicates that solute diffusion in and out of the pores is fast enough with respect to flow rate to maintain equilibrium solute distribution. Thermodynamic size exclusion is the fundamental basis common to all the SEC theories discussed in Section 2.5, where models for different-shaped solutes are considered in the quantitative prediction of the SEC calibration curve. The basic features of the thermodynamic theory of SEC retention are summarized in Table 2.2, which also shows the fundamental differences between SEC and other LC methods.

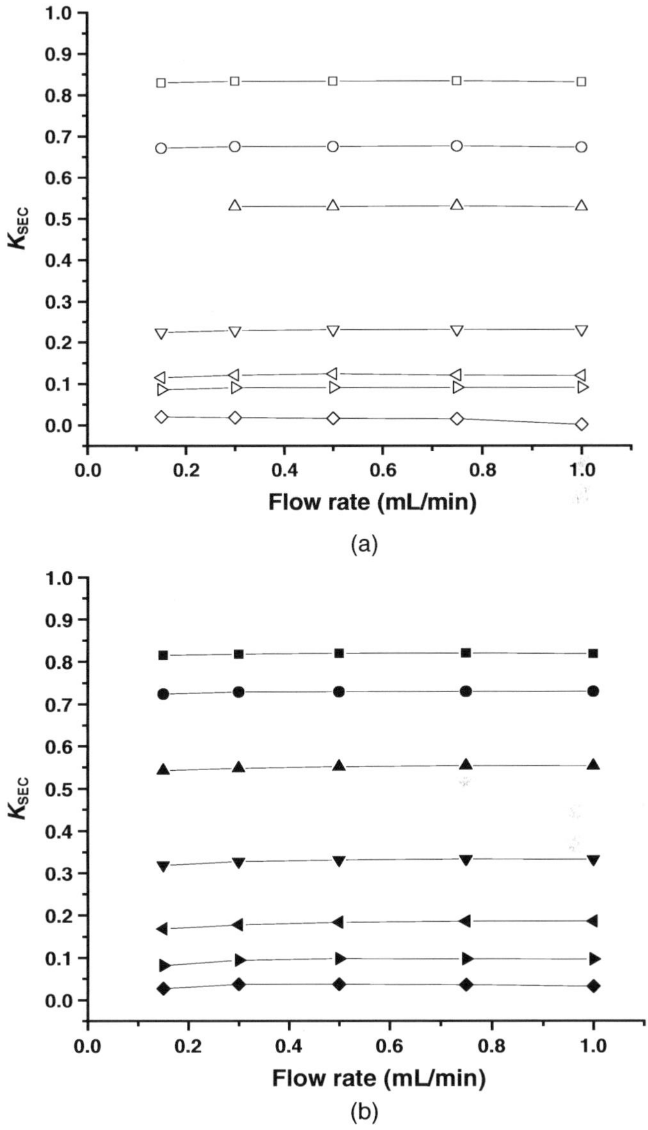

Figure 2.4 Flow rate independence of K_{SEC}. Narrow polydispersity linear standards in THF at 30°C: (a) polystyrene, with M_p in g/mol of (□) 925, (○) 8450, (△) 30,300, (▽) 189,000, (◁) 355,000, (▷) 500,000, (◇) 950,000; (b) poly(methyl methacrylate), with M_p in g/mol of (■) 1280, (●) 4910, (▲) 27,000, (▼) 107,000, (◀) 265,000, (▶) 467,000, (◆) 838,000. Results are averages of triplicate injections. In all cases, standard deviations are substantially smaller than data points and therefore are not shown. Solid lines are placed on graphs to guide the eye and are not meant to imply continuity between data points. Column, one 7.5 × 300 mm PLgel 10-μm particle size, 10^4-Å pore size. (Adapted from Ref. 10.)

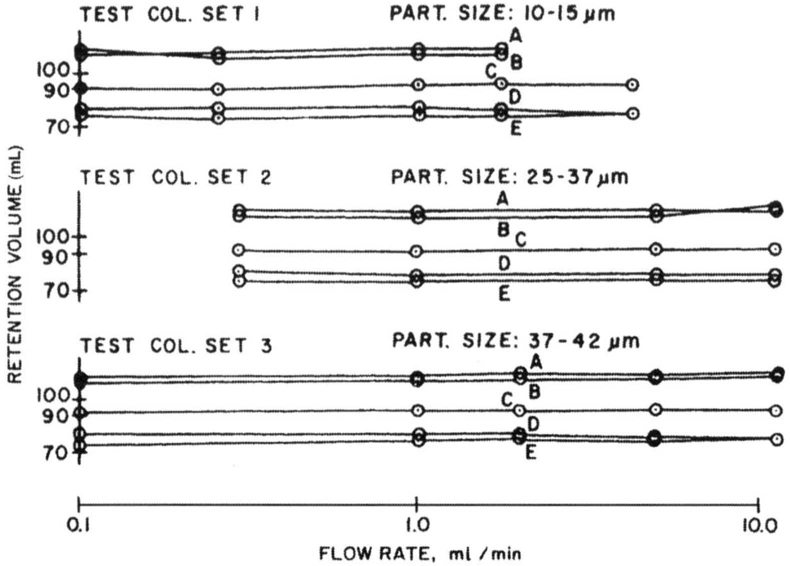

Figure 2.5 Independence of SEC retention on flow rate and particle size. A, 0.05% CH_3CN; B, 0.5% CH_3CN; C, 0.05% and 0.5% polystyrene M 19,850 g/mol; D, 0.5% polystyrene M 411,000 g/mol; E, 0.05% polystyrene M 411,000 g/mol (all tests under the same operating conditions). (Reprinted with permission from Ref. 11)

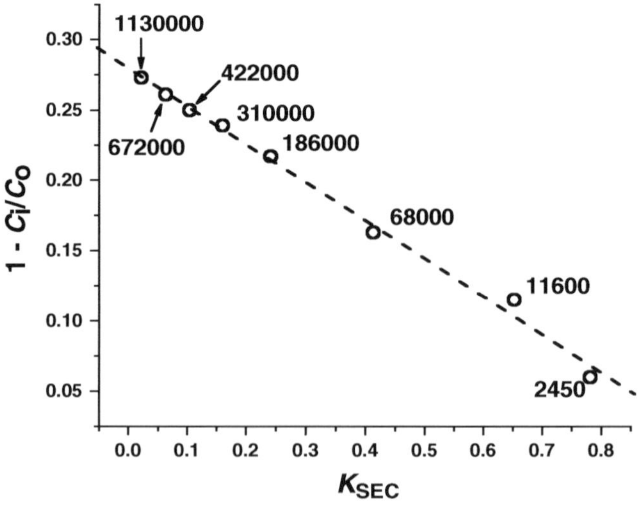

Figure 2.6 Static mixing data versus equilibrium solute distribution in SEC. Abscissa (K_{SEC}) values from flow-mode SEC experiment, ordinate ($1 - C_i/C_o$) values from static mixing experiment. PLgel 10-μm particle size, 10^4 Å pore-size column-packing material from the same manufacturer was used in both flow and static mixing experiments. 0.1% polystyrene in THF; 10 mL of solution with 2 g of porous stationary phase. Each point represents the average of triplicate measurements, with standard deviations along both axes substantially smaller than data points and therefore not shown. The dashed line represents a linear fit to the data ($r^2 = 0.994$). The numbers represent M_p, in g/mol, of each narrow polydispersity linear polystyrene standard. (Reprinted with permission from Ref. 13.)

Table 2.2 Thermodynamics of LC retention[a]

Size exclusion
$K_{\text{SEC}} = e^{-\Delta G^\circ/RT} \simeq e^{\Delta S^\circ/R}$
Entropy (S)-controlled process
ΔS° = negative for all solutes; S(stationary) $<$ S(mobile)
$K_{\text{SEC}} < 1$; solute elutes before solvent; k' = negative
Temperature independent
Other LC methods (partition, adsorption, ion exchange)
$K_{\text{LC}} = e^{-\Delta G^\circ/RT} \approx e^{-\Delta H^\circ/RT}$
Enthalpy (H)-controlled process
ΔH° = negative for most solutes; H(stationary) $<$ H(mobile)
$K_{\text{LC}} > 0$; solute elutes after solvent; k' = positive
Temperature dependent

[a] $\Delta G^\circ = \Delta H^\circ - T\,\Delta S^\circ$.

2.5 THEORETICAL MODELS OF SEC SEPARATION

The theoretical models described in this section are attempts to explain K_{SEC} and SEC calibration quantitatively as a function of the size and shape of the solute and the pore. The models are based on the equilibrium steric SEC theory described above. They are sometimes referred to as the *equilibrium theories of SEC separation*. Variations among the equilibrium theoretical models are related to the forms and structures of the solute molecules. For solutes of different conformation, K_{SEC} can have different physical significance such that different approaches to the theoretical interpretation are needed. The conformations of the pore structures are also important factors that affect only the value of K_{SEC}, not the nature of the size-exclusion effect. The hollow cylindrical pore shown in Figure 2.7 illustrates the exclusion effect of three

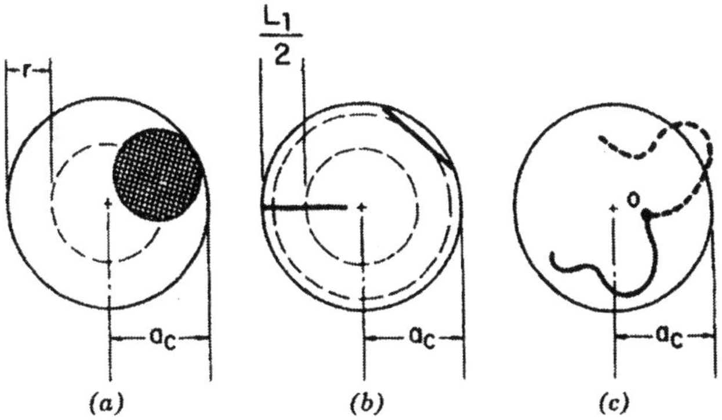

Figure 2.7 Exclusion effect in cylindrical void of radius a_c: (a) hard sphere of radius r; (b) thin rod of length L_1 in two orientations in the plane of the cross section; (c) random-flight chain with one end at point 0, showing allowed conformation (solid curve) and forbidden conformation (dashed curve). (Reprinted with permission from Ref. 16.)

types of solute molecules: the hard-sphere, rigid-rod, and random-flight coiled-chain models. The utility of the solute model varies depending on the true shape of the particular macromolecule of interest. The random-coil model is usually appropriate for synthetic polymers, whereas the rigid-rod and sphere models find applications primarily in biopolymer studies. SEC theories for the three solute models are discussed separately in this section.

2.5.1 Hard-Sphere Solute Model

The exclusion effect of hard spheres is illustrated in Figure 2.7a, which shows a spherical solute of radius r inside a cylindrical cavity of radius a_c. Here the exclusion process can be accounted for by straightforward geometrical considerations of the solute exclusion from the walls of the cavity. The center of the sphere (solute) cannot approach the cavity wall closer than a distance r away. Effectively, all the sphere sees is a smaller cylindrical volume of radius $a_c - r$ described by the dashed circle rather than the entire cavity volume of radius a_c. (It is assumed that the cylindrical cavity is infinitely long, that is, has negligible end effects.) The center of the hard sphere has free access to the space inside the dashed circle but cannot enter the space outside the dashed line. During solute distribution, the solute molecules can permeate only into the inner space of the cavity. At equilibrium, solute distribution there will be a step change of solute concentration across the dashed circle: The solute concentration inside this circle will be constant and equal to that in the open space outside the cavity, and the concentration outside the dashed circle to the cavity wall will be zero. Therefore, the average solute concentration of the entire cavity will be less than that outside the cavity. The ratio of the concentrations inside and outside the cavity is equal to the fraction of the cavity area inside the dashed circle. The situation is equivalent to one with a distribution coefficient:

$$K_e = \left(\frac{a_c - r}{a_c} \right)^2$$

or

$$K_e = \left(1 - \frac{r}{a_c} \right)^2 \qquad \text{(cylinder-shaped pores)} \qquad (2.14)$$

In this case the distribution coefficient is physically equivalent to the fraction of the pore volume accessible to the spherical solute molecules. The equilibrium solute distribution is represented here by K_e with subscript e to distinguish it from K_{SEC}, which is defined as the solute distribution coefficient in the SEC experiment. Thermodynamically, this exclusion process can be considered as the restriction of the solute spatial freedom inside the cavity due to the infinite energy barrier at the dashed circular line. Since the solute is geometrically symmetrical, considerations of configurational changes (rotational freedom) and conformational changes (intramolecular structural changes) are not necessary in this case.

Similarly, equations for K_e for spherical solutes with other simple pore shapes can also be derived using accessible pore volume considerations [17]:

$$K_e = \begin{cases} 1 - \dfrac{2r}{\bar{a}} & \text{(slab-shaped pores)} \quad (2.15) \\ \left(1 - \dfrac{2r}{3\bar{a}}\right)^3 & \text{(spherical pores)} \quad (2.16) \\ \left[1 - \left(\dfrac{2r}{\bar{a}}\dfrac{1}{1+P}\right)\right]\left[1 - \left(\dfrac{2r}{\bar{a}}\dfrac{P}{1+P}\right)\right] & \text{(rectangular pores)} \quad (2.17) \end{cases}$$

In the equations, P is the ratio of the long to the short side of the rectangular cavity, and $\bar{a}$ is defined as the effective radius:

$$\bar{a} \equiv 2 \times \frac{\text{pore volume}}{\text{pore surface area}} \quad (2.18)$$

The use of $\bar{a}$ to define the pore size of different pore shapes aids in the meaningful comparison of K_{SEC} for different pore geometries. This parameter, $\bar{a}$, is also a good fundamental quantity for describing chromatographic pore size. The following equivalences exist: $\bar{a}$ = radius of a cylinder; $\bar{a} = P/(1 + P)$ times the short side of a rectangle. (One finds that $\bar{a} = 2/s$, where s is defined as the surface area/unit pore volume, or the reciprocal of the hydraulic radius defined as the volume/surface area ratio [17]. Both $\bar{a}$ and s are experimentally measurable parameters regardless of pore shape.)

The plot of Equations 2.14 to 2.17 is shown in Figure 2.8. The dashed curve in the figure represents the separation of spherical solutes by a random-planes model suggested by Giddings et al. to describe the porous network structure of SEC packings [17]. The curve was calculated by the equation

$$K_e = \exp\left(-\frac{2r}{\bar{a}}\right) \quad \text{(random-planes pore model, spherical solute)} \quad (2.19)$$

A sketch of the random-planes pore model is shown in Figure 2.9. Pores in this model are formed randomly by intersecting planes.

The curves for different-shaped pores in Figure 2.8 differ considerably, except that as the solute radius decreases, all K_e curves approach unity along a common line. Further examination shows that all curves converge to $K_e = 1 - 2r/\bar{a}$ as the $r/\bar{a}$ value approaches zero. For sufficiently small spherical solutes, all elements of the pore inner surface appear as plane areas, and wall curvature and the corners of pores of different shapes become unimportant. The fact that the curves have a common convergence at small values of $r/\bar{a}$ strongly indicates that $\bar{a}$ (or s in the Giddings expressions [17]) is a fundamental chromatographic pore-size parameter. The curve for random planes has a more gradual change in slope and spreads over a wider range of $r/\bar{a}$ than the other curves. This is expected, since the pores of the assumed

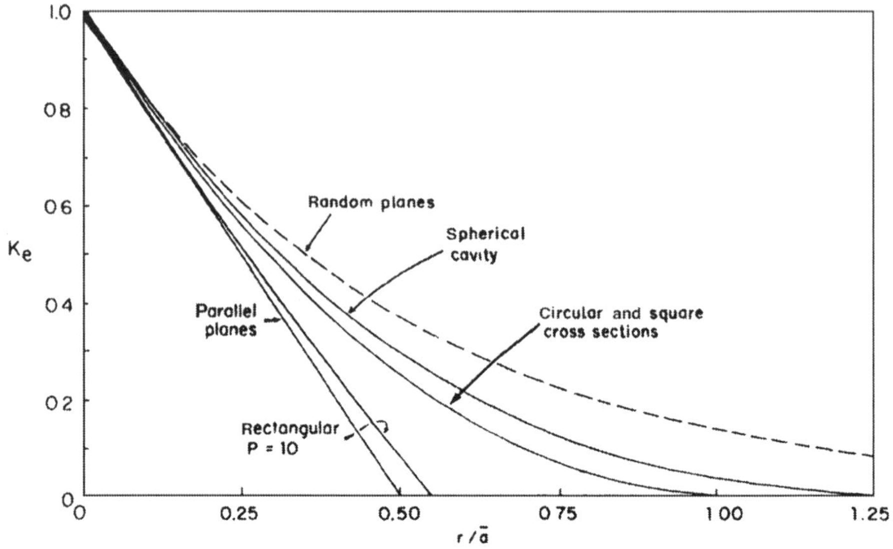

Figure 2.8 Distribution coefficient K_e for spherical molecules of radius r: K_e for various types of pores versus the ratio of r over the effective radius $\bar{a}$ of the pores. (Reprinted with permission from Ref. 17.)

random pore structure are not uniform in size, and the presence of a size distribution of pores tends to extend the $r/\bar{a}$ range. There have been other theories for hard-sphere solutes using different random porous network models, including the random-rod pore model [18,19] and the random-sphere pore model [20]. The random-rod model is used in the historical Laurent–Killander–Ogston theory of GFC retention. The

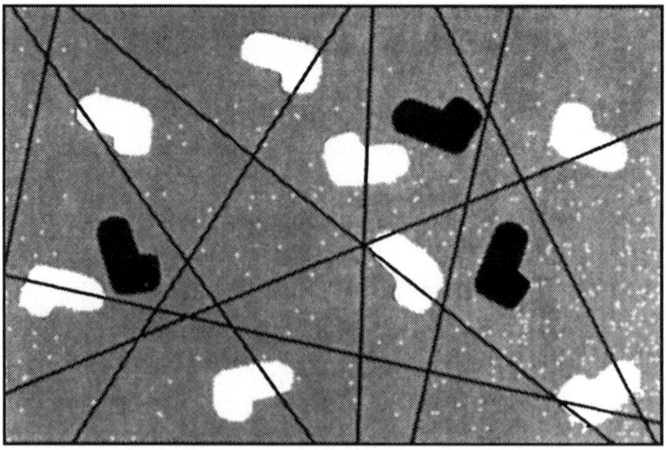

Figure 2.9 Size exclusion in random-planes pore structure. Unshaded bodies, excluded solute molecules; shaded bodies, permeating solute molecules. (Reprinted with permission from Ref. 17.)

random-sphere pore model approximates the pore shapes inside SEC packing particles by the voids between randomly arranged microspheres. The model is most suited for describing the porous silica microsphere (PSM) packings because of the expected similarity in pore structure. A random-pore geometry is more realistic than uniform-pore-shape models. Pore shapes in actual SEC packings are not uniform. Variations in pore shape and cross section are, in effect, a form of pore size distribution (PSD). Random-pore models account for these pore geometry variations.

2.5.2 Rigid Molecules of Other Shapes

Exclusion effects of rigid molecules is illustrated in Figure 2.7b, which shows a thin rod of length L_1 inside a cylindrical cavity of radius a_c. Quantifying the exclusion process here is much more complicated than for the hard-sphere model. For the rigid rod, the walls of the cavity restrain both the spatial and the rotational freedom of the rod. When the center of the rod is within the small dashed circle in the sketch (i.e., the rod is more than a distance $L_1/2$ away from the wall), the rod will have full freedom to rotate without touching the wall. As the center of the rod is moved closer to the wall, certain rotational angles in the plane of the cross section are no longer allowed because the ends of the rod hit the wall. Finally, as the rod reaches the position illustrated at the upper right corner of the sketch, it has *no* rotational or angular (configurational) freedom in the plane of cross section. The final theoretical account of this exclusion effect is complicated further by the necessity of considering rods situated not only in the plane of the cross section but also tilted out of the plane at all angles allowed.

This statistical problem for SEC solute distribution has been studied in detail by Giddings et al. [17]. The study suggested the following general expression for the statistical theory of equilibrium solute distribution:

$$K_e = \frac{\int e^{-u(q)/kT}\,dq}{\int dq} \tag{2.20}$$

where q represents the generalized coordinates indicating the solute position, orientation, and internal structural geometries that are needed to describe the changes of spatial, configurational, and conformational freedom of the solute molecules. The energy $u(q)$ is infinitely large when a geometric configuration (q) intersects with the wall of the cavity; $u(q)$ is equal to zero, otherwise. For rigid molecules, the conformational considerations are ignored, since there can be only one fixed solute conformation. (It may be recalled that in the case of a hard-sphere model, both the configurational and the conformational considerations were ignored.) The thesis of the statistical theory basically is a surface-overlapping phenomenon that forbids configurations that cause any part of the solute to intersect with the wall of the cavity.

Exact expressions for K_e for simple rods in cavities of even very simple shapes are quite complex. The equations of Reference 17 are not reproduced here, but the resulting curves are shown in Figure 2.10. The general shapes of the curves are similar to those in Figure 2.8. The main difference is that thin-rod curves have less

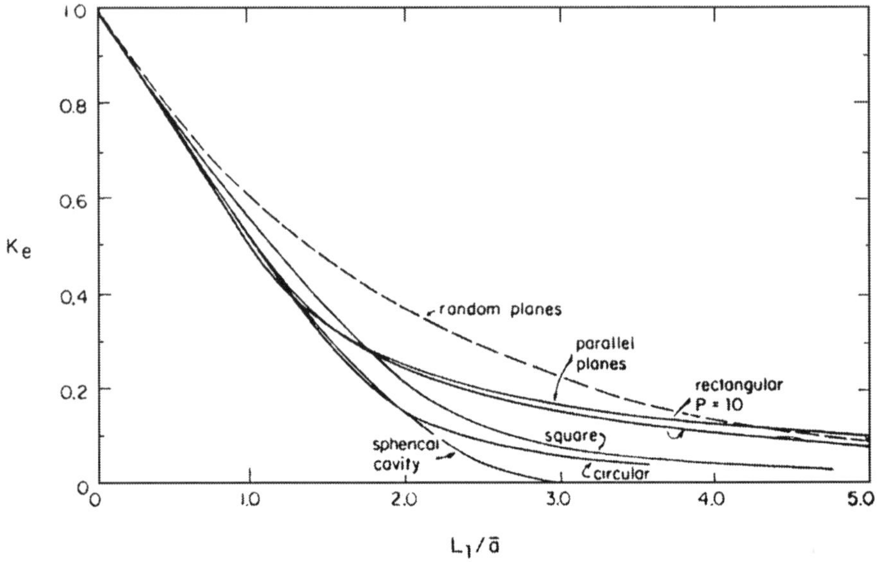

Figure 2.10 Distribution coefficient K_e for thin rods of length L_1: K_e for various pores versus the dimensionless parameter $L_1/\bar{a}$. (Reprinted with permission from Ref. 17.)

well defined exclusion limits than those for hard spheres. Except for the random-planes model, the curves in Figure 2.8 all intersect with the $r/\bar{a}$ axis at $K_e = 0$. On the other hand, the plots in Figure 2.10 take a gradual asymptotic approach to the exclusion limit $K_e = 0$, with the exception of the spherical-pore-shape curve. The small, yet finite K_e value at large $L_1/\bar{a}$ in Figure 2.10 can be attributed to the permeation of the finite number of rods being oriented in the direction of the long axis of the assumed infinitely long pores. The curve for the random-planes model in Figure 2.10 was calculated from

$$K_e = \exp\left(-\frac{\bar{L}}{\bar{a}}\right) \quad \text{(random-planes pore model, rigid solutes in general)} \tag{2.21}$$

where $\bar{L}$ is the mean external length of the solute, defined as the average length of the projection of the solute molecule along the axes of random orientations. For thin rod-shaped solutes, $\bar{L}$ in Equation 2.21 was replaced by $L_1/2$ in calculating the values of K_e.

The random-planes pore model illustrated in Figure 2.9 can be pictured as an initially free volume partitioned into pores by solid planes inserted at random location and orientation. If a molecule of given configuration in the free space is intersected by one or more of the inserted planes, that orientation of the solute molecule represents a forbidden state which is automatically excluded from the porous network [i.e., infinite $u(q)$ for that state]. In Figure 2.9 the randomly positioned bodies represent molecules initially in equilibrium in bulk fluid. Those molecules (unshaded) cut by the superimposed random surfaces are excluded from the hypothetical pore network created by the randomly oriented surfaces. The partition coefficient K_e is the ratio of the number of uncut (shaded) molecules to the total. This kind of statistical consideration leads to Equation 2.21. Mathematically, Equation 2.21 results from the more general distribution expression, Equation 2.20, after proper integration over the spatial coordinates. Equation 2.21 is generally applicable to rigid molecules of any shape, including ellipsoids, capsules, and doughnut-shaped solutes as well.

The SEC theory for rigid molecules is ideal for interpreting SEC separation of biological polymers or of small molecules, since these molecules lack the internal conformational degrees of freedom. An interesting case of a "once broken" (bent) rod has been examined [21], the result showing that K_e is rather insensitive to the presence of one universal joint at the center of an otherwise rigid rod. The random-coil model is more realistic in representing flexible synthetic polymers.

2.5.3 Random-Coil Solute Model

In the present discussion of the equilibrium SEC theory of random-coil polymers, we follow the explanations provided by Casassa [16,21–24,26]. Figure 2.7c illustrates two conformations of a flexible polymer chain with one end fixed inside the cylindrical cavity. Even with one end fixed, the chain can still assume many conformations. The presence of the wall makes some conformations no longer possible, however, as, for example, the dashed conformation shown in the sketch. This restraint of conformational freedom causes a decrease in both entropy and solute concentration inside the cavity. Calculation of K_e directly from Equation 2.20 is very difficult in this case. However, the problem can be solved with a second-order partial differential equation for a particle undergoing Brownian motion, subject to the boundary condition that at no time is the particle allowed to step out of the confines of the cavity wall [22]. The result of this approach for a cylindrical-pore model gives

$$K_e = 4\sum_{m=1}^{\infty} \beta_m^{-2} \exp\left[-\left(\frac{\beta_m R_G}{\bar{a}}\right)^2\right] \qquad (2.22)$$

where β_m is a numerical constant that equals the mth root of the Bessel function of the first kind of order zero and R_G represents the radius of gyration. The values of K_e from Equation 2.22 are shown in Figure 2.11 along with the plots for the slab and spherical pore models plotted as a function of $R_G/\bar{a}$ in the linear scale. The experimental points in Figure 2.11 are calculated using the apparent pore radius from mercury intrusion data.

The large quantitative discrepancy between the theoretical curves and experimental data shown in Figure 2.11 can be explained by the fact that mercury intrusion has underestimated the packing pore radius. A persistent hysteresis loop is usually observed in the mercury intrusion–depressurization cycles used in studying pore size [25]. The hysteresis observed, shown in Figure 2.12, suggests the presence of "ink-bottle" structures in the porous packing [28]. In Figure 2.12 it is possible that

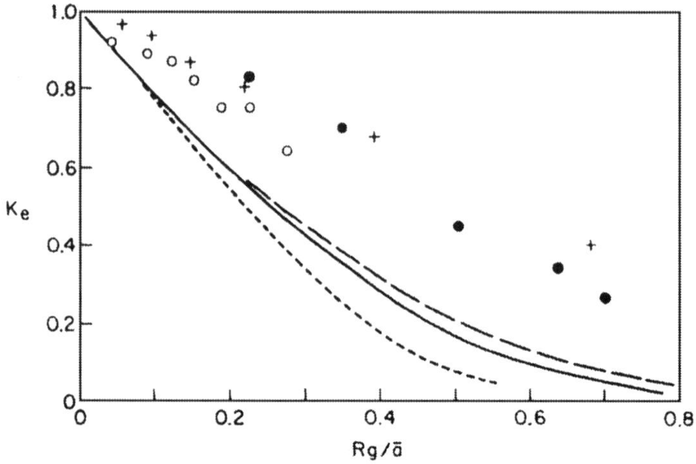

Figure 2.11 Dependence of the distribution coefficient on the molecular size/pore size ratio $R_G/\bar{a}$ for a linear flexible-chain polymers. The lower curves are theoretical results for random-coil solute in slab-shaped (short dashes), cylindrical (thick solid curve), and spherical (long dashes) cavities. The experimental points are the polystyrene data from Reference 27 (open and filled circles) and from Reference 25 (crosses). (Reprinted with permission from Ref. 26.)

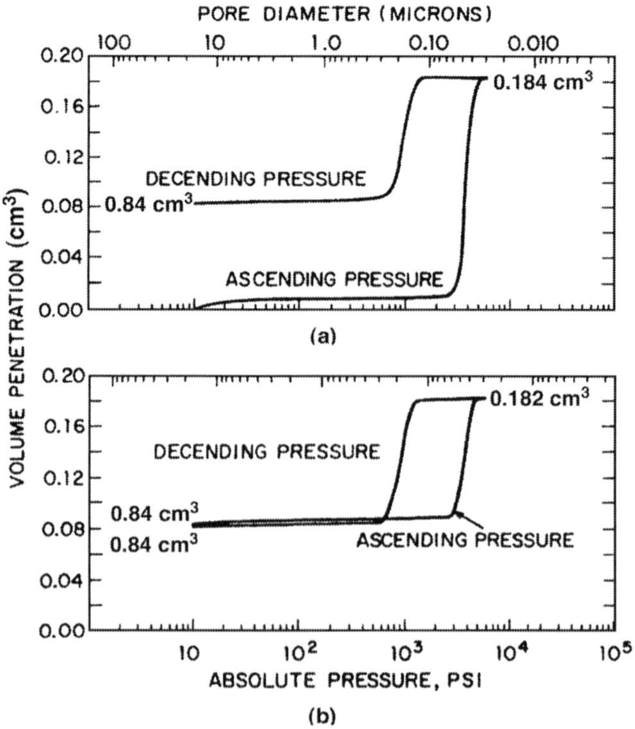

Figure 2.12 Mercury porosimetry curves for Bio-Glass 500 porous glass SEC packing: (a) first mercury intrusion-depressurizing cycle; (b) second consecutive mercury intrusion-depressurizing cycle. (Data from Ref. 29.)

the apparent pore radius of 210 Å for the mercury intrusion branch in both cycles may correspond to the narrow entrance of ink-bottle pores, and the method probably seriously underestimates the actual pore size of the packing. Calculation of the effective radius $\bar{a}$ from Equation 2.18 using the measured pore volume (e.g., mercury penetration) and the measured surface area (e.g., BET) gives $\bar{a} = 412$ Å. This new value of $\bar{a}$ brings about a much closer fit between Casassa's random-coil SEC theory and experiment. The good fit is shown in Figure 2.13, where the curves of K_e are now plotted versus $R_G/\bar{a}$ on a logarithmic scale. The success in explaining SEC separation using the value of $\bar{a}$ again verifies $\bar{a}$ as a basic SEC pore-size parameter. It should be noted that the same SEC data in Figure 2.11 from Reference 25 are used in Figure 2.13 to illustrate the concept.

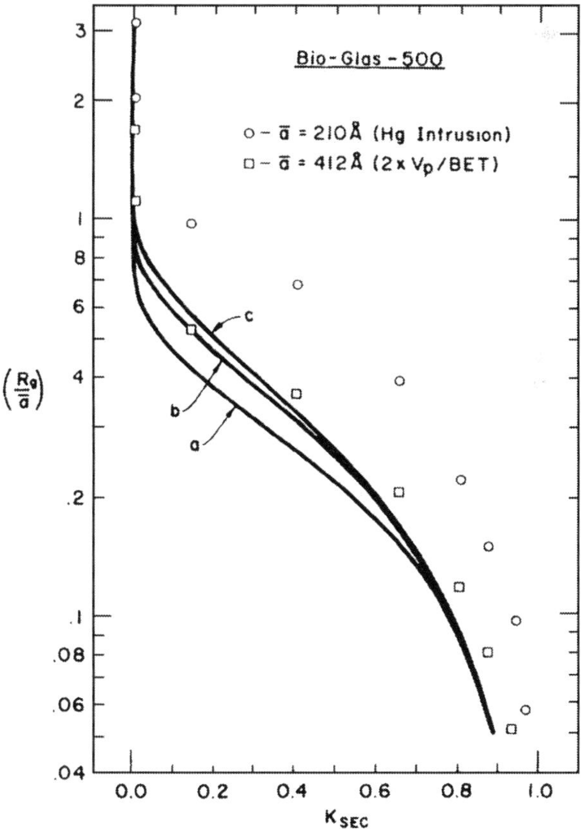

Figure 2.13 Single-pore-size SEC theoretical calibration curves: a, slab; b, cylindrical; and c, spherical pore models; circle and squares represent experimental data. (Reprinted with permission from Ref. 25.)

Comparison of the random-coil solute model in Figure 2.11 with the hard-sphere solute model in Figure 2.8 shows that for the cylindrical pore shape there is not much difference between the two corresponding curves. This indicates that the equivalent-sphere approximation of flexible polymers holds up quite well in interpreting SEC retention.

2.6 OTHER CONSIDERATIONS

2.6.1 Factors Influencing SEC Retention

All the theoretical SEC models discussed above express K_e or K_{SEC} as a function of the size ratio of the solute and pore (e.g., $R_G/\bar{a}$ for the case of coiled molecules). One can expect, therefore, that factors that affect either R_G or $\bar{a}$ will influence K_{SEC}. The chromatographic factors that affect K_{SEC} through their influence on $\bar{a}$ are pore size, pore shape, and pore-size distribution. (The pore volume of the packing affects SEC retention but not K_{SEC}.) Pore-size parameters can be utilized to combine columns effectively for optimizing SEC performance and accuracy of sample molar mass results (Sections 4.4, 7.9, and 8.3). Factors that affect K_{SEC} by their influence on R_G are solvent power, polymer branching, and copolymer composition. [The effects of flow rate and temperature on R_G are usually small and are not important considerations in optimizing SEC separation. However, these kinetic factors do affect SEC peak broadening and resolution, as discussed in Chapters 3 and 4. Only in a poor solvent is the polymer R_G appreciably affected by temperature (Table 2.3). In rare cases when SEC analyses in poor solvents are necessary, care needs be taken to control column temperature.]

There are also factors that interfere with SEC retention processes and perturb analytical information. For example, a successful SEC experiment should be free from surface interaction, polymer aggregation [31], and in situ shear degradation of the polymer in the columns [39]. Small gel particles should be filtered out of the polymer sample solutions. Concentration overloading should be kept to a minimum

Table 2.3 Effect of temperature on the radius of gyration of polystyrene[a] in cyclohexane

Temperature (K)	R_G (Å)
305.7	494
307.2	518
311.2	576
318.2	625
328.2	665
333.2	690

Source: Ref. 30.

[a]The molar mass of this sample was 3.2×10^6 g/mol.

[32]. The effects of many of these complicating factors are not well understood, and until they are, the best practice is to avoid them. These problems are also discussed in Chapter 7.

2.6.2 Failure to Define an Effective Polymer Radius

In accordance with the theory outlined in Section 2.4, the confinement entropy of the solute has been suggested as the fundamental operating parameter governing not only the separation in size-exclusion chromatography but also in hydrodynamic chromatography (HDC) and gel electrophoresis [33]. It was found, however, that under local equilibrium this entropy may not allow for the definition of an effective polymer radius, R_{eff}, that can correlate the elution behavior of different species. The relation between the solute distribution coefficient and the size of the analyte was examined for hard spheres, rigid rods, and Gaussian chains using a variety of internal (slit, capillary, spherical cavity) and external (planar slab, cylinders, spheres) pore models. An example of the results of this study is shown in Figure 2.14, where K_{SEC} is seen to vary as a function of solute size for spheres, rods, and chains. The figure depicts analyte behavior in a dilute, random array of cylinders. ε is the volume fraction of the cylinders, and the dimensionless coefficient V_d is defined as

$$K_{\text{SEC}} = 1 - V_d\varepsilon \tag{2.23}$$

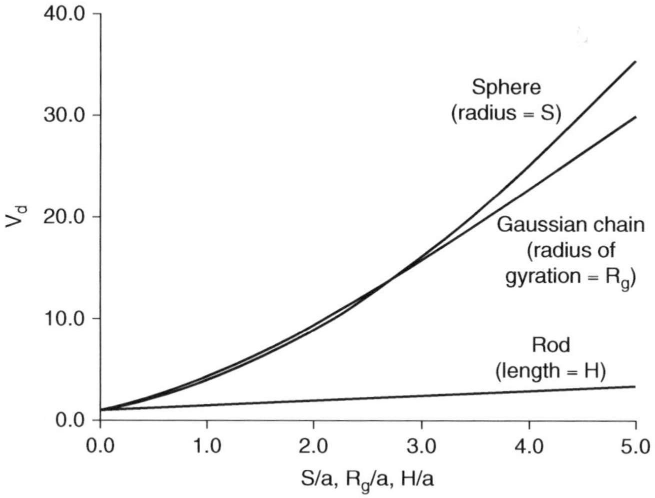

Figure 2.14 Dimensionless exclusion parameter as a function of dimensionless polymer size for spheres, rods, and Gaussian chains. Exclusion parameter V_d is for a dilute array of cylinders with radius *a*. Lower V_d implies less exclusion from matrix. (Reprinted with permission from Ref. 33.)

which applies in the limit of low ε. The abscissa normalizes polymer size (radius S for spheres, radius of gyration R_G for Gaussian chains, or length H for rigid rods) with cylinder radius a.

As can be observed from Figure 2.14, the different curves do not map onto one another except when the dimensionless solute size is small. In the small dimensionless size region, the cylinder surfaces may be considered planar and the various curves all collapse into one. In the more relevant regime of large dimensionless solute size, linear mappings fail (the sole exception being the case of a solute exposed to a single planar wall). This precludes the definition of an effective polymer radius when comparing to the pore radius a rather than to the effective pore radius $\bar{a}$. The failure is caused by the extra length scale characterizing the radius of curvature of the matrix. Representing Gaussian chains as spheres, for example, does not adequately mimic the behavior of the chains in the vicinity of cylinders, where the chains can wrap around the cylinders. As a result, Gaussian chains near cylinders are much less depleted than predicted using a spherical representation. This phenomenon becomes more significant with increasing ratio of chain size to cylinder radius.

More recently, Teraoka used numerical computations to compare the hydrodynamic radius, R_H, to the radius of gyration, R_G, as molecular dimensions determining the partition of polymers with different architectures in SEC (see Table 9.2 for a definition of the various macromolecular radii, including R_H and R_G) [34]. For a pore slit geometry of opening d, K_{SEC} for linear, star-branched, two-branch-point, and comb polymers nearly overlapped when plotted versus R_H/d. The agreement for plots of K_{SEC} versus R_G/d was poorer. Examples of these results are given in Figure 2.15.

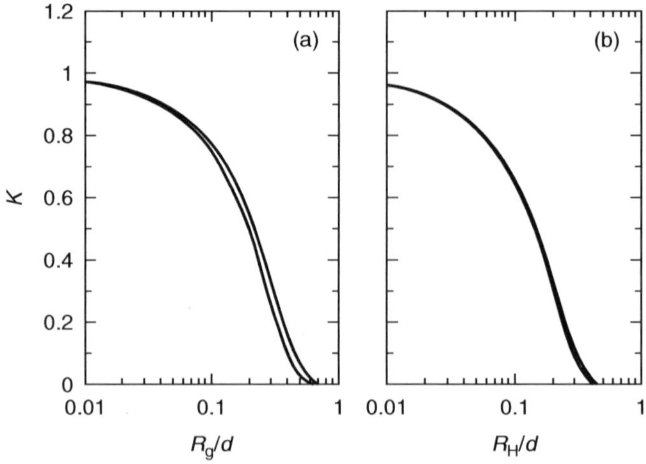

Figure 2.15 SEC partition coefficient, K for various architectures as a function of reduced polymer radii: (a) $R_G/d$and (b) R_H/d. From top to bottom, architecture is linear; symmetric three-arm star; symmetric two-branch-point, two-arm; and combination. (Reprinted with permission from Ref. 34.)

2.6.3 Hydrodynamic Chromatography Effects in SEC

Reports on the flow-rate dependence of elution volume in SEC have spawned a number of theories regarding the failure of the entropic model and/or asserting a flow-rate dependence of K_{SEC} [35]. It appears that all or most of the observations behind these theories can be explained on the basis of separation within the columns occurring by mechanisms alternative to size exclusion. Because most of the supposed "failures" of SEC have been reported for high molar mass species eluting at high flow rates, the two separation modes most likely to be responsible for the behavior observed are hydrodynamic chromatography and slalom chromatography. In this and the next section we consider these two modes of separation and their effects on SEC results.

Hydrodynamic chromatography (HDC) is a solution-based method that relies on the streamlines of flow in an open tube or in the interstitial volume of a packed column being preferentially sampled, in a size-dependent manner, by the dissolved/suspended particles [36,37]. Separation is due to the parabolic (Poiseuille) flow velocity profile in the open tube channel, which allows small particles to be close to the walls, where the flow is stagnant, while the larger particles remain nearer the center of the tube, where flow is fastest. Consequently, ideal HDC separation occurs under the influence of flow alone, unlike SEC, where separation occurs via a reverse-sieving mechanism based on the relative sizes of the analyte and the effective radius $\bar{a}$ of the column packing material. Elution order, however, is equal in both HDC and SEC, with larger analytes eluting earlier than smaller ones. Whereas in SEC this is due to the larger analytes sampling a smaller portion of the pore volume of the packing material than do the smaller analytes, in HDC elution order is due to the preferential sampling of faster streamlines by the larger particles.

Figure 2.16 shows the combined HDC–SEC separation of a series of narrow polydispersity, linear polystyrenes. The exclusion limit of the column set is approximately 50,000 g/mol and polystyrenes in the range 2,200 to 43,900 g/mol are separated from each other via size exclusion. Rather than eluting together at the total exclusion limit, polystyrenes ranging from 127,000 to 4,000,000 g/mol are actually separated from one another. This separation beyond the column exclusion limit occurs via a hydrodynamic chromatography mechanism whereby the larger analytes, which do not penetrate the pores of the column packing, sample the interstitial flow profile in a size-dependent manner.

Hydrodynamic chromatography effects in general, and on determination of the column void volume V_o and consequent determination of K_{SEC} in particular, may be several percent. These effects will depend on R_p/d_p, the ratio of the pore diameter to the diameter of the packing particles; on λ, the ratio of the solute radius to the radius of the flow channels; and on flow rate. A smooth transition from an SEC to an HDC mechanism appears to depend on the ratio R_p/d_p. A reduction in this ratio leads to a reduction in HDC effects, with the extrapolation to an infinitely small ratio leading to separation by a strict size-exclusion mechanism. Initial studies indicated that $R_p/d_p < 0.002$ led to negligible HDC effects [38].

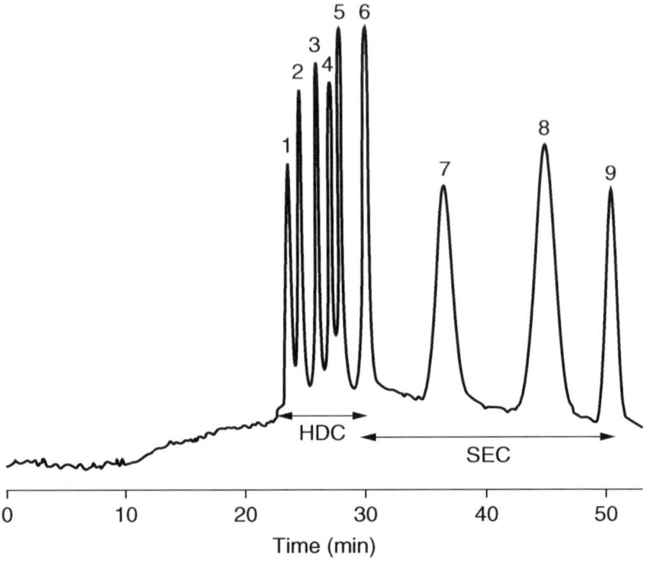

Figure 2.16 Hydrodynamic chromatography separation of polystyrenes on three Hypersil 45-cm length, 3-μm particle-size SEC columns, in THF at room temperature. M_w of peaks, in g/mol: 1, 4×10^6; 2, 2.2×10^6; 3, 7.75×10^5; 4, 3.36×10^5; 5, 1.27×10^5; 6, 4.39×10^4; 7, 1.25×10^4; 8, 2.2×10^3; 9, toluene. (Reprinted with permission from Ref. 38.)

Because most SEC columns will exceed an R_p/d_p ratio of 0.002, it is worth noting that once in HDC mode the selectivity of polymer separations is dependent on the ratio λ of the solute radius to the radius of the flow channels. Low values of λ lead to low chromatographic resolution. For example, $\lambda = 0.005$ leads to a velocity increase of only 1% as compared to infinitely small molecules. Conversely, when λ exceeds a maximum value (ca. 0.35), blockage of the column by the solutes may occur.

Avoiding hydrodynamic effects requires a balance between particle size and pore size, polymer size and channel size, and polymer relaxation time and flow rate. Reported inconsistencies in K_{SEC} can, in many cases, be explained by either inconsistencies in V_o as a result of HDC separation in the interstitial space or by critical deformation leading to distribution coefficients that appear to depend on flow rate but which, in actuality, do not.

High flow rates can extensionally deform large polymers, leading to early elution (flow-induced polymer degradation [39] is discussed in Section 7.2.3). However, this deformation may also be of a magnitude such that the highly extended polymer presents a smaller cross section relative to the direction of flow, permitting it to approach the packing particles more closely than in its unstreched state. This will allow for a more extensive sampling of the slower streamlines by the polymer and cause late elution. Flow-induced deformation has been explained successfully using the Deborah number, De, the ratio of hydrodynamic forces to Brownian forces, or the ratio of the longest relaxation time of the polymer to the convective time scale of the flow. For a flexible polymer in a good solvent,

$$\mathrm{De} = k\frac{\bar{v}}{d_p}\frac{6.12\Phi\eta_0 R_G^3}{RT} \tag{2.24}$$

where k is a constant that depends on the structure of the flow channels (usually, $k = 6$), $\bar{v}$ the superficial solvent velocity, Φ the value of Flory's constant after undergoing a correction for non-theta solvent–temperature conditions (see Section 9.6), η_0 the viscosity of the neat solvent, and all other symbols retain the same significance as before. At De < 0.1, polymer stretching can be considered insignificant and molecular size is the same as that at equilibrium conditions. Onset of deformation occurs at De $= 0.1$. At De $= 0.5$, critical deformation leads to highly extended, thread like structures. The latter leads to "abnormal" HDC separation, where the distribution coefficient appears to be flow-rate dependent, as large polymers elute later (higher K_{HDC}) at higher flow rates.

2.6.4 Slalom Chromatography Effects in SEC

In slalom chromatography (SC), ultrahigh molar mass polymers that have undergone critical flow-induced extension (at De ≥ 0.5) find themselves turning frequently around the column packing particles in their passage through the tortuous interstitial channels [40]. This repetitive and continuous turning retards elution, more so for longer polymer chains than for shorter ones. Elution order in SC is thus opposite to that in SEC or HDC, with longer polymers eluting later than shorter ones. Factors affecting the transition from SEC or HDC to SC behavior are the same as those affecting the Deborah number (De) above: flow rate, solvent viscosity, polymer size, and temperature. Column packing provides an additional factor in SC and in the SEC $\rightarrow$ SC and HDC $\rightarrow$ SC transitions. Failure to recognize that an SEC $\rightarrow$ SC transition has taken place can lead to the conclusion that nonentropic or nonequilibrium factors are involved in the SEC retention mechanism, when, in fact, the polymer is eluting by an entirely different mechanism, thereby obviating any comparison with SEC. An example of SC behavior for a polystyrene sample with M_w 20,500,000 g/mol eluting through a column packed with 15-μm nonporous particles is seen in Figure 2.17. At low flow rates (0.025 and 0.05 mL/min), the sample elutes by an HDC mechanism which shows the R_G distribution in the sample ranging from about 125 to 450 nm. At higher flow rates (>0.10 mL/min), a low-elution-volume plateau value of R_G is observed. This plateau R_G value is higher than the R_G of earlier-eluting chains and thus cannot be accounted for by an HDC mechanism. The elution order is due to the fact that the Deborah number for the largest chains has exceeded the critical value and a type of "coil-stretch" transition has occurred for the largest polymers [41]. At flow rates greater than 0.1 mL/min, these largest polymers elute by an SC mechanism, whereas their smaller, unstretched counterparts continue to elute by an HDC mechanism.

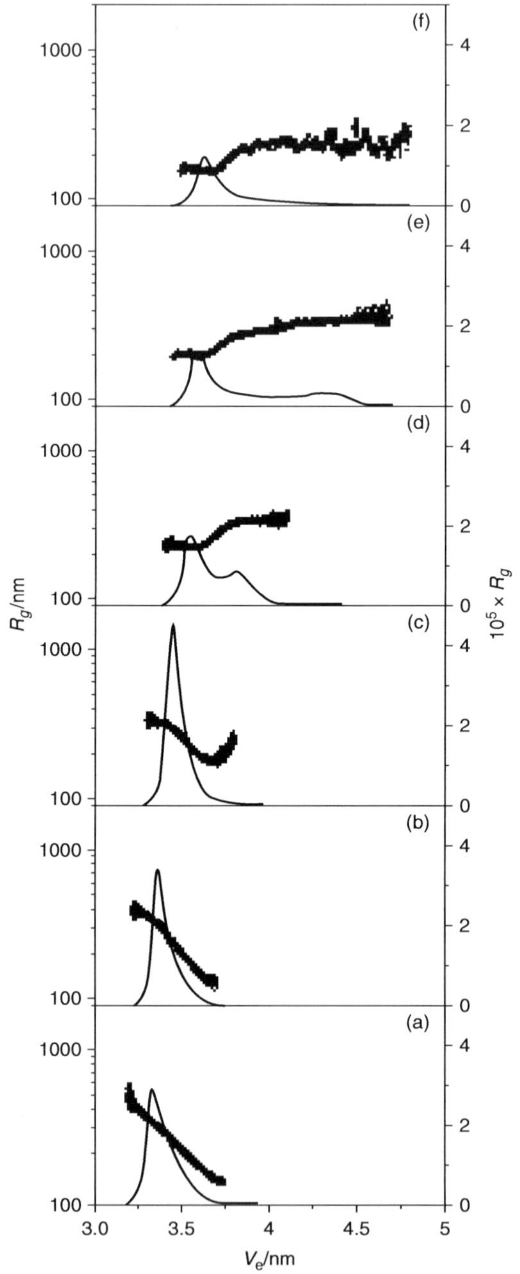

Figure 2.17 Transition from hydrodynamic to slalom chromatography: change in elution profile and R_G of a polystyrene with $M_w = 20{,}500{,}000$ g/mol, as a function of increasing flow rate. Abscissa, elution volume; right ordinate; static light scattering (16.8° angle) signal, tied to thin line; left ordinate, R_G, tied to thick line. Columns, two 300 × 4.6 mm, 15-μm particle size, nonporous PS-DVB columns; solvent, THF; temperature, 35°C. Flow rates, in mL/min: (a) 0.025; (b) 0.05; (c) 0.10; (d) 0.30; (e) 0.50; and (f) 1.00. (Reprinted with permission from Ref. 40.)

To summarize, HDC and SC effects become manifest in SEC mostly for high-*M* polymers at high flow rates, at De ≥ 0.1. With careful consideration in the choice of solvents (with respect to both viscosity and thermodynamic goodness or poorness), flow rate, and the size of the column packing material vis-à-vis the size of the polymer, HDC and SC mechanisms of separation can be avoided during an SEC experiment.

REFERENCES

1. L. R. Snyder and J. J. Kirkland, *Introduction to Modern Liquid Chromatography*, Wiley, New York, 1974, p. 339.
2. F. T. Gucker and R. L. E Seifert, *Physical Chemistry*, W.W. Norton, New York, 1966.
3. J. V. Dawkins, *J. Polym. Sci. A-2*, **14**, 569 (1976).
4. J. C. Giddings, *Unified Separation Science*, Wiley, New York, 1991.
5. A. M. Striegel, *J. Am. Chem. Soc.*, **125**, 4146 (2003). See the Erratum in *J. Am. Chem. Soc.*, **126**, 4740 (2004).
6. M. A. Boone and A. M. Striegel, *Macromolecules*, **39**, 4128 (2006).
7. M. A. Boone, H. Nymeyer, and A. M. Striegel, *Carbohydr. Res.*, **343**, 132 (2008).
8. M. A. Boone, H. Nymeyer, and A. M. Striegel, in preparation.
9. J. C. Moore, *J. Polym. Sci. A*, **2**, 835 (1964).
10. D. J. Richard and A. M. Striegel, in preparation.
11. J. N. Little, J. L. Waters, K. J. Bombaugh, and W. J. Pauplis, *J. Polym. Sci. A-2*, **7**, 1775 (1969).
12. W. W. Yau, C. P. Malone, and S. W. Fleming, *J. Polym. Sci.*, **6**, 803 (1968).
13. A. M. Striegel, *J. Chromatogr. A*, **1033**, 241 (2004).
14. W. W. Yau, H. L. Suchan, and C. P. Malone, *J. Polym. Sci. A-2*, **6**, 1349 (1968).
15. J. J. Hermans, *J. Polym. Sci. A-2*, **6**, 1217 (1968).
16. E. F. Casassa, *J. Phys. Chem.*, **75**, 3929 (1971).
17. J. C. Giddings, E. Kucera, C. P Russell, and M. N. Myers, *J. Phys. Chem.*, **72**, 4397 (1968).
18. T. C. Laurent and J. Killander, *J Chromatogr*, **14**, 317 (1964).
19. A. G. Ogston, *Trans. Faraday Soc.*, **54**, 1754 (1958).
20. M. E. Van Kreveld and N. Van Den Hoed, *J. Chromatogr.*, **83**, 111 (1973).
21. E. F. Casassa, *J. Polym. Sci. A-2*, **10**, 381 (1972).
22. E. F. Casassa, *J. Polym. Sci. B*, **5**, 773 (1967).
23. E. F. Casassa and Y. Tagami, *Macromolecules*, **2**, 14 (1969).
24. E. F. Casassa, *Sep. Sci.*, **6**, 305 (1971).
25. W. W. Yau and C. P. Malone, *Polym. Prepr.*, **12**, 797 (1971).
26. E. F. Casassa, *Macromolecules*, **9**, 182 (1976).
27. J. C. Moore and M. C. Arrington, International Symposium on Macromolecular Chemistry, Tokyo and Kyoto, 1966, paper VI-107.
28. J. W. McBain, *J. Am. Chem. Soc.*, **57**, 699 (1935).

29. W. W. Yau, J. J. Kirkland, and D. D. Bly, *Modern Size-Exclusion Liquid Chromatography*, Wiley-Interscience, New York, 1979.
30. C. Tanford, *Physical Chemistry of Macromolecules*, Wiley, New York, 1961, Chaps. 3 and 5.
31. A. H. Abdel-Almin and A. E. Hamielec, *J. Appl. Polym. Sci.*, **16**, 1093 (1972).
32. D. Berek, D. Bakos, L. Soltes, and T. Bleha, *J. Polym. Sci. B*, **12**, 277 (1974).
33. D. A. Hoagland, in *Strategies in Size Exclusion Chromatography*, ACS Symp. Ser. 635, M. Potschka and P. L. Dubin, eds., American Chemical Society, Washington, DC, 1996, Chap. 10.
34. I. Teraoka, *Macromolecules*, **37**, 6632 (2004).
35. J. H. Aubert and M. Tirrell, *Sep. Sci. Technol.*, **15**, 123 (1980).
36. S. S. Huang, in *Handbook of Size Exclusion Chromatography and Related Techniques*, 2nd ed., C.-S. Wu, ed., Marcel Dekker, New York, 2004, Chap. 23.
37. A. K. Brewer and A. M. Striegel, *Anal. Bioanal. Chem.*, **393**, 295 (2009).
38. G. Stegeman, J. C. Kraak, and H. Poppe, *J. Chromatogr.*, **550**, 721 (1991).
39. A. M. Striegel, *J. Liq. Chromatogr. Rel. Technol.*, **31**, 3105 (2008).
40. Y. Liu, W. Radke, and H. Pasch, *Macromolecules*, **38**, 7476 (2005).
41. C. D. DeLong and D. A. Hoagland, *Macromolecules*, **41**, 4887 (2008).

3

BAND BROADENING

3.1 INTRODUCTION

In column chromatography, a small volume of the sample solution is injected to form a band at the top of the column. As this band migrates downstream, its width increases. The sample solution in the band becomes increasingly more dilute as the band becomes more spread out in the direction of the flow, parallel to the axial (or longitudinal) direction of the column. Band broadening of a pure component can be used to measure the efficiency of the chromatographic system. (The term *band broadening* is commonly used with the implication that the solute band consists of a pure component.) Column band broadening is measured experimentally by the width of single chromatographic peaks such as those illustrated at the bottom of Figure 2.1. Gross overestimation of column band broadening can occur if the probe chromatographic species is not pure but contains partially separated components, or is a polymer species having an appreciable MMD.

The nomenclature used in reporting chromatographic band broadening in the literature is quite varied and sometimes confusing. Readers should be aware of the many near-synonyms for band broadening that appear in the literature, such as peak broadening; zone spreading; and instrumental, axial, longitudinal, or column dispersion.

All forms of band broadening are detrimental to chromatographic resolution. Basically, chromatographic separation is a demixing phenomenon, the maximization of separative transport relative to dispersive transport [1]. For example, in the LC

Modern Size-Exclusion Liquid Chromatography: Practice of Gel Permeation and Gel Filtration Chromatography, Second Edition
By André M. Striegel, Wallace W. Yau, Joseph J. Kirkland, and Donald D. Bly

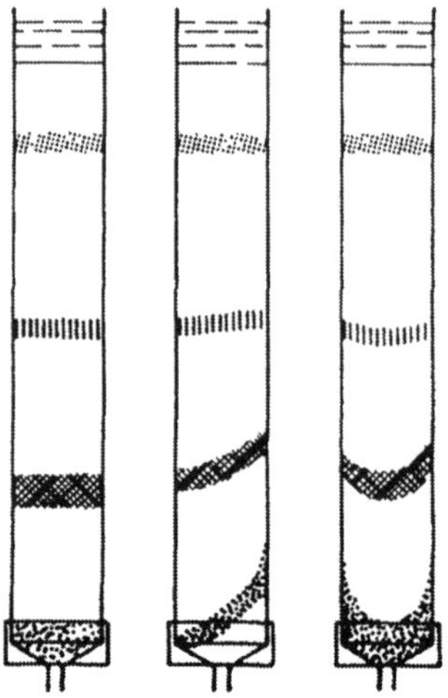

Figure 3.1 Effect of packing inhomogeneity on band distortion: (a) uniform migration of a band in a well-packed column; (b) band distortion due to uneven packing density across the column; (c) band distortion due to radial packing inhomogeneity.

analysis of small molecules, different molecular species in the original sample solution are demixed as they elute from the column in separated peaks. On the other hand, band broadening is a back-mixing or remixing phenomenon that causes the LC peaks to spread out and overlap. The effect is to make peak identification and peak-size analysis more difficult. Band broadening is also detrimental to SEC. In the analysis of broad MMD polymers by SEC, the effect of band broadening is to interfere with the integrity of the MMD information as displayed by the elution curve profile. A small distortion of the SEC curve shape by band broadening can cause large M errors in the SEC analysis, especially when using calibration curves. The problems of M accuracy and SEC resolution are discussed in detail in Chapter 4. Proper methods of correcting for the band-broadening effect in SEC calibration and calibration-based M computation procedures are discussed in Chapter 8.

Excessive peak broadening can result from poorly packed columns. A uniformly packed column is illustrated in Figure 3.1a, in which the entire band across the column is shown to migrate evenly through the column. In columns with large packing inhomogeneity as illustrated in Figure 3.1b and c, the solute band can become grossly distorted as the band migrates through the column. This band distortion is observed as excessive band broadening. Macroscopic channeling is another packing defect that can cause large peak broadening because of the fingering effect of the solute band in the packed bed. A detailed consideration of packing homogeneity and packing techniques is discussed in Chapter 6.

Gross band broadening can also result from the excessive extracolumn volume that is present in the chromatographic instrument. Large-volume elements such as a flow filter or pulse dampener must be installed before the sample injector to avoid excessive band broadening. Details for minimizing extracolumn band-broadening effects are discussed in Chapter 5.

In this and the next section, general LC band-broadening effects are discussed in detail. This basic information is not only useful to the understanding of band broadening of polymers in SEC, but is directly applicable to the practice of SEC separations of small molecules and oligomers.

3.1.1 Basic Column-Dispersion Processes

In a well-planned SEC experiment where well-packed columns and an efficient instrument are used, ultimate SEC column efficiency will depend on the inherent band-broadening processes occurring in the column. The uniform band broadening illustrated in Figure 3.1a is caused by molecular mass transfer processes and microscopic flow irregularities inherent in the column packing structure. These band dispersion effects constitute a large part of the overall band broadening observed in the usual SEC experiments. Accordingly, a large part of this chapter is devoted to a discussion of this important subject. An understanding of the basic column dispersion processes and their dependence on SEC operating variables is needed for making efficient SEC separations with the best compromises among time, accuracy, effort, and convenience.

The basic elements of SEC column dispersion are illustrated in Figure 3.2. Figure 3.2a represents a cross section of the solute band profile at the column inlet just after sample injection. Figure 3.2b shows one of the three fundamental processes leading to band broadening in SEC: *eddy diffusion*. This process arises because sample molecules take separate routes through the packed bed, as illustrated by the various arrows. Since the solute moves at different speeds in wide and narrow flow paths, some solute molecules move downstream faster than the others within a given time span. As a result of this eddy-diffusion phenomenon, a spreading of the solute molecules occurs from the initial narrow band in Figure 3.2a to a broader band in Figure 3.2b.

A second contribution to band broadening occurs as a result of the resistance of solute to *mobile-phase mass transfer* (Figure 3.2c). This broadening process is caused by the velocity gradient profile that exists in a single flow stream. Since liquid near the surface of the column packing particle moves relatively more slowly than the liquid at the center of the flow stream, solute molecules at the center migrate farther downstream than the others. Band broadening due to this dispersion process decreases with increasing lateral diffusion rate of the solute molecules between the fast- and slow-moving liquid regions. At times this dispersion process is called *mobile-phase lateral diffusion* or *extraparticle mass transfer*.

Band broadening caused by the resistance of the solute to *stationary-phase mass transfer* is illustrated in Figure 3.2d (for simplicity, a packing particle with a single pore instead of a complex pore structure is shown). This process of band broadening arises from the slow solute diffusion in and out of the pores of the packing particles. While some molecules are diffusing into the pores, others move with the solvent

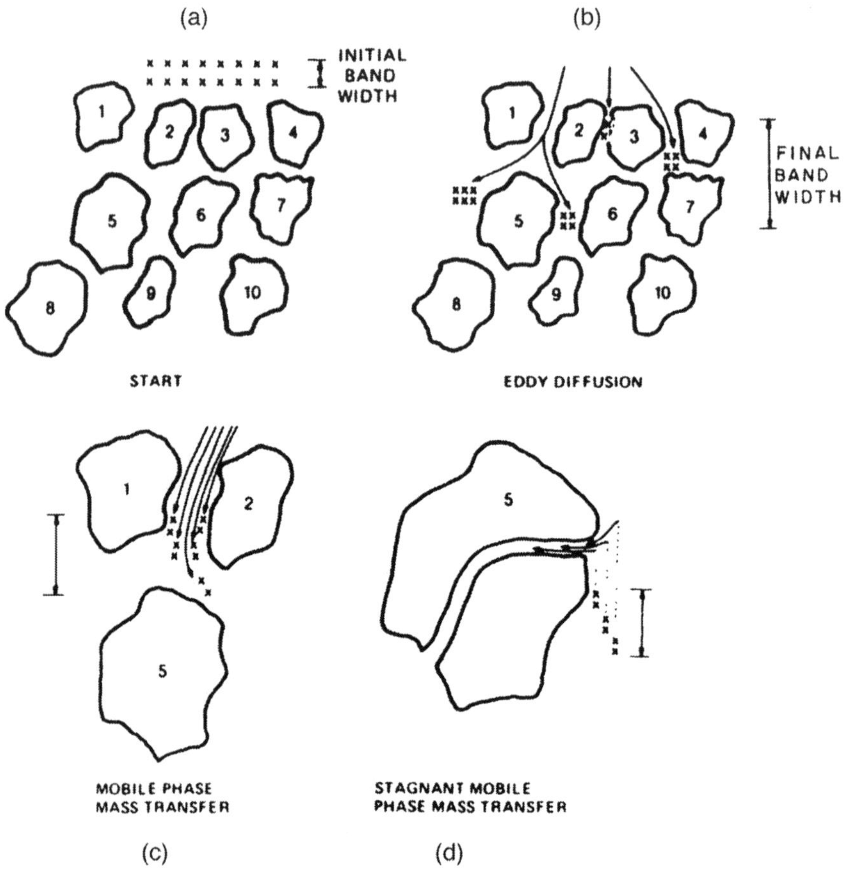

Figure 3.2 Basic peak dispersion processes.

farther downstream. For large solute molecules with low diffusion coefficients, this type of solute downstream migration will cause extensive band broadening. Therefore, this dispersion process, which has also been called *stationary-phase lateral diffusion*, *intraparticle mass transfer*, or *stationary-phase nonequilibrium mass transfer*, is the major contributor to band broadening in the SEC analysis of macromolecules. [For general LC methods other than SEC, this dispersion process is better known in terms of *stagnant mobile-phase mass transfer*, the term *stationary-phase mass transfer* being reserved to describe the dispersion effect due to the LC stationary phase (not shown in Figure 3.2).]

Longitudinal diffusion is another basic band-broadening process (not shown in Figure 3.2) in which the band is broadened along the column's axis parallel to the flow direction by molecular diffusion of the solute. This form of band dispersion is important in GC but is generally insignificant in large-molecule SEC because of the slow diffusion of macromolecules.

3.1.2 Peak Variance

The phenomenon of chromatographic band broadening is a random statistical process of solute mixing and is therefore subject to statistical analyses. In statistics, the fundamental parameter for describing the width of a statistical distribution is the variance of the distribution. This basic concept is adopted in chromatography, where the variance of single chromatographic peaks is the fundamental parameter for evaluating column band-broadening effects. Mathematically, peak variance in its most general form (Var) is defined as the second central moment of the peak (described in terms of a continuous distribution of normalized chromatogram height h):

$$\text{Var} \equiv \sigma_x^2 = \int_{-\infty}^{\infty} h(V - V_R)^2 \, dV \tag{3.1}$$

or defined as the mean-square deviation of V from V_R for the peak (described by discrete chromatogram heights h_i):

$$\text{Var} \equiv \overline{(V - V_R)^2} = \frac{\sum_i h_i (V_i - V_R)^2}{\sum_i h_i} \tag{3.2}$$

where σ_x is the standard deviation of a general statistical distribution, V and V_i the retention volume variable, V_R the retention volume of the chromatographic peak (see Figure 3.3), and the subscript i the sequence index for the discrete equally spaced data points used in the variance calculation.

A very important property of the statistical variance is the *additivity rule of the variances* [2], which states that the overall peak variance is the sum of the individual variances resulting from each of the independent band-broadening effects occurring in the chromatographic process. Thus, for *mutually independent dispersion*

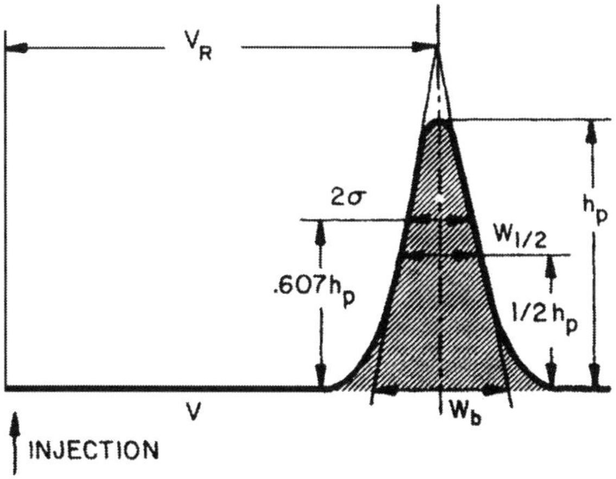

Figure 3.3 Band-broadening parameters; Gaussian peak model.

processes,

$$\text{Var} = \sum_i \text{Var}_i \tag{3.3}$$

where Var_i represents variance contributions from the various dispersion processes. Equation 3.3 is very useful in chromatographic band-broadening studies because it allows different dispersion effects and individual columns and volume elements to be evaluated separately or as an integral part of the chromatographic system. Specific forms of Equation 3.3 are used in later chapters to describe specific SEC dispersion effects.

The variance formulations of Equations 3.1 to 3.3 are universal expressions for chromatographic peaks in general, regardless of peak shape. These equations are the most accurate expressions for evaluating chromatographic band broadening. However, without a peak model for reference, it is difficult to visualize the physical significance of the peak variance concept and to establish a tie between the mathematic symbols in these equations and the observable parameters in the chromatographic experiment.

Conventional band-broadening parameters are developed from the Gaussian-peak-shape model, as shown in Figure 3.3. The contour of a Gaussian elution peak is described by the equation

$$h = \frac{A}{\sigma\sqrt{2\pi}} e^{-(V-V_R)^2/2\sigma^2} \tag{3.4}$$

where A is the area of the peak, σ the standard deviation of the Gaussian peak in retention volume units, and h, V, and V_R are as defined in Equations 3.1 and 3.2. (Peak standard deviation is sometimes reported in retention time units; however, this practice is not recommended. Unless flow rate is stated, the standard deviation in the time units gives only incomplete information about band broadening.) It can be shown that $\sigma = 0.43W_{1/2} = W_b/4$, where $W_{1/2}$ and W_b, also in volume units, are the peak width at half-height and at the base, respectively.

Substitution of h from Equation 3.4 into Equation 3.1 leads to the result that the variance of a Gaussian peak is equal to the square of the peak standard deviation:

$$\text{Var} = \sigma^2 = \left(\frac{W_b}{4}\right)^2 \qquad \text{(Gaussian peak)} \tag{3.5}$$

With Equation 3.5, the tie is established between the variance and the experimental quantities shown in Figure 3.3. According to Equation 3.5, peak variance increases linearly with the square of the peak width.

With the Gaussian peak model, the variance additivity rule in Equation 3.3 becomes

$$\sigma^2 = \sum_i \sigma_i^2 \tag{3.6}$$

Because band-broadening effects are summed according to the square of the σ values, the effect of one relatively large dispersion element is greatly magnified in the overall chromatographic band broadening. A single element with large dispersion in a system can dominate the total band broadening and damage the efficiency of the entire system. For example, the high-performance features of SEC columns cannot be realized (and thus the potential of the method will be wasted) if such columns are used with SEC instruments which generally exhibit large extracolumn band-broadening effects. To achieve high-quality system performance, care must be exercised to avoid the use of any element in the chromatographic system that causes excessive band dispersion.

Equations 3.5 and 3.6 derived from the Gaussian peak model provide good predictions of column performance with usual SEC experiments. The use of the Gaussian peak model to study band broadening is supported by both the plate and rate theories of band broadening (see the discussions of Equations 3.8, 3.10, and 3.30). However, for studying dispersion processes that cause large peak skewing, there will be errors in calculating variance using Equations 3.5 and 3.6 instead of Equations 3.1 to 3.3. Development of the band-broadening parameters with a skewed peak model is discussed in Section 3.5.

3.2 LC PLATE THEORY

There are two ways of approaching the theoretical interpretation of chromatographic band broadening. In the kinetic or *rate theory,* considered in the next section, band broadening is explained in terms of realistic models involving molecular diffusion and flow mixing. The other approach is the *plate theory*, which is a simplified, phenomenological approach. It explains band broadening by random fluctuations around the mean retention volume by a simulated partitioning model in a chromatographic column. Plate theory was first applied to LC studies by Martin and Synge [3], and many early advances in gas chromatography also owe a great deal of credit to development of this insight. Because of its simplicity, the plate theory will continue to be a useful, general model for studying chromatographic band broadening. The basic derivation of general plate theory can be found in many GC and LC books [4,5], and only a brief explanation is given below.

3.2.1 Basic Plate Theory

In the plate model the chromatographic column is pictured as being divided into N number of adjoining separation zones, with each zone having such a length that there can be complete equilibrium of the solute between the mobile and stationary phases within the zone. Each zone is called a *theoretical plate*, and its length in the column is called the *height equivalent to a theoretical plate* (HETP) or simply the *plate height*, H.

To illustrate the plate concept, a rudimentary five-plate column ($N = 5$) is shown in Figure 3.4, where the sequence of the plates is indexed by the serial number r. The feature of equilibrium partition in each plate is indicated in the figure by the

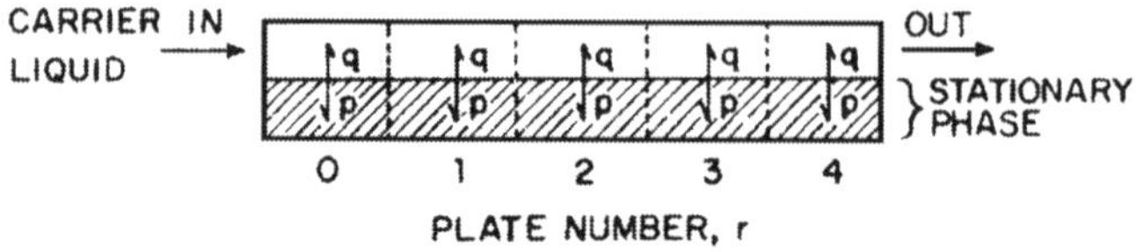

Figure 3.4 Hypothetical column of five theoretical plates.

balance between q and p, which are the fractions of the total solute in the mobile and the stationary phases, respectively, with $q + p = 1$. In this picture the flow of the carrier liquid is simulated by the sequential displacement of the entire top mobile-phase section to the right, one plate at a time. The number of times that this volume displacement has taken place following the introduction of a sharp band into the first plate is designated by the index number n. With each volume displacement, only a fraction q of the solute in each plate is carried to the next plate, leaving a fraction p behind. The solute in each plate reequilibrates in the new situation, and the displacement process repeats. This repetitive partition process leads to a solute distribution among many neighboring plates that follows the binomial distribution function. According to binomial statistics, the fraction of the original solute being in the rth plate following n displacements is

$$W(n,r) = \frac{n!}{r!(n-r)!} q^r p^{n-r} \tag{3.7}$$

In chromatography the solute concentration detector monitors the fraction q of the Nth (last) plate as a function of n. The elution curve is therefore described by $qW(n, N)$, where n is proportional to retention volume. For the usual large number of plates in chromatographic columns ($N > 50$), the binomial solute distribution becomes indistinguishable from the Gaussian distribution function [6]. With algebraic transformation, the Gaussian peak elution profile as predicted by the plate model can be expressed in terms of the experimental quantities of concentration c, retention volume V, peak retention volume V_R, sample weight W, and p, the fraction of solute in the stationary phase:

$$c = \frac{W}{\sqrt{2\pi p V_R^2/N}} e^{-N(V-V_R)^2/2pV_R^2} \tag{3.8}$$

By comparing Equation 3.8 with the general Gaussian function (Equation 3.4), one finds that

$$\sigma = V_R\sqrt{\frac{p}{N}} \quad \text{or} \quad N = \frac{pV_R^2}{\sigma^2} \tag{3.9}$$

As p approaches unity,

$$\sigma = \frac{V_R}{\sqrt{N}} \quad \text{or} \quad N = \frac{V_R^2}{\sigma^2} \tag{3.10}$$

Other results of the general plate theory are

$$H = \frac{L}{N} = L\frac{\sigma^2}{V_R^2} \tag{3.11}$$

and

$$H = \sum_i H_i = \frac{L}{V_R^2}\sum_i \sigma_i^2 \tag{3.12}$$

where L is the column length and H_i the individual plate height contribution of independent column dispersion effects. Equation 3.12 is derived directly from Equations 3.6 and 3.11.

In summary, the predictions resulting from the general plate theory are:

1. The peak shape is Gaussian (Equation 3.8).
2. Peak width increases linearly with retention volume (Equation 3.10).
3. Each peak in a chromatogram has approximately the same values of N and H (Equation 3.10).
4. N increases linearly with column length (Equation 3.11).

(Items 2 and 3 of these predictions are not observed in SEC. The SEC column dispersion has many unique features, as discussed in the next section.) The predicted dependence of band broadening on peak retention according to the general theory for GC and LC of small molecules is illustrated in Figure 3.5. Early peaks are tall and spikelike; later peaks are short and broad. The peaks in the figure were calculated from Equation 3.8 for a hypothetical column of 400 plates. Equal peak areas and $p = 1$ were assumed in the calculation.

The success of the plate theory can be attributed to the fact that experimental observations in GC and LC (other than SEC) are in good agreement with theoretical predictions. Approximate constancy of N and H for various probe peaks in a chromatogram is usually found experimentally. For peaks that are only very slightly retained ($p < 1$, Equation 3.9), the value of N can vary with V_R of the probe peak. In some cases experimental values of N calculated according to Equation 3.10, which

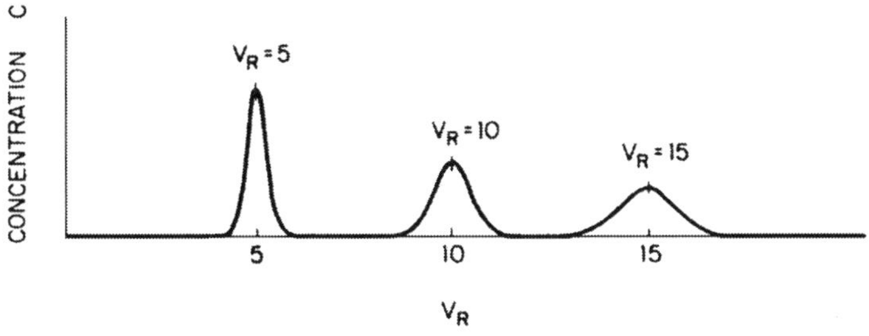

Figure 3.5 Theoretical peak shapes for a hypothetical column of 400 plates.

assumes that $p = 1$, are often somewhat larger for peaks of low retention [7]. For low-retention peaks, Equation 3.10 used in the experimental N-value calculation overestimates the true column plate count, because the actual value of p is less than the implied p value of 1 in the equation.

The number of theoretical plates N is a dimensionless quantity. The value of N is a fundamental measure of the system efficiency, independent of whether the chromatographic results are reported in retention volume or retention time units. The same is true for plate height H, which is in the units of column length; the same value of N is obtained whether it is calculated as $(V_R/\sigma)^2$ or as $(t_R/\sigma_t)^2$, where σ_t is the peak standard deviation in time. While N measures system efficiency, H measures the specific column efficiency. For systems with low extracolumn dispersion, H is a measure of the intrinsic efficiency of the column packing. In chromatography a plate is only a fictitious model, which does not actually exist for chromatographic columns. However, in practice, plate count N and plate height H are used as if they are real physical quantities.

Plate height equations derived from the basic plate height expression in Equation 3.12 permit critical evaluation of various dispersion processes in terms of their relative importance to system efficiency for different forms of chromatography. Each independent process is associated with a variance σ_i^2 and the corresponding plate height contribution H_i. The problem is to identify important dispersion processes and express the corresponding H_i contributions in terms of physical and experimental parameters. One approach to the problem is to use the random-walk model suggested by Giddings [8]. This model considers each dispersion process as being a random displacement of the solute molecules back and forth among flow streams of different velocities. In the random-walk model, the variance of each dispersion process can be expressed as

$$\sigma_x^2 = nl^2 \tag{3.13}$$

where n and l are the number and the mean characteristic length of the random steps, respectively. This semiempirical approach to derive the plate height equation can usually provide the correct functional dependence of H_i on important physical parameters. Since this model does not necessarily give a realistic description of the actual dispersion process, semiempirical adjustable constants are commonly included in the derived plate height expressions for explaining the experimental band-broadening data. The random-walk model is most useful for analyzing complex dispersion processes from complicated multichannel flow irregularities and mass transfer considerations. Many of these complex dispersion effects have been discussed in References 1 and 8.

3.2.2 The van Deemter Equation

For dispersion effects that involve simple flow and diffusion processes, exact expressions for H_i can be derived from rigorous mass transfer differential equations from the rate theory approach. A classical example is the theoretical work of van Deemter

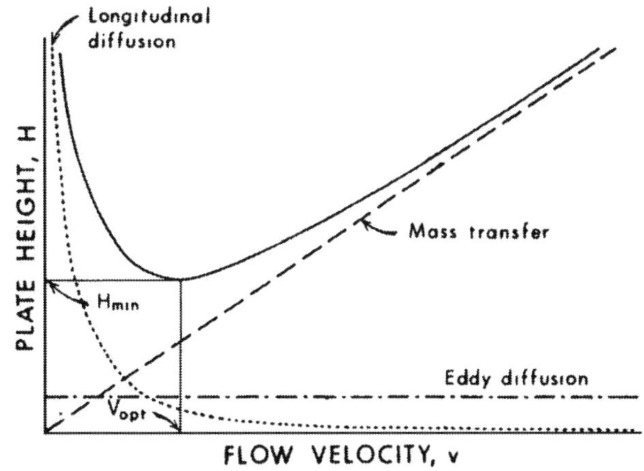

Figure 3.6 Theoretical van Deemter plot. (Reprinted with permission from Ref. 10.)

et al. [9], which led to the successful prediction of the dependence of GC column efficiency on carrier gas velocity. The now well-known van Deemter equation is highly instructive in illustrating basic peak dispersion processes.

For a general discussion of the effect of flow rate on plate height, the van Deemter equation can be simply represented by

$$H = A + \frac{B}{v} + Cv \tag{3.14}$$

where v is the flow velocity and the constants A, B, and C are associated with the plate height terms due to eddy diffusion, longitudinal diffusion, and mass transfer, respectively. A graphical representation of the parameters in Equation 3.14 is shown in Figure 3.6. Since Equation 3.14 is well known among chromatographers, reference to the three dispersion processes simply as A, B, and C terms is a commonly accepted practice. The dispersion process due to eddy diffusion (A term, Figure 3.2b) is a simple flow-splitting phenomenon that is not expected to vary with flow velocity. The value of the A term is largely dependent on particle size and the homogeneity of the packed column bed. Band broadening due to simple molecular diffusion in the long axis of the column is the B term. This term decreases (Figure 3.6) with increasing flow rate because a shorter time is available for longitudinal diffusion in a faster chromatographic separation. For the mass transfer or lateral diffusion processes (C term, Figure 3.2c and d), an increase in flow rate emphasizes the velocity differences between flow streams, which results in an increase in plate height. The magnitude of the C term is also dependent on the rate of diffusion of solute in and out of the pore structure. Therefore, larger, slower-diffusing molecules increase the value of the C term more than do smaller, faster-diffusing molecules. The solid line in Figure 3.6, which is the sum of all three dispersion processes, shows a minimum in plate height (H_{min}) which corresponds to the "optimum" velocity v_{opt}; at this velocity the column has maximum efficiency. In practice, flow rates somewhat higher than v_{opt} are often used for reasonably fast chromatographic separations. Band broadening in most LC and SEC separations is controlled by the mass transfer terms since the longitudinal effect (B term) is generally insignificant, and except for small molecules, H_{min} is not observed in SEC.

The C term in Equation 3.14 is the sum of plate height contributions from three possible processes: (1) the C_M term from the extraparticle effects, as illustrated in Figure 3.2c (this term is present even for nonporous solid packings), (2) the C_{SM} term from stagnant mobile-phase effects, as illustrated in Figure 3.2d (this is an important SEC term often called the "stationary" mass transfer term in SEC), and (3) the C_S term from conventional LC stationary-phase mass transfer effects involving the basic sorption–desorption processes.

Historically, all LC dispersion processes were considered as being independent of each other. This concept constitutes the classical interpretation of LC band broadening, as expressed in the expanded van Deemter equation,

$$H = A + \frac{B}{v} + C_M v + C_{\text{SM}} v + C_S v \tag{3.15}$$

This equation predicts a linear increase of plate height with increasing flow velocity at high-flow-rate regions, where the overall plate height is dominated by the C term. (This is expected to occur in LC at moderately high flow rates due to the relatively small A and B terms.) However, in practice, increase of plate height is found to taper off at high flow rates. A plausible explanation for this is provided by the Giddings coupling theory [1,8], which is discussed next.

3.2.3 Flow-Diffusion Coupling

The coupling concept is in contrast to the assumed independence of the eddy- and lateral-diffusion terms in the classical plate height theory (Equation 3.15). The coupling theory [1,8] maintains that both the eddy flow or stream-splitting effect and lateral diffusion can effectively move solute molecules from one flow stream to another. Thus, the combined effect of eddy and lateral diffusion provides more chances for each solute molecule to experience the different velocities in the various flow channels. The more frequently the individual molecules can sample the various flow velocities while traveling downstream in the column, the more likely it is that they can attain the same statistical mean velocity and can all elute from the column closer together.

The end result of coupling is reduced band broadening compared to that of eddy diffusion alone. A simplified explanation of these concepts is illustrated in Figure 3.7, where band broadening due to eddy diffusion alone is compared to that of coupled eddy-lateral diffusion. In this figure the locations of solute molecules are pictured in two time frames. The frames at the right (Figure 3.7b and d), taken a short time after the ones on the left (Figure 3.7a and c), show that the molecules have moved farther downstream with respect to the packing particles and formed a

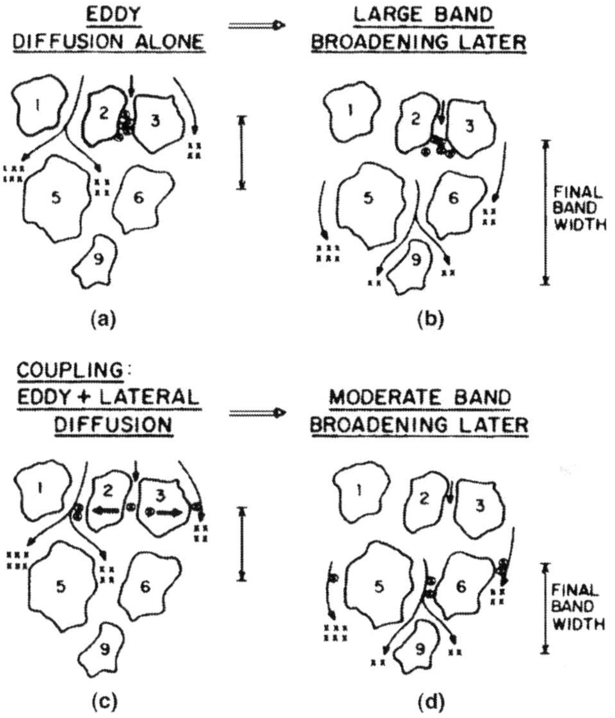

Figure 3.7 Reduced band broadening through coupling of eddy and lateral diffusion.

broader band. All the solute molecules in the figure are considered structurally identical, but the slower-moving molecules have been encircled for identification. The motions of these encircled molecules are the focus of this discussion. When eddy diffusion works alone (Figure 3.7a and b), these slow molecules lag far behind the others and contribute greatly to the overall band broadening. In coupling (Figure 3.7c and d), these molecules have a chance to escape from the slow flow stream via lateral diffusion around the packing particles (indicated by the two oppositely pointed arrows in Figure 3.7c). These diffusion-coupled molecules can thus follow faster-moving streamlines and elute closer to the other molecules, resulting in a reduced band width, as illustrated in Figure 3.7d. While Figure 3.7 shows how the diffusion-coupling effect of the microscopic flow irregularities can reduce band broadening, an analogous coupling situation can exist for band broadening due to nonuniform velocity profile over the column cross section [4,11]. Of course, the velocity profile contribution to plate height is less in well-packed columns.

Based on the random-walk model of the coupling concept, the combined mobile-phase plate height H_M can be expressed as [8]

$$H_M = \frac{1}{(1/A) + (1/C_M v)} \tag{3.16}$$

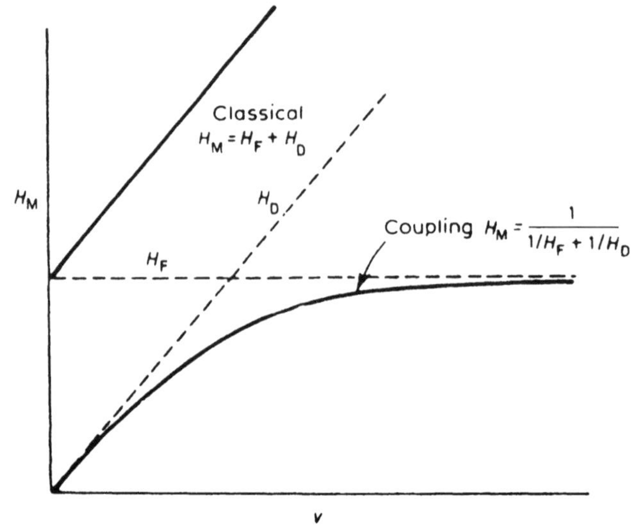

Figure 3.8 Extraparticle mobile-phase plate height contributions; classical versus coupling theory. (Reprinted with permission from Ref. 8.)

As shown in Figure 3.8, the plate height contribution of the coupled term calculated from Equation 3.16 is smaller than that of its individual component terms. At high flow rates, H_M approaches the eddy-diffusion term (the A or H_F term). The quantity H_D in Figure 3.8 reflects the C_M band-broadening term.

The plate height equation that incorporates the coupling concept can be expressed as

$$H = \frac{B}{v} + C_{\mathrm{SM}}v + C_S v + \frac{1}{(1/A) + (1/C_M v)} \tag{3.17}$$

The general curve shape of the H versus v plot predicted by this equation has been confirmed by many experimental studies and is also supported by data obtained on nonporous packings, as discussed in the next section. The coupling theory is generally considered theoretically sounder than the classical van Deemter expressions. However, it should be noted that the focus of the LC coupling theory is on extraparticle dispersion effects. This consideration is important to SEC only for small molecules. For the SEC of macromolecules, band broadening is dominated by the C_{SM} term, which is not subject to coupling.

Although the magnitude and relative importance of each plate height contribution from various dispersion mechanisms vary from one form of LC to another, the general functional dependence of each contribution to plate height on flow rate can be depicted by the plot shown in Figure 3.9. This figure shows the H versus v characteristics of each plate height component, and also the overall plate height:

$$H = H_L + H_{\mathrm{SM}} + H_S + H_M \tag{3.18}$$

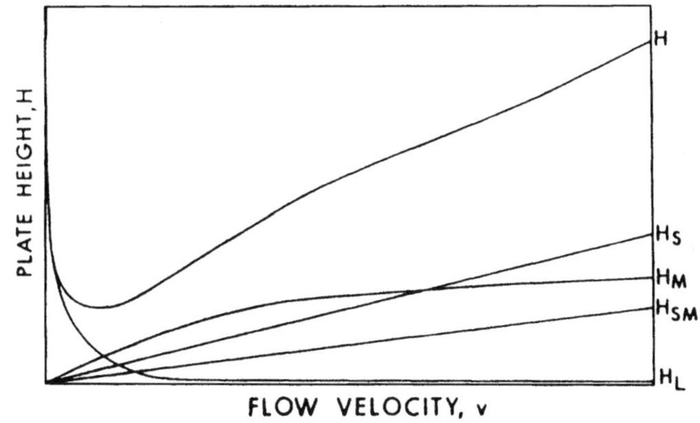

Figure 3.9 Dependence of plate height on mobile-phase velocity. (Reprinted with permission from Ref. 10.)

where H_L, H_{SM}, H_S, and H_M are the plate height contributions due to longitudinal-diffusion, stationary-mobile-phase, stationary-phase, and interparticle-mobile-phase mass transfer processes, respectively. They describe the corresponding terms in Equation 3.17.

The plate height factors given in Equation 3.18 represent a general rather than a comprehensive account of the column dispersion processes. The extracolumn dispersion effect, which is not included, is expected to behave much like the independent mass transfer terms, with its plate height contribution increasing linearly with flow velocity. The shape of the overall H versus v plot can vary greatly depending on the particular chromatographic technique used. When there is a single dispersive effect dominating in a particular chromatographic system, the shape of the overall H versus v plot will bear a resemblance to this component dispersion effect. In practice, it is desirable to have an experimental H versus v plot of the working chromatographic system. Such data can provide valuable insights into the relative importance of different plate height components and permit compromises in the experimental conditions to be made to obtain high resolution or separation speed.

For more elaborate chromatographic design considerations, a more detailed plate height equation with explicitly expressed dependence on packing particle size and solute-diffusion coefficients is more appropriate (8, 12):

$$H = b\frac{D_M}{v} + c_{\text{SM}}\frac{vd_p^2}{D_{\text{SM}}} + c_S\frac{vd_f^2}{D_S} + \frac{1}{(1/ad_p) + (D_M/c_M vd_p^2)} \tag{3.19}$$

where

d_p = particle diameter of the packing
d_f = film thickness of the LC stationary phase
D_M, D_{SM}, D_S = solute-diffusion coefficients corresponding to extraparticle, stagnant mobile phase, and stationary phase, respectively

with a, b, c_M, c_{SM}, and c_S being the coefficients of the respective dispersion terms in the plate height equation. The magnitudes of these coefficients are generally a function of the nature and the loading of the stationary phase, as well as the geometry of the packing and its pore structure. (The explicit expression for c_{SM} in the context of SEC band broadening is described in the next section.) Implicitly, the plate height is a function of many other operating variables, such as temperature, solvent viscosity, and so on, as discussed in Section 3.4 for the case of SEC.

3.2.4 Reduced Plate Height

The plate height equation can also be expressed in terms of dimensionless quantities, reduced plate height **h**, and reduced velocity **v** (8):

$$\mathbf{h} = \frac{H}{d_p} \tag{3.20}$$

$$\mathbf{v} = \frac{v d_p}{D_M} \tag{3.21}$$

The value of **v** is often several times larger in LC than in GC because of lower solute-diffusion rates in liquids (D_M values in liquids are on the order of about 10^{-5} cm^2/s). Even larger values of **v** are typical in SEC for macromolecules that have very small D_M values (about 10^{-7} cm^2/s). A typical value of **h** for a monomer with an efficient column is approximately 2 to 3.

To study flow-rate effects, the use of the reduced values **h** and **v** permits column efficiency data collected from different chromatographic studies to be compared effectively. An example of this is found in a band-broadening study of the extraparticle mobile-phase effects [13], where the Giddings coupling expression for the extraparticle effects ($H_L + H_M$ in Equation 3.18) is tested against several empirical equations to explain experimental data. Table 3.1 lists these equations with the original references. Each equation given in the table has a characteristic slope in the linear region of the log **h** versus log **v** plot. The slope predicted is 1 for equation (1), $\frac{1}{2}$ for equation

Table 3.1 Equations describing the plate height contribution of the extraparticle mobile-phase effects

Giddings [8]	$\mathbf{h} = \frac{b}{\mathbf{v}} + a\bigg/\left(1 + \frac{a}{c_M}\mathbf{v}^{-1}\right)$	(1)
Huber [14]	$\mathbf{h} = \frac{b}{\mathbf{v}} + a\bigg/\left(1 + \frac{a}{c_M}\mathbf{v}^{-1/2}\right)$	(2)
Horvath and Lin [13]	$\mathbf{h} = \frac{b}{\mathbf{v}} + a\bigg/\left(1 + \frac{a}{c_M}\mathbf{v}^{-1/3}\right)$	(3)
Done and Knox [15]	$\mathbf{h} = \frac{b}{\mathbf{v}} + a\mathbf{v}^{1/3} + c_M\mathbf{v}$	(4)
Snyder [16]	$\mathbf{h} = c_M\mathbf{v}^n,\ 0.3 \le n \le 0.7$	(5)

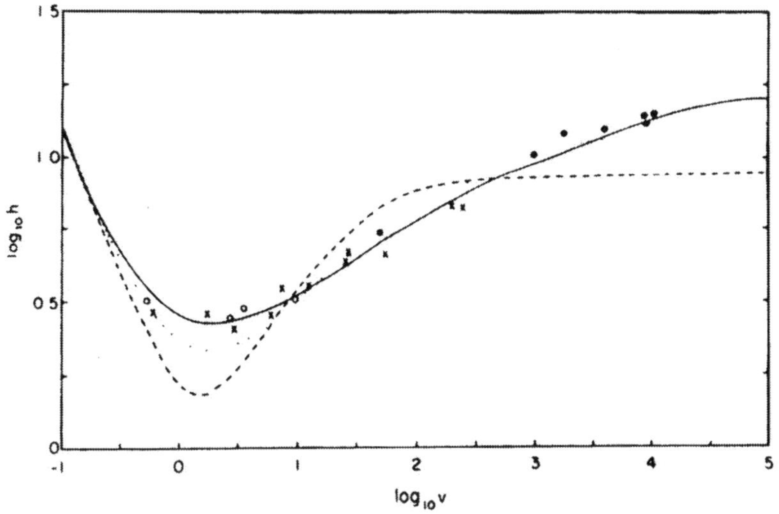

Figure 3.10 Plots of extraparticle mobile-phase effects. The data points were obtained with a single glass bead column by using acetone in *n*-hexane (×), benzene in *n*-hexane (○), and benzoic acid in ethylene glycol (•). The curves represent the equations given in Table 3.1 with the parameters that gave the best fit to the experimental data: dashed curve, equation (1); dotted curve, equation (2); solid curve, equation (3). (Reprinted with permission from Ref. 13.)

(2), and $\frac{1}{3}$ for equations (3) and (4) in Table 3.1. The slope of equation (5) is equal to the variable exponent n. The difference in the predicted slope is clearly seen among the theoretical curves shown in Figure 3.10, where equation (3) is chosen to illustrate the case for the slope of $\frac{1}{3}$. The curves in the figure were calculated from the values of a, b, and c_M selected to best-fit experimental data obtained from a column packed with solid glass beads. Clearly, the experimental data are best fitted by a slope of $\frac{1}{3}$ from equation (3) or (4). The rather poor agreement between equation (1) (Giddings' coupling theory) and the experimental data suggests a need for further theoretical development on the subject. Fortunately, lack of a quantitative theory poses much less of a problem to SEC than to other LC methods, because the extraparticle effect contributes little to the SEC plate height, especially for macromolecules. Equations (1), (3), and (4) from Table 3.1 were recently compared to the van Deemter equation (Equation 3.13). All four equations were found to model LC data comparably [47].

In the next section we consider the characteristic features of band broadening in SEC, which are mainly (1) the porous, nonsorptive nature of SEC packings, and (2) the slow, restricted, and molar-mass-dependent diffusion coefficient of macromolecules.

3.3 MECHANISM OF SEC BAND BROADENING

While the volume of the solvent inside the porous packing does not affect solute selectivity in other LC methods, it, in fact, serves as the stationary phase in SEC, in the sense that it causes the differential elution of solutes. Accordingly, while this liquid volume is described as the stagnant mobile phase in general LC discussions, it is called the *stationary phase* in SEC. This subtle difference in basic concept has caused much confusion and many inconsistencies between SEC and general LC terminology. Thus, a clarification of band-broadening terminology is presented here prior to discussion of the SEC band-broadening mechanism.

The meaning of the phrase *stationary-phase mass transfer* is different when used in SEC versus general LC discussions. The phrase means the H_{SM} term in SEC, but the H_S term in other LC methods (Equation 3.18). In a classical sense the LC stationary term H_S defines the dispersion effect of a distinct, separate LC stationary phase, but this does not at all apply to SEC separations involving nonsorptive packings. In SEC the primary concern is the H_{SM} term, which is called the stagnant-mobile-phase dispersion in LC discussions. Since the phrase *stagnant mobile phase* is somewhat confusing in SEC discussions, we have adopted the convention of calling H_{SM} the *stationary-phase* effect in the following discussions of SEC band broadening. Where conflict exists, the H_S term will be called the *LC stationary-phase* effect for distinction.

With regard to band broadening in SEC, the plate height contribution due to longitudinal diffusion, H_L, is minimal because the large solute molecules commonly encountered in SEC have very small diffusion coefficients (exceptions to this may be encountered in oligomeric SEC, where the diffusion coefficients are larger, and in SEC of ultrahigh molar mass polymers, which need to be analyzed at extremely low flow rates to prevent on-column, flow-induced degradation). With H_L and H_S dropped from Equations 3.18 and 3.19, we have, for SEC,

$$H = H_{\mathrm{SM}} + H_M \tag{3.22}$$

or

$$H = c_{\mathrm{SM}} \frac{v d_p^2}{D_{\mathrm{SM}}} + \frac{1}{(1/a d_p) + (D_M / c_M v d_p^2)} \tag{3.23}$$

Because the diffusion coefficients D_M and D_{SM} in Equation 3.23 are dependent on solute molar mass, band broadening is a function of sample molar mass. This poses a practical problem for the accurate interpretation of SEC data for broad-MMD samples.

3.3.1 Experimental Verification

The validity of Equation 3.23 is well substantiated by the data shown in Figures 3.11 to 3.15 [11,17,18]. In the studies cited, the H_M mobile-phase coupling term and the H_{SM} permeation term were successfully isolated for separate evaluation by using both porous and nonporous (nonpermeating) column packings in the experiments. The plate height data in Figure 3.11, obtained with nonporous packings, clearly show

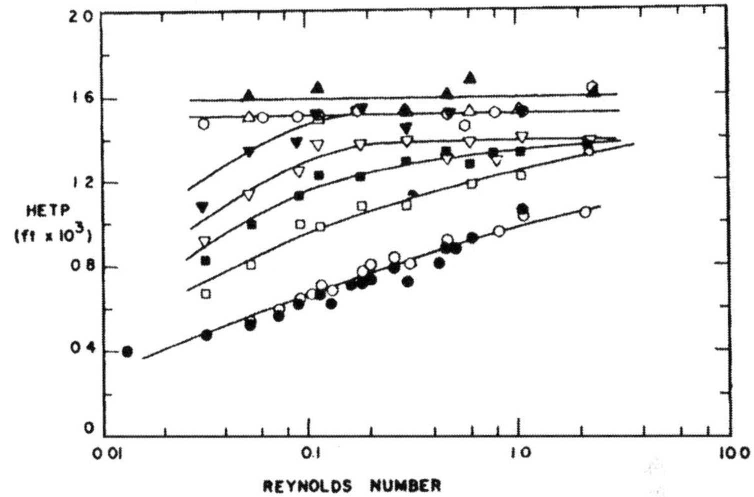

Figure 3.11 Plate height versus Reynolds number (vd_p/η_k) for 105- to 125-μm nonporous glass bead column. •, Hexane ○, cyclohexane; □, n-$C_{36}H_{74}$; ■, 2000 PS (polystyrene); ▽, 3600 PS; ▼, 10,300 PS; △, 97,200 PS; ▲, 160,000 PS. (Reprinted with permission from Ref. 11.)

the coupling characteristics of the mobile-phase dispersion effects, the second term in Equation 3.23. Here, plate height data are plotted against the Reynolds number, vd_p/η_k, where η_k is the kinematic viscosity (the Reynolds number may be thought of as the ratio of the inertial and viscous forces, the kinematic viscosity as the ratio of absolute fluid viscosity to fluid density). For small v or large D_M, the coupling term behaves much like C_M, the mobile-phase mass transfer term alone, and is expected to increase steadily with increasing flow rate. This effect is observed in Figure 3.11 for the monomer solutes: hexane, cyclohexane, and n-$C_{36}H_{74}$. With decreasing diffusion rate D_M and increasing flow velocity v, the chance for lateral solute exchange by diffusion is reduced, which brings out more of the eddy-diffusion characteristics (see the pictures of eddy and coupling effects in Figure 3.2). As shown in Figure 3.11 for polymer solutes, at the high flow rates the plate height observed approaches a constant value, which is the limiting eddy-diffusion plate height. For higher-molar-mass solutes, this limiting condition is reached at a lower flow velocity (lower Reynolds number), as expected. This definitive illustration of the extraparticle coupling effect is made possible through the use of polymer samples with large variations in diffusion coefficients. Actually, the polymer data in Figure 3.11 are more illustrative for demonstrating the LC mobile-phase coupling effects illustrated in Figure 3.10.

To illustrate the effect of particle size on the H_M term, the HETP data in Figure 3.11 were replotted in Figure 3.12 against the reduced velocity ($\mathbf{v} = vd_p/D_M$) to be compared with data obtained from a nonporous packing of much larger particle size. The figure shows the expected large increase in the plate height and slope of the plate height curve for the larger particle column packing.

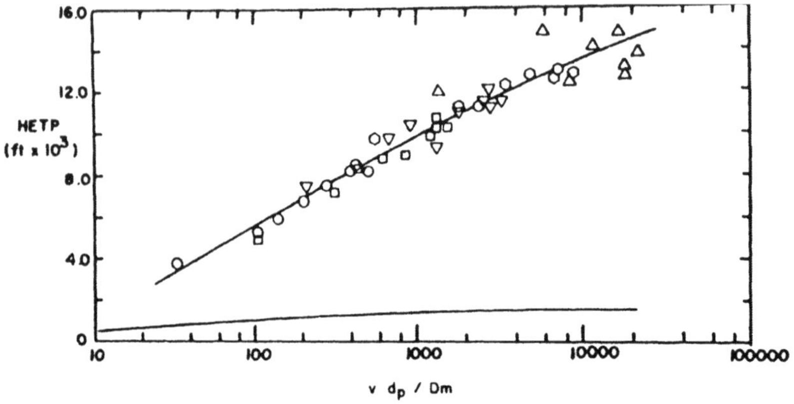

Figure 3.12 Plate height versus reduced velocity (vd_p/D_M) for 350- to 420-μm nonporous glass bead column. Symbols as in Figure 3.11. Data with 105- to 125-μm particles from Figure 3.11 are represented as a line near the bottom of the figure. (Reprinted with permission from Ref. 11.)

Under identical operating conditions, band broadening with a porous column packing is much larger than that with a nonporous packing. The additional band broadening is due to the SEC stationary mass transfer or permeation plate height contribution. This permeation contribution, which is the excess plate height of porous glass over nonporous glass of the same particle size, is shown in Figure 3.13 to

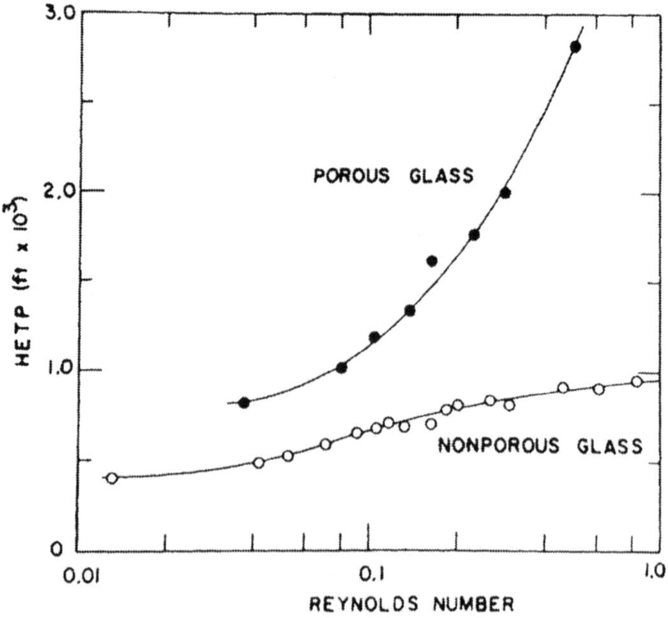

Figure 3.13 Comparison of band dispersion for porous and nonporous column packings. Particles, 105 to 125 μm; solute, cyclohexane. (Reprinted with permission from Ref. 17.)

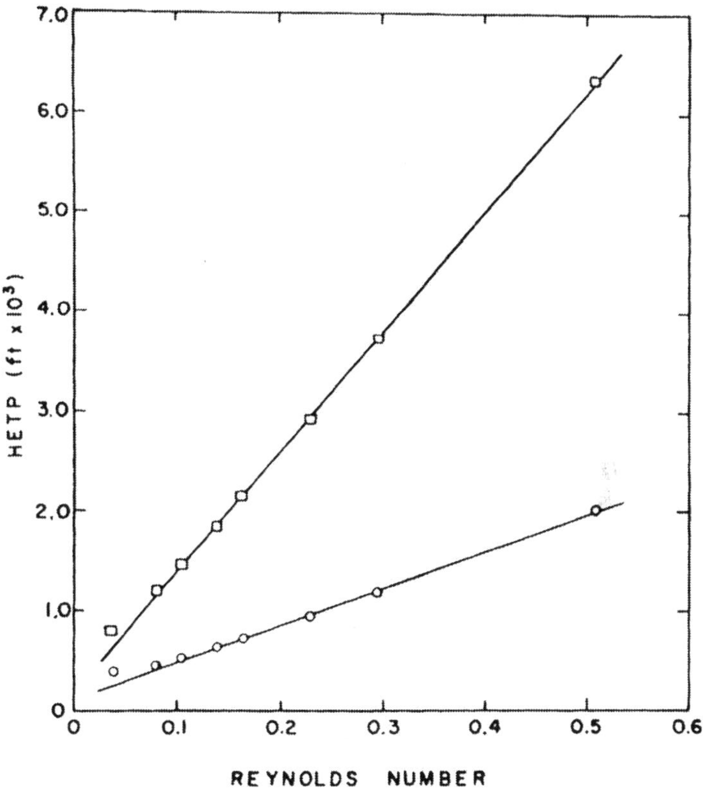

Figure 3.14 Effect of permeation on plate height as a function of Reynolds number (vd_p/η). Data with 105- to 125-μm Porasil A: ○ , cyclohexane; □ hexatriacontane. (Reprinted with permission from Ref. 17.)

increase steadily with Reynolds number, or flow rate. According to theory (the first term in Equation 3.23), this excess plate height due to permeation should vary linearly with flow rate, with the rate of increase being inversely proportional to the solute-diffusion coefficient. This is indeed observed experimentally, as illustrated in Figure 3.14. Note that the HETP curve for the larger solute (hexatriacontane, smaller D_{SM}) increases faster with flow rate than that of the smaller solute (cyclohexane, larger D_{SM}).

The drawback of the preceding method of extracting the permeation contribution from the SEC plate height is the assumption of equally well-packed columns. This assumption may not be realistic, since different columns never pack identically, especially those filled with porous versus nonporous packing materials. This potential problem can be obviated by using a nonpermeating species in the same column of porous packing to obtain the H_M term (i.e., to use a solute larger than the pores of the packing). The only dispersion experienced by bands of totally excluded solutes is due to the extraparticle mobile-phase effect. The successful use of nonpermeating

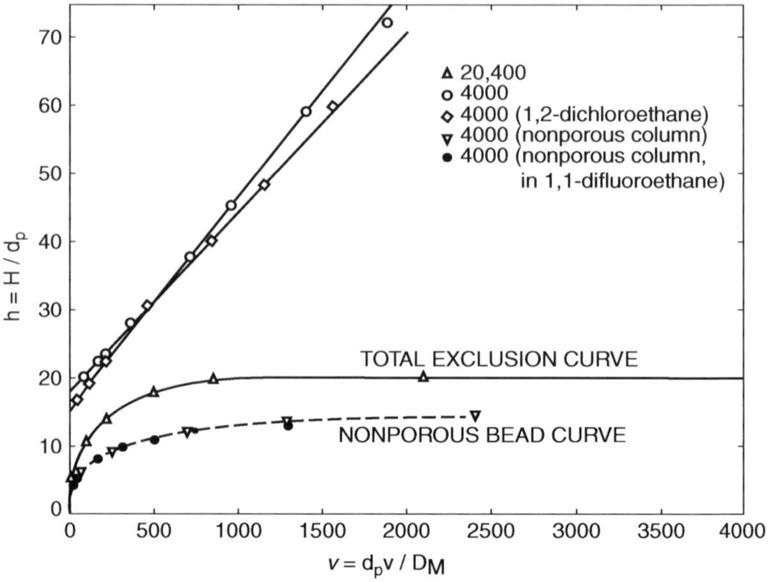

Figure 3.15 Effect of permeation on reduced plate height versus reduced velocity. Solvent, 1,1-dichloroethane, except as noted. (Reprinted with permission from Ref. 18.)

solutes to study the H_M term is illustrated in Figure 3.15, where, as expected, the total exclusion curve behaves much like the H_M curve of nonporous packings. Besides illustrating the large plate height and the flow-rate dependence of the permeation contribution, Figure 3.15 also shows that small chemical differences in the mobile phase have only secondary effects on the characteristics of SEC dispersion, provided that the different solvents are of comparable viscosity.

A different approach to determining interstitial band broadening, employing the same columns and analytes, involved using "wet" versus "dry" eluent. Benzene, 1,3-diphenylbutane, and a series of eight polystyrenes ranging from 10 to 200 kg/mol (designated as PSt 10 to PSt 200 in Figure 3.16) were analyzed using a column packed with 10-μm-diameter silica particles. Peak broadening was first measured using dry dichloromethane as eluent (water content less than 5 ppm). Subsequent to this, water-saturated (wet) dichloromethane, with a water content of about 2200 ppm, was used to fill the total pore volume of the column with water. Under wet conditions the pore volume is inaccessible to the polystyrenes. It is then possible to measure the interstitial band broadening for these samples. Results are shown in Figure 3.16, which plots the height equivalent of a theoretical plate, h, versus the linear velocity of a totally excluded inert sample, u_z. Figure 3.16a shows results using dry eluent. All curves are linear. The two totally excluded polystyrenes (PSt 111 and PSt 200) and the two monomers had the lowest and nearly identical values of h, virtually independent of u_z. Of note is the fact that the h values for PSt 10 are at least 15-fold greater than the values for benzene. When using wet eluent, however (Figure 3.16b), the difference between the h values of PSt 10 and benzene is, at most, a factor of 2. In wet

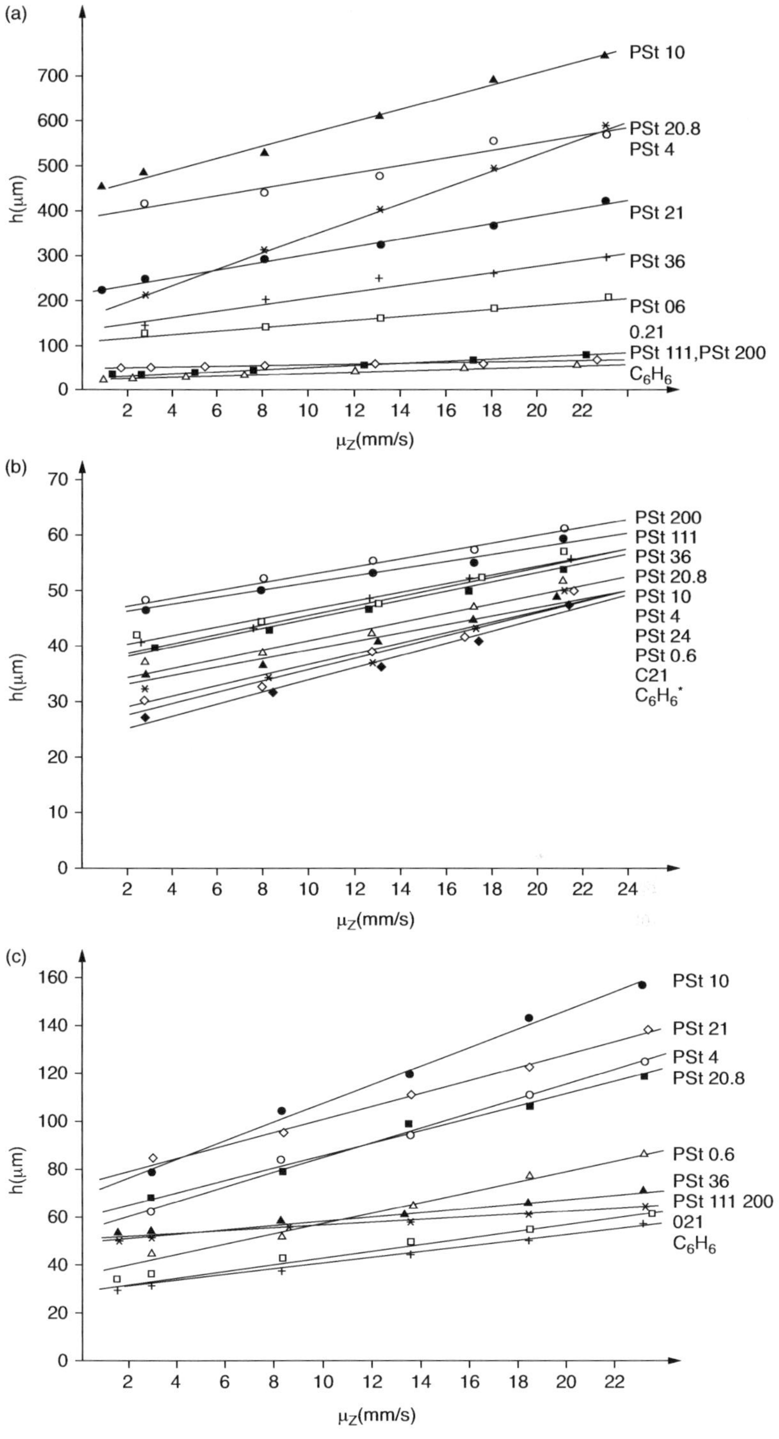

Figure 3.16 Effect of permeation on plate height versus velocity. Polystyrene ranging from 10 kg/mol (PSt 10) to 200 kg/mol (PSt 200), benzene (C_6H_6), and 1,3-diphenylbutane (0,21) analyzed on silica column using dichloromethane as eluent: (a) dry eluent (water content <5 pmm); (b) wet eluent (water content about 2200 ppm); (c) semiwet eluent (water content 0.2%). (Reprinted with permission from Ref. 19.)

eluent the values of h are much lower than for permeating species in dry eluent. Also, in wet eluent h increases linearly with molar mass at all flow velocities. Intermediate behavior is observed in semiwet eluent (water content 0.2%), where the column pores have been partially loaded with water, as shown in Figure 3.16c. In the latter case, however, it is important to note that not only the pore volume but also the mean pore size has changed. In each case, dry, wet, and semiwet, the pore porosity (ε_p) was measured as the ratio $V_{p,\max}/V_k$, with $V_{p,\max}$ being the pore volume available to the smallest inert sample and V_k the empty column volume. With dry dichloromethane, $\varepsilon_p = 0.382$; with wet eluent, $\varepsilon_p = 0.010$; at semiwet conditions, $\varepsilon_p = 0.100$.

In using Equation 3.23 for SEC band broadening, the longitudinal diffusion contribution to dispersion is ignored. This contention is generally supported by results from interrupted flow SEC experiments. As seen in Figure 3.17a, two nearly superimposable elution curves were obtained for a 280,000-g/mol poly(ethyl methacrylate) sample, PEMA, with one curve eluted immediately after sample injection and the other curve obtained after holding the sample on the column for 16 hours with the solvent flow stopped. This is a good illustration of the very slow longitudinal diffusion rate of polymer molecules in packed columns. This means that in the usual high-polymer SEC experiments, the increase in plate height at low flow rates, like that shown by the theoretical curves in Figure 3.9, is of little practical concern. Longitudinal diffusion can become an issue in oligomeric SEC (Chapter 13), however, and/or when using the types of stop-flow probes favored in some two-dimensional chromatographic setups (Chapter 14) and in certain SEC-NMR experiments (Chapter 10). An example of this is shown in Figure 3.17b. Here, a 2000-g/mol polystyrene sample shows considerable band broadening, corresponding to an approximately 40% reduction in reduced peak height (**h**), after undergoing the same type of stop-flow treatment as the high-molar-mass PEMA sample in Figure 3.17a. Table 3.2 gives the results of the 16-hour stop-flow experiments for narrow and broad polydispersity polymers covering a variety of chemistries over a wide molar mass range. Under the conditions of the experiment, which included only linear polymers, the effects of longitudinal diffusion became negligible above approximately 30,000 g/mol.

The data presented in Figures 3.11 to 3.17 provide general support for the validity of Equation 3.23 and also provide the following insights. In the useful SEC separation range where partial solute permeation is expected, the SEC peak broadening is more dominated by the permeation process itself than by mobile-phase effects. When studying SEC band broadening, the emphasis must therefore lie heavily on understanding the permeation term in Equation 3.23. To explain the total exclusion curve in Figure 3.15, one has to assume that c_{SM} goes to zero (as $K_{\mathrm{SEC}} = 0$) at total exclusion; that is, the coefficient c_{SM} in the permeation term must be a function of the extent of permeation (K_{SEC}, Chapter 2). In addition, the quantity D_{SM} in the permeation term, which represents the solute-diffusion coefficient in the pores of SEC packings, can vary greatly with solute molar mass. Because solutes in SEC analyses are often of sizes comparable to those of the pores, the "restricted diffusion" effect must be considered in interpreting D_{SM} (see the discussion in Section 3.4). The rate theories of SEC dispersion described below provide more nearly quantitative understanding for some of these effects.

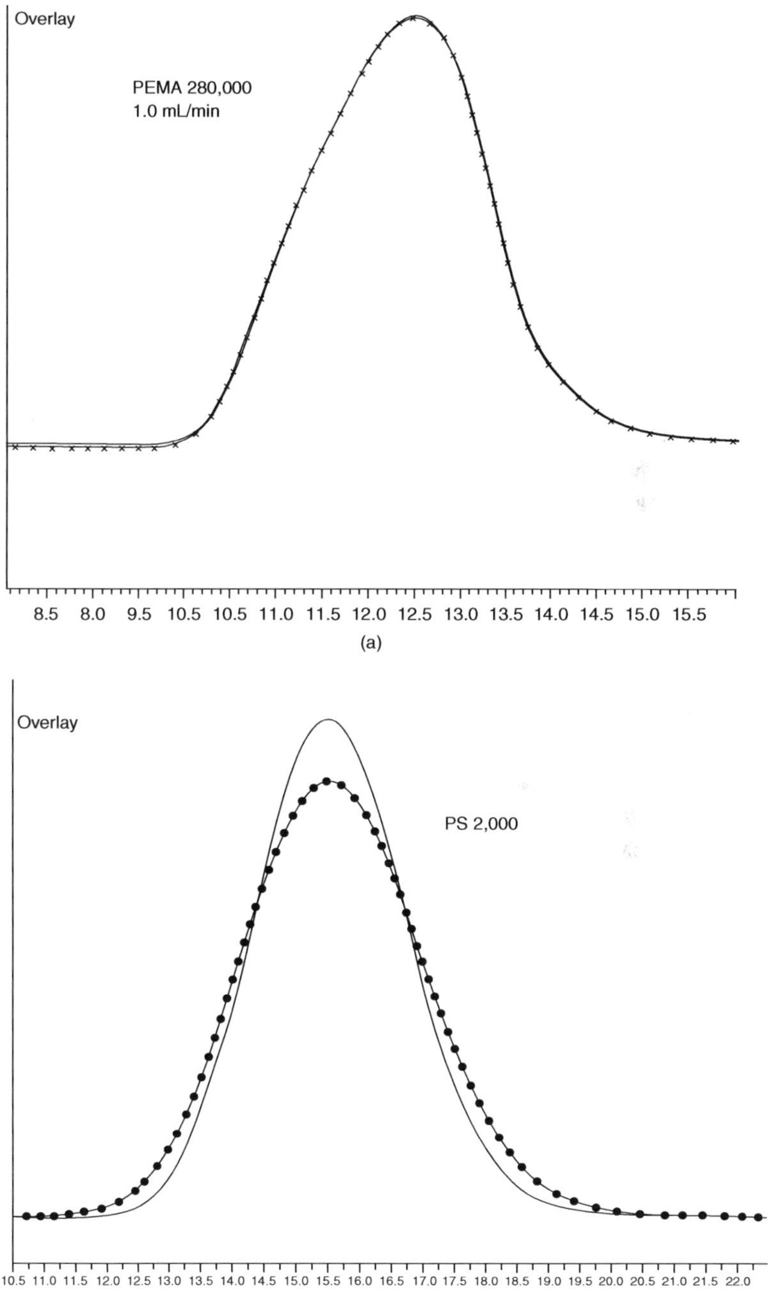

Figure 3.17 Effect of interrupted flow on SEC chromatograms. Retention volume vs. DRI response. Solid curves correspond to normal injection and elution, curves with crosses or filled circles correspond to elution after holding sample on-column for 16 hours. (a) 280,000 g/mol poly(ethyl methacrylate); PEMA; (b) 2,000 g/mol polystyrene. Solvent, THF; temperature, 35°C; flow rate during elution, 1 mL/min; columns, PLgel 5-μm Mixed-C. (Adapted from Ref. 20.)

Table 3.2 Influence of molar mass, polydispersity, and analyte chemistry on longitudinal diffusion

Sample[a]	M_r (g/mol)	M_w/M_n	ΔP^b	$\Delta W_{0.5}{}^c$	$\Delta W_{0.1}{}^d$	$\Delta \mathbf{h}^e$
PS	2,000	1.05	+ +[f]	+ + +	+ + +	41 ± 2
PS	5,000	1.04	+	+ + +	+ + +	42 ± 2
PMMA	7,800	1.14	0	+ +	+ + +	22 ± 2
PS	11,600	1.04	0	+ +	+ + +	33 ± 3
PVF	10,000–15,000	5.62	0	+ +	+ +	25 ± 2
PC	20,000–25,000	2.74	0	+	+	14 ± 4
PMMA	27,000	1.11	0	0	+ +	3 ± 2
PVB	36,000	2.40	0	0	0	1
PVC	68,000	2.88	0	0	0	0
PBD	100,000	2.18	0	0	0	0
PMMA	107,000	1.10	0	0	0	0
PS	170,000	1.04	0	0	0	0
PEMA	280,000	2.43	0	0	0	0
PMMA	400,000	1.14	0	0	0	0
PS	470,000	1.06	0	0	0	0
PS	1,130,000	1.06	0	0	0	0

Source: Ref. 20.
[a] PS, polystyrene; PMMA, poly(methyl methacrylate); PVF, poly(vinyl formal); PC, polycarbonate; PVB, poly(vinyl butyral); PVC, poly(vinyl chloride); PBD, polybutadiene; PEMA, poly(ethyl methacrylate).
[b] Denotes decrease in peak height.
[c] Denotes increase in peak width at 50% peak height.
[d] Denotes increase in peak width at 10% peak height.
[e] Denotes percentage change in reduced peak height. Where no standard deviation (std) is given, std < 1.
[f] $0 = \Delta 0$ to 5%; $+ = \Delta 6$ to 10%; $+ + = \Delta 11$ to 15%; $+ + + = \Delta$ 16 to 20%.

3.3.2 Rate Theory

In developing a rate theory, differential equations are derived to describe solute mass balance in a differential column section such as that illustrated in Figure 3.18 [9,21,22]. The four classical, independent mass transfer processes are shown in their partial differential form: longitudinal diffusion ($D_M \partial C_m/\partial x$), eddy diffusion ($D_E \partial C_m/\partial x$), stationary mass transfer ($D_{SM} \partial C_s/\partial r$), and mobile-phase lateral diffusion ($D_M \partial C_m/\partial y$), where x is the column-length variable along the column axis, y the distance variable in the lateral direction, and C_m and C_s the solute concentrations in the mobile and stationary phases, respectively. The quantity D_E is the eddy-diffusion coefficient, which is expected to be proportional to d_p and v [23].

A rate theory for SEC dispersion employs the partial differential equations (3.24) to (3.26) (Figure 3.18). For stationary mass transfer (permeation),

$$\frac{\partial C_s}{\partial t} = D_{SM}\left(\frac{\partial^2 C_s}{\partial^2 r} + \frac{2}{r}\frac{\partial C_s}{\partial r}\right) \tag{3.24}$$

where r is the radial distance from the center of the spherical porous particle. For

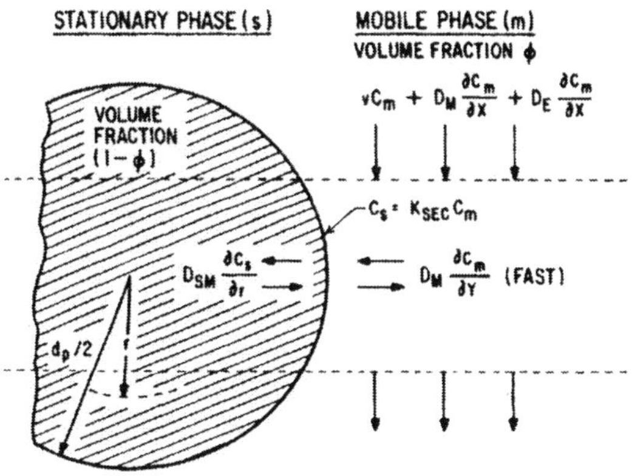

Figure 3.18 Mass transfer of solute in a thin section of SEC column.

mobile-phase mass transfer,

$$\frac{\partial C_m}{\partial t} + v\frac{\partial C_m}{\partial x} - D_M\frac{\partial^2 C_m}{\partial^2 x} = -\sigma D_{\mathrm{SM}}\left(\frac{\partial C_s}{\partial r}\right)_{r=d_p/2} \tag{3.25}$$

where $\sigma = 6(1-\phi)/d_p\phi$, with ϕ being the volume fraction of the extraparticle solvent volume. For the boundary condition, at $r = d_p/2$,

$$C_s = K_{\mathrm{SEC}}C_m \tag{3.26}$$

In these expressions, coupling of eddy diffusion to mobile-phase transfer and longitudinal diffusion effects is neglected. These assumptions are appropriate for SEC, since overall mobile-phase dispersion is usually small compared to permeation.

Although the complete solution of these differential equations (Equations 3.24 and 3.25) under the specific boundary conditions is not known, an approximate solution can be obtained for a limiting case where near-equilibrium solute distribution exists between phases. This limiting condition closely approximates most SEC experiments and predicts a near-Gaussian elution peak shape.

A numerical solution to the differential equations for SEC dispersion can be obtained [25]. The elution curves computed are shown in Figures 3.19 and 3.20 to demonstrate the predicted effect of packing-particle diameter (d_p) and flow rate on SEC peak shapes. In Figure 3.19a, the simulated monomer SEC peak becomes increasingly broader without changes in peak symmetry as the packing particle size increases. Figure 3.19b shows the particle-size effect on an earlier eluted SEC peak of a polydisperse high-molar-mass polymer. Owing to the lower diffusion coefficient, the polymer peak is more sensitive to the increase in d_p. With increasing d_p, the polymer peak becomes broader and, because of polydispersity, more skewed, with its peak maximum leaning more toward the direction of low retention volume.

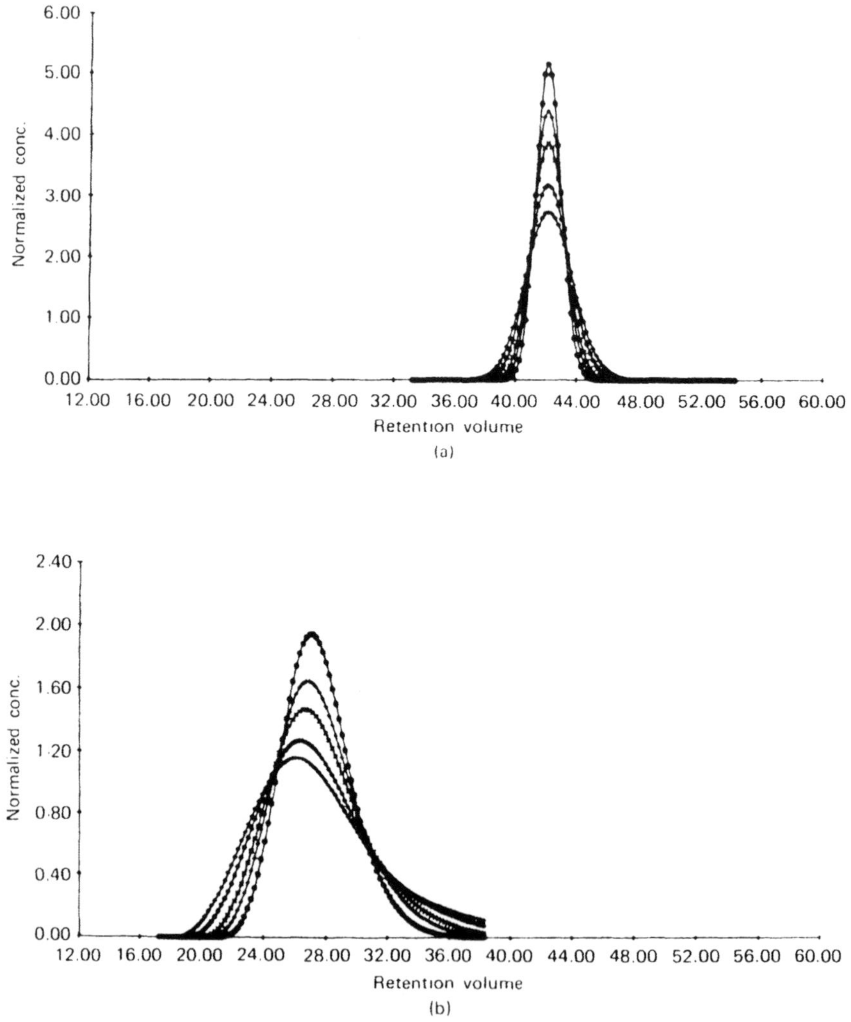

Figure 3.19 Predicted effect of particle diameter on SEC peak shapes: (a) ethylbenzene peak; (b) peak for polystyrene, 160,000 g/mol. The d_p values (in units of μm). ○, 30; +, 40; ★, 50; Ø, 70; ▲, 90. (Reprinted with permission from Ref. 25.)

Figure 3.20 shows the effect of flow rate on the theoretical chromatogram of a polydisperse polymer sample. As the flow rate increases, the width as well as the skewness of the peak also increases, with the peak maximum again shifting to lower retention volume. All the theoretical predictions described above are quite commonly observed in SEC experiments, again demonstrating that SEC dispersion is a permeation-rate-limited process. [The curves in Figures 3.19 and 3.20 were computed assuming that $D_M \propto (M)^{-\alpha_m}$ and $D_{SM} \propto (M)^{-\alpha_s}$ with $\alpha_m = 0.6$ and

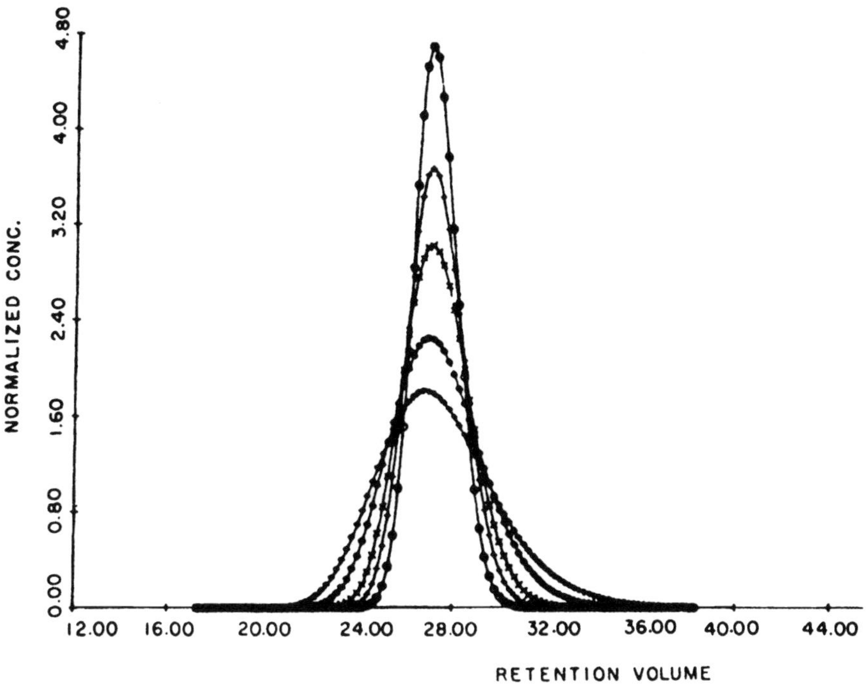

Figure 3.20 Predicted effect of flow rate on SEC peak shape. Particle diameter d_p, 60 μm; solute, polystyrene, 160,000 g/mol. Flow rates: ○, 2 mL/min; +, 3 mL/min; ×, 4 mL/min; •, 6 mL/min; ▲, 8 mL/min. (Reprinted with permission from Ref. 25.)

$\alpha_s = 1.0$. The molar mass (M) dependence of diffusion coefficients is considered in Section 3.4.]

Explicit expressions of peak shape in terms of statistical moments [2] can be derived from the original differential equations by using Laplace transformations without actually having to solve the equations or make any assumptions about the limiting cases [24]. The results obtained for the permeation dispersion process are described below. As an approximation, for the first three moments, μ_1 represents mean retention, μ_2 represents peak variance, and μ_3 represents peak skewness. These terms are expressed quantitatively as

$$\mu_1 \equiv V_R = \left(1 + \frac{1-\phi}{\phi} K_{\text{SEC}}\right) V_o = V_o + K_{\text{SEC}} V_i \tag{3.27}$$

$$\mu_2 \equiv \overline{(V - V_R)^2} = \frac{V_o V_i}{30L} K_{\text{SEC}} \frac{v d_p^2}{D_{\text{SM}}} = \frac{1}{30} K_{\text{SEC}} \, V_i \frac{F d_p^2}{D_{\text{SM}}} \tag{3.28}$$

$$\mu_3 \equiv \overline{(V - V_R)^3} = \frac{1}{420} K_{\text{SEC}} V_i \left(\frac{F d_p^2}{D_{\text{SM}}}\right)^2 \tag{3.29}$$

where ϕ, V_o, V_i, L, and F are the void-volume fraction, void volume, internal pore volume, column length, and volume flow rate, respectively.

Since $H = L/N = L\mu_2/\mu_1^2$ (Equation 3.11), the exact expression for the H_{SM} term (SEC stationary-phase dispersion) in Equations 3.22 and 3.23 can be obtained by combining Equations 3.27 and 3.28 to give the expression

$$H_{SM} = c_{SM}\frac{vd_p^2}{D_{SM}} = \frac{K_{SEC}V_i/V_o}{30(1 + K_{SEC}V_i/V_o)^2}\frac{vd_p^2}{D_{SM}} \tag{3.30}$$

3.3.3 Theoretical Inferences

Many important features of SEC can be explained by the rate theory (Equations 3.27 to 3.30). For example, Equation 3.27 indicates that the peak retention volume V_R, when defined as the first moment or the center of gravity of the peak, is not a function of flow rate. This prediction, demonstrated by the theoretical elution curves in Figure 3.20, is also verified experimentally by the data in Figure 3.21. Although the retention volume at the peak apex of a polydisperse polymer sample decreases with increasing flow rate, the average retention volume or center of gravity of the peak is unchanged. At low flow rates, the retention volume at peak maximum approaches that of the first moment (or center of gravity) of the peak when the peak becomes more symmetrical. An example for a limiting case of Equation 3.30 is found in the earlier discussion of H_M coupling (see Figures 3.13 and 3.15), where the H_{SM} term is forced to zero by using either a nonporous packing ($V_i = 0$) or a nonpermeating solute ($K_{SEC} = 0$).

Equation 3.28 predicts that the SEC peak dispersion increases linearly with flow rate, and the slope of an H versus v plot increases with solute M (except for

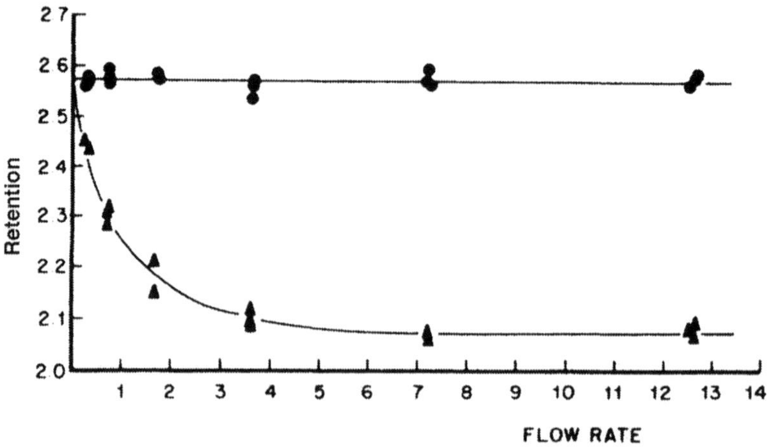

Figure 3.21 Effect of flow rate on peak apex and first moment position. Sample, polystyrene, 51,000 g/mol; • first moment (V_R); ▲ peak maximum. Retention and flow rate are reported in terms of the weight of solvent eluted. (Reprinted with permission from Ref. 26.)

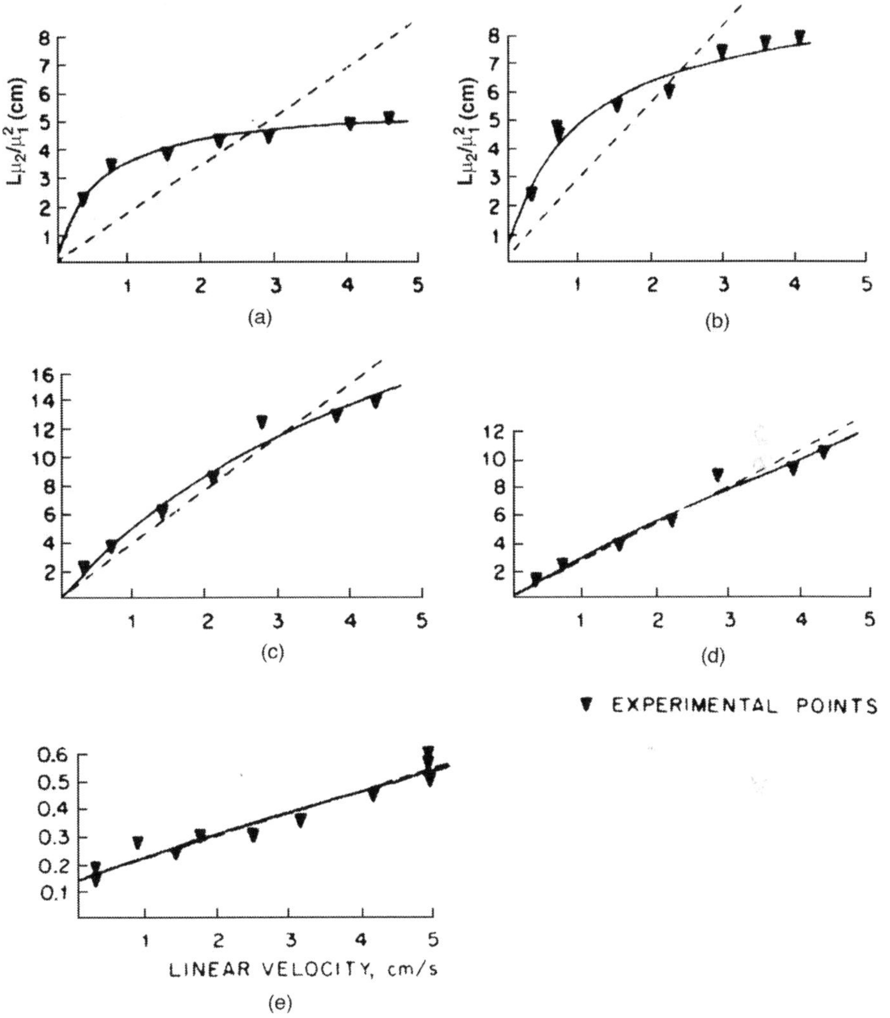

Figure 3.22 Dependence of plate height ($L\mu_2/\mu_1^2$) on flow rate. Samples, polystyrene standards of following *M*, in g/mol: (a) 160,000; (b) 97,200; (c) 51,000; (d) 20,400. For (e), the solute is toluene. Abscissa, velocity in cm/s. (Reprinted with permission from Ref. 26.)

nonpermeating species). This is generally observed in SEC, except in those experiments that involve very high flow rates and high-molar-mass solutes of low diffusion coefficient D_{SM}. Deviations from the predicted linearity follow a unique pattern, as shown in Figure 3.22. The characteristic shape of the H versus v curve for high-molar-mass solutes shown in a and b of the figure suggests the possibility of flow-diffusion interaction within the pore structure of the SEC packing. In fact, the data can be fitted quantitatively by an intraparticle interaction expression, including diffusion and an empirical intraparticle flow, or convection velocity [26]. This empirical interaction term provides the solid curves plotted in Figure 3.22. No rigorous interpretation of this intraparticle convection is yet available, although it may be attributed either to eddy currents or to flow through all or part of the pore structure [27]. This flow-diffusion interaction term, required to explain the data in Figure 3.22, is not to be confused with the Giddings coupling term (Equation 3.16), which only plays a role in the extraparticle mobile phase involving plate height values orders of magnitude smaller than those of concern in the present discussion.

The predicted effect of M and flow rate on the skew of SEC peaks as shown in Figures 3.19 and 3.20 can be studied quantitatively by using peak moments from Equations 3.28 and 3.29. Peak skew γ_{SM}, which is due to SEC stationary-phase lateral diffusion, can be expressed as [1]

$$\gamma_{\mathrm{SM}} = \frac{\mu_3}{\mu_2^{3/2}} = \left(\frac{15}{98 K_{\mathrm{SEC}} V_i} \frac{F d_p^2}{D_{\mathrm{SM}}} \right)^{1/2} \tag{3.31}$$

Both Equation 3.31 and the theoretical curves in Figure 3.20 predict that peak skewing increases with increasing flow rate, particularly for large solutes. Skewing contributes (along with other band-broadening processes) to the underestimation of the number-average molar mass, M_n, and overestimation of the weight- and z-average molar masses, M_w and M_z, respectively, which leads to an overestimation of the molar mass polydispersity, M_w/M_n, and to distorted molar mass distributions, MMD. This is especially true when employing calibration-curve-based methods for determining the MMD and corresponding M averages of polymers.

Since the stationary-phase mass transfer term H_{SM} is the most important dispersion factor in SEC, further implications of Equations 3.30 and 3.31 are considered in the next section. Many of the fundamental concepts discussed in this chapter serve as the foundation for optimizing columns (Chapter 6) and operating variables (Chapter 7).

3.4 INFLUENCING FACTORS

In this section we direct attention to experimental parameters that can affect peak dispersion in SEC stationary phases. Other dispersion processes are not considered in this section since they are relatively unimportant in SEC, even for small-molecule SEC analyses. Adequate general information about these other dispersion processes has been given in earlier sections. Specifically, in this section, Equations 3.30 and 3.31 are considered more carefully in terms of the effects of (1) column parameters: V_i/V_o, K_{SEC}, and d_p; (2) kinetic factors: v, F, and D_{SM}; and (3) other experimental parameters: temperature, solvent viscosity, sample concentration, and so on.

3.4.1 Column Parameters

The column parameter V_i/V_o, the ratio of pore volume to void volume, is basically an SEC retention parameter that is directly proportional to the porosity of SEC packings. Although the plate height contribution H_{SM} increases with V_i/V_o (Equation 3.30), SEC column packings with large porosity are still generally preferred because of their better separating ability. Large values of V_i/V_o mean more useful pore volume available for the molar mass (M) separation and better overall SEC resolution. Because V_i/V_o is independent of column dimensions, so is H_{SM}, according to Equation 3.30. Thus, neither column length nor column diameter variations should have much effect on the SEC plate height contribution H_{SM} (except as they lead to packing differences).

The quantity K_{SEC}, the SEC distribution coefficient, is also a retention parameter that is dependent on the size of the solute molecules relative to the packing pore size. Like V_i/V_o, K_{SEC} is usually optimized for molar mass selectivity and SEC resolution rather than for peak broadening. The retention volume dependence of SEC plate height approximated by H_{SM} is the result of the combined effects of the extent and rate of permeation determined by K_{SEC} and D_{SM}. A small K_{SEC} value means low solute retention V_R, large solute M, and small solute-diffusion coefficient D_{SM}. Since D_{SM} changes more than K_{SEC} as a function of V_R, the diffusion coefficient (D_{SM}) usually dominates SEC band broadening (H_{SM}) in the majority of the SEC separation range. Only when K_{SEC} approaches zero (the total exclusion limit) will H_{SM} reverse the trend and start to decrease. This reversal predicts a maximum in the H versus K_{SEC} plot, as verified experimentally by the data in Figure 3.23 [28–30]. The solid line in the figure describes the theoretically predicted curve [31].

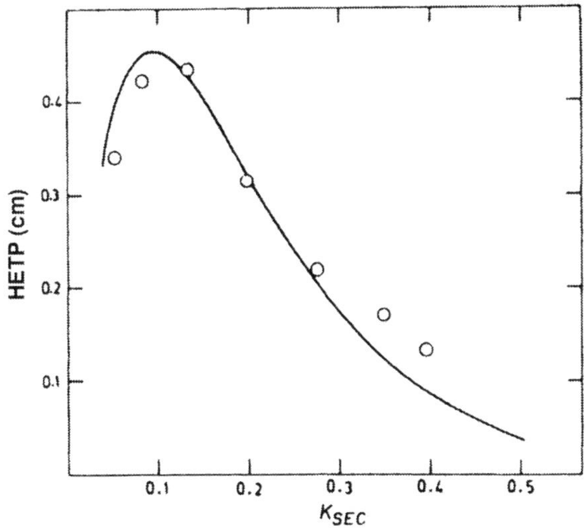

Figure 3.23 Variation of SEC plate height as a function of K_{SEC}. Points, data from Ref. 30; curve, semiempirical prediction by Kubìn [31]. (Reprinted with permission from Ref. 31.)

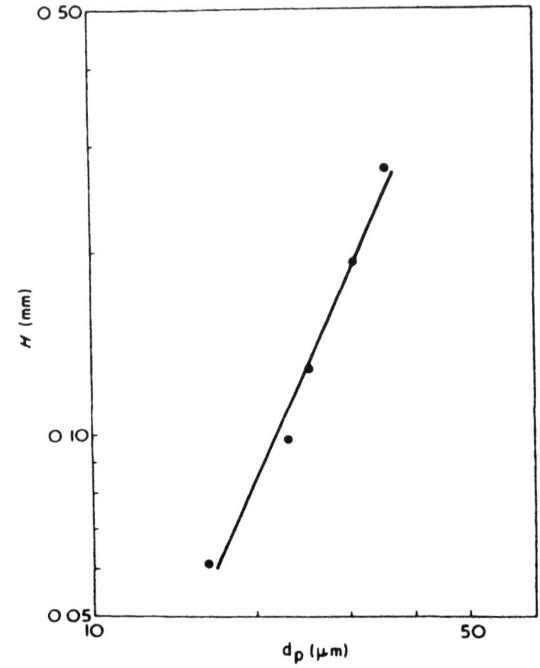

Figure 3.24 Dependence of SEC plate height on particle diameter. Sized fractions of the same SEC gel packing particles are used. (Reprinted with permission from Ref. 32.)

Packing particle diameter d_p is the most influential of all experimentally adjustable parameters affecting chromatographic band broadening. The strong d_p dependence of H_{SM} is predicted by the squared term d_p^2 in Equation 3.30. This square dependence on particle size also predicts a line with a slope of 2 for the H versus d_p data in a log-log plot, as verified in Figure 3.24 [32]. The data in the figure suggest a slight decrease in the slope at small d_p values, which may indicate more poorly packed columns at small d_p, or, alternatively, sufficiently small H_{SM} terms that become less dominating over the overall SEC plate height. Practical implications of the effect of d_p on SEC efficiency are discussed in Sections 6.4 and 7.3.

In SEC, peak skewing due to H_{SM} peak dispersion is reduced substantially with the use of small d_p packings. Equation 3.31 indicates that peak skew should increase with decreasing K_{SEC} and V_i, but decrease with decreasing d_p. For a 10-μm particle and the usual SEC conditions (assume that $F = 1$ mL/min, $K_{SEC}V_i = 10$ mL), the estimated γ_{SM} values for peak skew according to Equation 3.31 are very small: $\gamma_{SM} = 0.05$ for a polymer solute of a diffusion coefficient of $D_{SM} = 10^{-7}$ cm^2/s, and $\gamma_{SM} = 0.005$ for a monomer solute of $D_{SM} = 10^{-5}$ cm^2/s. This calculation indicates that the peak skew observed in SEC experiments must be due to restricted diffusion in the pore structure so that D_{SM} actually becomes much smaller than 10^{-7} cm^2/s.

The discussion above has considered the effect of totally porous particles on peak dispersion. However, superficially porous particles (solid core, porous shell) have also been used for SEC, to reduce the influence of H_{SM} peak broadening. For superficially porous packings, d_p is approximated by the thickness of the porous surface layer. For larger polymer solutes that are prone to flow-induced degradation, the use of large superficially porous packings of 10 to 20 μm with a 2- to 3-μm porous layer is preferred to achieve column efficiency that would otherwise require small porous particles ($d_p < 5$ μm) and high shear forces in the flow streams.

3.4.2 Kinetic Factors

Among kinetic parameters, the effect of solute diffusion coefficient (D_{SM}) on band broadening is not as well studied as that of solvent velocity (v) or flow rate (F). Classically, flow-rate studies are used in LC and SEC as a practical tool for studying column dispersion and for evaluating the performance of chromatographic systems. Since flow-rate effects were discussed in detail in the preceeding section, the focus of the following discussion is on D_{SM}.

The slow permeation in SEC is sometimes called a nonequilibrium process. This is only to indicate a microscopic nonequilibrium state of solute distribution between phases in the column. Since most of the local solute nonequilibrium is averaged out during the course of solute migration through the entire column, this effect results only in band broadening without affecting solute retention.

The molar mass dependence of D_{SM} is the main cause of M dependence of the SEC plate height, as pointed out in the discussion of Figure 3.23. The reason that D_{SM} varies with solute M is twofold: (1) solute size increases with M, causing D_M (diffusion coefficient in open space) to decrease, and (2) restriction of diffusion inside the SEC pore structure increases with increasing solute size [i.e., D_{SM} is a function of M; $D_{SM} = D_M f(M)$]. A rigorous expression is available to explain the influence of solute size on D_M [33]:

$$D_M = \frac{RT}{6\pi \eta_0 N_A} \left(\frac{10\pi N_A}{3K} \right)^{1/3} M_\nu^{-(1+a)/3} \tag{3.32}$$

where R is the gas constant, T the absolute temperature, η_0 the solvent viscosity, N_A is Avogadro's number, and K and a are constants of the Mark–Houwink viscosity equation $[\eta] = K M_\nu^a$, where $[\eta]$ is the solute intrinsic viscosity and M_ν is the viscosity-average molar mass of the solute (see Sections 1.4 and 8.2.3). The usual value of a for random-coil polymers varies from 0.5 in a poor solvent to 0.8 in a good solvent. Therefore, D_M is expected to vary with $M^{-\alpha_m}$, with α_m falling between 0.5 and 0.6.

No rigorous expression for the restricted diffusion of molecules in real pores is available. This is understandable in view of the difficulty of defining a realistic pore model to account for the restriction of solute diffusion in the irregular pores of SEC packings. However, some attempts have been made to describe restricted diffusion using simple models. For instance, the following expression was derived from considerations of wall friction on the motion of a solid sphere in a cylindrical pore [34]:

$$D_{\mathrm{SM}} = \frac{D_M}{\tau}(1 - 2.104\lambda + 2.09\lambda^3 - 0.95\lambda^5) \quad (3.33)$$

with λ the ratio of solute diameter to pore diameter and τ the tortuosity factor (ca. 2.1 to 2.4). For polymers, the size parameter λ is a function of solute M [i.e., $\lambda \propto (M)^{0.5\text{–}0.6}$]. Diffusion data for simple organic solutes in fine-pore structures suggest the following simple empirical equation [34]:

$$\log_{10} \frac{D_{\mathrm{SM}}\tau}{D_M} = -2\lambda \quad (3.34)$$

3.4.3 Experimental Factors

Equations 3.33 and 3.34 both indicate that D_{SM} is directly proportional to D_M. Therefore, the restricted diffusion coefficient D_{SM} should increase as D_M increases with increasing temperature T according to Equation 3.32. The result should be improved SEC resolution. This prediction is verified by the experimental data given in Figure 7.9 [35,36]. The same equations also predict that D_{SM} should increase in solvents of lower viscosity η_0 or small Mark–Houwink exponent a (poor solvent for the polymer). However, solvent choices for many important polymers are often limited. Therefore, optimization of SEC efficiency via solvent selection is not a common practice. The use of a poor solvent to gain SEC efficiency is generally not recommended; very poor solvents may increase the chance of solute adsorption on a packing surface. Many of these considerations are discussed in Chapters 7 and 8.

Solute concentration overloading can add to band broadening. The data show that polymer solutes at higher concentrations elute later, suggesting a more compact polymer conformation at the overloading concentrations. This overloading effect poses an SEC precision problem and an upper limit to the sample concentration that can be used to improve the signal/noise ratio of SEC analyses (Section 7.4). An upper limit on sample concentration is given by the critical overlap concentration, c^*, which represents a boundary between near-infinitely dilute and semidilute solutions (Figure 7.10). Four different definitions of c^* exist, given in turn by the radius of gyration (R_G), the hydrodynamic or Stokes radius (R_H), the second virial coefficient (A_2), and the intrinsic viscosity ($[\eta]$) [37]. The most consistent choice for c^* is that based upon A_2, given as

$$c^*_{A_2} = \frac{1}{A_2 M_w} \quad (3.35)$$

A much simpler approximation, however, is that based on $[\eta]$, which with the

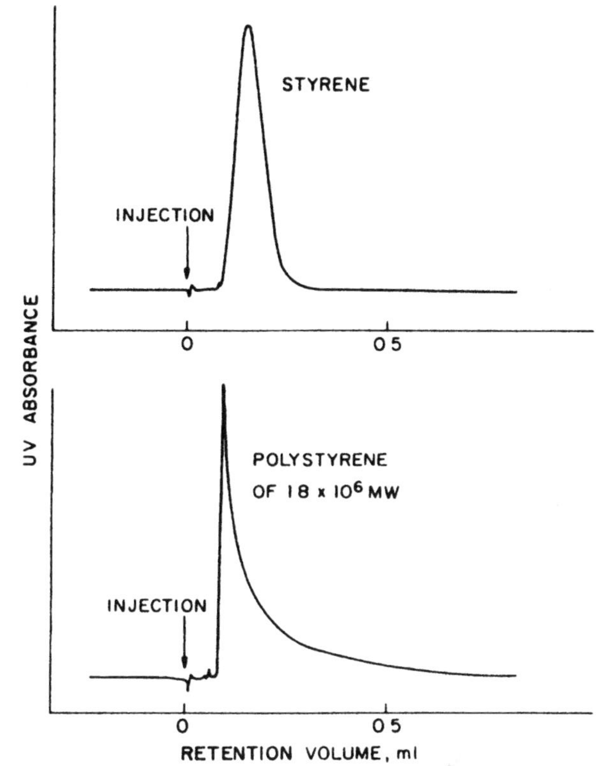

Figure 3.25 Extracolumn peak dispersion in SEC. Straight tubing, 0.018 in. i.d., 1 m long; solvent, chloroform; flow rate, 0.5 mL/min. (Reprinted with permission from Ref. 29.)

popularity of online viscometers (Section 9.5), has become a conveniently measurable parameter:

$$c_{\eta}^{*} = \frac{1}{[\eta]} \tag{3.36}$$

When using Equation 3.36, care must be taken to calculate c^* based on the intrinsic viscosity of the largest species in the distribution.

Extracolumn dispersion effects are of special importance to the overall performance of SEC, because of the high-efficiency (low-dispersion) of SEC columns. Since the dispersion effect of each element in an SEC instrument is different [38], the subject is discussed in more detail in Chapter 5, with special attention given to individual effects. In general, these dispersion effects have characteristics similar to those of highly nonequilibrium lateral mass transfer processes, as illustrated by the highly skewed elution peak profile in Figure 3.25 for a polymer sample eluted from long tubing.

While the retention parameters void volume V_o, pore volume V_i, and distribution coefficient K_{SEC} determine the extent of solute permeation and therefore control solute retention and separation, they are rather unimportant factors in SEC band broadening. The opposite is true for the quantities d_p, v, F, and D_{SM} and such D_{SM}-related factors as T, η_0, and a. Although the latter parameters (d_p, v, etc.) can affect the rate of solute permeation into the pores and control the column dispersion, they have no or very little effect on SEC solute retention and separation. This is evident in the expression for the first moment of the peak (Equation 3.27), which contains none of the latter parameters.

3.5 EXPERIMENTAL METHODS

3.5.1 Plate Number

Because chromatographic separations can be described as stepwise processes, a larger number of available steps corresponds to better separations. This number of steps is defined by the plate number, N. There are several methods available for measuring N, defined as V_R^2/σ^2. First, N should be determined using a monodisperse sample of low M. The slow diffusion coefficients of macromolecules lead to much lower plate counts. Also, the peak width of polydisperse macromolecules is a combination of band-broadening processes and sample polydispersity. If the peaks are symmetrical and close to Gaussian-shaped (Equation 3.4), Equation 3.2 can be expressed in variables that are easily measured experimentally. The most often used approximations include [39]

$$N = 16\left(\frac{V_R}{W_b}\right)^2 \tag{3.37}$$

where W_b is the baseline width formed by the tangents of the peak intersecting the baseline (approximately 4σ; see Figure 3.3),

$$N = 5.54\left(\frac{V_R}{W_{1/2}}\right)^2 \tag{3.38}$$

where $W_{1/2}$ is the peak width at one-half the peak height, and

$$N = 2\pi\left(\frac{h_p V_R}{A}\right)^2 \tag{3.39}$$

where h_p is the peak height and A the peak area. Although Equations 3.37 to 3.39 are widely used as a measure of chromatographic performance, it is well recognized that serious errors in the calculation can result if the peaks of interest are not close to a Gaussian shape [42].

The error in the calculation of plate count by the simplified methods of Equations 3.37 to 3.39 was determined quantitatively in a computer simulation study using the skewed peak model of the exponentially modified Gaussian (EMG) function [40]. The peak contour of this model is described by the convolute integral between a Gaussian constituent having the standard deviation σ and an exponential modifier having the time constant τ:

$$h = \frac{A}{\tau\sigma\sqrt{2\pi}} \int_0^{\infty} \exp\left[-\left(\frac{V - V_R - V'}{\sqrt{2}\sigma}\right)^2 - \frac{V'}{\tau}\right] dV' \qquad (3.40)$$

The variance of the peak (σ_x^2) equals $\sigma^2 + \tau^2$, and peak skew or tailing increases with the ratio τ/σ:

$$\text{peak skew} = \gamma \equiv \frac{\mu_3}{\mu_2^{3/2}} \simeq \gamma' \equiv \frac{2(\tau/\sigma)^3}{[1 + (\tau/\sigma)^2]^{3/2}} \qquad (3.41)$$

Figure 3.26 shows several synthesized chromatographic peaks with varying τ/σ ratios in which the value of σ for the peaks was held constant and the value of τ was

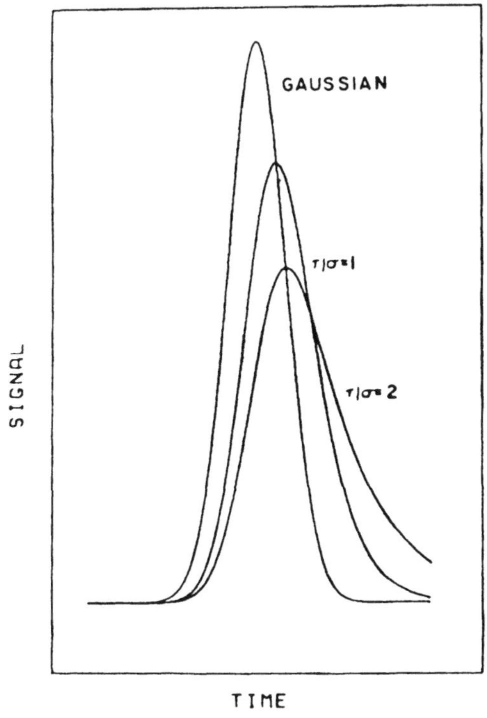

Figure 3.26 Effect of τ/σ on peak shape. (Reprinted with permission from Ref. 41.)

Table 3.3 Plate count calculation (constant σ^2, increasing τ^2)

		Plate Count				
τ/σ	Peak Skew, γ	True Value	Moment Method (Eq. 3.2)	Area Method (Eq. 3.37)	Tangent Method (Eq. 3.35)	Peak Half-Width Method (Eq. 3.36)
0.1	0.002	7670	7690	7680	7330	7630
0.5	0.18	6200	6210	6440	6230	6410
1.0	0.71	3870	3880	4740	4470	5060
1.5	1.15	2380	2410	3580	3850	4010
2.0	1.43	1550	1620	2760	3060	3200

Source: Ref. 42.
From peaks in Figure 3.26.

increased. Table 3.3 shows the results of the plate count calculations obtained on the series of peaks with constant σ and increasing τ. With values $\tau/\sigma > 1$ (peak skew > 0.7; peak asymmetry > 1.2; Equation 3.41), methods other than the moment method give significant positive errors in plate count. The exponentially modified Gaussian peak model has been used in a more precise method (compared to the statistical moments calculations, Equation 3.2) to isolate σ and τ constituents for characterizing the variance and the skewing of experimental chromatographic peaks [43]. It is also the basic model of the GPCV3 calibration method described in Section 8.3. Peak skew values of the σ–τ model are related to the more practical peak asymmetry factors shown in Figure 3.27 and described further in Section 6.4.

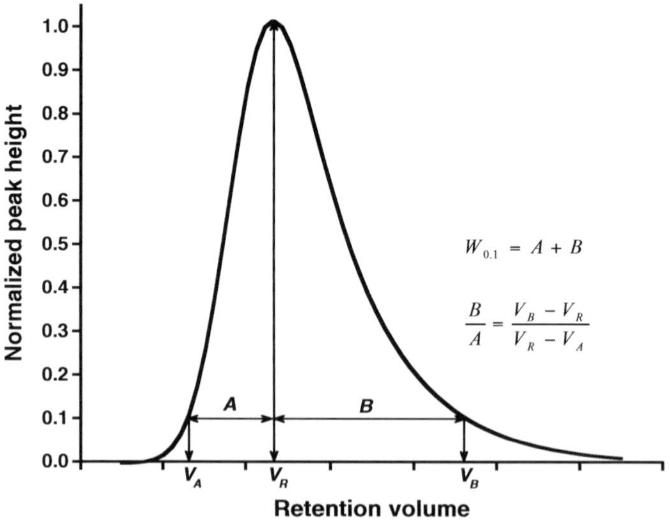

Figure 3.27 Peak asymmetry factors.

One of the most accurate methods for calculating column efficiency has been shown to be the Foley–Dorsey equation (Equation 3.42), which is also based on the EMG skewed peak model and the asymmetry parameters shown in Figure 3.27 [44]:

$$N = \frac{41.7(t_R/W_{0.1})^2}{B/A + 1.25} \tag{3.42}$$

For Equation 3.27 to work, the asymmetry factor B/A must be greater than unity, corresponding to a tailing peak. For fronting peaks, the asymmetry factor must be inverted (i.e., the ratio A/B should be used instead of B/A).

It should be noted that in SEC, many peaks, even those of narrow standards, will have a non-Gaussian shape due to non-band-broadening factors such as molar mass polydispersity and the type of distribution (e.g., Poisson-like) produced by the polymerization mechanism employed in making the polymer.

The actual variance of a column series is often larger than that predicted from the sum of the variances of the individual columns. This error arises because Equations 3.37 to 3.39, which do not account for peak skewing, are often used with tailing peaks to estimate the variance of columns. The additivity of plate count for a column set is described more fully in Section 7.10.

3.5.2 Column-Dispersion Calibration

Column dispersion is a major factor that causes inaccuracy in quantitative SEC interpretations, because it distorts the elution curve and affects the calibration and the molar mass calculations derived from the calibration curve. Compensation for the dispersion effect in SEC calibration and molar mass calculation is considered in Section 8.3. To account for column dispersion using the calibration methods developed in Section 8.3, one needs first to know how much peak broadening has been imposed on the experimental SEC elution curve. Unfortunately, it is difficult to determine true SEC column dispersion, because all polymer SEC peaks except a totally excluded polymer peak are somewhat broadened by molar mass separation as well as by column dispersion. (An example of the broadening due to molar mass separation, even for narrow-MMD polystyrene standards, can be seen in Figure 7.5. The relatively flat H versus v plot of the PS 3600 molar mass standard suggests that most of the band width of this standard must be due to molar mass separation, as the plate height contribution due to column dispersion is expected to change with changing flow rate, whereas that due to molar mass separation is not.) Peak broadening caused by molar mass separation can vary from one polystyrene standard to another, so it is not possible to distinguish peak broadening due to molar mass separation from broadening caused by column dispersion. This situation causes problems in the accurate characterization of SEC column dispersion over the entire separation volume range. Although there are two techniques that can be used to solve this problem, the reverse-flow experiment [28,30] and the recycle technique [45,46], these are rather complicated methods, used only to obtain very accurate calibration for SEC column dispersion or values of polydispersity M_w/M_n for narrow-MMD polymer standards.

In the reverse-flow technique, the polymer sample is injected in the normal way, but when the sample peak is halfway through the column, the flow is reversed. The molar mass separation processes are now reversed, but band broadening due to dispersion effects continues. When the peak reaches the detector, now located at the top of the column, it reflects only the band broadening due to dispersion processes. Molar mass separation has been completely canceled by the flow reversal, assuming equal elution time each way. The results of such an experiment are shown in Figure 3.23; such data can be used to obtain the σ-calibration curve for a column. Once determined, this curve is expected to be independent of the nature of the polymer sample and can be used directly in the SEC calibration methods described in Section 8.3 to compensate for the column-dispersion effect. The recycle technique of characterizing SEC column dispersion is described in Section 15.3.

REFERENCES

1. J. C. Giddings, *Unified Separation Science*, Wiley-Interscience, New York, 1991.
2. W. Feller, *An Introduction to Probability Theory and Its Applications*, 2nd ed., Wiley, New York, 1957, p. 216.
3. A. J. P. Martin and R. L. M. Synge, *Biochem. J.*, **35**, 1358 (1941).
4. A. B. Littlewood, *Gas Chromatography*, 2nd ed., Academic Press, New York, 1970, Chaps. 5 and 6.
5. S. Dal Nogare and R. S. Juvet, Jr., *Gas–Liquid Chromatography*, Wiley, New York, 1962, Chap. 3.
6. C. S. G. Phillips, *Gas Chromatography*, Academic Press, New York, 1956, p. 95.
7. L. R. Snyder and J. J. Kirkland, *Introduction to Modern Liquid Chromatography*, Wiley, New York, 1974, p. 29.
8. J. C. Giddings, *Dynamics of Chromatography*, Marcel Dekker, New York, 1965.
9. J. J. van Deemter, F. J. Zuiderweg, and A. Klinkenberg, *Chem. Eng. Sci.*, **5**, 271 (1956).
10. B. L. Karger, L. R. Snyder, and C. Horvath, *An Introduction to Separation Science*, Wiley, New York, 1973, Chap. 5.
11. R. N. Kelley and F. W. Billmeyer, Jr., *Anal. Chem.*, **41**, 874 (1969).
12. R. J. Hamilton and P. A. Sewell, *Introduction to High Performance Liquid Chromatography*, Wiley, New York, 1977, Chap. 2.
13. C. Horvath and H. J. Lin, *J. Chromatogr.*, **126**, 401 (1976).
14. J. F. K. Huber, *J. Chromatogr. Sci.*, **7**, 85 (1969).
15. J. N. Done and J. H. Knox, *J. Chromatogr. Sci.*, **10**, 606 (1972).
16. L. R. Snyder, *J. Chromatogr. Sci.*, **7**, 352 (1969).
17. R. N. Kelley and F. W. Billmeyer, Jr., *Anal. Chem.*, **42**, 399 (1970).
18. J. C. Giddings, L. M. Bowman, Jr., and M. N. Meyers, *Macromolecules*, **10**, 443 (1977).
19. R. Groh and I. Halász, *Anal. Chem.*, **53**, 1325 (1981).
20. A. M. Striegel, *J. Chromatogr. A*, **932**, 21 (2001).
21. L. Lapidus and N. R. Amundson, *J. Phys. Chem.*, **56**, 984 (1952).

22. P. R. Kasten, L. Lapidus, and N. R. Amundson, *J. Phys. Chem.*, **56**, 683 (1952).
23. A. Klinkenberg and F. Sjenitzer, *Chem. Eng. Sci.*, **5**, 258 (1956).
24. J. J. Hermans, *J. Polym. Sci. A-2*, **6**, 1217 (1968).
25. A. C. Ouano and J. A. Barker, *Sep. Sci.*, **8**, 673 (1973).
26. M. E. van Kreveld and N. van den Hoed, *J. Chromatogr.*, **149**, 71 (1978).
27. C. M. Guttman and E. A. DiMarzio, *Macromolecules*, **3**, 681 (1970).
28. L. H. Tung and J. C. Moore, in *Fractionation of Synthetic Polymers*, L. H. Tung, ed., Marcel Dekker, New York, 1977, Chap. 6.
29. W. W. Yau, C. P. Malone, and H. L. Suchan, *Sep. Sci.*, **5**, 259 (1970).
30. L. H. Tung and J. R. Runyon, *J. Appl. Polym. Sci.*, **13**, 2397 (1969).
31. M. Kubìn, *J. Chromatogr.*, **108**, 1 (1975).
32. J. V. Dawkins, T. Stone, and G. Yeadon, *Polymer*, **18**, 1179 (1977).
33. A. Rudin and H. K. Johnston, *J. Polym. Sci. B*, **9**, 55 (1971).
34. C. N. Satterfield, C. K. Colton, and W. H. Pitcher, Jr., *Am. Inst. Chem. Eng. J.*, **19**, 628 (1973).
35. G. Trenel, M. John, and H. Delleweg, *FEBS Lett.*, **2**, 74 (1968).
36. J. Y. Chuang, A. R. Cooper, and J. F. Johnson, *J. Polym. Sci. C*, **43**, 291 (1973).
37. W. Burchard, *Adv. Polym. Sci.*, **143**, 113 (1999).
38. J. C. Sternberg, in *Advances in Chromatography*, Vol. **2**, J. C. Giddings and R. A. Keller, eds., Marcel Dekker, New York, 1966, p. 205.
39. A. T. James and A. J. P. Martin, *Analyst*, **77**, 915 (1952).
40. E. Grushka, *Anal. Chem.*, **44**, 1733 (1972).
41. R. E. Pauls and L. B. Rogers, *Sep. Sci.*, **12**, 395 (1977).
42. J. J. Kirkland, W. W. Yau, H. J. Stoklosa, and C. H. Dilks, Jr., *J. Chromatogr. Sci.*, **15**, 303 (1977).
43. W. W. Yau, *Anal. Chem.*, **49**, 395 (1977).
44. J. P. Foley and J. G. Dorsey, *Anal. Chem.*, **55**, 730 (1983).
45. J. L. Waters, *J. Polym. Sci. A-2*, **8**, 411 (1970).
46. Z. Grubisic-Gallot, L. Marais, and H. Benoit, *J. Polym. Sci. A-2*, **14**, 959 (1976).
47. K. M. Usher, C. R. Simmons, and J. G. Dorsey, *J. Chromatogr. A*, **1200**, 122 (2008).

4

RESOLUTION

4.1 INTRODUCTION

4.1.1 Chromatographic Resolution

Traditionally, chromatographic column performance has been expressed in terms of the number of theoretical plates N (Equations. 3.11 and 3.37 to 3.39), the plate height H, or the *column resolution* R_s:

$$R_s = \frac{2(V_{R2} - V_{R1})}{W_{b1} + W_{b2}} \tag{4.1}$$

where V_R is the peak retention volume; W_b is the chromatogram peak width formed by intersection of the tangents to the curve inflection points with the baseline in retention volume units, $W_b = 4\sigma$ (Figure 3.3); and σ is the peak standard deviation (proportional to peak width) caused by column dispersion and expressed in volume units (e.g., milliliters). The subscripts 1 and 2 serve to identify two closely eluting solutes. The plate height H (or HETP, height equivalent to a theoretical plate) is equal to L/N, where L is the column length (Section 3.2).

Equation 4.1 may also be written as

$$R_s = \frac{V_{R2} - V_{R1}}{2(\sigma_1 + \sigma_2)} \approx \frac{\Delta V_R}{4\sigma} \tag{4.2}$$

Modern Size-Exclusion Liquid Chromatography: Practice of Gel Permeation and Gel Filtration Chromatography, Second Edition
By André M. Striegel, Wallace W. Yau, Joseph J. Kirkland, and Donald D. Bly

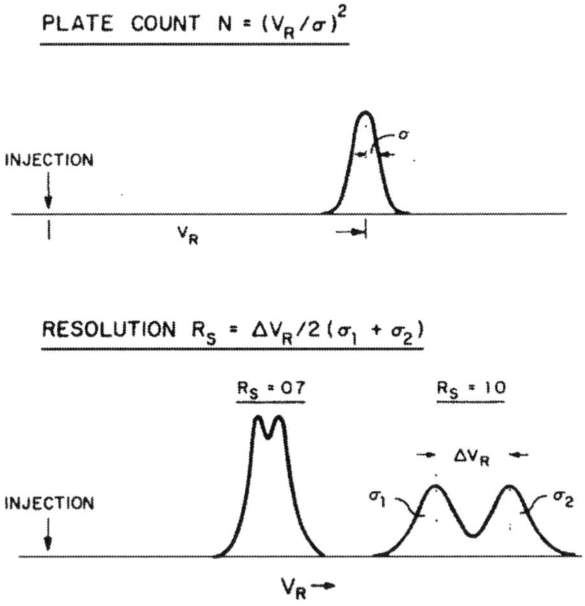

Figure 4.1 Traditional column performance parameters.

The values of σ are determined experimentally from the chromatograms of single molecular species (see Section 3.5), and to a first approximation, $\sigma_1 = \sigma_2 = \sigma$. The resolution factor R_s is a more meaningful column performance parameter than plate count N. R_s accounts for peak broadening (σ, N) as well as the selectivity of the column (ΔV_R). In LC separations other than SEC, the value of R_s calculated by Equation 4.2 depicts how well peaks are resolved. An illustration of this is shown in Figure 4.1 for the cases $R_s = 0.7$ and 1.0. However, the R_s factor is still not a totally adequate general column performance parameter, because its value varies with the particular choice of peaks. Since R_s is a dimensionless quantity, the resolution of a particular pair of solute peaks has the same value of R_s whether the peak elution is recorded in retention volume or in retention time units. As for the calculation of plate count N (Sections 3.1 and 3.2), the resolution relationships (Equations. 4.1 and 4.2) implicitly assume the Gaussian (symmetrical) peak shape.

To provide a visualization of resolution units, standard resolution curves calculated for theoretical Gaussian peaks are shown in Figures 4.2 and 4.3. In Figure 4.2 all the component peaks are of equal height, simulating equal concentrations of component solute species. The evidence of a double peak begins at $R_s = 0.5$, which is sometimes called 2σ resolution because it corresponds to $\Delta V_R = 2\sigma$ in Equation 4.2. At $R_s = 1.0$ (4σ separation), the peaks are reasonably well resolved. Complete peak separation to baseline resolution occurs at $R_s = 1.5$ (6σ separation).

Actual solute overlapping or mixing between the elution peaks of equal size is not as extensive as it might appear from casual observation. At $R_s = 0.5$, there is only

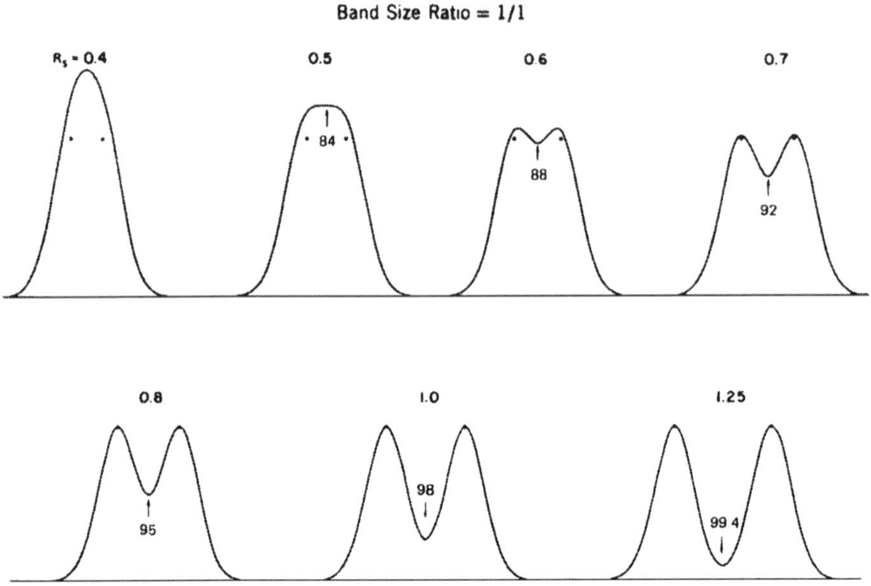

Figure 4.2 Standard resolution curves for a band-size ratio of 1 : 1. Values of R_S, 0.4-1.25 (Reprinted with permission from Ref. 1.)

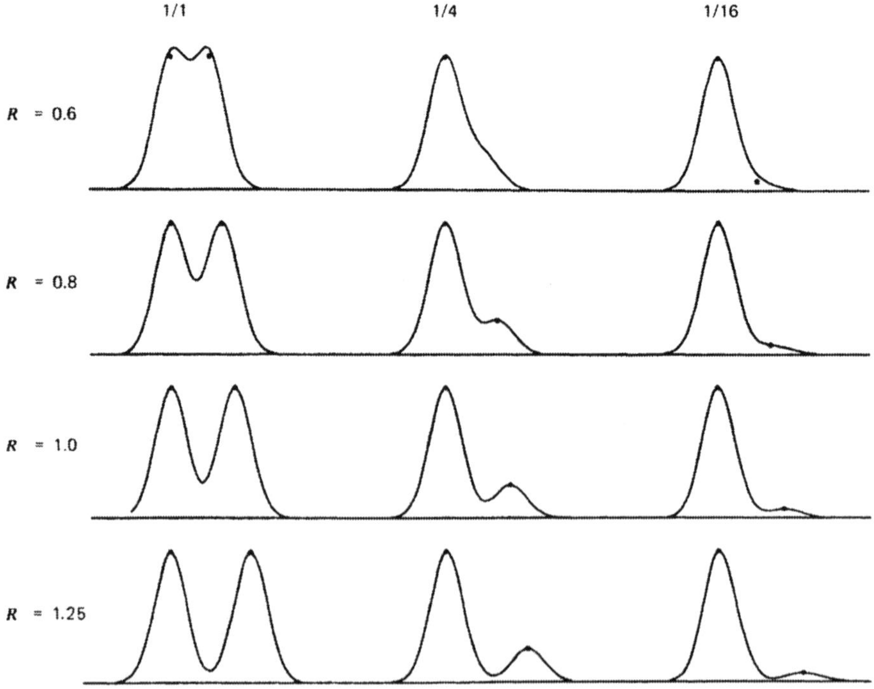

Figure 4.3 Separation as a function of R_S and relative band concentration. (Reprinted with permission from Ref. 1.)

a 16% actual solute overlap. The overlap of one solute on the other is only 2% at $R_s = 1.0$. In other words, even at the low-resolution case of $R_s = 0.5$, fractions of each component species of 84% purity can be recovered at the equal purity cut point indicated by the arrows in Figure 4.2. At $R_s = 1.0$, the two recovered fractions are 98% pure for equal-height peaks. As a result of curve overlapping, the peak apexes of the composite chromatogram at low resolution (e.g., $R_s = 0.6$) are taller and closer to each other than those of the component peaks. The apexes of the original component peaks are indicated by the dots shown in Figures 4.2 and 4.3.

Similar sets of the theoretical standard resolution curves are available for other band concentration ratios [1,2]. A selection of such curves is shown in Figure 4.3. Such standard reference curves are very useful for estimating the values of R_s of experimental peaks. With a recollection of the various R_s curve shapes, a quick estimate of the value of R_s can be made on the spot by glancing at the features of the experimental chromatogram. Since the expression for R_s in Equation 4.2 is independent of the individual peak heights, the same value of R_s can correspond to resolution curves very different in shape, depending on band ratios. As shown in Figure 4.3, as the band ratio increases, the features of the smaller peak are less distinguishable because of the increased interference of the larger peak. This effect makes the quantitative detection of smaller peaks on the tailing edge of larger peaks more difficult. Under these circumstances, the standard R_s curves can be very helpful for comparison with experimental chromatograms to detect the presence and estimate the areas of the smaller peaks, or to locate the proper cut point if a fraction of the smaller peak of a certain purity is desired. For a pair of peaks with a large band concentration difference, the equal-purity cut point shifts toward the smaller peak, since the solute molecules of the larger peak spread more into the smaller peak. For a further discussion of the use of standard resolution curves, see References 1 and 2.

Since column dispersion and thus N for polymer solutes in SEC varies as a function of retention volume (Sections 3.3 and 3.4), the traditional LC resolution and peak-capacity expressions described below are of little use in SEC of polymers. However, they are generally applicable to SEC separations of small molecules. Special considerations are required for studying SEC resolution in polymer MMD analyses, as described in Section 4.2.

For LC methods other than SEC, experimental values of plate count are nearly independent of the retention volumes of the individual solutes; that is, according to Equation 3.11, peak width increases linearly with increasing retention volume. This is illustrated in Figure 4.4 by the unshaded peaks in the region marked "other LC." Constant values of N mean $\sigma \propto (1 + k')$, where $k' = (V_R - V_t)/V_t$, with k' being the usual LC peak-capacity factor described in Section 2.1 and V_t being the retention volume of the total permeation peak (often called the unretained peak in discussing conventional LC separations). In terms of basic LC retention parameters, the resolution in Equation 4.2 can be expressed as

$$R_s = \frac{\Delta k' \sqrt{N}}{4(1 + k')} = \frac{(\alpha - 1)\sqrt{N}}{4} \frac{k'}{1 + k'} \tag{4.3}$$

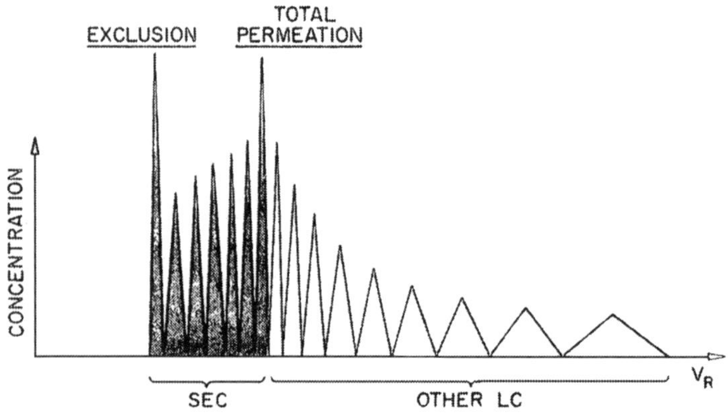

Figure 4.4 Characteristics of column dispersion and peak capacity in SEC and the other LC methods.

where α is the separation factor, which equals the k' ratio of the two adjacent peaks (i.e., $\alpha = k_2'/k_1'$), and the plate count N is assumed constant. Equation 4.3 is very useful for the design and optimization of LC methods other than SEC, since the resolution of LC peaks can be controlled by independently changing separation selectivity α, efficiency N, or capacity k' [2]. Reference 17 provides a review of the various resolution equations used in the different column chromatographic techniques.

4.1.2 Peak-Capacity Concept

For small molecules, the quality of separation can also be described effectively in terms of *peak capacity* n. This term is defined as the maximum number of peaks that can be resolved within a specified range of retention volume. For cases in which solute peaks having the same plate count are to be separated with 4σ resolution, it has been shown that [3]

$$n = 1 + \frac{\sqrt{N}}{4} \Delta \ln V_R \tag{4.4}$$

where $\Delta \ln V_R$, the difference between the logarithms of the retention volumes, specifies the retention range of interest. The relationship described by Equation 4.4 for constant N is illustrated in Figure 4.4 by the closely spaced peaks beyond the total permeation volume.

As with Equation 4.3, a constant plate number N must also apply in Equation 4.4 to all the peaks of interest in a chromatogram. This is usually the case in LC methods other than SEC. As a result of the significant decrease in solute diffusion with increasing solute size, the earlier peaks in SEC actually suffer more band broadening due to column dispersion. This is a trend directly opposite to that of the other LC methods (see Figure 4.4 for illustration and Sections 3.3 and 3.4 for further discussion).

The term *peak resolution* is not commonly used in SEC because it does not fit properly in the context of describing SEC column performance. A major use of SEC is not to resolve and identify species but to retrieve MMD information from the chromatogram. Special concepts of SEC resolution and M accuracy are required to define SEC column performance in polymer analyses. However, these concepts are derived from the general LC resolution considerations discussed above.

4.2 RESOLUTION CONCEPT IN SEC OF POLYMERS

A quantitative expression of SEC resolution for polymer MMD analyses is needed to determine where the performance of SEC stands relative to conventional SEC and with respect to theoretical and instrumental performance limits, and to determine practical goals for SEC performance relative to cost and time.

Because the dependence of SEC peak separation ΔV_R on solute molar mass is known via the SEC calibration curve (e.g., Figure 4.5), a unique opportunity exists in SEC for eliminating the dependence of the resolution factor R_s on the probing solutes [4–6]. The useful portion of the SEC calibration curve can be approximated by a straight line of slope D_2 and intercept D_1:

$$M = D_1 e^{-D_2 V_R} \tag{4.5}$$

By taking the natural logarithm and rearranging, Equation 4.5 becomes

$$V_R = \frac{\ln D_1 - \ln M}{D_2} \tag{4.6}$$

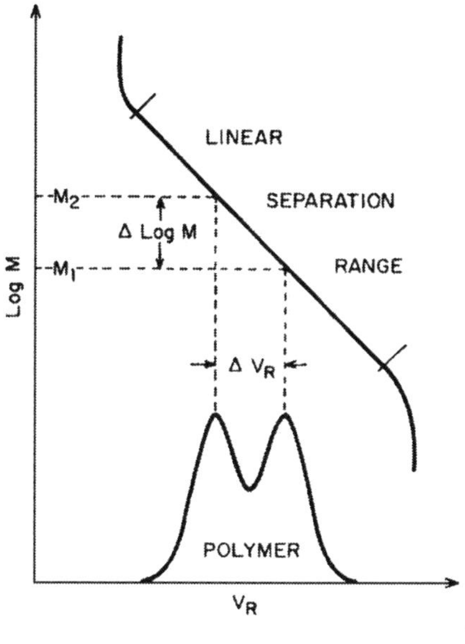

Figure 4.5 Dependence of SEC peak separation on solute M.

or

$$\Delta V_R = \frac{\Delta \ln M}{D_2} = \frac{\ln(M_2/M_1)}{D_2} \tag{4.7}$$

Substitution of Equation 4.7 into the basic resolution expression, Equation 4.2, gives

$$R_s = \frac{\ln(M_2/M_1)}{2D_2(\sigma_1 + \sigma_2)} \simeq \frac{\Delta \ln M}{4\sigma D_2} \tag{4.8}$$

This equation describes how well the SEC column can distinguish between two molecules of the same polymer type but differing by a molar mass factor M_2/M_1. This description of SEC resolution is useful but too specific to allow the data of different columns or different laboratories to be compared. However, in SEC we are interested in the resolution pertained in the elution curve as a whole, not so much that between specific pairs of eluted fractions. To provide a general measure of SEC resolution, the concept of *specific resolution* R_{sp} has been developed [4]. Dividing Equation 4.8 by $\Delta \log M$ leads directly to the expression for SEC specific resolution:

$$R_{sp} = \frac{R_s}{\Delta \log M} = \frac{0.58}{\sigma D_2} \tag{4.9}$$

[the conversion between the natural and the base 10 logarithm expressions (i.e., ln $M = 2.303 \log M$) is accounted for in Equation. 4.9]. Note that in Equation 4.9 the explicit dependence of SEC resolution on sample molar mass is now eliminated in the expression for R_{sp}. Specifically, Equation 4.9 states that the resolution factor R_{sp} in the linear calibration region is equal to the usual chromatographic resolution R_s (Equation 4.1) for a pair of peaks having a decade of molar mass difference (Section 7.10). Experimental values of R_{sp}, which are not expected to vary much with the selection of probe samples, can be used as a SEC column performance parameter for evaluating and comparing SEC columns or column sets.

To provide a performance factor for comparison of different column packings, the expression for R_{sp} must be compensated for column length. Since D_2 is proportional to the reciprocal of the column length L, and σ is proportional to the square root of L, Equation 4.9 can be normalized for column length to give the packing resolution factor, which is equivalent to R_{sp} for a 1-cm column:

$$R^*_{sp} = \frac{0.58}{\sigma D_2 \sqrt{L}} \tag{4.10}$$

The advantage of using the resolution factors R_{sp} and R^*_{sp} instead of plate count N in evaluating SEC columns and column packings is further illustrated by the experimental results in Tables 4.2 and 4.3.

4.3 MOLAR MASS ACCURACY CRITERION

The quality of the SEC results in polymer analyses should be assessed in terms of the accuracy of the final calculated values of M. It is important that the molar mass accuracy of the SEC polymer analyses can be predicted from measurable column parameters. The resolution concept in SEC still does not provide the same utility as in GC and the other LC methods, where a resolution value can unequivocally define the system efficiency as well as the quality of the final results. A simple resolution value simply does not provide the desired molar mass accuracy information about a system for polymer analyses. Fortunately, a relationship exists between the SEC resolution and the M accuracy, as described below.

In SEC the elution curve is broadened by column dispersion as illustrated in Figure 4.6. The SEC-M accuracy problem resulting from column dispersion is related directly to the differences between the experimental, $F(V_R)$, and the true, $W(V_R)$, elution curves. The elution curves F and W are related by a convolution integral [7]:

$$F(V_R) = \int_{-\infty}^{\infty} W(y)G(V_R - y)\,dy \tag{4.11}$$

where $G(V_R - y)$ is an instrument-column-dispersion function which describes the weight fraction of a solute that should have been eluted at the retention volume y but is actually dispersed and detected at the retention volume V_R. The true values of M_w

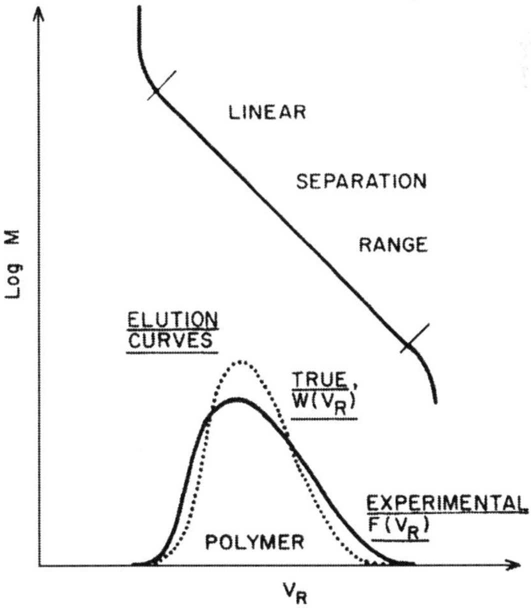

Figure 4.6 Effect of instrumental band broadening on SEC elution-curve shape.

and M_n of a polymer sample for linear calibration (Equation 4.5) can be written as [8–11]

$$(M_n)_{\text{true}} = \frac{1}{\sum_{V_R} \left[W(V_R)/D_1 e^{-D_2 V_R}\right]} \tag{4.12}$$

and

$$(M_w)_{\text{true}} = \sum_{V_R} W(V_R) D_1 e^{-D_2 V_R} \tag{4.13}$$

On the other hand, observed molar mass averages are calculated from the experimental elution curves $F(V_R)$ instead of $W(V_R)$:

$$(M_n)_{\text{exp}} = \frac{1}{\sum_{V_R} \left[F(V_R)/D_1 e^{-D_2 V_R}\right]} \tag{4.14}$$

and

$$(M_w)_{\text{exp}} = \sum_{V_R} F(V_R) D_1 e^{-D_2 V_R} \tag{4.15}$$

The true and experimental molar mass averages can be related to each other directly by a single correction factor X [8–11]:

$$M_{\text{true}} = (X) M_{\text{exp}} \tag{4.16}$$

Equation 4.16 represents an important theoretical advance in modern SEC data reduction. The values of X for various molar mass averages derived from two instrument dispersion functions, one for symmetrical and one for skewed peak shapes, are summarized in Table 4.1, where M_υ and a are the viscosity-average molar mass and

Table 4.1 SEC correction factors for various instrument dispersion functions[a]

	Delta Function (Hamielec)	Gaussian Function (GPCV2)	Exponentially Modified Gaussian Function (GPCV3)	
M_n	1	$e^{(D_2\sigma)^2/2}$	$e^{(D_2\sigma)^2/2}\left(\frac{1}{1-D_2\tau}e^{-D_2\tau}\right)$	for $D_2\ \tau < 1$
M_w	1	$e^{-(D_2\sigma)^2/2}$	$e^{-(D_2\sigma)^2/2}(1 + D_2\tau)e^{-D_2\tau}$	for $D_2\ \tau > -1$
M_z	1	$e^{-3(D_2\sigma)^2/2}$	$e^{-3(D_2\sigma)^2/2}\left(\frac{1+2D2\tau}{1+D2\tau}e^{-D_2\tau}\right)$	for $D_2\ \tau > -\frac{1}{2}$
M_υ	1	$e^{-\mathbf{a}^2(D_2\sigma)^2/2}$	$e^{-\mathbf{a}^2(D_2\sigma)^2/2}(1 + aD_2\tau)e^{-\mathbf{a}D_2\tau}$	for $D_2\ \tau > -\mathbf{a}$

[a] $M_{\text{true}} = (X)M_{\text{exp}}$.

the exponent constant, respectively, for the Mark–Houwink viscosity–molar mass relationship (Section 2.4; see Reference 18 for published values of a). These correction factors, which are the same as those used in the linear calibration methods (i.e., Hamielec, GPCV2, and GPCV3), are noted in the column headings of the table. The delta function in the second column simply indicates that zero column dispersion is assumed. The Gaussian function in the third column simulates symmetrical peak dispersion. The exponentially modified Gaussian function used to develop the fourth column is the same generalized skewed peak model, as previously described in Section 3.5 [12,13].

Based on Equation 4.16, a molar mass error expression can now be derived in terms of column parameters only. The general expression of this molar mass error, normalized by the value of the molar mass average, is

$$\frac{M_{\text{exp}} - M_{\text{true}}}{M_{\text{true}}} = \frac{1}{X} - 1 \qquad (4.17)$$

Using Equation 4.17 and Table 4.1, fourth column, the errors in M_w and M_n can be predicted for various band-broadening situations (different values of σ and τ). The molar mass errors calculated are plotted in Figure 4.7. It is shown that molar mass errors increase with increasing column dispersion as measured by the term $\sigma^2 + \tau^2$ (Section 3.5). The effect of increased peak skewing as measured by τ/σ is to cause more peak tailing into the longer retention volume region and larger error in experimental values of M_n.

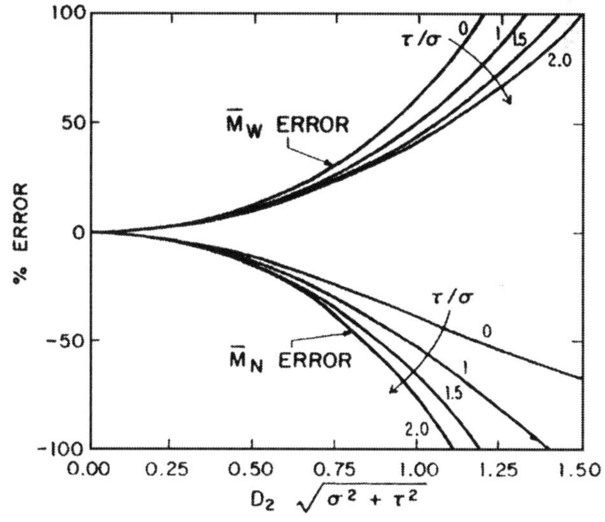

Figure 4.7 Predicted SEC-M error due to column dispersion. Curves for τ/σ of 0, 1, 1.5, and 2.0 are calculated according to Equation 4.17 using the correction factor X in Table 4.1, fourth column. (Reprinted with permission from Ref. 9.)

The level of molar mass error or inaccuracy directly reflects the performance of SEC columns. This fact underlies the basic concept of the molar mass accuracy criterion for SEC column performance. Specifically, the molar mass accuracy criterion is defined as M^*, the molar mass error averaged for M_w^* and M_n^*, which are derived from the Gaussian dispersion function:

$$M_n^* = e^{-(1/2)(\sigma D_2)^2} - 1 \tag{4.18}$$

and

$$M_w^* = e^{(1/2)(\sigma D_2)^2} - 1 \tag{4.19}$$

Note that the value of M_n^* is always negative and the value of M_w^* is always positive according to Equations 4.18 and 4.19. The absolute value of M_n^* is to be used for the M^* calculation. These equations have practical utility, since they serve to predict molar mass accuracy directly from experimental column parameters σ and D_2. Also, these equations can be used to specify the values of column σ and D_2 required to achieve a desired SEC-M accuracy. Both σ and D_2 are positive quantities in SEC analysis. Familiarity with the basic properties of D_2 and σ in these equations (see Table 4.4) is needed to make the best practical use of the molar mass accuracy criterion.

It is important to note that both the SEC specific resolution (R_{sp}) and molar mass accuracy (M^*) factors are defined uniquely by the value of σD_2 for the chromatographic system. Therefore, the product of σ and D_2 is by itself a fundamental SEC column performance parameter. Inherently, columns of different individual values of σ and D_2 can perform equally well as long as they have the same combined value of σD_2. In practice, SEC systems with *small* values of σD_2 are sought to achieve high resolution and M accuracy. Note also that the values of R_{sp}, M^*, and σD_2 are all dimensionless and are therefore valid for studying SEC systems in general, regardless of whether retention volume, syphon counts, or retention time is used in defining the SEC calibration and elution curves. These performance parameters provide the interesting feature that they are independent of sample MMD. Sample MMD is not used in the derivation and does not appear in the final expression for these parameters. Therefore, values of R_{sp}, M^*, and σD_2 reflect properties of the column alone and should be nearly the same for a particular column set, regardless of differences in probe sample MMD (whether it is a single, bimodal, broad, or narrow distribution).

4.4 APPLICATIONS OF COLUMN PERFORMANCE CRITERIA

The validity of the SEC performance concept above is in practice dependent on the basic premise that σD_2 is reasonably constant and independent of solute molar mass and retention volume. The experimental value of D_2 can be calculated from Equation 4.7 if narrow-MMD polymer standards are available, or by the broad-MMD standard calibration methods described in Section 8.3. By connecting columns of different pore-size packings, the value of D_2 (or the slope of the SEC calibration curve) of the assembled column set can be made essentially invariant over a wide molar mass separation range (see the discussion in Section 4.5 and in Sections 7.9 and 8.6 for a bimodal-pore-size column set). Near the exclusion and the total permeation volumes, D_2 approaches infinity, which forces the SEC resolution to zero. Therefore, in evaluating SEC performance, polymer standards that elute too close to the exclusion or the total permeation volume should be avoided.

The value of σ or column dispersion is determined, to a first approximation, as the experimental value of σ for a very narrow MMD polymer standard. The value of σ for a monomer peak should not be used, since it usually grossly underestimates the true column dispersion. To obtain more accurate values of column σ, special SEC experiments such as recycle (Section 15.3) and reverse-flow techniques are required (Section 3.5).

Usually, a constant value of σ is not observed experimentally for use in the σD_2 resolution concept. The value of column σ in SEC is dependent on the retention volume (Figure 3.23). However, in practice, the value of σ used is the average of the smaller σ values as determined for narrow-MMD polystyrene standards. It would be most accurate to account for this σ variation. However, this is difficult to accomplish and the dependency of σ on V_R is small relative to the total magnitude of σ.

One way to test the molar mass accuracy criterion (Equations 4.18 and 4.19) is to vary the value of σ of a column set by changing the solvent flow rate and then to compare the observed and predicted values of M^*. The results of such an experiment are shown in Figure 4.8, where the experimental values of M^* (open circles) are shown to correspond closely to the theoretical values (dashed curves) that are calculated

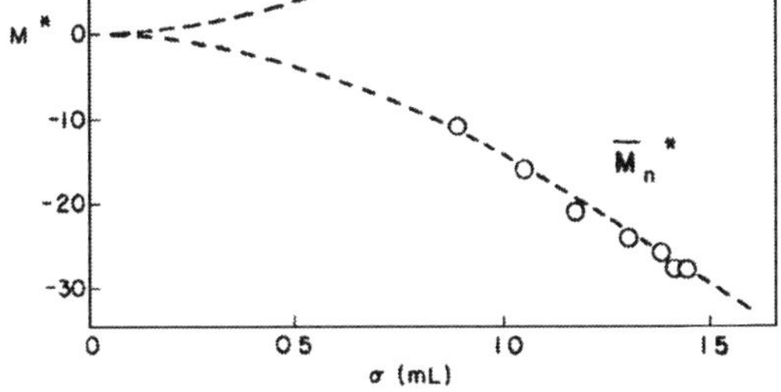

Figure 4.8 Effect of column dispersion on M accuracy. Columns, Vit-X column set (see Table 4.2); sample, polystyrene standard, 97,200 M. ○, experimental values, – – –, theoretical values of M_w^* and M_n^*. (Reprinted with permission from Ref. 4.)

Table 4.2 Performance comparison of several column sets using various SEC column-packing materials

Column Packing	Number of Columns	Total Length (cm)	Flow Rate (mL/min)	Sample Analysis Time (min)	Particle Size (μm)	Linear Calibration Range M (g/mol)	Plate Count N, Toluene	σD_2[a]	R_{sp}	M^*[b] (%)
Styragel	4	488	1	180	50	10^3–10^6	7,500	0.45	1.27	11
Porasil	4	488	1	180	75–150	2×10^4–10^6	2,700	0.37	1.56	7
Vit-X	4	200	2	15	30	5×10^3–10^6	3,500	0.59	0.97	18
μ-Styragel	4	120	2	15	10	2×10^3–10^6	13,000	0.50	1.14	13
LiChrospher	4	100	1.5	15	10	5×10^3–10^6	5,800	0.23	2.50	3
PSM	5	60	1.25	15	7	10^3–2×10^6	24,500	0.21	2.72	2

Source: Ref. 4.

[a] σ measured with 97,200 g/mol polystyrene.

[b] M^* is obtained from Equations 4.18 and 4.19 using the measured values of σ and D_2.

from various values of σ and the measured value of column D_2. Experimental values of M_w and M_n for the test polystyrene standard were calculated directly by the usual point-by-point summation of the elution curves observed at different flow rates (Equations 4.14 and 4.15). These values are compared, as in Equation 4.17, to the "true" value of M of the standard supplied by the vendor to calculate the experimental M_w^* and M_n^* errors. The particular column set used in this experiment was chosen arbitrarily, for illustration. However, similar results were obtained for column sets of other packing materials, listed in Table 4.2. They also support the general utility of the SEC-*M* accuracy criterion.

Table 4.2 also verifies that plate count N, measured by the value of σ for a monomer peak (Section 3.2), is a poor indicator of SEC column performance in terms of resolution or polymer molar mass accuracy. For example, for the 2700-plate Porasil column set, the molar mass error (M^*) caused by column dispersion is 7%, compared with 11% for a Styragel column set of $N = 7500$. The better molar mass accuracy of the Porasil column set in this case is due partly to its smaller value of D_2 as compared to the Styragel column set. These data support the contention that column plate count measured from a monomer peak does not accurately reflect the capability of the SEC system for polymer molar mass analyses and that the R_{sp} and M^* accuracy values are more useful quantitative criteria for SEC column performance.

The performances of the various column sets in Table 4.2 are compared directly in Figure 4.9, which represents a master plot of M^* versus σD_2. This plot, the basic SEC-*M* accuracy criterion, can be used universally for comparing the performances of all SEC column sets. The data in Table 4.2 and Figure 4.9 show, for example, that the PSM column set studied exhibits an excellent level of molar mass accuracy of 2% for a 15-minute analysis. Column-packing particle size is the most significant factor

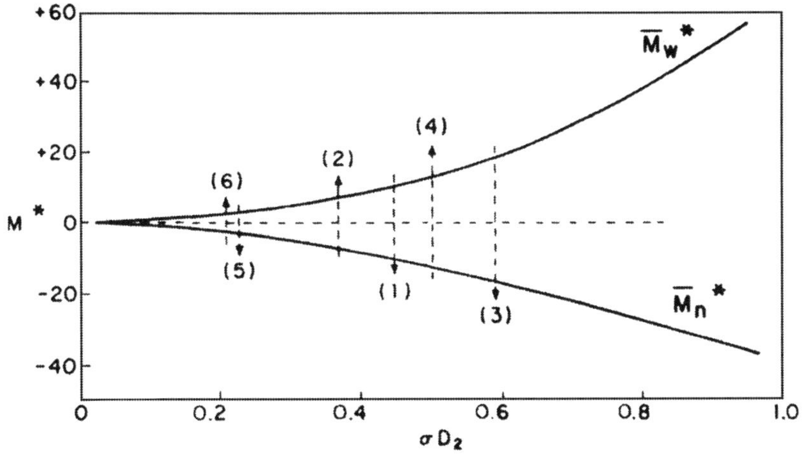

Figure 4.9 Comparison of column sets using the molar mass criterion M^*. 1, Styragel; 2, Porasil; 3, Vit-X; 4, μ-Styragel, 5, LiChrospher; 6, PSM. (Reprinted with permission from Ref. 4.)

Table 4.3 Comparison of unmodified LiChrospher and PSM packings for SEC

	Pore Size (Å)	Volume Porosity (%)	Particle Diameter (μm)	Linear M Fractionation Range (g/mol)	D_2	σ (mL)			R^*_{sp}	
						Toluene	PS[a]	M (g/mol)	Toluene	PS[a]
LiChrospher 100	100	72.5	10	3×10^3–5×10^4	1.28	0.087	0.229	5,000	1.05	0.39
PSM-500	125	45.8	9	5×10^3–4×10^4	2.17	0.067	0.147		0.80	0.36
LiChrospher 500	500	63.7	10	1.5×10^4–1.5×10^5	1.32	0.107	0.275	51,000	1.01	0.32
PSM-800	300	48.3	6	6×10^3–2×10^5	2.25	0.054	0.118		0.95	0.43
LiChrospher 1000	1000	63.7	10	3×10^4–2×10^6	1.98	0.096	0.142	97,000	0.61	0.41
PSM-1500	750	50.2	9	4×10^4–2×10^6	4.56	0.030	0.068		1.30	0.60
LiChrospher 4000	4000	63.7	10	$10^5 - > 7 \times 10^6$	3.84	0.092	0.144	390,000	0.33	0.21
PSM-4000	3800	53.0	6	$7 \times 10^4 - > 7 \times 10^6$	5.76	0.052	0.136		0.50	0.19

Source: Reference 14.

[a]PS, polystyrene.

differentiating the 15-minute "high-performance" SEC from the 3-hour conventional SEC analysis.

The utility of the R_{sp}^* factor (Equation 4.10) for comparing SEC packings is demonstrated in Table 4.3. Note that R_{sp}^* is independent of column dimensions and that LiChrospher and PSM particles have comparable values of R_{sp}^* for polymer analyses. The data also verify that values of R_{sp}^* for solutes at total permeation (toluene in this case) are not very useful for defining the performance of SEC packings for polymer analyses. While LiChrospher packings show greater selectivity (smaller D_2), presumably because of a generally larger porosity, PSM packings have the advantage of higher efficiency (smaller H and σ) because of smaller particles and narrower particle and pore-size ranges.

It should be noted at this point that Equations 4.18 and 4.19 (the M^* criterion) predict the accuracy of the values of M_w and M_n, not the accuracy of the entire MMD curve. The requirement of *MMD accuracy* on column performance is more stringent than that of the *molar mass accuracy*. Predictions of satisfactory accuracy of the average molar mass do not necessarily mean acceptable accuracy for the entire MMD curve observed. The value of M^* is the error caused by column dispersion alone and does not include errors in values of M assigned to polymer standards, errors due to flow-rate variation, operator errors, and so on. In actual polymer sample analyses, molar mass errors due to column dispersion can be corrected by using the appropriate SEC calibration and molar mass calculation method (Section 8.3). Methods for correcting column dispersion in the MMD curve are discussed in Section 8.7. It should be emphasized that the validity of the SEC resolution calculations depends on the accuracy of the following approximations: the linearity of the calibration curve, the appropriateness of using a Gaussian instrument function, and a constant value of column σ.

4.5 PORE GEOMETRY AND OPERATIONAL EFFECTS

Properties of SEC separating systems that are of great practical importance can be deduced from theoretical insights into σ and D_2. For clarity, we discuss only the conclusions and practical guidelines in this section. Detailed discussions of properties of D_2 and σ may be found in Chapters 2 and 3, respectively.

4.5.1 Connecting Columns

The total value of σ for a column series can be calculated from those for the individual columns according to the additive property of peak variance (Section 3.1):

$$\sigma^2 = \sum_i \sigma_i^2 \tag{4.20}$$

On the other hand, it can be shown from the additivity property of peak retention that [15]

$$\frac{1}{D_2} = \sum_i \left(\frac{1}{D_2}\right)_i \tag{4.21}$$

or

$$C_2 = \sum_i (C_2)_i \tag{4.22}$$

where C_2 is the linear calibration constant when the SEC calibration is expressed as $V_R = C_1 - C_2 \log M$, where $C_2 = \ln 10/D_2$ (Section 8.3). Note that for solute molar mass values outside the linear separation region, the value of D_2 for the individual columns approaches infinity (value of C_2 approaches zero). As predicted by Equation 4.20, one poor column with an exceptionally large value of σ_i can dominate the value of σ of a column set and degrade column performance as a whole (Section 7.10).

4.5.2 Separation Capacity of Single Pores

By the nature of the size-exclusion mechanism, there is a finite minimum slope to the calibration curve, that is, a lower limit to the value of D_2 (smaller D_2 means better resolution) even when there is no pore-size distribution (PSD) in the SEC column packing [15]. It often is mistakenly assumed that a broad spectrum of pore sizes is required for the SEC packing to effectively fractionate broad-MMD polymers. In fact, however, pores of a single pore size are capable of fractionating polymer molecules over a substantial molar mass range (1.5 to 2 decades for random-coil polymer solutes). Figure 4.10 illustrates simply how a single pore can separate solute molecules of differing sizes by means of a solute-to-wall exclusion effect inside the pore. Because of steric interference, the centers of large incoming solute molecules are kept away from the interior walls of the pore, as illustrated by the inner dashed

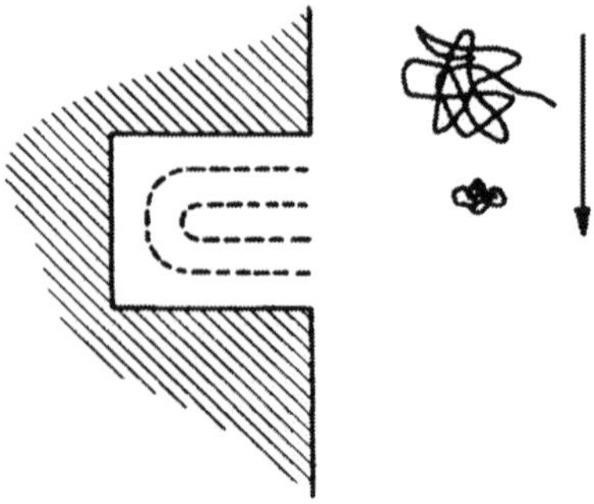

Figure 4.10 Size-exclusion effect in a single pore. Larger solute sees smaller effective pore volume.

line. However, smaller molecules can approach closer to the wall, as represented by the outer dashed line in the figure. Thus, a larger fraction of the pore volume is accessible to smaller molecules than to larger molecules. The progression from total permeation to total exclusion does not occur abruptly (if it did, it would produce a horizontal SEC calibration line with $D_2 = 0$) but it takes place gradually (with a finite value of D_2 over a substantial size range for solute molecules).

The limiting values of D_2 for single pores can be predicted from the basic retention theory presented in Section 2.4. For random-coil solutes (Equation 2.22),

$$\text{limiting } \mathrm{D}_2 \simeq \frac{1}{3 \times \text{pore volume}} \tag{4.23}$$

Thus the limiting value of D_2 is inversely proportional to the column pore volume and, therefore, to the internal porosity of the SEC packing particles.

The effect of pore shape on the limiting D_2 value is small. The large shape differences between the cylindrical and slab pore models cause only a 20% difference in D_2 (Figure 2.11). Particles with equal pore volume but different size have identical values of D_2 according to the SEC retention theory (Section 2.4). A change in pore size only shifts the SEC calibration curve up or down along the molar mass scale, without affecting its slope or value of D_2.

On the other hand, the limiting value of D_2 of a single pore is strongly dependent on the shape of the solute molecule. For a certain chemical structure and molar mass, the more extended the conformation of a macromolecule, the more it will be excluded from the pores of SEC packings. Therefore, SEC calibration curves for different solute conformations behave like those illustrated in Figure 4.11. The more open structure of the rigid-rod shape elutes before the random-coil structure of the same molar mass, with the calibration curve of the rod molecule falling below that of the coiled molecule. On the other hand, the more compact structure of the hard-sphere type will elute after the coiled molecules of the same molar mass, producing a calibration curve that lies above that for the coiled molecule.

The dependence of solute size on molar mass varies with solute conformation. Solute size varies in proportion to the molar mass raised to a power of about 1, $\frac{1}{2}$, and $\frac{1}{3}$ for the rod-like, the coil-like, and the sphere-like solutes, respectively (Section 11.4). It is to be expected, therefore, that the calibration curve will be the steepest for the sphere-like solutes, with its value of D_2 being $\frac{3}{2}$ that of the coiled solute. The curve for the rod-like solutes has the lowest slope, with its value of D_2 being only $\frac{1}{2}$ of that of the coiled solute. The molar mass separation range expected for a single pore is about one decade for rod-like molecules and three decades for spheres, as compared to the usual approximately two-decade M separation range for random-coil solutes.

4.5.3 Effect of Packing Pore-Size Distribution

Because the pores in actual SEC packings have irregular cross sections and finite pore-size distributions (PSDs), the observed value of D_2 and molar mass separation

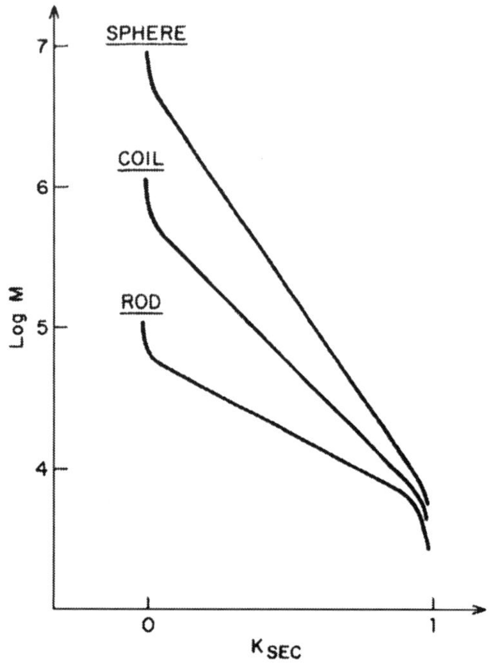

Figure 4.11 Effect of solute geometry on SEC-M calibration curve slope. Rigid rod, $R_g \propto M$; flexible coil, $R_g \propto M^{\alpha}$, $\alpha \simeq \frac{1}{2}$; solid sphere, $R_g \propto M^{1/3}$.

range are always larger than predicted by the theoretical limits even for the single-pore-size columns. However, the effect of pore geometry on D_2 and molar mass range has often been overestimated. It has sometimes been mistakenly assumed that the shape of the SEC calibration curve is dictated entirely by the PSD curve of the packing, leading to the misconception that the SEC separation capacity (value of D_2) can be greatly improved by the use of packings with a very narrow PSD. This fallacy is caused by a failure to recognize the theoretical limit on the value of D_2 as described by Equation 4.23. The value of D_2 and molar mass range of SEC columns of a single pore size are usually only 30 to 60% higher than the theoretical limits. However, the theoretical limits are based on a simplified model, and in practice irregularities in pore cross sections are unavoidable, so it is not possible to recover this 30 to 60% loss in SEC separation capacity by minimizing the PSD of the SEC packing.

Because SEC separation capacity is limited by the available pore volume of the column packing (Figure 4.4), the design of an SEC experiment involves a trade-off between resolution and versatility. With SEC columns of only one pore size, all the SEC separation capacity is concentrated in a narrow molar mass range to give a minimum D_2 (or maximum resolution). However, the linear molar mass range of a single-pore-size column is too narrow to provide accurate analyses for broad-MMD polymers in general. For example, the MMD of a typical condensation polymer as described by the Flory MMD curve is quite broad, extending over two decades in molar mass. The narrow molar mass range of single-pore-size columns can force the wings of a Flory MMD curve into the nonlinear calibration region, causing distortion of the polymer elution curve and an error of more than 20% in molar mass values calculated [15]. A preferable approach to the SEC analysis of polymers is the use of an SEC column set of different pore sizes to provide a wide molar mass separation range (Section 7.9). The increased convenience and versatility can usually justify the use of a wide-molar-mass-range column set for general-purpose SEC. A column set with a wide linear molar mass separation range when used in conjunction with broad standard linear calibration methods (Section 8.3) gives good molar mass accuracy in SEC analyses.

Proper SEC column selection is a compromise between two goals: *wide-molar-mass calibration range* for convenience and versatility, and a calibration curve with *good linearity* for maximum accuracy in *M* determination. The best compromise is obtained by using columns with packings of only two pore sizes (i.e., the *bimodal PSD approach*) [15].

By simulating the conventional method of connecting columns of many similar pore sizes, Figure 4.12 shows how the SEC-*M* calibration curve broadens in range as

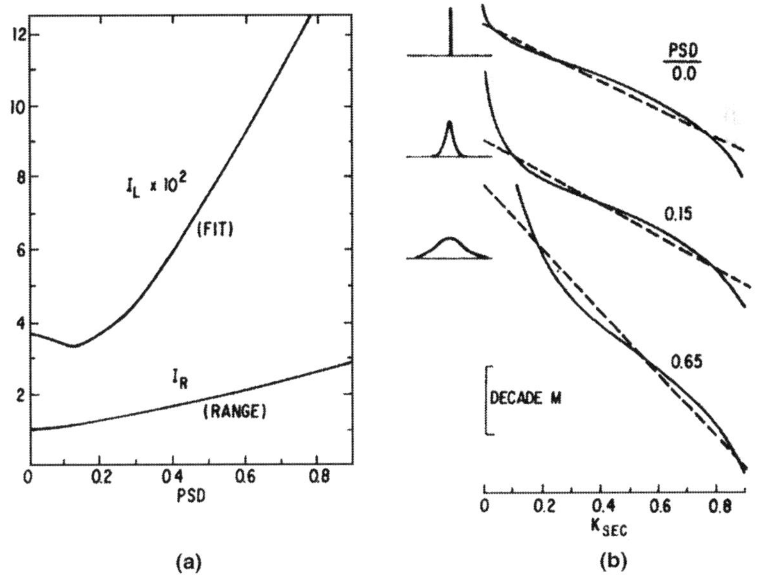

Figure 4.12 Effect of pore-size distribution on calibration linearity and molar mass range for SEC: monomodal. I_R and I_L in units of decades of radius of gyration. (Reprinted with permission from Ref. 15.)

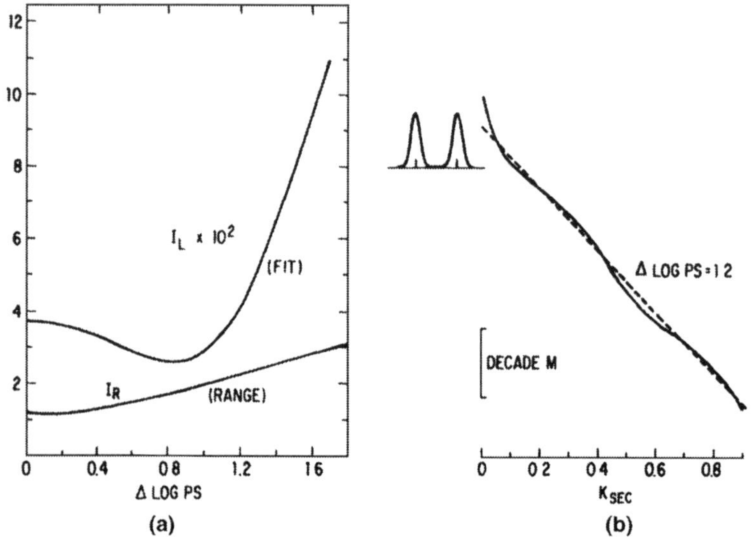

Figure 4.13 Effect of pore-size distribution on calibration linearity and molar mass range: bimodal. Pore-size distribution, 0.15; pore volume ratio, 1.0 units as for Figure 4.12. (Reprinted with permission from Ref. 15.)

the PSD of the packing increases from zero (single pore size) to 0.15 and 0.65. Here PSD is expressed as the standard deviation of the log-normal PSD curve. The dashed lines in the figure are the linear approximation of the calibration curves. Figure 4.12a shows that the separation range (I_R) increases with increasing PSD; however, the linear fit (I_L) rapidly becomes poorer beyond a PSD of 0.15. The separation range, I_R, is given in decades of the solute radius of gyration, R_G. The value of I_R is calculated as the difference in the logarithm of the limiting values of R_G near exclusion and total permeation. The "goodness" of the linear fit between the dashed line and the calibration curve in Figure 4.12b is measured by the root-mean-square derivation of the fit I_L, in the same units as I_R. The situation is much improved in the case of the bimodal PSD approach. Figure 4.13 shows that as the difference in pore size increases (increasing $\Delta \log \mathrm{PS}$), I_R increases steadily. However, I_L goes through a minimum at $\Delta \log \mathrm{PS} = 1$ (with two pores of about one decade difference in size), representing the best linear calibration fit. The calibration curve in Figure 4.13b has a four-decade range of molar mass with an excellent linear fit. (See Section 7.9 for selecting bimodal PSD columns.)

4.5.4 Effect of Operating Parameters

The dependence of SEC column dispersion on retention is quite complex, as discussed in Section 3.3. Since SEC peak dispersion is a mass-transfer-limited process, it is very sensitive to most experimental parameters, including packing particle size,

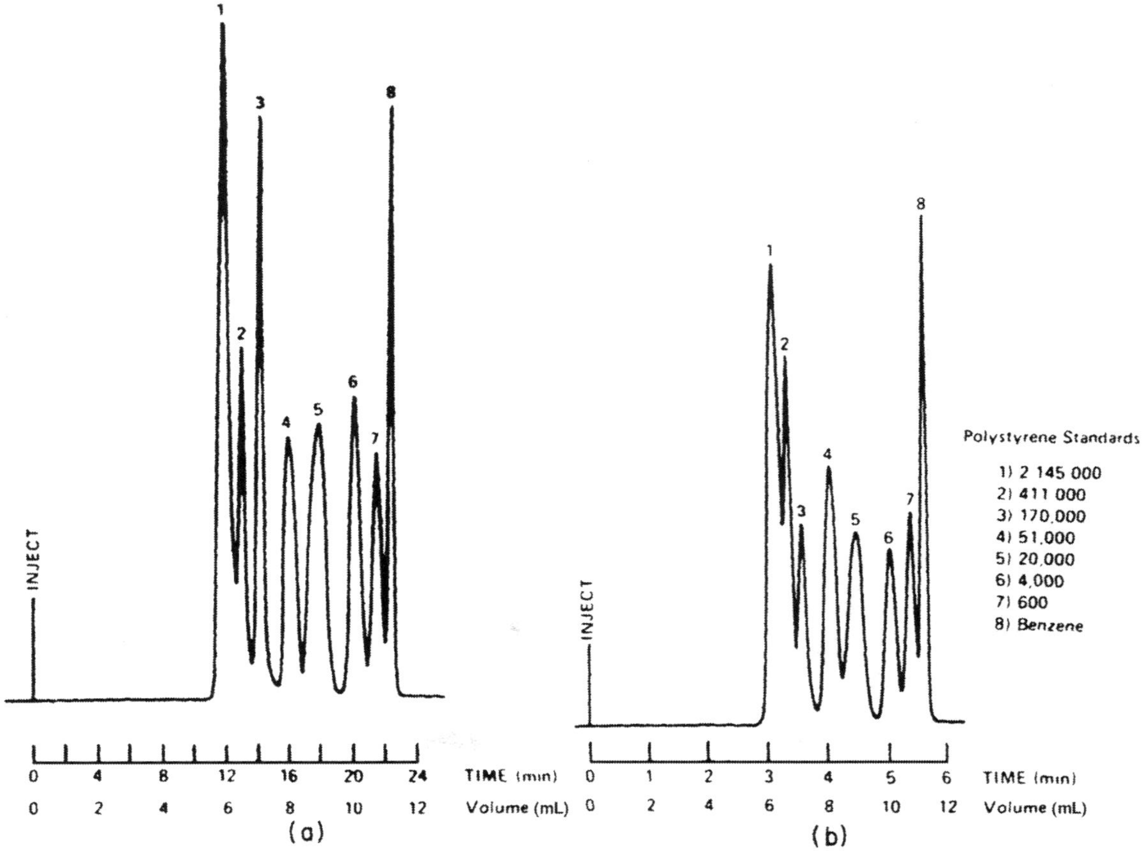

Figure 4.14 Effect of flow rate on resolution: separation of polystyrene standards on μ-Bondagel columns. Columns, 125, 300, 500, and 1000 Å; mobile phase, methylene chloride. Flow rate: (a) 0.5 mL/min; (b) 2.0 mL/min. (Reprinted with permission from Ref. 16.)

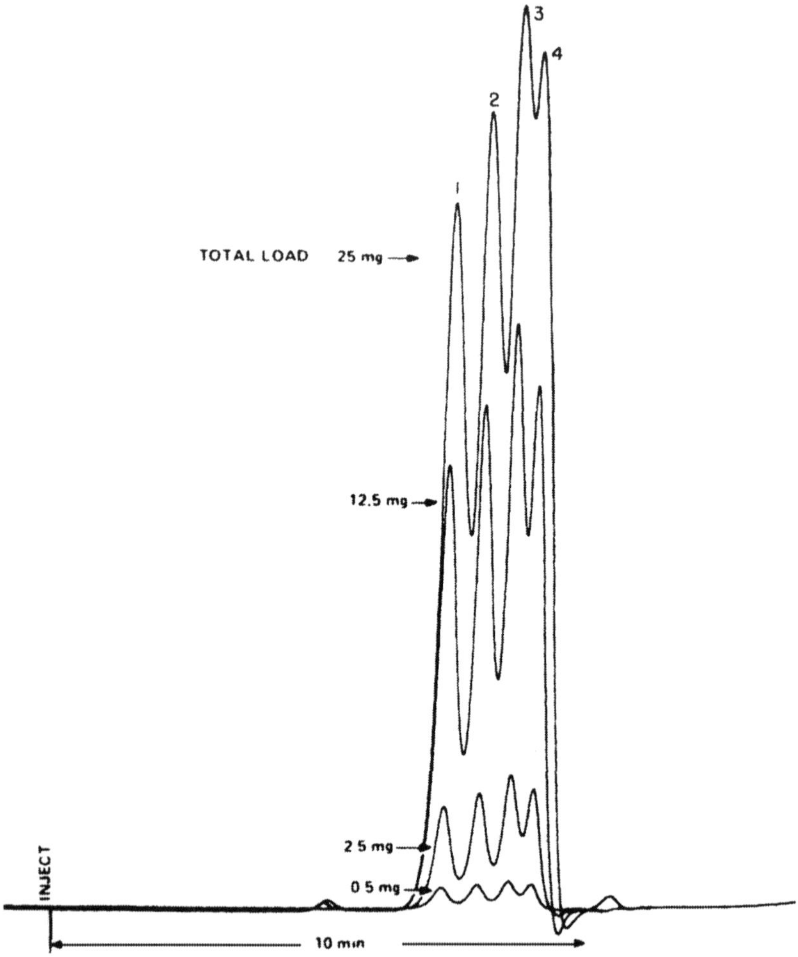

Figure 4.15 Effect of sample load on resolution. Column, two μ-Styragel 100 Å; mobile phase: THF; flow rate, 20 mL/min. Solutes: 1, dioctylphthalate; 2, dibutylphthalate, 3, diethylphthalate; 4, dimethylphthalate. (Reprinted with permission from Ref. 16.)

flow rates, solvent viscosity, sample concentration, extracolumn effects, and packing inhomogeneity (Section 3.4). Because of the dependence on all these factors, values of column σ should be determined each time a change in experimental conditions is made (Section 3.5). Illustrations of the loss of resolution by increasing flow rate and sample concentration are shown in Figures 4.14 and 4.15, respectively. Since SEC resolution depends on the product of σ and D_2, any operating parameter that affects either σ or D_2 will affect the resolution. The expected effects of some common SEC experimental parameters are summarized in Table 4.4.

In summary, both experience and theory have shown that σ, D_2, R_{sp}, and M^* are the most accurate and effective terms for expressing SEC column performance.

Table 4.4 Influence of operating parameters on SEC performance[a]

Parameter	D_2	σ	R_{sp}, M^*
Column volume	++	++	++
Particle size	−	++	++
Particle porosity	+	−	+
Particle shape	−	−	−
PSD	+	−	+
Pore size	−	−	−
Pore shape	−	−	−
Solute conformation	++	+	++
Flow rate	−	++	++
Solvent viscosity	−	+	+
Temperature	−	+	+

[a]−, Negligible (or unsubstantiated) effect; +, Moderate effect; ++, Large effect.

REFERENCES

1. L. R. Snyder and J. J. Kirkland, *Introduction to Modern Liquid Chromatography*, 2nd ed., Wiley, New York, 1979, Chap 2.
2. L. R. Snyder, *J. Chromatogr. Sci.*, **10**, 200 (1972).
3. J. C. Giddings, *Anal. Chem.*, **39**, 1027 (1967).
4. W. W. Yau, J. J. Kirkland, D. D. Bly, and H. J. Stoklosa, *J. Chromatogr.*, **125**, 219 (1976).
5. D. D. Bly, *J. Polym. Sci. C*, **21**, 13 (1968).
6. A. E. Hamielec, *J. Appl. Polym. Sci.*, **14**, 1519 (1970).
7. L. H. Tung, *J. Appl. Polym. Sci.*, **13**, 775 (1969).
8. W. W. Yau, H. J. Stoklosa, and D. D. Bly, *J. Appl. Polym. Sci.*, **21**, 1911 (1977).
9. W. W. Yau, H. J. Stoklosa, C. R. Ginnard, and D. D. Bly, 12th Middle Atlantic Regional Meeting, American Chemical Society, Apr. 5–7, 1978, paper PO13.
10. A. E. Hamielec and W. H. Ray, *J. Appl. Polym. Sci.*, **13**, 1319 (1969).
11. T. Provder and E. M. Rosen, *Sep. Sci.*, **5**, 437 (1970).
12. E. Grushka, *Anal. Chem.*, **44**, 1733 (1972).
13. W. W. Yau, *Anal. Chem.*, **49**, 395 (1977).
14. J. J. Kirkland, *J. Chromatogr.*, **125**, 231 (1976).
15. W. W. Yau, C. R. Ginnard, and J. J. Kirkland, *J. Chromatogr.*, **149**, 465 (1978).
16. R. V. Vivilecchia, B. G. Lightbody, N. Z. Thimot, and H. M. Quinn, *J. Chromatogr. Sci.*, **15**, 424 (1977).
17. J. P. Foley, *Analyst*, **116**, 1275 (1991).
18. M. Kurata and Y. Tsunashima, in *Polymer Handbook*, 4th ed., J. Brandrup, E. H. Immergut, and E. A. Grulke, eds., Wiley-Interscience, New York, 1999, Chap. VII/1.

5

EQUIPMENT

5.1 INTRODUCTION

To provide high-quality results, SEC equipment must be designed according to many of the criteria listed in Table 5.1. Although to meet various goals (e.g., high analysis speed), particular equipment designs are required, analytical accuracy necessitates the greatest range and control of operating parameters. Thus, an apparatus that provides good analytical accuracy often will meet the design requirements of any SEC method. An apparatus constructed to meet all the criteria listed in Table 5.1 should be useful for any separation. A general schematic for equipment is shown in Figure 5.1. Additional components may be needed for specialized analyses.

Whether to choose modular equipment (i.e., equipment assembled from components) or a completely integrated apparatus depends on the application anticipated. If great versatility or range of applicability (e.g., operation at higher temperatures) is not required, or if there are budget limitations, simple modular equipment may well be adequate. On the other hand, integrated commercial instruments generally provide better convenience and are particularly attractive when methods are to be exchanged between laboratories.

In this chapter we describe in some detail the components required for an effective SEC system. The reader is also informed of the advantages and disadvantages of various instrumental designs to permit the choice of equipment to satisfy a particular need [1,2]. However, before describing the apparatus, a general discussion of extra-column effects is needed to aid in understanding this important parameter.

Modern Size-Exclusion Liquid Chromatography: Practice of Gel Permeation and Gel Filtration Chromatography, Second Edition
By André M. Striegel, Wallace W. Yau, Joseph J. Kirkland, and Donald D. Bly

Table 5.1 Criteria for SEC equipment

Equipment Design Feature	Goal: Analytical Accuracy	Retention Reproducibility	Analysis Speed	Separation Versatility
Precise flow rate	×	×		
Temperature control	×	×		×
Precise sampling	×	×		
Stable detection	×			
High-signal/noise detection	×		×	×
Fast detection	×		×	
High-pressure pumping	×		×	×
Efficient columns	×		×	
Automatic data handling	×		×	
Low-dead-volume system	×		×	
Flow-rate sensing	×			
Range of column packings				×
Chemically resistant				×
Variety of detectors	×			×

5.2 EXTRA-COLUMN EFFECTS: GENERAL

In addition to the inherent band broadening that occurs within the chromatographic column (Chapter 3), additional broadening also occurs outside the column. This extra-column band broadening results from the sample injection and from other elements of the apparatus, such as the detector cell, column-end fittings, connectors,

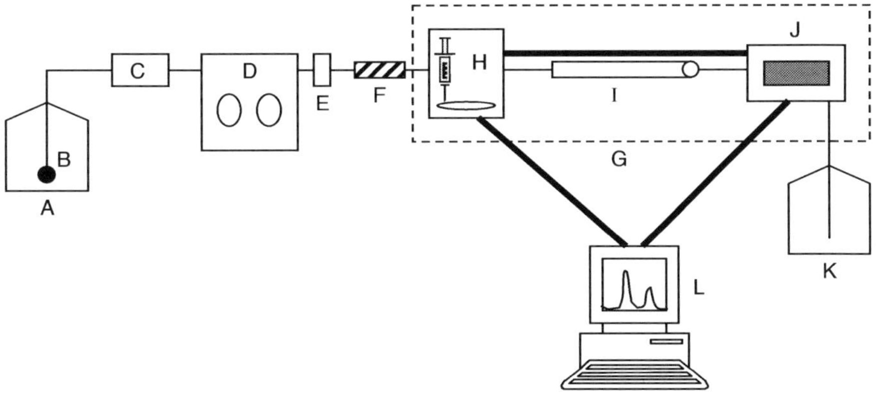

Figure 5.1 Typical SEC apparatus. A, inlet reservoir; B, inlet ("plunger") filter; C, degasser; D, pump; E, in-line filter; F, pulse dampener; G, thermostatted oven; H, sample injector and autosampler; I, chromatographic column; J, detector; K, waste reservoir; L, data acquisition and handling. Thin lines correspond to plumbing connections, thick lines to electronic connections.

and so on. Thus, the total band width observed, W_t, is a function of the sample band volume (width at the baseline, about 4σ) due to column dispersion W_c and the sample injection volume W_i, plus the extra-column band broadening that occurs within the SEC apparatus. This peak broadening relationship may be expressed in terms of peak volume:

$$W_t^2 = W_c^2 + W_i^2 + W_d^2 + W_j^2 + W_x^2 \quad (5.1)$$

The quantities W_d, W_j, and W_x represent the increased peak widths (volumes) associated with the extra-column effects in the detector, end fittings, and connecting tubing, respectively. Extra-column band broadening should be kept to a minimum so that the peak volume W_t observed closely approximates the actual peak volume W_c. As a rule of thumb, this means that the total of the injected volume and the other extra-column peak volumes should be less than one-third of W_c for a monomer peak in the chromatogram. This then limits the increase in W_t to about 10%.

Because the volume W_c of a band can be quite small (e.g., $\leq$40 μL in extreme cases), it is particularly important that extra-column effects be minimized. Particular attention must be placed on the design of all equipment components to ensure that these cause insignificant broadening of the true band width. As discussed in Section 4.3, significant band broadening can cause large errors in molar mass results. The origin of extra-column effects for the various equipment components is discussed in the following sections.

5.3 MOBILE-PHASE RESERVOIRS, INLET FILTERS, AND DEGASSERS

Because flow rates in SEC typically are 1 to 3 mL/min, and separations are usually completed in a half-hour or less, the volume of mobile phase used for a single analysis is relatively small. As a result, the total volume used in a workday is moderate, and reservoirs typically hold about 1 L. For preparative applications involving large-diameter columns, larger volume reservoirs are needed (e.g., several liters). A typical set of runs, employing a flow rate of 1 mL/min, consumes 1440 mL in 24 hours. As a result, solvent reservoirs with a capacity greater than 1 gallon are rarely necessary. The exception is preparative applications, where solvent consumption is substantially greater. Reservoirs are usually made of stainless steel or glass, but should be inert to the mobile phase and not easily broken. Tubing on the solvent inlet side can be fitted with an inlet or "plunger" filter. These filters help prevent particulate contamination from reaching the pump and injector. Inlet filters are usually stainless steel, have 2-μm pores, generate negligible backpressure or cavitation due to their large surface area, and can easily be replaced or cleaned by sonicating or backflushing.

Some reservoirs are designed so that the mobile phase may be degassed in situ to prevent bubbles from forming in the detector during the separation. Elimination of oxygen is also required to prevent reaction with certain samples. Dissolved oxygen can also lead to baseline drift, reduced sensitivity and signal-to-noise ratios when using electrochemical and fluorescence detectors. Oxygen can also be a problem with UV detection, because solvated oxygen complexes absorb significantly in the region 190 to 260 nm. To facilitate in situ degassing, reservoirs may be equipped with a vent valve, a heater, a stirring mechanism (e.g., magnetic stirring bar), and separate inlets for vacuum, nitrogen, or helium purge. Degassing can also be achieved effectively by thoroughly purging the mobile phase with helium, which has a very low solubility in virtually all liquids. After initial sparging by a fast flow of helium for a few minutes, a slow purge of helium is then used to maintain the mobile phase. A helium purge also prevents oxygen from redissolving in a sensitive mobile phase after degassing and improves safety by preventing accidental ignition of flammable vapors.

Online degassers are usually equipped with multiple ports and are placed between the inlet reservoir and the pump. These units are able to handle several solvent lines simultaneously. As the solvent flows through the narrow-bore tubing within the online degasser, dissolved gases are vacuum-filtered across a gas-permeable membrane. State-of-the-art units can handle flow rates as high as 10 mL/min with pressure drops of 0.06 kg/cm^2, achieving degassing efficiencies of 0.6 ppm at the channel outlets.

An alternative means of degassing solvent is by ultrasonication; however, this is a short-term solution as redissolution of air will usually occur.

5.4 SOLVENT-METERING SYSTEMS (PUMPS)

Providing a constant, reproducible supply of mobile phase to the column is the most important function of the solvent-metering system. Relatively high pump pressures are needed to overcome resistance to flow offered by the small particles used in the columns (Section 7.3). The general operational requirements for a solvent-metering system in SEC are listed in Table 5.2. Constant-flow reciprocating pumps are most widely used, because constant-pressure pumps are susceptible to flow variation with temperature and to other changes that affect column backpressure (see Section 5.4.2).

Most modern solvent-metering systems for chromatography are constructed of stainless steel, PEEK [poly(ether ether ketone)], or Teflon, for a maximum resistance to chemical attack. Pump seals made from virgin or filled Teflon resist most solvents

Table 5.2 General requirements for pumps

Deliver mobile-phase volume flow rate with an overall precision of better than 0.2% and an overall accuracy of better than 1%
Have a pressure output of at least 6000 psi
Provide pulse-free or pulse-dampened output, with pressure pulsations of less than 1% at 1 mL/min
Provide flow rates in the range 0.1 to 3 mL/min and, preferably, extending from 0.01 to 10 mL/min in either 0.1- or 0.01-mL/min increments
Be chemically resistant to a wide range of mobile phases
Have small hold-up volume for rapid solvent changes and recycle operation (desirable but not essential)

that have been used for SEC. However, parts made from Teflon should generally not be subjected to pressures of about 2000 psi. Alternatively, PEEK is strong, relatively inert, can be machined into different, often complex shapes, and can be used at pressures up to about 6000 psi. Sapphire pistons are most often used to generate the pumping pressure required.

5.4.1 General Pump Specifications

The solvent-metering or pumping system can often be the limiting factor for accurately determining the performance of the chromatographic separation, particularly when molar mass information is desired. Constancy of flow rate is especially important, as elaborated in Section 7.2. Certain specifications become dominant when considering solvent-metering systems: (1) pump resettability, (2) short-term precision, (3) pump pulsation or "noise," (4) drift, and (5) flow-rate accuracy. By *resettability* (or *repeatability*), we mean the ability to reset the pump to the same flow rate repeatedly. Short-term precision is a measure of the reproducibility of the volume output by the pump over a few minutes. Pump "noise" or pulsation arises from flow changes as result of operational functions such as piston movement and check valve operation. Drift is a measure of a generally continuous increase or decrease in the pump output over relatively long periods (e.g., hours). Pumping accuracy relates to the ability of pumps to deliver exactly the flow rate indicated by a particular setting. Although all the foregoing considerations regarding pumps are important, pump resettability and drift are usually the most critical (Section 7.2). In addition to these performance features, operational convenience, durability, and serviceability should also be considered when selecting a pump.

Commercially available pumps can be classified into three groups: reciprocating, positive displacement, and constant pressure. Here we focus on reciprocating pumps, as they are the most widely used.

5.4.2 Reciprocating Pumps

Reciprocating pumps are the most widely used pumps because of their generally satisfactory performance. Models with output pressures up to about 10,000 psi (680 bar) and maximum volumetric outputs of 10 to 20 mL/min are typical of pumps used for analysis. Various commercial models are distinguished by the techniques used to minimize pulsating flow output and by the mode of compensation used for solvent compressibility and flow-rate changes.

Simple, single-head reciprocating pumps, such as those shown schematically in Figure 5.2, are relatively inexpensive. However, with these simple types, variations in solvent or column backpressure can cause minor flow-rate changes because of solvent compressibility changes. Pulsations are also greatest with simple reciprocating pumps, resulting in increasing detector noise, which increases with increasing flow-rate delivery. A pulse damper generally is used to minimize this noise, the extent of dampening observed being a function of the detector type. Ultraviolet photometers are more tolerant to pulses than is a refractometer. Pulse-damping devices

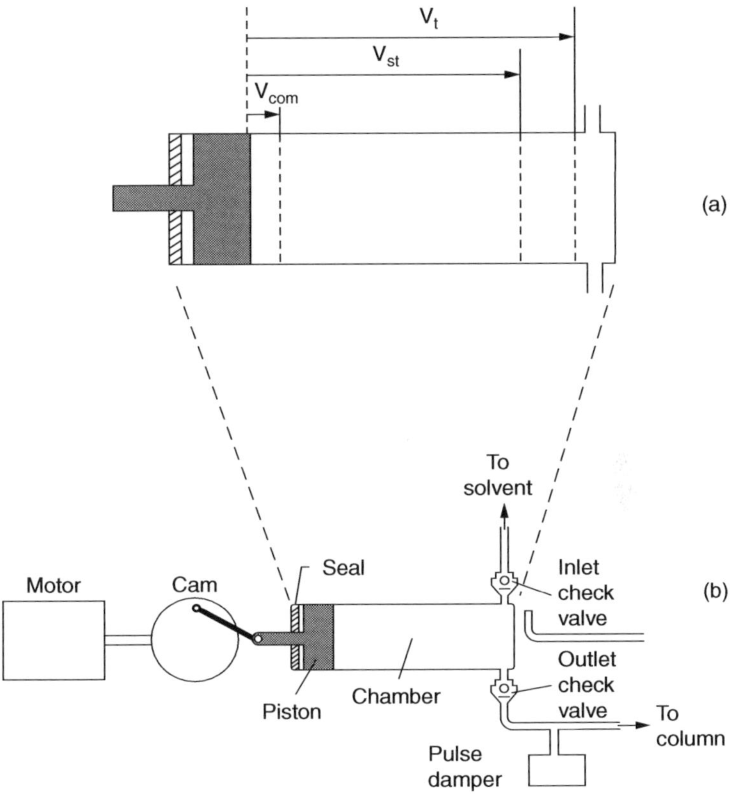

Figure 5.2 Simple reciprocating pump: (a) pump chamber; (b) single-head reciprocating pump. (Reprinted with permission from Ref. 2.)

represent a compromise with convenience, however, since they increase the volume of the system between the pump and the chromatographic column and require additional purging when changing the mobile phase.

More sophisticated single-head pumps utilize a sinusoidal cam to drive the pump piston in the pumping and refill cycles, so that pulsations are minimized. One approach uses a circuit design that recognizes the approaching end of the pumping stroke so that the motor driving the piston suddenly speeds up to deliver extra liquid in anticipation of the upcoming fast (200 ms) refill stroke when no liquid flows. The motor torque, which is proportional to the volume output of the pump, is monitored so that the motor speed is returned to the level operating before the back-fill stroke. As a result of such special devices, sophisticated single-head pumps generally exhibit lower pumping noise and improved pumping accuracy compared to simple types.

Dual-head pumps with pistons controlled by circular cams operated at 180° out-of-phase produce reduced-flow pulsations (Figure 5.3b) relative to single-head circular cam pumps (Figure 5.3a) but are more expensive. With these pumps, one chamber fills while the other provides flow to the column. Additional reduction in pump

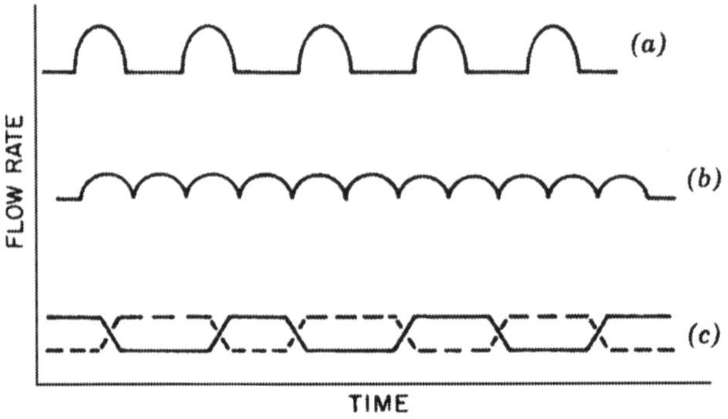

Figure 5.3 Reciprocating pump output patterns: (a) Simple single-head reciprocating pump; (b) dual-head, circular-cam pump; (c) dual-head, sinusoidal-cam output (with changeover ramp).

pulsation is obtained with dual-head pumps driven by sinusoidal cams designed to produce a linear displacement of the piston. If the cam-activated strokes are perfectly matched, an essentially pulse-free output flow can result (Figure 5.3c). However, in practice, some mismatch occurs, which produces a slight pulse at the end of the changeover points in the pumping cycle. Pulses are minimized by arranging a piston-driving cycle of slightly more than 180°, to include gradual takeover periods of one pump head relative to the other. This type of pump is often quite satisfactory and represents one of the most widely used types at present.

Several manufacturers provide flow-feedback systems to adjust and correct the imperfect flow of a reciprocating pump. The general approach is the continuous measurement of flow rate by an appropriate transducer, which produces a signal when the flow rate varies from the preset value. The signal is then used to adjust the pump electronically to deliver more or less solvent to maintain an essentially constant flow of solvent relatively pulse free. For example, in one method for flow-rate control, a differential pressure transducer measures the pressure drop across a restricter. The pressure (and therefore the flow) is maintained constant by controlling the rate of the pump motor electronically, which controls solvent output. For highest accuracy, this particular method requires individual calibration for each solvent.

The use of special systems to compensate for flow-rate variations resulting from mobile-phase compressibility and pulsations is illustrated schematically in Figure 5.4. Output (a) represents the uncorrected flow output from a dual-head reciprocating pump as a result of the cam rotation driving the piston against the column backpressure; mobile-phase pulsations are significant. However, against relatively high column backpressures, the average of this uncorrected flow rate is lower than the set value desired. Output (b) is compensated for solvent compressibility, so the average flow rate now is correct compared to the set value; however, the output is still pulsating. In trace (c) the pump output is corrected with pressure feedback; flow rate is correct and pulsations have been greatly decreased.

Diaphragm reciprocating pumps are similar to piston pumps except that a flexible stainless steel or Teflon diaphragm is in contact with the mobile phase. This

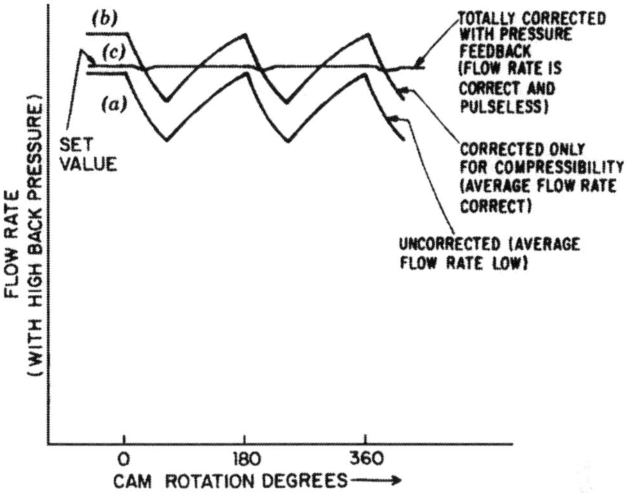

Figure 5.4 Effect of pressure feedback and compressibility correction on pumping. (Adapted with permission of Spectra-Physics, Inc.)

diaphragm is actuated by a piston working on an oil cavity which on each stroke of the piston flexes the diaphragm to produce a pulsating solvent output.

A general advantage of reciprocating pumps is that solvent delivery is continuous; therefore, there is no restriction on the size of the reservoir that is used or the length of time that a pump is operating. These pumps are particularly useful in equipment for automatic operation (e.g., overnight). A specific advantage of piston-reciprocating pumps is that their internal volume can be made small, and this type is particularly useful for recycle chromatography (Section 15.3).

The newest generation of dual-head piston pumps relies on independently driven pistons and a dual-pressure transducer feedback loop. These pumps, shown schematically in Figure 5.5, use a serial flow design with primary and accumulator piston chambers. Through a "first in, first out" principle, the primary head receives the solvent and delivers it to the accumulator head. Because the pistons are driven independently of each other, the fill rates of both heads need not be identical. Solvent delivery from the primary head is also controlled independently so that this head can supply solvent at a rate sufficient to replenish the accumulator and maintain system flow and operating pressure. For gradient elution, the system is equipped with programmable piston volume and a gradient proportioning valve for solvent proportioning. Figure 5.6 traces the piston velocities through one delivery cycle.

5.5 SAMPLE INJECTORS AND AUTOSAMPLERS

The method of introducing the sample onto the column can be a significant factor in determining SEC performance. As discussed in Section 5.2, the sample should

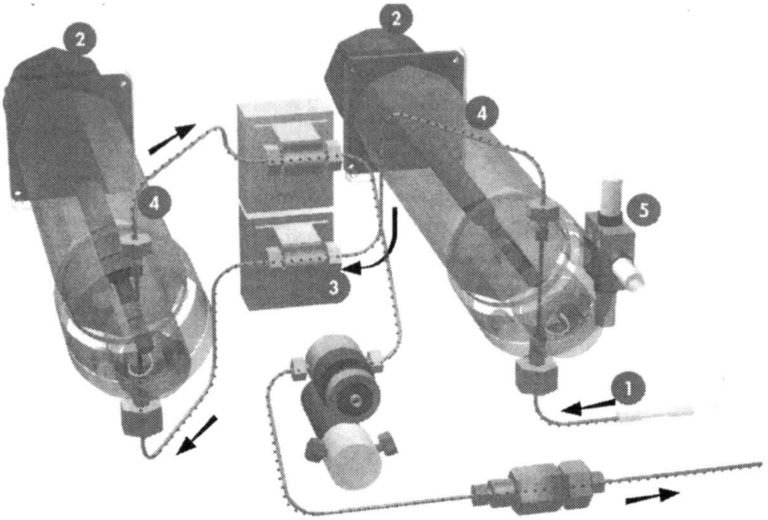

Figure 5.5 Dual-head pump with independently driven pistons. 1, Serial flow; 2, independent piston drive; 3, dual pressure transducers; 4, programmable piston volume; 5, gradient proportioning valve. (Courtesy of Waters Corp.)

be introduced onto the column in a sufficiently narrow band so that peak broadening from this cause is negligible. Ideally, the sample injector should introduce sharp plugs of a wide variety of samples into the columns with insignificant band broadening. Injectors should be convenient to use, reproducible, and operable against high

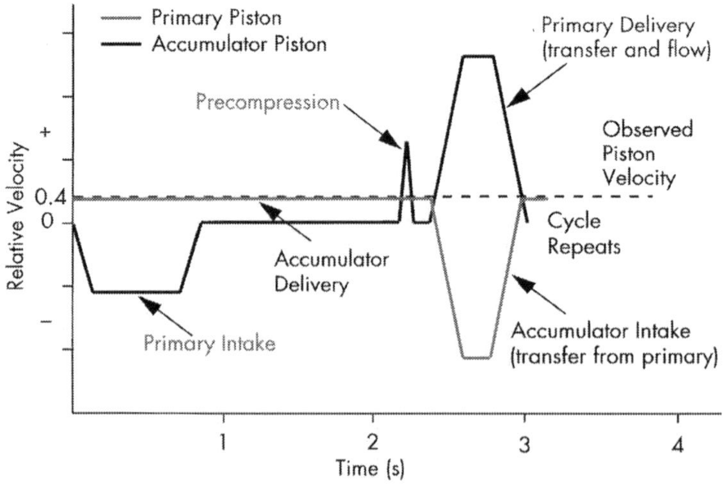

Figure 5.6 Piston velocities profile for dual-head pump with independently driven pistons, through one cycle of the delivery process. (Courtesy of Waters Corp.)

backpressures. Some sample types require injection at elevated temperatures to meet solubility requirements.

The most generally useful sampling device is the injector valve. These special valves permit the sample to be introduced reproducibly into pressurized columns without significant interruption of solvent flow, even at higher temperatures. Figure 5.7 shows schematic drawings of a six-port, plug-type valve in which the sample is contained in an external loop. [Long, narrow loops are preferred over shorter, wider-inside-diameter (i.d.) loops when large sample volumes are required.] The loop of appropriate volume is filled at low pressure by flushing it thoroughly with the sample solution, using an ordinary syringe (Figure 5.7a). A clockwise rotation of the valve rotor places the sample-filled loop into the mobile-phase stream with subsequent injection of the sample onto the top of the column (Figure 5.7b). Other valve types use sample cavities which consist of annular rings in a sliding rod that can be thrust into the flowing stream.

The particular advantage of valve injection is the rapid, reproducible delivery of large volumes (e.g., up to several milliliters at $\ll 1\%$ error) with pressures to 7000 to 10,000 psi. These large-volume injections are required when performing preparative SEC (see Section 15.2). High-performance valves deliver appropriate sample volumes without significant extra-column band broadening. These valves are only moderately expensive, delivery volumes are essentially operator independent, and valves can be obtained in automated versions. One minor disadvantage is that the sample loop must be changed to obtain various sample volumes. (However, with the device shown in Figure 5.7, it is possible to inject variable sample sizes into a given loop, using a specially designed syringe.) A special advantage of valves is that they can be located within a controlled-temperature environment, such as an oven, for use with samples that require dissolution and injection at elevated or controlled temperatures (e.g., up to 220°C).

Low-volume, high-pressure switching valves are also available for use in special techniques such as recycle SEC (Section 15.3). These valves come in a variety of configurations and can be operated at pressures up to 10,000 psi; some can be used at elevated temperatures but at lower pressures.

Automatic sampling devices are commercially available, so that large numbers of samples may be analyzed routinely without the need for operator intervention. The type of device shown schematically in Figure 5.8 allows samples contained in small vials to be pressurized consecutively into a sampling valve for injection. Automatic loop flushing is part of the sampling cycle, and multiple (up to 99) injections from the same vial can be made with total control of the chromatographic cycle. Automation even allows for different injection volumes and/or run times for samples in different vials. Injection volumes ranging from a few microliters to several milliliters are possible, although often a change of injection loop is necessary. Precision of $\pm 0.5\%$ has been achieved over the injection volume range of 0 to 2000 μL, with sample carryover from injection to injection of less than 0.1%. Temperature control allows for both subambient (as low as 4°C) and high-temperature (up to 220°C) operation, although not usually in the same apparatus. Additional features of modern autosamplers include options for sample filtration, variable-speed mixing and shaking, and

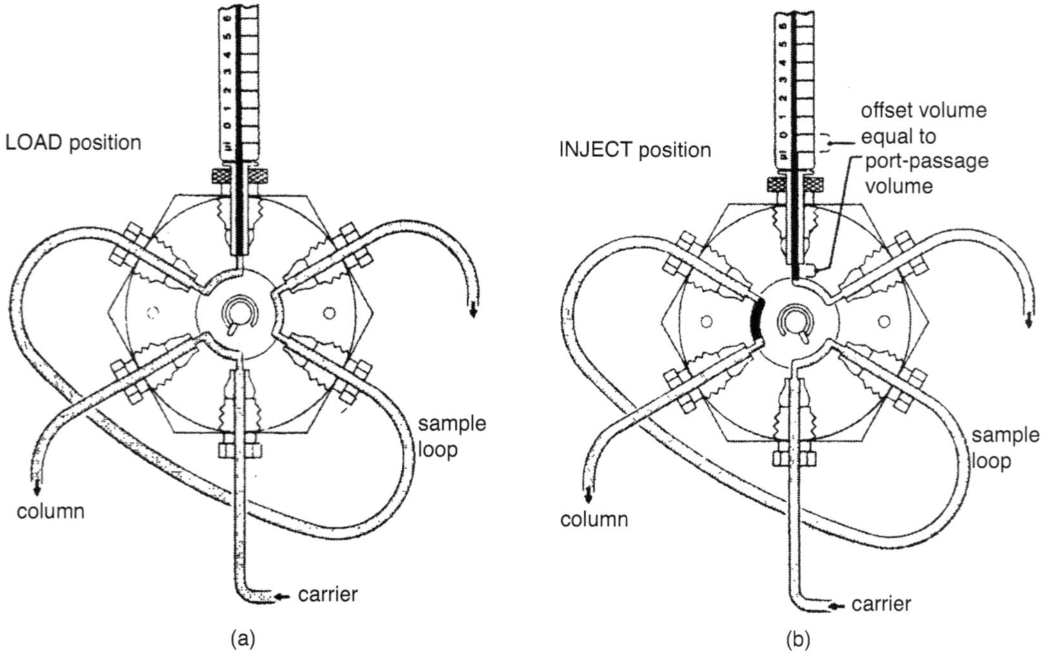

Figure 5.7 Six-port microsampling valve for SEC. (Reprinted with permission of Valco Instruments Co.)

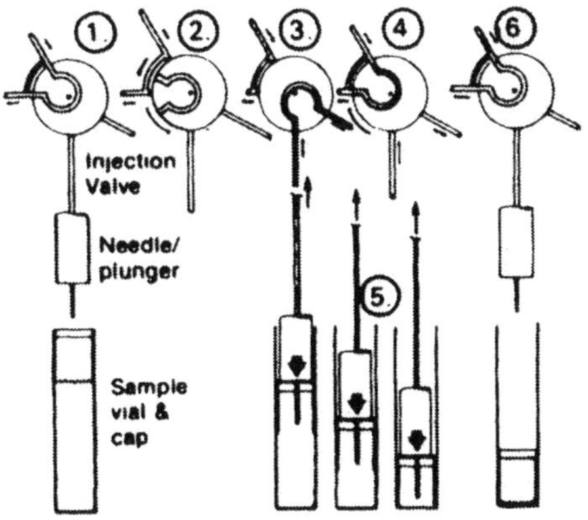

1. Vial is positioned beneath needle for injection.
2. Motorized injection valve rotates to "load" position as solvent diversion valve opens to allow continuing flow of mobile phase around sample loop.
3. With motorized injection valve in "load" position, hollow needle penetrates polyethylene vial cap and displaces sample into injector loop.
4. To make injections, motorized valve rotates to allow reverse displacement of sample onto column.
5. Valve action in 3 & 4 repeat for additional injections.
6. After injection completion, needle is withdrawn.

Figure 5.8 Positive-displacement automatic sampler. (Reprinted with permission of Micromeritics Instruments Corp.)

needle wash. Autosamplers can also accommodate different sample vial carousels; each carousel is designed to handle vials of a different volume. Care must be taken to note the minimum fill-level requirements for vials. This is due to two facts:

1. The injector needle does not descend all the way to the bottom of the sample vial. Therefore, a vial that is filled only partway may result in air being injected into the system if the needle, at its most extended position, still does not reach the solution in the vial.
2. The hole in the injector needle through which sample aspiration and injection occur is not at the bottom of the needle (as this could easily result in plugging of the hole as the needle penetrates the vial cap liner or septum), but on the side of the needle at a position somewhat higher than the needle bottom. Again, for partway-filled vials the needle hole must descend into the solution far enough to aspirate the desired injection volume of solution. If multiple injections are being conducted from the same vial, the operator should calculate the amount of solution needed, taking into account the minimum height required of the solution that remains in the vial subsequent to all injections. In some newer systems, needle height can be programmed through the on-board computer, as can be the injector and needle purges necessary when changing operating solvents.

5.6 MISCELLANEOUS HARDWARE

Line filters should be used between the pump and the sample injector to prevent particulates from clogging the column inlet. Most commercial instruments use stainless steel frits or filters of 2-μm porosity. However, experience has shown that with columns of particles below 10 μm, it is advantageous to use 0.5-μm porosity filters. The volume of these devices should not be large, to facilitate solvent changeover.

Pressure monitors (e.g., transducers) are desirable as diagnostic tools to indicate system plugging or leaks. These devices are available with high- and low-pressure alarms of cutoff circuits to protect the chromatographic system.

Pulse dampers are required by certain pumping systems. (The effects of a pulsating mobile phase are discussed in Section 5.4.) Many modular and most integrated commercial instruments containing reciprocating pumping systems are equipped with pulse-damping devices. An effective damping system for homemade equipment is a combination of about 5 m of 0.25-mm-i.d. capillary tubing and an associated diaphragm or Bourdon-type gauge. The capillary tubing acts as a flow restrictor and the gauge as a capacitor, so that the combination of these two components usually reduces pulsations of simple reciprocating pumps to manageable levels. Pulse-dampening devices increase the volume between the pump and the sample injector, and decrease solvent changeover convenience. Detector output may be affected by mobile-phase pulsation, but these pulses have no effect on column efficiency.

All fittings and connectors between the sample injector and the detector should be designed to be cleanly swept and with a minimum dead volume. Extraneous volumes act as mixing chambers, which significantly contribute to extra-column band broadening (Section 5.2). Comparison of ordinary compression fittings with "zero-dead-volume" fittings for use with columns shows that low-volume fittings must be used between the sample injector and detector to minimize extra-column band broadening.

Thermostats are needed to control above-ambient column temperatures. Use of circulating-air baths is convenient and generally preferred, since $\pm 1^\circ$C is easily maintained around the column. This usually results in a variation of no more than about $\pm 0.2^\circ$C in the temperature of the column packing. Column air baths in liquid chromatographs are very similar to those used in gas chromatographs and usually consist of high-velocity air blowers and electronically controlled thermostats. Some instrument thermostats use contact heaters for controlling the column temperature. These sometimes are more convenient, but variations in the temperature within the column can be problematic. Alternatively, columns may be jacketed and the temperature controlled by circulating a fluid through the jacket system from a constant-temperature bath. This approach is generally less flexible but is practical for routine analyses.

As discussed in Section 7.2, mobile-phase flow rates must be precise, since retention times are often used in conjunction with calibration curves to develop sample molar mass information. Since flow-rate variations can result from even minor failure of the pumping system, it is important that techniques be available for careful monitoring of the flow rate during sample analysis. Flow rates can either be determined manually or may more conveniently be measured automatically with any one of the several devices described below.

Volumetric measurements of flow rate are most commonly used. The mobile phase simply is collected for a measured time in a calibrated vessel, usually a small volumetric flask. Flow-tube methods are sometimes used. Typically, an air bubble is introduced into the detector eluent stream, which passes through a transparent, volume-calibrated tube. The bubble is then timed while it travels between two volume markers on the tube. With this approach the flow rate can be measured quickly with a precision of about 1%.

Fraction collectors are not used routinely in analytical SEC, but are employed in preparative SEC (Section 15.2) and in the "heart cuts" approach in two-dimensional chromatography, with SEC as one of the separation dimensions (Chapter 14). Modern devices collect fractions based on time, peak, mass, or combinations of these and also allow for manual collection. For analytical-scale separations, flow rates as high as 10 mL/min may be used, while for preparative-scale separations, fraction collectors can handle flow rates as high as 100 mL/min. Pooling of collections from multiple injections of the same sample is possible, as is Peltier thermostatting to protect thermally labile samples. Collection can be done from and into wellplates, vials, and test tubes, and several fraction collectors may be used in parallel to increase throughput.

5.7 LABORATORY SAFETY

The general aspects of solvent handling are described in Section 7.6.5. Specifically, instruments should be operated in well-ventilated areas, and although a hood is usually not required for instrument operation, it is recommended for preparing samples. Most commercial instruments possess built-in safety cutoff devices (e.g., vapor sensors) to protect from hazards that might result from solvent spills or leaks.

REFERENCES

1. C. F. Poole, *The Essence of Chromatography*, Elsevier, Amsterdam, 2003.
2. L. R. Snyder and J. J. Kirkland, *Introduction to Modern Liquid Chromatography*, 2nd ed., Wiley-Interscience, New York, 1979.

6

THE COLUMN

6.1 INTRODUCTION

As discussed in Section 3.3, better column efficiencies and separations are obtained with small particle packings and solutes having high diffusion rates. Plate height is essentially dependent on the square of the particle diameter (d_p^2) but is a linear reciprocal function of the solute diffusion coefficient (D_m). Thus, the effect of particle size is most important for macromolecules that diffuse slowly, and the use of columns with very small, totally porous particles is particularly favored in SEC.

In Section 3.1 we have described the band broadening that is inherent in the SEC method. What can be done to prepare columns to minimize these band-broadening effects? Eddy diffusion can be reduced by preparing homogeneously packed beds. This generally is accomplished more readily with spherical particles rather than with irregular particles. Both mobile-phase and stagnant-mobile-phase mass transfer are improved significantly by using very small particles. As suggested in Figure 6.1, movement of solute molecules in and out of stagnant mobile phases is much faster in very small particles than in larger totally porous particles with deeper pools of stagnant mobile phase.

6.2 COLUMN PACKINGS

A variety of porous packing materials is available for SEC. Semirigid organic gels and rigid solids are available, and from these materials must be chosen the packing

Modern Size-Exclusion Liquid Chromatography: Practice of Gel Permeation and Gel Filtration Chromatography, Second Edition
By André M. Striegel, Wallace W. Yau, Joseph J. Kirkland, and Donald D. Bly

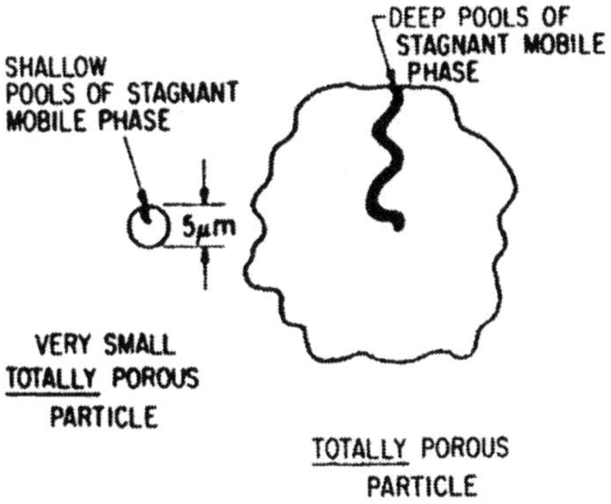

Figure 6.1 Stagnant mobile phase in large and small porous particles.

best suited for the particular application. Optimum performance of an SEC packing material involves high resolution and low column backpressure; good mechanical, chemical, and thermal stability; minimal shear degradation of high-molar-mass fractions in macromolecular samples; minimal specific interactions between analyte and column packing material; and minimal hindrance to solute diffusion. A combination of these desirable properties allows a column to be used at high resolution with different solvents over a range of flow rates and temperatures.

Originally, most SEC analyses of synthetic organic polymers were made using cross-linked, semirigid polystyrene gel packings. Later, small rigid inorganic particles (e.g., silica) became available which had several significant experimental advantages over the organic gels. Rigid particles are relatively easily packed into homogeneous columns which are mechanically stable for long times. A much wider range of mobile phases can be used, providing greater versatility and increased convenience in application. The rigid packings equilibrate rapidly with new solvents, so that solvent changeover is rapid. Columns with rigid packings are stable with the high-temperature solvents required for characterizing some synthetic macromolecules, while organic gels of particles smaller then 5 μm are often not usable under these conditions. Rigid particles can also be used in aqueous systems for separating high-molar-mass, water-soluble solutes. A potential disadvantage of the rigid inorganic particles is adsorption or degradation of solutes (e.g., denaturing of proteins). However, siliceous particles can often be easily modified with certain organic functional groups to effectively eliminate these difficulties for most applications [1–3].

Since soft particles (e.g., agarose) collapse at high inlet pressures, they are not utilized in packed-column SEC. While soft gel packings traditionally have been used for separating high-molar-mass, water-soluble substances, such aqueous SEC separations are now being carried out at high pressures using columns of deactivated rigid particles. In the following section, a brief introduction is provided to the processes for making semirigid organic gels and rigid inorganic particles.

Table 6.1 summarizes most of the commercially available column lines based on semirigid organic gels and rigid inorganic particles, along with the different types

Table 6.1 Some commercially available SEC columns and packings

Manufacturer	Product Name	Solvent Compatibility[a]	Molar Mass Range (g/mol)	Packing Material[b]
Agilent	Zorbax PSM	A and O	10^3–10^6 (proteins) 10^2–10^7 (synthetics)	Unmodified, trimethylsilane-modified, and diol-modified silica
	Zorbax Bio-Series GF	A	10^2–10^6	Zirconia-stabilized silica, modified with hydrophilic organosilane
Bio-Rad	Bio-Sil and Bio-Select	A and O	10^3–10^6 (proteins)	Hydrophilic bonded-phase silica
Jordi	Jordi Gel	A and O	10^2–10^6	DVB and derivatized DVB[c]
Phenomenex	Phenogel	O	10^2–10^7	S/DVB
PSS	SDV	O	10^2–10^7	S/DVB
	HEMA Bio and SUPREMA	A	10^2–10^8	Methacrylic ester
	HEMA and SUPREMA Basic	A and O	10^2–10^6	N/A
Sepak	Nanofilm and CNT	A and O	10^2–10^6 (proteins)	Coated silica
Shodex	K series	O	10^2–10^8	S/DVB
	OHpak	A	10^2–10^7	Polyhydroxymethacrylate
	Protein K series	A	10^2–10^5	Polyhydroxylated silica
	Sugar K series	A	10^2–10^7	Sulfonated polystyrene
SynChrom	SynChropak HPC		10^3–10^7 (proteins) 10^2–10^7 (synthetics)	Porous silica bonded with γ-glycidoxypropylsilane
	SynChropak CATSEC		10^2–10^6	Porous silica with polyamine bonded phase
Tessek	Separon	A and O	10^2–10^6	HEMA/EDMA
TosoHaas	TSK-GEL H	O	10^3–10^8	S/DVB
	TSK-GEL SW	A	10^3–10^7 (proteins) 10^2–10^5 (dextran)	Ethylene glycol/ methacrylate
Varian/Polymer Laboratories	PLgel	O (up to 10% A)	10^2–10^7	S/DVB
	PL Aquagel-OH	A (up to 50% methanol)	10^2–10^7	S/DVB

Table 6.1 *(Continued)*

Manufacturer	Product Name	Solvent Compatibility[a]	Molar Mass Range (g/mol)	Packing Material[b]
Malvern/ Viscotek/ Tosoh	ViscoGEL HR	O	10^2–10^7	S/DVB
	ViscoGEL PWXL	A (up to 50% O)	10^2–10^6	Polymethacrylate
Waters	Protein-Pak	A	10^3–10^5	Diol-derivatized silica
	Styragel	O	10^2–10^8	S/DVB
	Ultrahydrogel	A (up to 20% O)	10^2–10^7	Hydroxylated polymethacrylate

[a] A, aqueous; O, organic.
[b] DVB, divinylbenzene; S/DVB, styrene/divinylbenzene; HEMA/EDMA, hydroxyethyl methacrylate/ethylene dimethacrylate.
[c] Derivatizations include fluorination, glucosidation, hydroxylation, polyaminodation, and sulfonation.

of packing material, the molar mass range that is covered, whether the columns are meant for use with aqueous or organic solvents, and the current manufacturers. For columns used with aqueous mobile phases, exclusion limits are usually determined with globular proteins, polysaccharides such as dextran or pullulan, or PEO/PEG. For columns used with organic mobile phases, exclusion limits are usually determined with linear polystyrene. Molar mass ranges given in the table include the lowest and highest limits available in each product line (i.e., normally a single column will not cover the entire range).

Table 6.2 lists commercial offerings for soft organic gel packings, along with manufacturer, packing material, and molar mass range covered based on globular proteins, dextrans, or both. Due to their extreme sensitivity to high pressures, these materials are normally available in bulk, not in the form of packed columns.

Table 6.2 Some commercial soft organic gel SEC packings available in bulk

Manufacturer	Product Name	Molar Mass Range (g/mol)[a]	Packing Material
Bio-Rad	Bio-Gel	10^2–10^8 (p)	Polyacrylamide
GE Healthcare	Sephacryl	10^3–10^7 (d) 10^3–10^6 (p)	Allyl dextran and *N,N′*-methylene bisacrylamide
	Sephadex	10^2–10^5 (p/d)	Dextran and epichlorohydrin
	Sepharose	10^3–10^7 (p/d)	2,3-Dibromopropanol and agarose derivatives
	Superdex	10^2–10^5 (p/d)	Agarose with covalently attached dextran
	Superose	10^3–10^6 (proteins)	Agarose and epichlorohydrin
Merck	Fractogel	10^3–10^6	Polyethylene glycol dimethacrylate

[a] Molar mass range for globular proteins (p), dextrans (d), or both proteins and dextrans (p/d).

In addition to the analytical columns and soft gel packings listed, a variety of specialty columns are also commercially available. These include columns for ultrahigh molar mass analysis, for oligomeric analysis, for high-temperature analysis, and for use with specialty solvents (e.g., hexafluoroisopropanol), cationic and polar columns, guard columns, narrow-bore columns, and preparative columns.

6.2.1 Semirigid Organic Gels

Beaded polymeric supports are usually manufactured via a suspension polymerization process consisting of three stages: (1) droplet formation, (2) droplet stabilization, and (3) droplet hardening and control of particle size and pore size [4].

6.2.1.1 Droplet Formation. *Droplet formation* usually proceeds by a two-phase suspension polymerization process where liquid microdroplets are converted into solid microbeads. For water-insoluble monomers (e.g., styrene, divinylbenzene), an oil-in-water (o/w) suspension is used with direct conversion of monomer droplets to polymer beads. For water-soluble monomers (e.g., acrylamide), a water-in-oil (w/o) suspension (*inverse suspension*) is used instead.

In suspension systems there is a series of collisions and redivisions of oil droplets. As the polymerization increases, however, redivision becomes more difficult because of the concomitant viscosity increase. Once redivision becomes almost impossible (*sticky period*), at about 25 to 75% conversion, depending on the composition of the monomer mixture, continued droplet coalescence leads to coagulation of the entire bulk of the monomer phase. Once individual droplets are hardened, at the end of the sticky period, the droplets will not coalesce in the event of a collision.

6.2.1.2 Droplet Stabilization. To prevent coagulation during the sticky period, the surface tension of the droplet needs to be controlled and the collision force between droplets minimized. The collision force is controlled via stirring speed and reactor design. The stirring speed can be decreased but must be kept high enough to prevent aggregation and separation of the droplets during the sticky period.

To reduce surface tension, a small amount (0.15 to 1%) of droplet stabilizer is added as a coagulation inhibitor. For oil-in-water suspensions, stabilizers include inorganic salts such as calcium sulfate, calcium phosphate, and benzonite, or organic polymers such as 75 to 98% hydrolyzed poly(vinyl alcohol), poly(vinyl pyrrolidone), methyl cellulose, gelatin, or other natural gums. Polymeric stabilizers are usually preferred, because they are easier than inorganic salts to remove from the bead surface.

6.2.1.3 Control of Particle and Pore Size. Particle size depends on reactor design, stir rate, the ratio of monomer phase to water (or to oil, for inverse suspensions), the viscosity of both phases, and the type and concentration of droplet stabilizer used. The size distribution of beads obtained from a two-phase suspension process depends on the configuration of the reactor as well as on the shape of the stirrer and on the stirring speed. Stirring speed, in particular, provides a convenient means for controlling particle size. Normally broad particle-size distributions (e.g., 1 to 50 μm) are obtained. Large (>50 μm) particles can be separated by dry sieving, 20- to 50-μm particles by wet sieving, and particles of less than 20 μm can be separated by wet sedimentation, counterflow settling, counterflow centrifugation, or air classification.

The pore size, pore-size distribution, and surface area of organic polymeric supports can be controlled by precipitation processes during suspension conversion. Pore size is controlled by the ratio of solvating and nonsolvating diluents in the monomer mixture. A higher amount of nonsolvating diluents increases pore size, and vice versa. For example, adding divinylbenzene (DVB) to a polystyrene suspension produces beads with wide porosities and pore sizes, depending on the ratio of DVB to styrene monomer. The porosity of beads is controlled by the ratio of diluent ("porogen") to monomer in the organic phases, a higher ratio corresponding to higher porosity. Macroporous polymer beads are produced using inert linear organic polymers [e.g., polystyrene, alkyl celluloses, poly(vinyl ether)] or inorganic polymers (e.g., silica microbeads) as porogen. Subsequent to polymerization, the porogens are removed by solvent extraction or by hydrolysis with strong alkali, leaving formed macropores in the polymer beads.

Porosity and surface area can be measured by nitrogen absorption–desorption, mercury intrusion, or low-angle x-ray analysis. Visual evidence of pore size and pore-size distribution can be obtained by electron microscopy.

6.2.2 Rigid Inorganic Packings

Totally porous siliceous particles for SEC are made by several methods [13,14]. For example, porous silica with pores of a specified size can be formed from small (80 to 1000 Å) silica sol ultraparticles, agglutinated to form microspheres. Silanol groups on the surface of untreated porous silica packings can cause problems in SEC by adsorbing the solute. This biases the desired size-exclusion mechanism, so that the desired relationship between retention volume and molecular size (or molar mass) will not be obtained. Mixed retention can be evidenced by tailing chromatographic peaks, lowered column efficiency, and retention beyond the total permeation volume. Fortunately, silica surfaces are altered by adding certain organic functional groups to effectively eliminate this disadvantage for most applications. By proper selection of the organic functionality, the surface of particles can be modified for both organic and aqueous mobile-phase applications. Porous silica is also made by gellation with a porogen, by spray drying, and other methods.

One approach used to eliminate solute adsorption in organic solvents is to maximize conversion of surface silanol groups on silica by endcapping: for example, by converting to their trimethylsilyl derivatives [1,4,5]. This hydrocarbon-modified packing can be prepared by refluxing a large molar excess of a short-chain reactive silane (e.g., chlorotrimethylsilane) with the siliceous support, the surface of which has previously been fully hydrolyzed (e.g., heating at 90 to 100°C at pH 9 in aqueous systems). With this approach, unreacted silanol groups (which constitute about one-half of the total silanol concentration) become shielded by an "umbrella" of tightly packed trimethylsilyl organic groups. As long as the trimethylsilyl groups are at a sufficiently high concentration on the surface (>3.5 μmol/m^2), the residual silanol groups remain essentially unavailable for unwanted adsorptive interactions. Reaction of surface Si–OH groups with chlorosilane reagents to high yields is promoted by (1) using a large excess of reactant, (2) conducting the reaction in the neat liquid reactant or in a dry solvent, (3) mechanically removing the volatile reaction product during the reaction (e.g., volatilization) [6], or (4) by using an appropriate acid acceptor such as pyridine [7]. Trimethylsilation of both small-pore (60 Å) and large-pore (750 Å) porous silica microspheres causes no significant change in the molar mass calibration plots for untreated particles, as indicated in Figure 6.2.

If desired, untreated silica packing may be silanized by in situ reaction with chlorotrimethylsilane [8]. This approach is useful to resilanize a set of columns that have become somewhat adsorbing because of loss of deactivating bonded organic groups, but is less convenient than the general procedure described above for silanizing larger quantities of bulk packings.

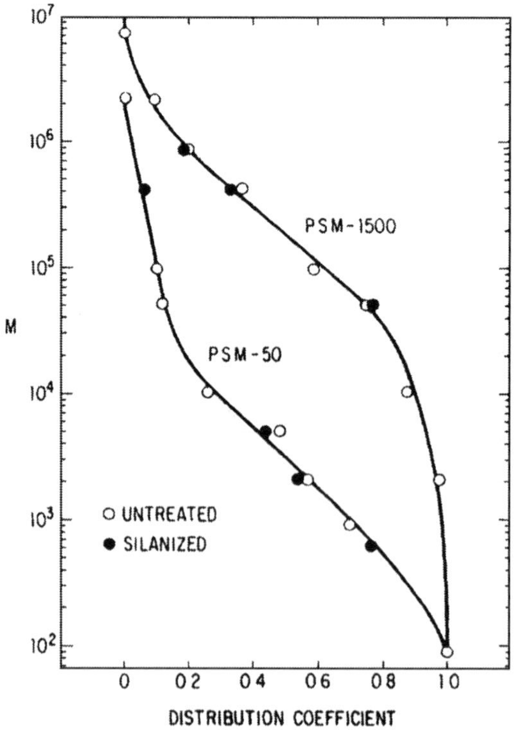

Figure 6.2 Effect of particle silanization on molar mass calibration curves. Polystyrene standards; mobile phase, THF; 22°C; flow rate, 2.5 mL/min; pressure, 925 psi; UV detector at 254 nm; sample, 25 μL, 0.25%; 60-cm set of porous silica microsphere columns 60 to 3500 Å. (Reprinted with permission from Ref. 1.)

6.3 COLUMN-PACKING METHODS

6.3.1 Particle Technology

As indicated in Section 3.3, particle size is an important factor in the preparation of efficient SEC columns. Both plate height and column permeability decrease approximately as the square of the particle diameter. Thus, when using smaller particles to gain higher column resolution, higher column inlet pressures are required. On balance, columns of smaller particles are generally worth the increased cost and higher column pressures to gain increased resolution or decreased analysis time. Preferred particle sizes for most polymeric analyses are 5 and 10 μm. For linear polymers with ultrahigh-M ($M > 10^6$ g/mol), 20-μm packing particles are recommended in order to decrease the risk of on-column, flow-induced degradation of the macromolecules (see Section 7.2.3) [15]. In oligomeric analysis (see Chapter 13), the separation of individual components in a multicomponent mixture is often paramount, but the possibility of on-column, flow-induced analyte degradation is rarely an issue. In this case, particles of less than 5 μm generally provide the needed resolution, although lower flow rates (e.g., <1 mL/min) may be needed both to optimize resolution and also because of the high backpressures generated when multiple columns of sub-5-μm particle size are coupled to one another.

Spherical particles are preferred over irregular materials, because reproducibly packed, high-efficiency columns are prepared more easily from the former. In addition, columns packed with spherical particles tend to have higher permeability for the same apparent average particle size. Irregularly shaped particles with a wide particle-size range cause particular difficulties in packing homogeneous column beds.

For best results the particle-size range in a given packing should be relatively narrow, for example, $\pm 20\%$ from the average, as illustrated for the porous silica microspheres in Figure 6.3. A wider range of particle sizes makes packing homogeneous columns more difficult because of the tendency of the particles to size across the column during the packing process. Segregation of particles during the packing procedure produces a distribution of flow velocities across the column. This difference in the packing structure at the column wall and at the center of the column will cause band distortion. Particle agglomeration during packing makes these flow effects more serious. Inhomogeneous beds result in lower column plate numbers and column permeabilities, as compared to homogeneous bed, narrow-size-range packings of the same average size. Thus, optimum column packings for SEC favor spherical particles with a narrow size range.

Good mechanical stability of particles is required for the preparation of high-efficiency columns, particularly by the high-pressure slurrying-packing technique described below. The shear force imposed on particles during this packing process is relatively large. If particles are not sufficiently strong, they will fragment or compress, resulting in significantly reduced column performances and greatly increased column backpressure. The higher strength of rigid siliceous packings permits higher mobile-phase pressures and velocities during packing and subsequent use (i.e., higher column input pressures) compared to organic gels. The excellent

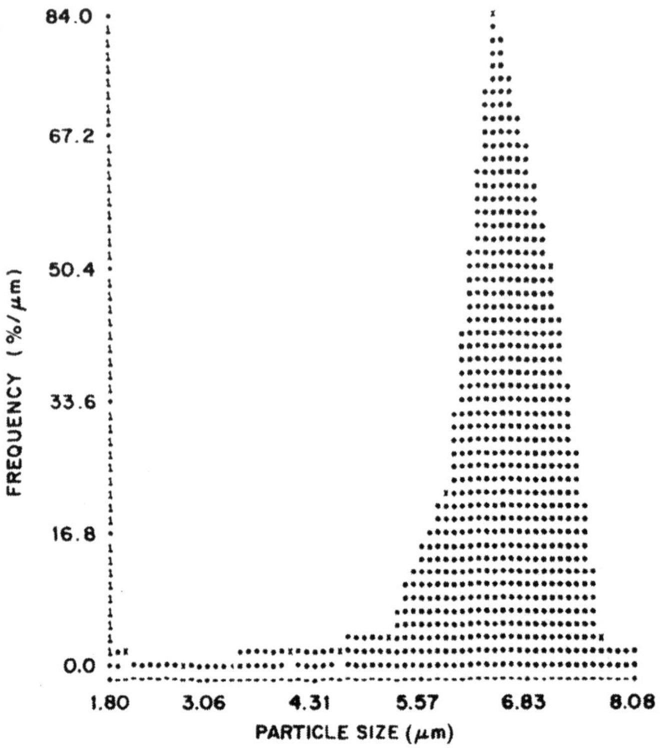

Figure 6.3 Particle-size range of porous silica microspheres. Size range weighted by volume; Quantimet analysis.

mechanical rigidity of these particles also allows rapid changing of mobile phase without the changes in swelling associated with organic gels.

6.3.2 Basis of Column-Packing Techniques

There is no one "best" column-packing method for all packings, since the optimum procedure is determined by the particle size and the nature of the material [9]. The prime goal is to pack the column uniformly without channels or particle sizing within the column. Rigid solids and semirigid, hard gels generally are packed as densely as possible without fracturing. Columns of rigid solids may be made by dry-packing or slurry-packing techniques, depending on particle size; dry packing is normally used with particles larger than 20 μm. Any packing technique that rapidly establishes a dense, stable structure is applicable. Several techniques work satisfactorily, and good columns can usually be produced with a little care.

As discussed previously, higher column efficiency results from more homogeneous structures in the packed bed. The desired homogeneous bed structure is not obtained by the classical technique of simply pouring particles into the column until full; with use, a void will usually develop at the column inlet, and a serious decrease in column performance will result. The purpose of all column-packing techniques is to produce a compact, homogeneous, and stable packing structure.

While dry-packing techniques have been used in high-performance liquid chromatography for some time [10], SEC columns of high efficiency are difficult to achieve as the particle size is decreased, because small particles have very high surface energies relative to their mass. Small particles tend to form larger aggregates, producing an effect analogous to that obtained in packing a column with a very wide particle-size range. The incidence of particle agglomeration increases as the particle diameter decreases, and totally porous particles smaller than about 20 μm cannot be dry-packed easily into homogeneous beds. Because dry packing of SEC columns is rarely performed nowadays, we do not delve deeper into this subject here.

Since particles smaller than about 20 μm are difficult to form into high-efficiency columns by dry-packing techniques, high-pressure "wet"-packing or slurrying techniques are normally used for this purpose [9]. A suitable liquid is utilized to wet the particles, reduce surface energy, and eliminate aggregation. With the proper procedure, the small packing particles can be dispersed homogeneously in the liquid by vigorous agitation [11]. It is important that the suspension liquid thoroughly wet the packing. Polar silica or hydrophilic organic-modified silicas require relatively polar liquids, while most low-surface-energy packings, such as trimethylsilyl-modified silicas, need relatively nonpolar media. Table 6.3 lists the properties of fluids that have been used for high-pressure slurry-packing techniques. The column-filling process,

Table 6.3 Properties of some slurry-packing solvents

	Density, ρ (g/mL)	Viscosity, η (cP, 20°C)
Diiodomethane (methylene iodide)	3.3	2.9
1,1,2,2-Tetrabromoethane[a]	3.0	—
Dibromomethane (methylene bromide)	2.5	1.0
Iodomethane (methyl iodide)	2.3	0.5
Tetrachloroethylene (perchloroethylene)[a]	1.6	0.9
Carbon tetrachloride	1.6	1.0
Chloroform	1.5	0.6
Trichloroethylene	1.5	0.6
Bromoethane (ethyl bromide)	1.5	0.4
Dichloromethane (methylene chloride)	1.3	0.4
Water	1.0	1.0
Pyridine	1.0	0.9
Tetrahydrofuran	0.9	0.5
n-Butanol	0.8	3.0
n-Propanol	0.8	2.3
Ethanol	0.8	1.2
Methanol	0.8	0.6
Cyclohexane	0.8	1.0
n-Heptane	0.7	0.4
Isooctane	0.7	0.5

[a] Most halogenated solvents are somewhat toxic, but this one is particularly toxic.

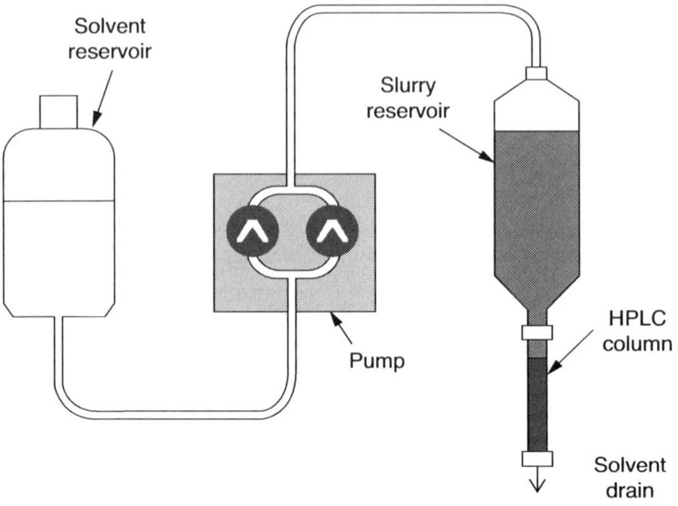

Figure 6.4 Column filling process using slurrying technique. (Reprinted with permission from Ref. 9.)

using the slurrying technique, is shown in Figure 6.4. The design of both analytical and preparative columns is shown in Figure 6.5 (see Section 15.2 for a discussion of preparative SEC).

In addition to the aforementioned packing material requirements of sphericity and narrow particle-size distribution, a combination of physical, chemical, and mechanical conditions that lead to optimal column performance and stability include the following [9]:

1. The column-loading apparatus needs to be designed to permit suitable flow patterns to maintain the same internal diameter from the bottom outlet of the slurry reservoir into the column inlet.
2. The slurry reservoir must be of sufficient volume to permit slurry concentrations in the solids range 7 to 15%.
3. Column tubing must have mirror-finish walls, to minimize column wall effects and to avoid formation of fines during loading.
4. The packing retainer devices (e.g., frits) should be thin and homogeneous.
5. Column-loading pressure should be as high as possible for forming homogeneously packed beds of high stability. Typically, the column loading pressure should be half again higher that the maximum pressure used in applications.
6. The choice of slurry liquid is critical. For best results, this liquid should be of low viscosity and a good energy match for the chemical surface of the packing particles.

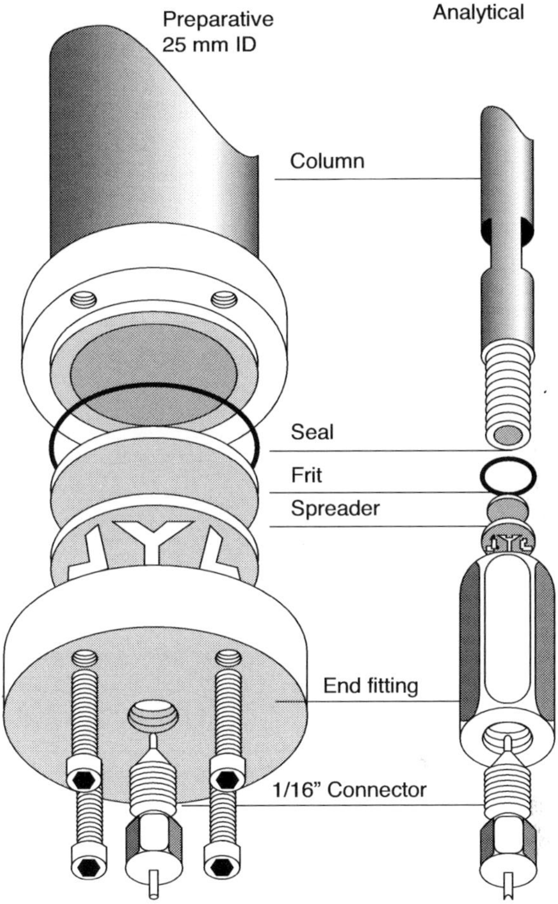

Figure 6.5 Analytical and preparative column design. (Courtesy of Varian/Polymer Laboratories.)

Regarding the mechanical aspects of column stability, the following statements can be made:

1. The exposure of a particle-packed bed to high-pressure flow during column filling and in normal operation causes elastic compression of the particles.
2. The walls of SEC columns can expand during the pressure-filling process, then collapse back when the pressure is relieved, causing further compression of the packing particles. When subjected to pressures of several kilobar, the stress distribution and dimensional changes that occur in both stainless steel and silica chromatographic column tubes are the subject of recent studies [9,12].
3. The part of the column that packs first is able to retain some of the packing compression due to particle bridging and wall interference.

4. The last part of the column to fill is not able to maintain any of the packing compression because of the opportunity for the packing to extrude from the open end of the packed column.
5. For column bed stability, it is important to close a freshly packed column quickly. This serves to prevent the column from completely relieving the packed-bed compression that is initially located at one end of the column but which quickly redistributes throughout the packed-bed after closing both ends of the column.
6. Retaining some internal packed-bed compression created during column filling is important, because it helps compensate for later stresses imparted by column operation at high pressures and elevated temperatures.

6.4 COLUMN PERFORMANCE

Column manufacturers now supply a great deal of information regarding care, use, and in-house testing of purchased columns. This includes information about the type of tubing and connectors to use for installation, maximum flow rates and backpressure tolerances, recommended sample concentration and injection volume, and suggestions for short- and long-term storage. Regarding eluent use, for columns with aqueous mobile phases the maximum amount of allowed organic modifier is usually given as well as compatible buffer solutions. For columns with organic mobile phases, a maximum water tolerance is usually given. In addition, manufacturers often provide eluent transfer guidelines for getting from the solvent in which the column is shipped to the solvent in which the user wishes to work (this may involve initial transfer to a solvent with polarity and/or miscibility intermediate to those of the shipped and end-use solvents). Other specifications may include typical operating backpressure, as well as limiting operating ranges for pH and temperature.

Quantitative and graphical information on column efficiency is also usually provided, giving plates per meter based on different tests and on one or several test compounds, peak symmetry (see Figure 3.27), and peak width. An example of the type of test data sheet that is provided along with a purchased column is shown in Figure 6.6. Some manufacturers also provide a small amount of the test material. In this way, the user may test the column under the operating conditions of choice, and note differences in plate number, peak symmetry, and so on, using exactly the same test compounds as used by the manufacturer.

A new column, whether purchased or packed in the laboratory, should always be tested before use to ensure that it satisfies the intended need. The user should also make sure that the column meets the proper specifications as given in the performance data sheet included with the column or as specified by the user to the manufacturer prior to purchase. Performance can be compared to that of columns previously used or to published data. Also, this testing provides reference performance data for future comparison. In the event that the performance of a column becomes doubtful, information on the initial pressure drop and plate count is important. Comparison of

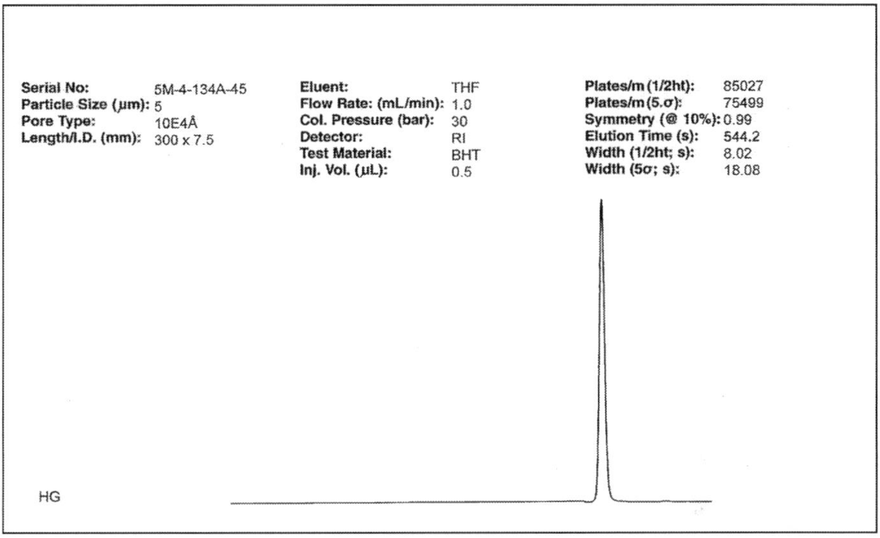

Figure 6.6 Column performance data sheet shipped with column. Test data for a 300 × 7.5 mm, 10-μm particle size, 10^4-Å pore size PLgel PS/DVB column. BHT, butylated hydroxytoluene.

column performance should only be made when pertinent variables are maintained constant: flow rate, mobile phase, solute, temperature, and apparatus.

Vibration and extremes of temperature should be avoided. While no change in column performance occurs when columns are used and stored with some neat organic mobile phases, aqueous mobile phases should be used only in the pH range 2 to 8 to prevent degradation of silica-based column packing. It is unwise to allow aqueous systems with the pH at the extremes of this range to stand in a column for long periods of time. Of particular importance when using aqueous mobile phases (e.g., buffer solutions) is bacterial growth, which can form within the column and cause pluggage. It is suggested that the column be purged with absolute methanol (or another pure, aprotic organic solvent, such as acetonitrile) before storage for long periods. The drying out of columns of semirigid gels must be avoided. Closing both ends of all columns with compression-fitting caps is desirable for storage.

In some situations, adsorbed sample residues can affect retention-time reproducibility and column efficiency or cause non-Gaussian peaks. In unusual situations such as this, it may be necessary to purge the column periodically with a solvent strong enough to remove tightly held components that have collected at the inlet. This condition generally does not occur when proper choice of the combination of mobile phase and stationary phase has been made.

REFERENCES

1. J. J. Kirkland, *J. Chromatogr.*, **125**, 231 (1976).
2. K. K. Unger, R. Kern, M. C. Ninou, and K. F. Krebs, *J. Chromatogr.*, **99**, 435 (1974).

3. F. E. Regnier and R. Noel, *J. Chromatogr. Sci.*, **14**, 316 (1976).
4. M. J. Lu, in *Column Handbook for Size Exclusion Chromatography*, C.-S. Wu, ed., Academic Press, San Diego, CA, 1999, Chap. 1.
5. K. K. Unger and P. Ringe, *J. Chromatogr. Sci.*, **9**, 463 (1971).
6. J. J. Kirkland, *Chromatographia*, **8**, 661 (1975).
7. I. Halász and I. Sebestian, *Chromatographia*, **7**, 371 (1974).
8. J. J. Kirkland and P. E. Antle, *J. Chromatogr. Sci.*, **15**, 137 (1977).
9. J. J. Kirkland and J. J. DeStefano, *J. Chromatogr. A*, **1126**, 50 (2006).
10. L. R. Snyder and J. J. Kirkland, *Introduction to Modern Liquid Chromatography*, 2nd ed., Wiley-Interscience, New York, 1979, Chap. 5.
11. S. Bakalyar, J. Yuen, and R. H. Henry, *Spectra-Physics Chromatography Technical Bulletin* 114–76, 1976.
12. F. Chen, E. C. Drumm, and G. Guiochon, *J. Chromatogr. A*, **1083**, 68 (2005).
13. K. K. Unger, *Porous Silica,* Elsevier, Amsterdam, 1979.
14. U. D. Neue, *HPLC Columns*, Wiley-VCH, New York, 1997.
15. A. M. Striegel, *J. Liq. Chromtogr. Rel. Technol.*, **31**, 3105 (2008).

7

EXPERIMENTAL VARIABLES AND TECHNIQUES

7.1 INTRODUCTION

Some considerations of the operational variables in SEC are different from those normally encountered in other LC methods. In SEC the substrate (the porous packing) primarily determines retention and resolution. As a result, the mobile phase is chosen primarily for sample solubility and to eliminate unwanted solute or substrate effects. With large molecules, diffusion is slow, and the need to maintain high column efficiency by using small particles and low-viscosity solvents is an important consideration. In this chapter we discuss the general aspects of mobile phase, substrate, and sampling effects in SEC. We also consolidate information from other chapters and summarize the adjustment of these variables for optimum operation. Finally, we describe experimental laboratory procedures involving these variables.

7.2 SOLVENT EFFECTS

7.2.1 Sample Solubility

The dissolution process for polymers is, in some respects, rather different from that for low-molar-mass substances. Polymer dissolution is preceded by swelling of the bulk solid phase, typically a slow process. In a second stage the molecules of the

Modern Size-Exclusion Liquid Chromatography: Practice of Gel Permeation and Gel Filtration Chromatography, Second Edition
By André M. Striegel, Wallace W. Yau, Joseph J. Kirkland, and Donald D. Bly

swollen polymer phase disentangle and enter into solution. The rate-controlling step is the solvent diffusion rate within the swollen polymer. Warming the mixture will reduce solvent viscosity and may speed up diffusion and dissolution.

Polymer dissolution rate is affected by the crystallinity of the sample. In the case of highly crystalline polymers (e.g., linear polyethylene), the first step toward dissolution is to melt the crystalline region by heating to facilitate solvent permeation into the polymer. While the physical state of the polymer sample affects the rate of dissolution, the chemical structure of the polymer dictates the total solubility. Generally, polymer–solvent structural similarity favors enhanced polymer solubility. Dissolution occurs because the net attractive forces between the solvent and the solute outweigh those between pairs of solvent and solute molecules, respectively. As a general rule, solubility decreases with increasing solute M.

Solvents that provide rapid dissolution rates are usually small, compact molecules. However, solvents promoting high rates of dissolution do not necessarily possess the favorable thermodynamic properties that control the quantity of polymer that is dissolved. Some liquids dissolve polymers quickly to relatively high concentrations. Interestingly, for the dissolution of some high-M crystalline polymers solvent diffusion into the bulk polymer many not follow *Fick's second law*, which is

$$\frac{\partial C}{\partial t} = D\frac{\partial^2 C}{\partial x^2}$$

where C is the concentration of the solution, t the time, D the diffusion coefficient of the analyte in solution, and x the distance. This *non-Fickian diffusion* (also termed *type II transport*) is thought to be due to the fact that as solvent penetrates into the bulk polymer, the polymer chains relax into a more randomly coiled conformation than the highly ordered conformation encountered in the crystalline solid. Because this relaxation can be slower that the solvent diffusion process, overall dissolution is controlled not by Fick's law but by polymer relaxation kinetics [1].

The dissolution process occurs if the free energy of mixing, ΔG, is negative:

$$\Delta G = \Delta H - T\,\Delta S \tag{7.1}$$

The entropy of mixing, ΔS, is positive for polymer solutions. If the enthalpy of mixing, ΔH, is negative (meaning that there is a net positive attraction favoring solvent–solute pairs), dissolution will occur at any temperature. When the polymer is in a poor solvent (e.g., ΔH is positive), the thermodynamics of dissolution depends on the temperature. Dissolution can occur at high temperatures because the negative $T\,\Delta S$ term in Equation 7.1 becomes dominant. Conversely, as temperature decreases, the $T\,\Delta S$ term is less important and the solvent becomes thermodynamically poorer (i.e., ΔG becomes less negative). Finally, with some systems a consolute temperature, T_c, is reached below which polymer and solvent are no longer miscible in all proportions and the mixture separates into two phases.

Table 7.1 Theta-temperature data for polymer–solvent systems

Polymer	Solvent	Θ (K)
Polystyrene	Octadecanol	474
	Cyclohexanol	358.4
	Cyclohexane	307.2
	Ethylcyclohexane	343.2
Polyethylene	Nitrobenzene	503
Polyisobutene	Diisobutyl ketone	331.1
Poly(methyl methacrylate)	4-Heptanone	305
Polydimethylsiloxane	Phenetole	358
	Butanone	298.2
Cellulose tricaprylate	Dimethylformamide	413
	3-Phenyl-1-propanol	323
Poly(acrylic acid)	Dioxane	302.2
Polymethacrylonitrile	Butanone	279

Source: Ref. 3.

The dependence of T_c on polymer M can be expressed by [2]

$$\frac{1}{T_c} \simeq \frac{1}{\Theta}\left(1 + \frac{C}{M^{1/2}}\right) \tag{7.2}$$

where C is a constant and Θ is the theta temperature for the specific polymer–solvent system [2]. (Both Θ and T_c are in Kelvin.) The theta temperature is the critical miscibility temperature at "infinite" molar mass. By convention, a solvent whose theta temperature is close to room temperature is called a "poor" solvent, since the polymer solution is then close to precipitation. Similarly, a "good" solvent is one whose theta temperature is well below room temperature. Table 7.1 shows the theta temperatures of some polymer–solvent systems. Also by convention, theta solvents are designated as those in which the theta temperature is approximately room temperature, or specified for whatever temperature is to be employed. As such, "good," "poor," and "theta" are terms that actually denote a specific set of solvent–temperature conditions, reminding us of the intimately linked roles that solvent and temperature play in polymer science.

In a thermodynamically good solvent, where polymer–solvent interactions are favored, a polymer coil in solution is more extended than in a poor solvent. The radius of gyration of a polymer molecule (i.e., R_G; see Section 9.3 and Table 9.2) in solution is expanded by a factor α times that in the theta condition. Above the theta temperature the following relationship holds [2]:

$$\left(\alpha^5 - \alpha^3\right) \propto \left(1 - \frac{\Theta}{T}\right) M^{1/2} \tag{7.3}$$

At $T = \Theta$, $\alpha = 1$, but α becomes larger with increasing T (in Kelvin); at the same time, α becomes more molar-mass dependent. At large α values, $\alpha^5 \propto M^{1/2}$, or the limiting molar mass dependence of $\alpha \propto M^{0.1}$. These relationships explain why R_G increases with temperature in a poor solvent, as illustrated in Table 2.3. (The opposite effect, that R_G decreases with increasing temperature, can occur in a good solvent.) The predicted molar mass dependence on the expansion factor α is in agreement with observations [2] that

$$R_G \propto M^{0.5-0.6} \tag{7.4}$$

and

$$[\eta] \propto M^{0.5-0.8} \tag{7.5}$$

In SEC experiments, good solvents help to avoid possible packing surface adsorption effects. Near Θ conditions, the solvent is less effective in preventing polymer solute molecules from adsorbing to the packing surface.

Another measure of polymer–solvent interaction is provided by the solubility parameter. When only dispersion forces are involved, the enthalpy of mixing can be calculated by [4]

$$\Delta H = \phi_s \phi_m (\delta_s - \delta_m)^2 \tag{7.6}$$

where ϕ_s and ϕ_m are the volume fractions of solvent and macromolecule, respectively, and δ_s and δ_m are the respective solubility parameters of the solvent and macromolecule, defined by

$$\delta = \left(\frac{\Delta E}{V}\right)^{1/2} \tag{7.7}$$

where $\Delta E/V$ is the energy of vaporization per unit volume, or the cohesive energy density. For ΔH to be minimal, δ_s and δ_m should be similar. This solubility parameter approach was originally devised for small molecules, but it has been applied successfully to predict polymer solubility [5,6]. Equation 7.6 predicts that $\Delta H = 0$ if $\delta_s = \delta_m$, so that two substances with equal solubility parameters should be mutually soluble because of the negative entropy factor. This is in accordance with the general rule that chemical and structural similarity favors solubility ("like dissolves like"). As the difference between δ_s and δ_m increases, the tendency toward dissolution decreases. As with the theta-temperature approach, the solubility parameter method is used to study polymers in poor solvents (i.e., positive ΔH). For cases where ΔH is negative, the polymer is so easily solubilized that evaluation of solubility has no real meaning. Polymer and solvent are miscible in any proportion at any temperature.

Whereas δ_s can be measured directly, δ_m cannot. Polymers do not vaporize, and therefore δ_m has to be measured indirectly through swelling or viscosity experiments using solvents of known δ_s. Typical values of δ_m and δ_s for common polymers and solvents are shown in Table 7.2 and Figure 7.1. The values of "δ calc." in the last

Table 7.2 Experimental and calculated values of δ_m for some polymers[a]

Polymer	δ_m Expt. Range ($J^{1/2}/cm^{3/2}$) From:	To:	δ_m calc. ($J^{1/2}/cm^{3/2}$)
Polyethylene	15.8	17.1	16.0
Polypropylene	16.8	18.8	17.0
Polyisobutylene	16.0	16.6	16.4
Polystyrene	17.4	19.0	19.1
Poly(vinyl chloride)	19.2	22.1	19.7
Poly(vinyl bromide)	19.4	—	20.3
Poly(vinylidene chloride)	20.3	25.0	20.6
Poly(tetrafluoroethylene)	12.7	—	11.7
Poly(chlorotrifluoroethylene)	14.7	16.2	15.7
Poly(vinyl alcohol)	25.8	29.1	—
Poly(vinyl acetate)	19.1	22.6	19.6
Poly(vinyl propionate)	18.0	—	18.8
Poly(methyl acrylate)	19.9	21.3	19.9
Poly(ethyl acrylate)	18.8	19.2	19.2
Poly(propyl acrylate)	18.5	—	18.7
Poly(butyl acrylate)	18.0	18.6	18.3
Poly(isobutyl acrylate)	17.8	22.5	18.7
Poly(2,2,3,3,4,4,4-heptafluorobutyl acrylate)	13.7	—	15.8
Poly(methyl methacrylate)	18.6	26.2	19.0
Poly(ethyl methacrylate)	18.2	18.7	18.6
Poly(butyl methacrylate)	17.8	18.4	17.9
Poly(isobutyl methacrylate)	16.8	21.5	18.3
Poly(*t*-butyl methacrylate)	17.0	—	18.0
Poly(benzyl methacrylate)	20.1	20.5	19.3
Poly(ethoxyethyl methacrylate)	18.4	20.3	18.6
Polyacrylonitrile	25.6	31.5	25.7
Polymethacrylonitrile	21.9	—	22.8
Poly(α-cyanomethyl acrylate)	28.7	29.7	23.8
Polybutadiene	16.6	17.6	17.5
Polyisoprene	16.2	20.5	17.4
Polychloroprene	16.8	18.9	19.2
Polyformaldehyde	20.9	22.5	20.5
Poly(tetramethylene oxide)	17.0	17.5	17.6
Poly(propylene oxide)	15.4	20.3	18.9
Polyepichlorohydrin	19.2	—	20.1
Poly(ethylene sulfide)	18.4	19.2	18.9
Poly(styrene sulfide)	19.0	—	19.6
Poly(ethylene terephthalate)	19.9	21.9	20.5
Poly(8-aminocaprylic acid)	26.0	—	25.7
Poly(hexamethylene adipamide)	27.8	—	28.0

Source: Ref. 6.

[a]Conversion factor: $J^{1/2}/cm^{3/2} = 0.49\ cal^{1/2}/cm^{3/2}$.

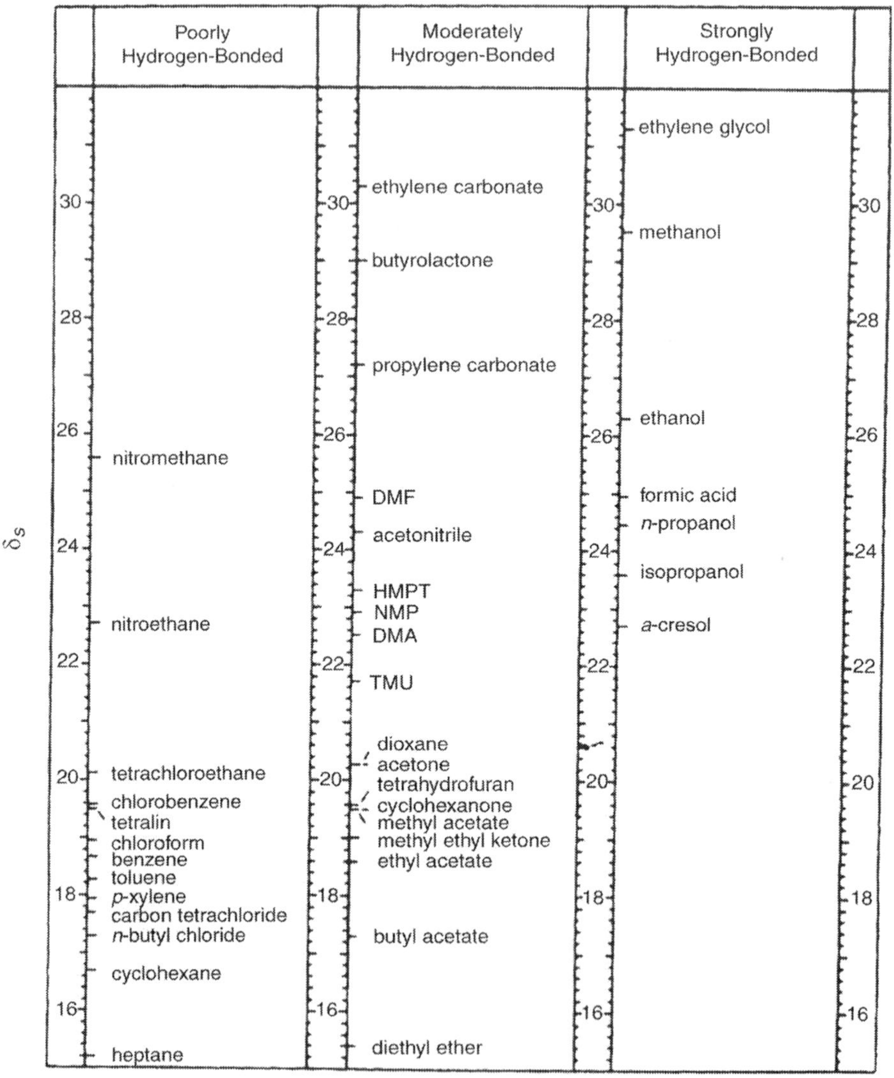

Figure 7.1 Solubility parameter and hydrogen-bonding tendency of solvents. DMA, *N,N*-dimethylacetamide; DMF, *N,N*-dimethylformamide; HMPA, hexamethylphosphoramide; NMP, *N*-methylpyrrolidone; TMU, tetramethyl urea. (Reprinted with permission from Ref. 6.)

column of Table 7.2 were obtained by summing contributions to δ from structural groups in the polymer [6]. The validity of using structural group summation to estimate δ_m is illustrated by the good agreement shown for experimental and calculated values of δ for polymers. Figure 7.1 shows solvents in three groups representing different hydrogen-bonding tendencies. The arrangement of the figure is convenient to use, since mutual solubility only occurs if the degree of hydrogen bonding between the polymer and solvent is approximately matched (see also Table 7.7). The values of δ in Table 7.2 and Figure 7.1 have the dimensions $J^{1/2}/cm^{3/2}$. Values of δ expressed in $cal^{1/2}/cm^{3/2}$ have become familiar quantities for many investigators. Conversion of $cal^{1/2}/cm^{3/2}$ to $J^{1/2}/cm^{3/2}$ requires multiplication by a factor of 2.046.

The effect of δ_h (hydrogen bonding) is significantly different from those of δ_d (dispersion) and δ_p (polar forces). Therefore, it is advantageous to classify the ability of solvents to dissolve solutes by separating the contributions of δ_h, δ_d, and δ_p. Utilization of the parameter $\delta_v = (\delta_d^2 + \delta_p^2)^{1/2}$ leads to a degree-of-solubility diagram involving δ_v versus δ_h, as illustrated in Figure 7.2 for the interaction between

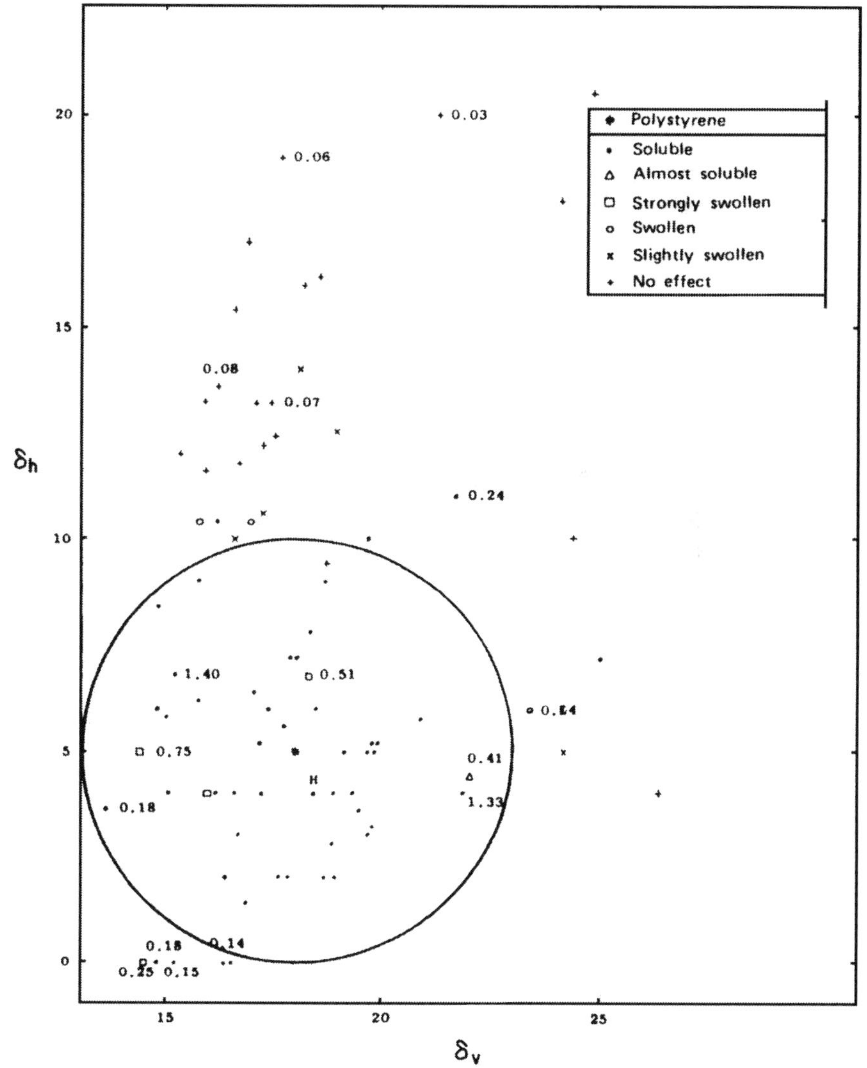

Figure 7.2 Solubility of polystyrene in various solvents. Numbers represent volume polystyrene/volume solvent. (Reprinted with permission from Ref. 6.)

polystyrene and a number of solvents [7]. The majority of the points for good solvents fall in a single region, which can be defined approximately by a circle. The center of the circle is indicated by the symbol * in Figure 7.2 and has the coordinate values $\delta_v = 18$, $\delta_h = 5$. Thus, in the (δ_h, δ_v) diagram the degree of solubility (volume of polymer per volume of solvent) can be indicated by a number. The solubility region can be defined approximately by a circle with a radius of about 5 units of δ. It can be seen that the solubility generally decreases as the distance from the center increases. As a general rule, polystyrene is soluble in solvents for which

$$\sqrt{(\delta_v - 18)^2 + (\delta_h - 5)^2} < 5 \tag{7.8}$$

The (δ_h, δ_v) diagram is an efficient way to represent polymer–solvent interactions.

Solubility limits of a given polymer are closely related to the theta temperatures of the polymer in various solvents, as shown in the (δ_h, δ_v) diagram for polystyrene in Figure 7.3. Thus, the closer the data points are to the center of the circle in Figure 7.3, the better is the solubility of the system and the lower are the values of the theta temperature. Typical values of δ_d, δ_p, and δ_h for common polymers and solvents are listed in Tables 7.3 and 7.4.

Until now, our discussion of polymer solubility has been limited to the use of a single solvent, as mixed solvents are rarely used in SEC. The reason for this is the preferential solvation of the macromolecule by one solvent over another and the differential basis of most of the detection techniques used. For example, a polymer may dissolve in a mix of solvents A and B. While the A/B ratio in the mix may be 50 : 50, the ratio within the hydrodynamic volume of the polymer may well be different from this (e.g., 60 : 40). Most physical detectors are of the differential type (e.g., differential refractometer, differential viscometer, static light scattering; see Chapter 9), in which the detector response for the polymer itself is obtained by subtracting the response for the solvent from the response for the polymer solution. In a neat solvent (or a solvent with a modest amount of additive), the solvent composition within the hydrodynamic volume occupied by the polymer will be the same as the composition outside this volume, and subtraction of the solvent response (baseline subtraction) is straightforward. In mixed solvents; however, the solvent baseline does not accurately represent the solvent contribution to the detector response for the polymer solution.

Many chemical detection methods (e.g., mass spectrometry; see Section 10.2), and even certain concentration-sensitive detectors (e.g., evaporative type; see Section 9.2.3), can handle mixed solvents and solvent gradients. Within these detectors, desolvation occurs by thermal or pneumatic means, or by a combination of both. Because of the increased use of these types of detection methods, a discussion of mixed solvents is warranted.

In *mixed solvents* the solubility relationships become more complicated, but nevertheless, can be formulated entirely from considerations of intermolecular forces.

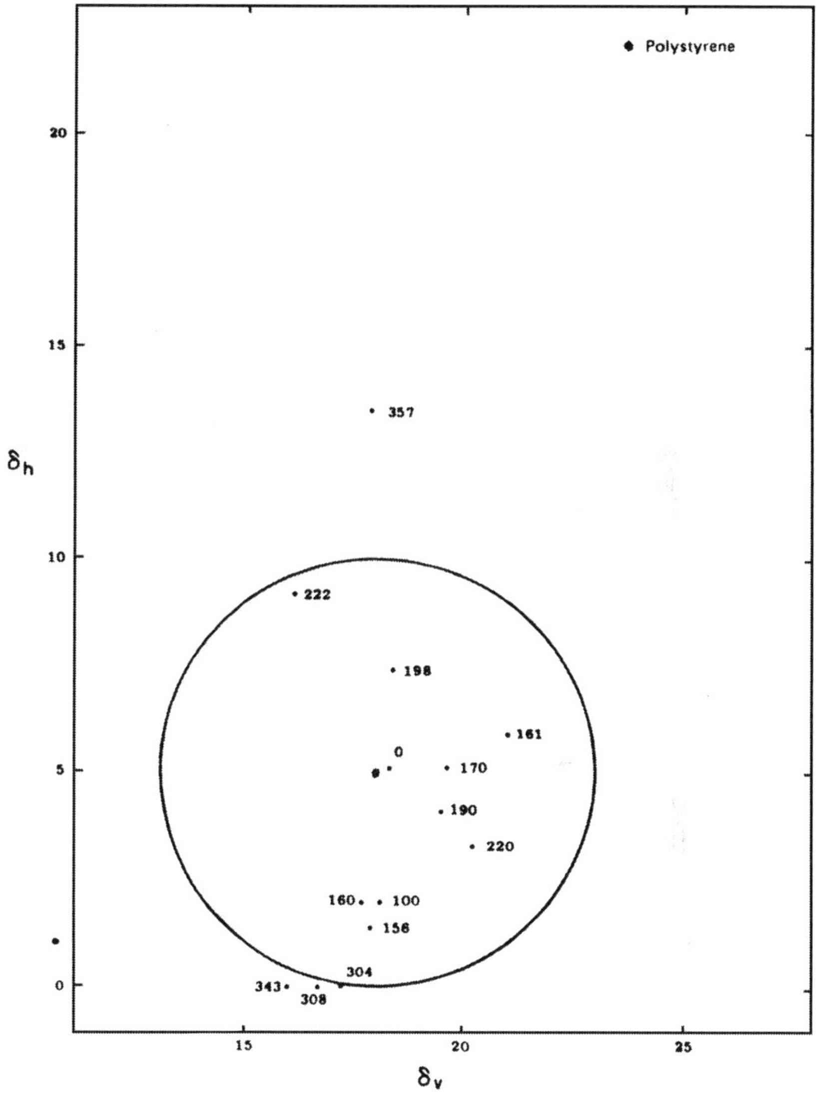

Figure 7.3 Theta temperature of polystyrene in various solvents. Numbers represent theta temperatures. (Reprinted with permission from Ref. 6.)

In Table 7.5 some examples are compiled for the solubility of polymers in mixtures of two liquids, where neither liquid by itself is a solvent for that polymer. These solubility phenomena can be explained as follows [9].

Polystyrene does not dissolve in acetone because the intermolecular association between the acetone molecules through polar forces is too strong. Polystyrene is also insoluble in nonane, since the intermolecular dispersion forces between the

Table 7.3 Solubility parameters specified for some polymers[a]

Polymer	δ	δ_d	δ_p	δ_h
Polyisobutylene	17.6	16.0	2.0	7.2
Polystyrene	20.1	17.6	6.1	4.1
Poly(vinyl chloride)	22.5	19.2	9.2	7.2
Poly(vinyl acetate)	23.1	19.0	10.2	8.2
Poly(methyl methacrylate)	23.1	18.8	10.2	8.6
Poly(ethyl methacrylate)	22.1	18.8	10.8	4.3
Polybutadiene	18.8	18.0	5.1	2.5
Polyisoprene	18.0	17.4	3.1	3.1

Source: Ref. 6.
[a] All δ in $(\mathrm{J/cm^3})^{1/2}$.

polystyrene molecules (because of high π-electron polarizability) are stronger than the intermolecular forces between polystyrene and nonane. However, in the mixture of acetone and nonane, polystyrene dissolves easily at 20°C to form a homogeneous solution, because the intermolecular acetone associations that inhibit strong solvation are broken up by the nonane molecules. The single acetone molecules that result are more able to solvate polystyrene molecules through polar forces to an extent sufficient for dissolution.

Similar arguments define the solubility behavior of poly(vinyl chloride) in acetone/carbon disulfide, poly(vinyl chloride) in nitromethane/trichloroethylene, polychloroprene in acetone/hexane, and Buna S in methyl acetate/pentane. In systems of poly(vinyl acetate) + ethanol/carbon tetrachloride and cellulose–tribenzyl ether + ethanol/carbon tetrachloride, dissociated alcohol molecules hydrogen bond to the polymer oxygen-containing groups. With phenol alone, the intermolecular hydrogen bonding is so strong that polystyrene does not dissolve. However, if the phenolic hydroxyl groups are made partially inactive for intermolecular association through complexation by the addition of acetone, polymer dissolution is possible.

Propanol is a nonsolvent for poly(methyl methacrylate) because of steric hindrance. However, if the propanol contains water, associations form between propanol and water and the "solubility arm" of the alcohol is lengthened. In this way the macromolecules are solvated adequately. A similar situation is also true for poly(vinyl acetate) + ethanol–water and other such systems.

With random copolymers, solubility in general is greater than that of the respective homopolymers, because the irregular arrangements of groups responsible for the intermolecular forces in the copolymer cause a lower degree of interaction between adjacent molecules. Therefore, random copolymers can be solvated more easily than the corresponding homopolymers [9,37].

So far we have dealt primarily with dissolving synthetic polymers in organic solvents. However, dissolution of biopolymers in aqueous solvents can also be troublesome at times, particularly if the material is labile. Because the hydrogen bonding of water molecules largely dictates the mode and extent of solvation, the various solubility forces discussed above do not generally operate in aqueous

Table 7.4 Three-dimensional solubility parameters of solvents[a]

Solvent	δ_o	δ_d	δ_p	δ_h
Acetic acid	10.5	7.10	3.9	6.6
Acetic anhydride	10.30	7.50	5.4	4.7
Acetone	9.77	7.58	5.1	3.4
Acetonitrile	11.9	7.50	8.8	3.0
Acetophenone	9.68	8.55	4.2	1.8
Aniline	11.04	9.53	2.5	5.0
Benzaldehyde	10.40	9.15	4.2	2.6
Benzene	9.15	8.95	0.5	1.0
α-Bromonaphthalene	10.25	9.94	1.5	2.0
1,3-Butanediol	14.14	8.10	4.9	10.5
n-Butanol	11.30	7.81	2.8	7.7
2-Butoxyethanol	10.25	7.76	3.1	5.9
n-Butyl acetate	8.46	7.67	1.8	3.1
n-Butyl lactate	9.68	7.65	3.2	5.0
Butyric acid	9.2(?)	7.30	2.0	5.2
γ-Butyrolactone	12.78	9.26	8.1	3.6
Butyronitrile	9.96	7.50	6.1	2.5
Carbon disulfide	9.97	9.97	0	0
Carbon tetrachloride	8.65	8.65	0	0
Chlorobenzene	9.57	9.28	2.1	1.0
1-Chlorobutane	8.46	7.95	2.7	1.0
Chloroform	9.21	8.65	1.5	2.8
m-Cresol	11.11	8.82	2.5	6.3
Cyclohexane	8.18	8.18	0	0
Cyclohexanol	10.95	8.50	2.0	6.6
Cyclohexanone	9.88	8.65	4.1	2.5
Cyclohexylamine	9.05	8.45	1.5	3.2
Cyclohexylchloride	8.99	8.50	2.7	1.0
Diacetone alcohol	10.18	7.65	4.0	5.3
o-Dichlorobenzene	9.98	9.35	3.1	1.6
2,2-Dichlorodiethyl ether	10.33	9.20	4.4	1.5
Diethyl amine	7.96	7.30	1.1	3.0
Diethylene glycol	14.60	7.86	7.2	10.0
Diethylene glycol monobutyl ether	8.96	7.80	3.4	5.2
Diethylene glycol monomethyl ether	10.72	7.90	3.8	6.2
Diethyl ether	7.62	7.05	1.4	2.5
Diethyl sulfide	8.46	8.25	1.5	1.0
Diisobutyl ketone	8.17	7.77	1.8	2.0
Dimethylformamide	12.14	8.52	6.7	5.5
Dimethyl sulfoxide	12.93	9.00	8.0	5.0
Dioxane	10.0	9.30	0.9	3.6
Dipropylamine	7.79	7.50	0.7	2.0
Dipropylene glycol	15.52	7.77	9.9	9.0
Ethanol	12.92	7.73	4.3	9.5
Ethanolamine	15.48	8.35	7.6	10.4

(continued)

Table 7.4 Three-dimensional solubility parameters of solvents[a] (*Continued*)

Solvent	δ_o	δ_d	δ_p	δ_h
Ethyl acetate	9.10	7.44	2.6	4.5
Ethylbenzene	8.80	8.70	0.3	0.7
2-Ethylbutanol	10.38	7.70	2.1	6.6
Ethylene chloride	9.76	9.20	2.6	2.0
Ethylene glycol	16.30	8.25	5.4	12.7
Ethylene glycol monoethyl ether	11.88	7.85	4.5	7.0
Ethylene glycol monoethyl ether acetate	9.60	7.78	2.3	5.2
Ethylene glycol monomethyl ether	12.06	7.90	4.5	8.0
2-Ethylhexanol	9.85	7.78	1.6	5.8
Ethyl lactate	10.5	7.80	3.7	6.1
Formamide	17.8	8.4	12.8	9.3
Formic acid	12.15	7.0	5.8	8.1
Furan	9.09	8.70	0.9	2.6
Glycerol	21.1	8.46	—	—
Hexane	7.24	7.23	0	0
Isoamyl acetate	8.32	7.45	1.5	3.4
Isobutyl isobutyrate	8.04	7.38	1.4	2.9
Isophorone	9.71	8.10	4.0	3.6
Mesityl oxide	9.20	7.97	3.5	3.0
Methanol	14.28	7.42	6.0	10.9
Methylal	8.52	7.35	0.9	4.2
Methylene chloride	9.93	8.91	3.1	3.0
Methyl ethyl ketone	9.27	7.77	4.4	2.5
Methyl isoamyl ketone	8.55	7.80	2.8	2.0
Methyl isobutyl carbinol	9.72	7.47	1.6	6.0
Methyl isobutyl ketone	8.57	7.49	3.0	2.0
Morpholine	10.52	9.20	2.4	4.5
Nitrobenzene	10.62	8.60	6.0	2.0
Nitroethane	11.09	7.80	7.6	2.2
Nitromethane	12.30	7.70	9.2	2.5
2-Nitropropane	10.02	7.90	5.9	2.0
1-Pentanol	10.61	7.81	2.2	6.8
n-Propanol	11.97	7.75	3.3	8.5
Propylene carbonate	13.30	9.83	8.8	2.0
Propylene glycol	14.80	8.24	4.6	11.4
Pyridine	10.61	9.25	4.3	2.9
Styrene	9.30	9.07	0.5	2.0
Tetrahydrofuran	9.52	8.22	2.8	3.9
Tetralin	9.50	9.35	1.0	1.4
Toluene	8.91	8.82	0.7	1.0
1,1,1-Trichloroethane	8.57	8.25	2.1	1.0
Trichloroethylene	9.28	8.78	1.5	2.6
Water	23.5	6.0	15.3	16.7
Xylene	8.80	8.65	0.5	1.5

Source: Ref. 8.
[a] All δ in $(\text{cal/cm}^3)^{1/2}$.

Table 7.5 Dissolution of polymers in mixtures of two nonsolvents for the polymer

Polymer	Nonsolvent 1	Nonsolvent 2
Polystyrene	Acetone	Nonane
	Methyl acetate	Nonane
	Phenol	Acetone
Buna S	Methyl acetate	Pentane
Buna N	Dimethyl malonate	*p*-Cymene
Gel rubber	Benzene	Butanol
Poly(vinyl acetate)	Water	Ethanol
	Ethanol	Carbon tetrachloride
Poly(methyl methacrylate)	Propanol	Water
Poly(vinyl alcohol) with 30% acetyl groups	Methanol	Water
Poly(vinyl isobutyral)	Benzene	Ethanol
Poly(vinyl chloride)	Acetone	Carbon disulfide
	Nitromethane	Trichloroethylene
Polychloroprene	Acetone	Hexane
Cellulose tribenzyl ether	Ethanol	Carbon tetrachloride

Source: Ref. 9.

systems. However, biopolymer solubility can be significantly altered by adjusting salt concentration and pH levels. For example, at low salt concentrations (e.g., $<0.01\ M\ PO_4^{3-}$), a strong increase in the solubility of nonapeptides occurs with increasing salt concentration (salting-in effect) [10]. At higher salt concentrations (e.g., $>0.05\ M$), this increase reaches a maximum and then decreases (salting-out effect). Thus, ionic strength or salt concentration significantly affects solute conformation and size through solvation effects [11]. As with synthetic polymers, if possible biopolymers should also be dissolved in the mobile phase used for the separation. However, provided that denaturization or precipitation is not a problem, sometimes it is feasible to dissolve a sample at the ionic strength required to obtain a high sample concentration, then chromatograph this solution in a mobile phase at another ionic strength for optimum separation.

The adjustment of pH for sample dissolution is limited by solute stability. Some solutes (e.g., enzymes) often are unstable outside pH 5 to 9. Also, some solutes (e.g., DNA polymers) are less soluble at low pH. In the latter case it may be desirable to raise the solution to pH 9 with dilute sodium or ammonium hydroxide solution and then neutralize carefully after the polymer has dissolved.

Sometimes it is difficult to wet a freeze-dried polymer due to gel formation at the surface. Agitation on a vortex mixer is useful in these cases, but ultrasonic dissolution should not be performed because of possible "shear degradation," an example of which is shown in Figure 7.8. (The term *shear degradation* is a misnomer. As explained in Section 7.2.3, what actually occurs during sonication is a manifestation of transient elongational flow-induced degradation). Heating some biopolymers at 40 to 50°C for 5 to 10 minutes to increase dissolution rate is permitted. However, solutions of very unstable compounds may have to be kept cool at all times (5°C for best protection) and chromatographed as soon as possible to ensure minimum decomposition.

7.2.2 Other Solvent Effects

Unlike other LC methods, the mobile phase in SEC is not varied to control resolution. Rather, the mobile phase is limited to the solvents that can dissolve the sample macromolecule. If permitted, a solvent of low viscosity at the temperature of separation (to improve mass transfer; see Section 3.4) is preferred, to ensure high column plate number. To maintain high resolution, mobile phases that have boiling points only about 25 to 50°C higher than the column temperature should be used. In such a case, the viscosity of the mobile phase will usually be below 1 cP, and solute diffusion rates will be relatively high. Of course, in the case of difficultly soluble samples, the solvent must be selected primarily to provide sufficient solubility, and viscosity considerations are then secondary.

A particular mobile phase also has to be selected on the basis of its compatibility with the solute detector. Thus, if a differential refractometer is to be used, the refractive index of the mobile phase should be as different as possible from that of the sample. Or the mobile phase must have a much lower absorption than the solute at the wavelength of detection with an ultraviolet (UV) or infrared photometric detector.

An occasional problem in SEC is molecular association of the sample. For example, in some solvents, ionic surfactants form micelles which produce asymmetrical bands and solute retentions that change with sample size. However, by using a more dilute sample solution or a mobile phase that is a better solvent for the sample, this type of association can usually be eliminated.

Controlled sample association can be advantageous for certain separations. For instance, carboxylic acids dimerize in nonbasic solvents, such as benzene, but are solvated with more basic solvents, such as tetrahydrofuran and remain monomeric. The retention of a carboxylic acid varies with its apparent molecular size and decreases in nonbasic solvents. Association between solvent and sample molecules (e.g., certain alcohols plus tetrahydrofuran) can also be used to control certain separations. As shown in Figure 7.4, 1,8-octanediol and 1-octanol coelute in chloroform, but with tetrahydrofuran complete resolution of these solutes is obtained. Presumably, bonding with tetrahydrofuran increases the size of the difunctional alcohol relative to that of the monofunctional, resulting in improved resolution. Changes in retention due to bonding with tetrahydrofuran are usually small. However, by adding a larger associating solvent to the mobile phase, the molecular size of a particular solute may sometimes be artificially increased through association, to carry out an advantageous separation (e.g., association of benzoic acid with aniline in *N,N*-dimethylformamide mobile phase).

The effect of the solvent on the various SEC packings must also be considered. For example, the cross-linked polystyrene gels can tolerate a moderate range of organic solvents, but acetone, alcohols, and other highly polar solvents cannot be used (Section 7.9). Aqueous systems outside a pH range of about 2 to 8 degrade siliceous packings. Strong bases such as NaOH and tetramethylammonium hydroxide should be avoided, but organic amines (e.g., triethylamine) are well tolerated [12]. Salts in

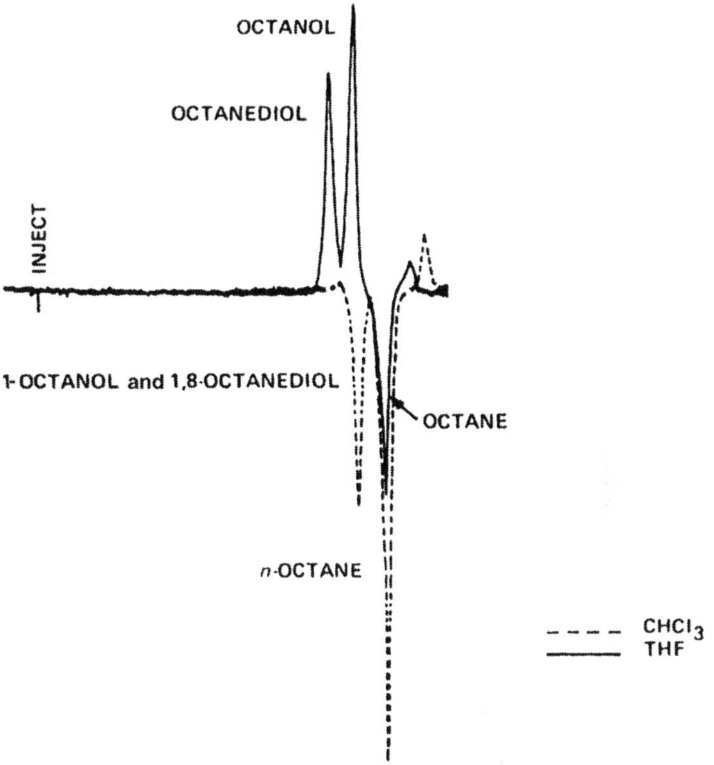

Figure 7.4 Sample/solvent association effects. Column, four 30 × 0.78 cm μ-Styragel 100 Å; mobile phase, chloroform or THF; flow rate, 2.0 mL/min; sample, 1,8-octanediol, 1-octanol, *n*-octane; detector, refractive index. (Reprinted with permission from Waters Associates.)

general appear to hasten the degradation of silica packings, particularly at temperatures above ambient, the effect being a direct function of pH and ionic strength. As discussed in more detail in Sections 6.2 and 7.9, solvents that cause the collapse of organic gels must be avoided. With certain very polar solvents such as dimethylformamide, a salt must often be added to the solvent to reduce solute adsorptive effects and maintain a constant ionic strength. However, large changes in salt concentration can cause an organic gel to collapse. Salt concentration also affects molecular size in some cases. The manufacturer's literature should be consulted for each packing to determine which mobile phases are allowed.

7.2.3 Flow-Rate Effects

Flow-rate level has a significant influence on the efficiency and the resolution of an SEC column. Figure 7.5 shows the effect of mobile-phase velocity on plate height for a monomer and a series of polystyrene standards with a set of porous silica microphere columns. For lower-molar-mass solutes, this set of columns exhibits a relatively small increase in plate height with increased mobile-phase velocity. This

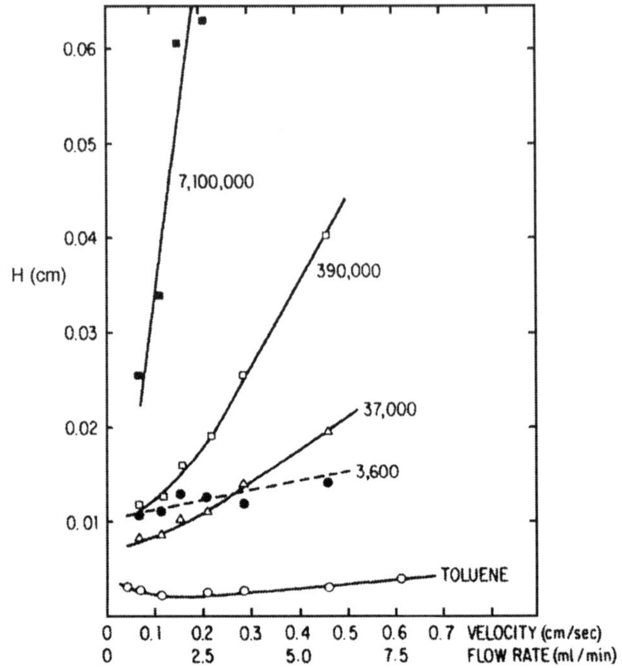

Figure 7.5 Effect of mobile-phase velocity and flow rate on column plate height. Columns, porous silica microspheres: 10 cm, 60 Å; 15 cm, 125 Å, 10 cm, 300 Å, 10 cm, 750 Å; 15 cm, 3500 Å (60 cm total, all columns 0.78 cm i.d.). Mobile phase, THF; temperature, 22°C; sample volume, 25 μL; polystyrene standard solutions. (Reprinted with permission from Ref. 13.)

is characteristic of the relatively rapid solute equilibration associated with small (<10 μm) porous particles. In this case, mobile-phase velocity can be increased substantially without significant sacrifice in the resolving power because of the excellent mass transfer characteristics of smaller molecules. The shape of the plate height versus velocity plot for the totally permeating solute toluene in Figure 7.5 agrees with plate height theory (Section 3.2). The plate height minimum, at a mobile phase velocity of about 0.15 cm/s, represents about 25,000 theoretical plates for small molecules. With this column set, a 97,200-g/mol polystyrene standard exhibited an apparent plate count of 10,000 at the same mobile-phase velocity.

The polymer standards in Figure 7.5 show the anticipated increase in plate height with increasing velocity; however, the less steep slope of the 390,000- and 37,000-g/mol polystyrene plots at low velocities does not agree with theory. The most plausible explanation for this artifact is peak broadening as a result of partial fractionation of these polymer standards by the high-efficiency system. This trend is particularly apparent for the 3600-g/mol polystyrene (dashed line in Figure 7.5), which shows an anomalous, almost flat plate height versus velocity plot, presumably because of significant molar mass fractionation. These data indicate that because of partial fractionation of polymer molecular sizes present in the standards, measured plate height (and

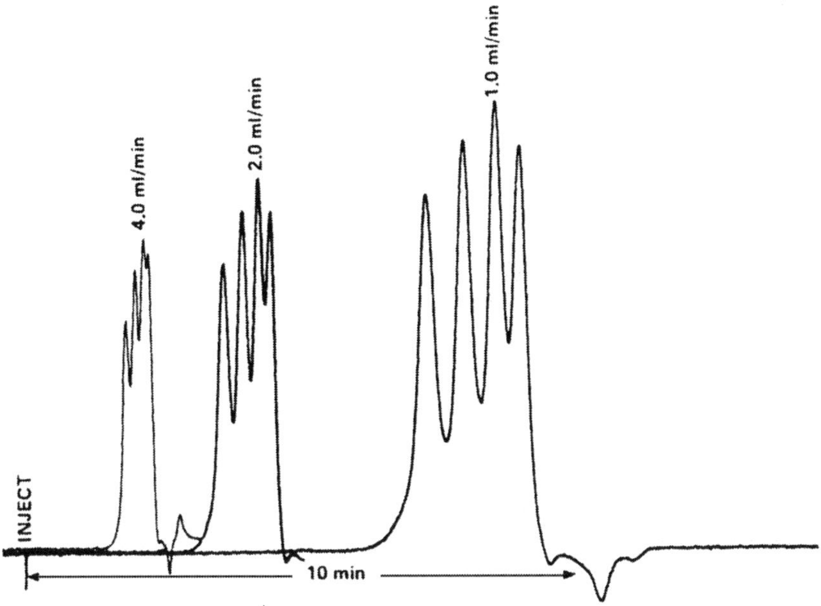

Figure 7.6 Effect of mobile velocity on separation of small molecules. Column, 30 × 0.78 cm μ-Styragel 100 Å; mobile phase, THF; sample, dioctyl phthalate, dibutyl phthalate, diethyl phthalate, and dimethyl phthalate; detector, UV. (Reprinted with permission of Waters Associates.)

σ values) may be significantly larger than the true values for these materials. This effect often causes the apparent resolution of SEC columns to be less than actual.

The practical effect of mobile-phase velocity on column efficiency is illustrated in Figure 7.6. When optimum resolution is required and high separation speed is of secondary importance, SEC columns are normally operated at mobile-phase velocities of 0.2 cm/s or less (measured with a totally excluded solute). Typically, flow rates of less than 2 mL/min are used with a 0.78-cm-i.d. column, and 1 mL/min is often preferred as a compromise between resolution and speed.

Very high molar mass polymers are degraded at high mobile-phase velocities [13,14,38], although the exact mechanism of degradation is not known. For permeating species, chain scission is probably due to a combination of flow-induced degradation in the interstitial medium and at the pore boundary, and caused by steady-state elongational flows. These conclusions were arrived at through two sets of experiments, one comparing degradation strictly in the interstitial medium to degradation in both the interstitial medium and at the pore boundary (intrapore degradation appears extremely unlikely, due to the fact that the solvent within the pore is considered stagnant), the second comparing these results to degradation resulting from exposure to transient elongational flow fields [15]. Both sets of experiments are explained in more detail in what follows.

Figure 7.7 shows the on-column flow-induced degradation of a narrow polydispersity, ultrahigh molar mass polystyrene with $M_p = 1.92 \times 10^6$ g/mol and $M_w/M_n = 1.03$. Figure 7.7a overlays chromatograms of this PS, at various flow rates, analyzed

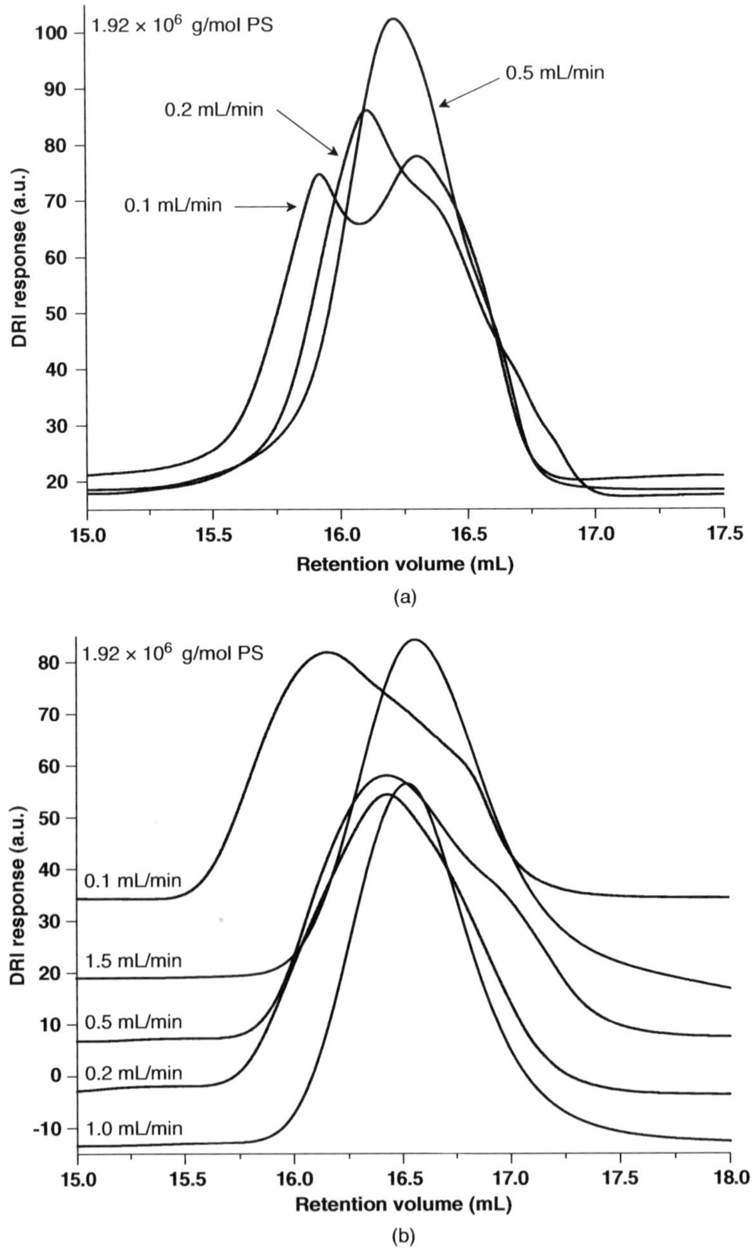

Figure 7.7 On-column flow-induced degradation in SEC: (a) Degradation in the interstitial medium only; (b) degradation in both the interstitial medium and at the pore boundary. Flow rates corresponding to each chromatogram are given in the figures. Sample, PS with $M_p = 1.92 \times 10^6$ g/mol, $M_w/M_n = 1.03$; solvent, THF; temperature, 35°C; detector, DRI; concentration, 0.1 mg/mL. Columns: (a) PLgel 5-μm particle size, 50-Å pore size, exclusion limit 2×10^3 g/mol; (b) PLgel 10-μm particle size, Mixed B, exclusion limit 10×10^6 g/mol. In (b), chromatograms are offset from each other on the *y*-axis for viewing convenience. (Adapted from Ref. 15.)

using a column with an exclusion limit of 2×10^3 g/mol. As the flow rate increased from 0.1 to 0.5 mL/min, the early elution volume (high M) species are seen to degrade, as evidenced by the chromatogram shifting to larger elution volumes. Degradation in Figure 7.7a occurs strictly in the interstitial medium of the column, as the analyte is too large to penetrate the pores of the packing material. (The bimodality observed at 0.1 mL/min is probably due to the finite polydispersity of the sample in combination with a hydrodynamic chromatography mechanism operating in the interstitial medium; see Section 2.6.2.) Figure 7.7b overlays chromatograms of the same PS chromatographed at flow rates varying from 0.1 to 1.5 mL/min, using a column with an exclusion limit of 10×10^6 g/mol. In this case, the analyte samples both the interstitial volume and the pore volume. Again flow-induced degradation of the analyte is seen to occur, although the patterns in the chromatograms are different from those in Figure 7.7a, indicating that a different or, more likely, an additional mechanism of degradation is present at the pore boundaries compared to degradation solely in the interstitial medium.

The second set of experiments dealing with on-column flow-induced degradation has to do with the types of flow fields that degrade polymers in SEC columns. Figure 7.8 overlays chromatograms of the same 1.92×10^6 g/mol PS as shown in

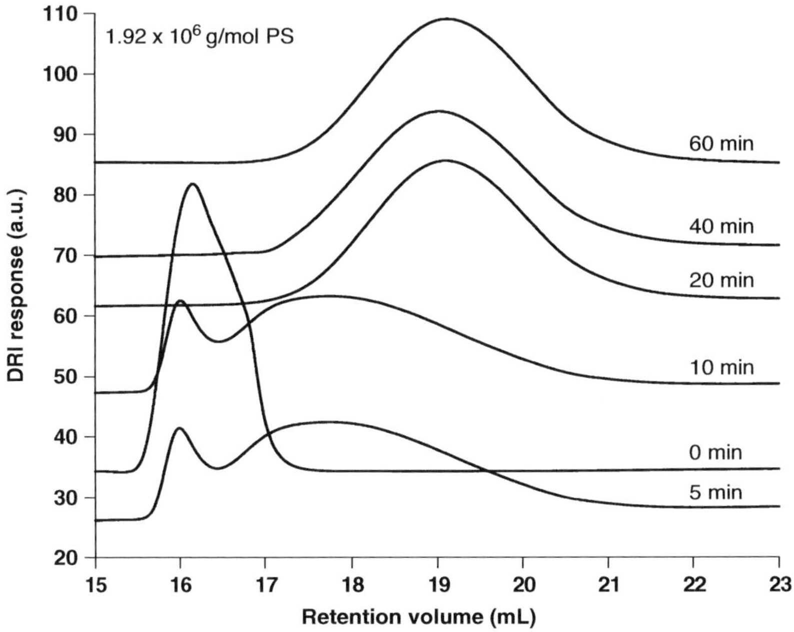

Figure 7.8 Ultrasonic (transient elongational flow) polymer degradation. Analyte and experimental conditions same as in Figure 7.6b. Flow rate, 0.1 mL/min. Times above each chromatogram correspond to times of exposure to ultrasonic irradiation at 47 kHz and 185 W. Chromatograms are offset from each other on the y-axis for viewing convenience. (Adapted from Ref. 15.)

Figure 7.7. This time, solutions of the PS were exposed to ultrasonic irradiation for varying time lengths. The cavitational bubble collapse characteristic of ultrasonication produces elongational flow fields that are termed *fast transient* or *transient elongational.* In transient elongational flow, the velocity time scale of a volume element of fluid is several orders of magnitude greater than the relaxation time of the polymer. In "steady-state" elongation flows, the velocity of a volume element of fluid is of the same order of magnitude as the polymer's relaxation time. Transient elongational flow fields are also produced during turbulent flow, so that from both a fluid mechanics and a kinetics point of view, ultrasonic degradation resembles degradation caused by turbulence. The chromatograms shown in Figure 7.8 were obtained at the same solvent and temperature conditions as those in Figure 7.7, using the same column as that used to obtain Figure 7.7b. The flow rate for Figure 7.8 was 0.1 mL/min, where no or negligible on-column flow-induced degradation was observed (this conclusion was reached by analyzing the sample at flow rates below 0.1 mL/min and noting that these chromatograms appeared identical to chromatograms obtained at 0.1 mL/min). This PS degrades more and more as a function of increasing exposure to ultrasonication, again as shown by the shift in the primary mode of the elution profiles to larger retention volumes. The pattern observed in the overlays in Figure 7.8 is quite different from the patterns observed in Figure 7.7. This difference indicates that on-column flow-induced degradation, either solely in the interstitial medium or in a combination of interstitial medium and pore boundary, is not caused exclusively, or even primarily, by transient elongational flow fields. If it were, the patterns in Figure 7.7 would be expected to resemble those in Figure 7.8. Degradation can also occur during sample filtration, as discussed in Section 7.11.2 and shown in Figure 7.16.

In SEC molar mass calculations based on calibration curves, instrumental flow-rate fluctuations can cause large errors in calculated molar mass averages and molar mass distributions. In high-performance systems, solute retention volumes are relatively small, and errors associated with flow-rate variations can be significant. An accurate, constant flow rate is required not only during a given experiment, but across the entire time span, from the calibration runs to the sample analyses. Toluene and acetone are often used as *flow-rate markers.* Addition of a small amount (a few microliters) of either of these substances usually produces a sharp, distinct peak far away from any analyte peaks. If toluene (or acetone) is added to each of the calibration standards, the retention volume of the toluene peak can be averaged over all injections of all standards. Comparing the retention volume of toluene in an individual injection of a calibration standard to the average value over all injections of all standards, the retention volumes of the polymer standards can be corrected relative to the average. A calibration that is internally self-consistent is thus generated. Addition of toluene (or acetone, as the case may be) to each sample solution then allows for correction of minor flow-rate fluctuations, due to pumping instabilities, and so on, during sample analysis. Most commercial SEC data-processing software packages allow for retention volume corrections using a flow-rate marker. It is important to realize that large (more than a few percent) fluctuations in the flow-rate marker peak are probably indicative of a problem with the solvent delivery system, with the columns, or of a leak somewhere in the system. At this point, flow-rate corrections should be forsaken in favor of troubleshooting the system.

Retention volumes of macromolecules injected at low concentrations in small volumes are essentially independent of flow rate for practical SEC systems, as shown in Figures 2.4 and 2.5 [16,17]. In Figure 2.5, flow rates of 0.1 to 12.5 mL/min were used with 0.85-cm-i.d. columns. The rate of mass transfer associated with permeation into the pores is fast compared to the SEC experiment, so solute distribution between phases is not affected by flow rate (see also Section 2.4). No change in retention volume occurs for either large molecules (polystyrene $M = 411{,}000$ g/mol; diffusion coefficient, $D_s = 22 \times 10^{-8}$ cm^2/s) or small molecules (acetonitrile, $M = 41$ g/mol; $D_s = 1 \times 10^{-5}$ cm^2/s). This effect is observed even with columns packed with larger particles (37 to 42 μm) which have much wider interstitial spaces and deeper pore networks. It appears that most macromolecules have time to equilibrate fully with the column porous network at flow rates normally employed for SEC. The variations in elution volume with changes in flow rate that have been reported [18–20] can be attributed to the apparent shifting of peak maxima because of degradation-induced changes in peak shape at high flow rates. Alternatively, such shifts could be due to non-SEC mechanisms (e.g., hydrodynamic chromatography) than can manifest themselves at high flow rates (see Sections 2.6.3 and 2.6.4).

7.2.4 Temperature Effects

In SEC, temperature is increased to enhance sample solubility or to improve column efficiency by decreasing solvent viscosity. However, for convenience, many SEC separations are carried out at room temperature. The characterization of higher-M polyolefins (and polyamides in some solvents) requires temperatures greater than 100°C, since these materials are difficult to dissolve at lower temperatures (Section 16.2). If the sample is sufficiently stable, SEC using an aqueous mobile-phase can also be employed at higher temperatures (e.g., 50 to 70°C), to improve column efficiency and resolution [21]. Figure 7.9 shows the influence of temperature on the separation of some carbohydrates by SEC. Table 7.6 shows the effect of temperature on the plate number for a series of polystyrene standards with a silica microparticle column.

Although temperature can have a significant effect on column resolution, its influence on the slope and position of the molar mass calibration curve is relatively minor as long as the macromolecule is dissolved in a good solvent (Section 2.4). Figure 7.10 shows the dependence of retention volume on column temperature for polystyrenes chromatographed on a set of Bio-Glas columns with 1,2,4-trichlorobenzene. A slight shift toward smaller retention volumes is observed for this system with increasing temperature.

Although adsorption of the solute on the column packing plays no part in an ideal SEC separation, in actual practice limited adsorption does sometimes occur. Adsorption is more noticeable with lower-molar-mass solutes, because permeation exposes the solutes to a much higher surface area in the packing. Decrease of adsorption with increasing temperature can be expected in some situations with low-molar-mass

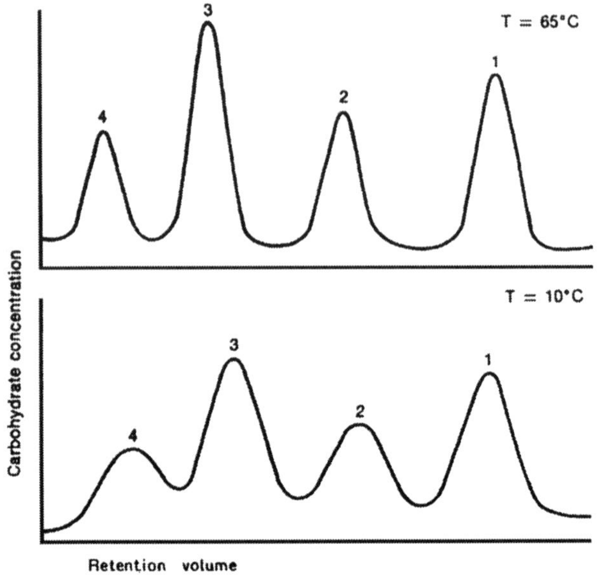

Figure 7.9 Effect of temperature on a SEC separation. Column, 100 × 2.5 cm Bio-Gel P-2; mobile phase, water; sample, 20 μL of a solution containing 320 μg glucose (1), 240 μg lactose (2), 360 μg raffinose (3), and 200 μg stachyose (4); detector, DRI. (Reprinted with permission from Ref. 22.)

materials. Adsorption sometimes occurs when analyzing biomolecules (proteins, peptides, DNA, plasmids, etc.) using silica-based particles in aqueous systems (Chapter 12).

Operation of SEC columns at high temperatures can also have a significant effect on some detector outputs, as discussed in Section 5.2. Therefore, detectors based on the bulk solution properties (e.g., refractive index and viscosity, Chapter 9) must be carefully thermostated to maintain the required baseline stability.

Table 7.6 Effect of temperature on column efficiency[a]

	Plate Number			
M (g/mol)	297 K	323 K	373 K	417 K
411,000 (excluded)	2601	2315	2100	2427
173,000	130	141	230	235
98,000	151	212	235	298
51,000	220	242	329	329
19,000	363	389	460	465
5,000	1126	1163	1214	1238
600	3125	3571	3677	3968
106 (ethylbenzene)	6757	6803	6898	6945

Source: Ref. 23.

[a] Analytes, PS standards; column, 25 × 0.4 cm porous silica 70 to 600 Å, 10 μm; mobile phase, tetrahydrofuran; velocity, 0.1 cm/s; detector, UV at 254 nm.

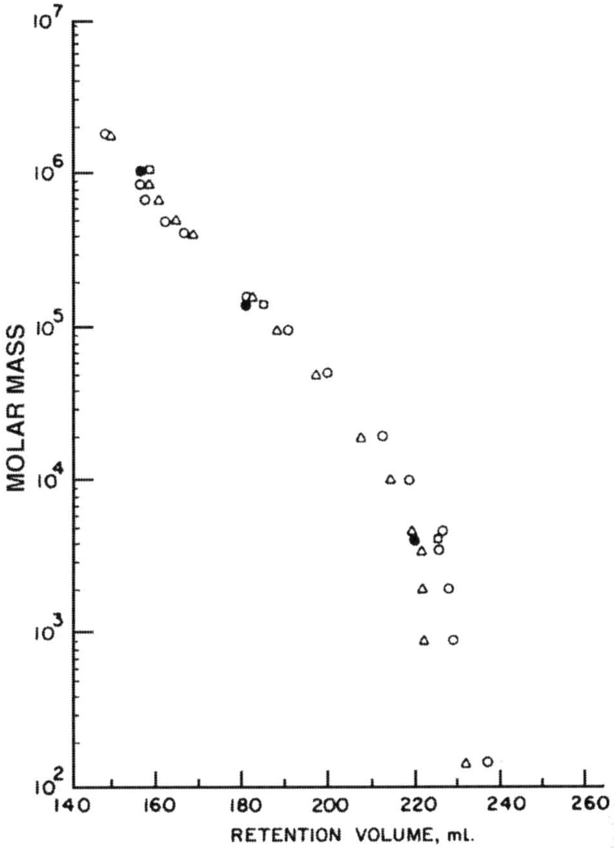

Figure 7.10 Effect of temperature upon SEC retention calibration. Retention volume/molar mass relationships for polystyrene and polyisobutene eluted from treated Bio-Glas columns with 1,2,4-trichlorobenzene: ○, polystyrene, 25°C, △, polystyrenes, 150°C; □, polyisobutenes, 25°C; ●, polyisobutenes, 150°C. (Reprinted with permission from Ref. 24.)

7.3 SUBSTRATE EFFECTS

As discussed in Chapter 2, pore structure of the substrate (the packing material) largely determines retention in SEC. Consequently, the first step in an SEC separation is the selection of an appropriate packing to cover the expected molecular-size range of the sample (Chapter 6). The physical characteristics of the substrate have considerable effect on the operation of the SEC system. The general effect of particle size on column performance was discussed in Section 3.3 and Chapter 6. The plate number advantage of smaller particles is particularly evident at the higher mobile-phase velocities needed for high-speed SEC separations. A narrow particle-size distribution (e.g., not more than ±20% of the average size) is desired to produce high-efficiency columns with the best permeability (Chapter 6). The internal pore volume or the specific porosity of a packing should be as large as possible for optimum resolution and highest molar mass accuracy when using calibration curves (Section 2.5). The method of particle preparation largely determines particle porosity (Section 6.2). As discussed in Section 7.9, pore-volume constancy for the various pore-size packings is particularly important to obtain wide linearity in the molar mass calibration by the bimodal pore-size approach. Pore volume for each pore size used in the column set generally should vary by no more than about 20% and preferably, by less than 10% to allow useful linearity of the molar mass calibration [25,26]. Packing pore-size distribution and pore geometry are also important SEC variables that can significantly influence resolution (Chapter 4).

Improper selection of the packing pore size can lead to false double peaks in SEC. When part of a polymer sample is totally excluded (i.e., has a molar mass or, more accurately, a size greater than the exclusion limit), that part elutes at the total exclusion volume. Therefore, the presence of a sharp polymer band at the total exclusion volume should always be considered suspect in molar mass interpretation. In such cases, the sample should be rechromatographed on a column with a larger exclusion limit if accurate molar mass is required. If the sharp band at the exclusion limit from the first separation is eliminated, the false exclusion peak in the first experiment is identified. However, if the initial band is still apparent in the second separation with the larger exclusion limit but no longer at the total exclusion volume, a true bimodal molar mass distribution exists.

Retention of the solute by mechanisms other than size exclusion greatly complicates the process of obtaining sample molar mass information from SEC chromatograms. Adsorption or "matrix" effects involving a form of partition or adsorption can be superimposed on size exclusion, resulting in excessive retention so that the desired relationship between retention volume and molecular size is not obtained. As discussed in Section 6.2, exhaustive reaction of silica particles with chlorotrimethylsilane reduces the adsorption tendency of silica columns for SEC with organic eluents. In aqueous SEC, the modification of siliceous surfaces is made with hydrophilic groups (e.g., ether or diol) to avoid unwanted interactions between the solute and substrate. In rare instances, an improper combination of solute and mobile phase can also result in excessive retention.

Adsorption or matrix effects can often be minimized by utilizing the most polar mobile phase permitted by sample solubility. For example, water is an effective mobile phase with unmodified siliceous particles for some separations. At $pH > 4$, some acidic SiOH groups on silica are ionized and can function as ion-exchange sites. Aqueous phases containing buffers or salts can be used effectively to eliminate undesired ion-exchange interaction of certain solutes (e.g., proteins) with unmodified siliceous surfaces. Where permitted by sample properties, working at $pH < 3$ often will also eliminate unwanted ion-exchange effects. Highly polar solvents such as hexafluoroisopropanol and dimethyl sulfoxide have proven to be useful in eliminating the adsorption of polar polymers to surfaces of unmodified silica particles.

The mobile phase can affect the surface or pore characteristics of organic gel columns, and these columns are limited to a few solvents (see Sections 6.2 and 7.9). Improper mobile phases can collapse organic gel structures and destroy the

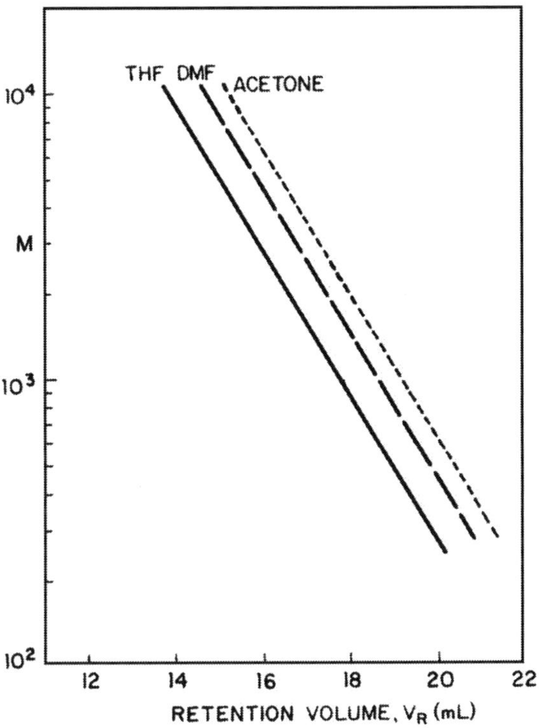

Figure 7.11 Effect of mobile phase on polystyrene molar mass calibration. Column, 100 × 0.62 cm porous silica microspheres, 60 Å (silanized); calibrations in solvents, acetone, dimethylformamide, tetrahydrofuran; flow rate, 1.5 mL/min; sample, 25 μL; detector, UV with THF, DRI with DMF and acetone. (Reprinted with permission from Ref. 27.)

column. On the other hand, columns of rigid particles (e.g., silica) can be used with a wider range of solvents without harm. Different solvents can also cause substantial changes in the calibration curves for organic gel columns because of swelling differences. Conversely, changes in the solvent result in only minor variations in the calibration curves for columns of rigid particles. Figure 7.11 shows the effect of various mobile phases on polystyrene calibration curves obtained on porous silica microsphere columns optimized for oligomeric SEC (Chapter 13). The slopes of the calibration curves are constant, and the small changes in the intercept are probably due to changes in the hydrodynamic volume of the polymer standards with different solvents.

The stability of the substrate can also be influenced by other aspects of the mobile-phase environment. For instance, with silica-based particles the mobile phase should be in the pH range 2 to 8, or particle degradation can occur. With columns of organic gels or organic-modified rigid particles, it is desirable to exclude oxygen from the mobile phase to improve long-term column stability. This is particularly the case for silica-based particles with glycol–ether or ether-bonded organic-modifying films that could form peroxides.

7.4 SAMPLE EFFECTS

7.4.1 Sample Volume

Sample injection variables can lead to significant band broadening in SEC, which ultimately reduces resolution and causes inaccurate molar mass measurements. Extra-column variance caused by the volume of injected sample is accurately predictable, whether syringe or valve sampling is used. For plug sampling, $\frac{1}{12}V_{\text{inj}}^2$ (where V_{inj} is the injection volume) is added to the peak variance [28,29]. Band broadening due to sample injection is independent of flow rate, but it is part of the extra-column effect that differs from the band broadening associated with the detector, tubing, and connectors, whose variances are flow-rate dependent. Band broadening caused by injected sample volume is not separable from the eddy-diffusion term in the chromatographic plate height equation (Section 3.2).

For a constant sample volume, improved performance can be expected for larger-diameter columns; sample injection volume variance σ_{inj}^2 is fixed, but peak volume variance σ_v^2 increases with increasing column diameter. Therefore, for larger-volume columns, column efficiency increases because of the reduced contribution of the sample volume and other extra-column effects to band broadening.

7.4.2 Sample Weight or Concentration

Contrary to what is found with the other LC methods, the retention volumes of polymers have been observed to increase with increased sample concentration. This type of column overloading appears to be due to a change in the effective dimensions (i.e., radius of gyration) of the macromolecular coils with concentration, the effect increasing with polymer molar mass and decreasing as the solvent approaches the theta condition [30]. Column overloading effects are only a function of the weight of sample injected ($V_{\text{inj}} \times$ concentration). To obtain highest accuracy in determining molar masses using calibration curves, the concentration dependence of retention volumes should be minimized by working at the lowest possible constant sample concentration consistent with the required detector signal/noise ratio. When using molar-mass-sensitive detectors such as the static light-scattering photometer or the differential viscometer, accurate calculation of M and/or of parameters such as the intrinsic viscosity ($[\eta]$) of a polymer is dependent on the polymer solution being at near-infinitely dilute concentration. To achieve this, and to avoid column overloading, sample concentrations below c^*, the critical overlap concentration, should be used. For random coil polymers, c^* can be calculated using Equations 3.35 and 3.36. For rodlike polymers, calculating c^* is a bit more involved but can be done using [31]

$$c^* = \frac{M}{L_c^3 N_A} \tag{7.9}$$

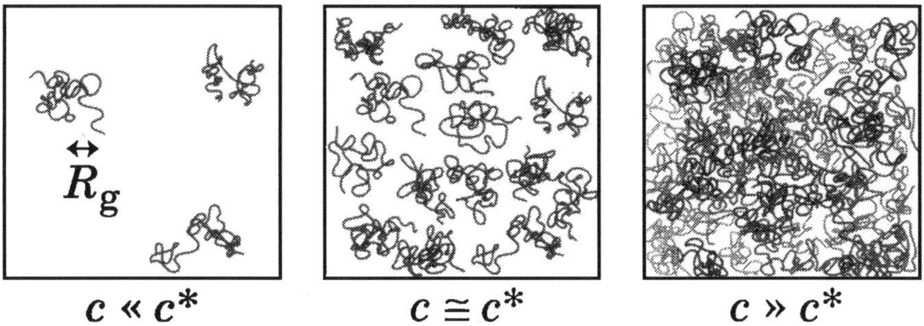

Figure 7.12 Concentration regimes for solution of coils. Concentration regimes for solutions of random coil polymers below, at, and above the critical overlap concentration, c^*. (Reprinted with permission from Ref. 31.)

where L_c is the contour length of the polymer chain (see Section 11.6). The concentration regimes below, at, and above c^* are shown schematically in Figures 7.12 and 7.13 for random coils and rods, repectively.

Sample size in SEC also is sometimes limited by sample viscosity. At high sample concentrations, increased solution viscosity can cause significant band broadening due to viscous streaming or "viscous fingering" on the trailing side of the solute band. Viscous fingering occurs at the interface of two fluids of different viscosities as the fluids move through a porous bed. An example of this effect is shown in Figure 7.14. Extra care should be taken in preparative SEC experiments (Section 15.2), where high sample concentrations are normally used. To prevent problems of this type, a rough guide is that an injected sample solution should have a viscosity no greater than twice that of the mobile phase [32]. For high-molar-mass polymers, concentrations of $\leq 0.1\%$ are often required to eliminate the undesirable effects of concentration on both molecular coil dimensions and sample viscosity.

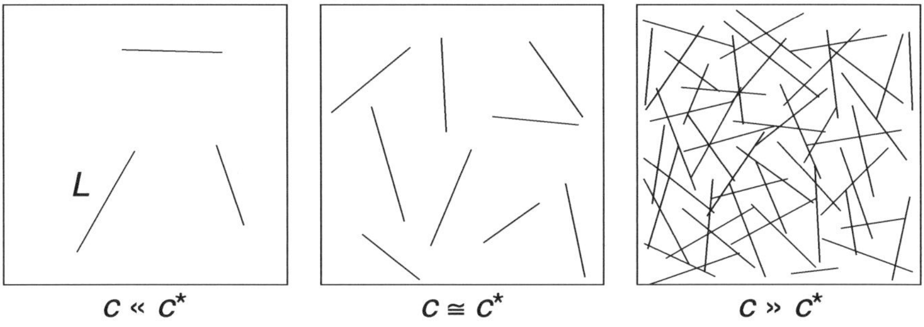

Figure 7.13 Concentration regimes for solution of rods. Concentration regimes for solutions of rodlike polymers below, at, and above the critical overlap concentration, c^*. (Reprinted with permission from Ref. 31.)

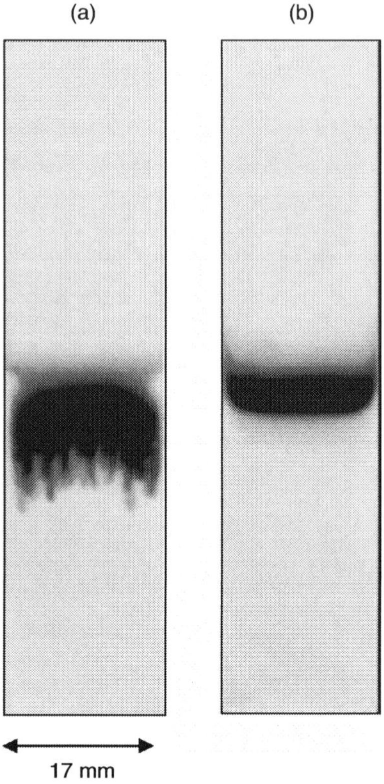

Figure 7.14 Viscous fingering effect: (a) viscous and (b) nonviscous fingering bands. In (a) the difference between sample and solvent viscosities is about 0.48 cP, whereas in (b) the difference is about 0 cP. (Reprinted with permission from Ref. 33.)

7.5 LABORATORY TECHNIQUES

The rest of this chapter provides the reader with a checklist of options and guidelines for optimizing SEC experiments. Besides "how-to" information, typical values for operating parameters are provided as well as experimental suggestions and caveats.

Prior to performing an experiment, the problem objectives must be understood clearly. A qualitative survey scan may be sufficient to establish that there are several components in a sample or that molar mass distribution differences exist between two samples. Such survey scans can probably be made without close attention to detail or to optimizing resolution. On the other hand, if small differences in molar mass distribution must be discerned or absolute molar mass values are required, careful attention must be given to experimental details to optimize resolution and obtain accurate data. It should also be determined if the experiment is to be a single run or a repetitive analysis. Generally, a large number of analyses justify the cost of more sophistication and automation. Preparative SEC experiments require yet another approach (see Section 15.2).

7.6 SOLVENT SELECTION AND PREPARATION

Solvent selection for SEC involves a number of considerations, including convenience, sample type, column packing, operating variables, safety, and purity. These are discussed next in terms of how they affect the SEC process.

7.6.1 Convenience

The most convenient solvent is that already in the instrument. The greatest expenditure of time in preparing SEC experiments often occurs in solvent changeover and in obtaining detector baseline stability with a new solvent. If a solvent change can be avoided, analyses of different sample types can be continued without delay.

7.6.2 Sample Type

For samples of any type, the solvent used for the mobile phase in SEC must satisfy the following criteria:

1. The solvent must dissolve the sample completely; if not, the sample may be partially fractionated by the solvent according to molar mass, crystallinity, or composition.
2. The solvent must permit adequate detection of solute in the eluent: Typically with a differential refractometer, the solvent refractive index (RI) must differ from the sample RI by ± 0.05 units or more. With a UV detector, the solvent should transmit more than 10% of the incident energy at the wavelength chosen.
3. The solvent must not degrade the sample during dissolution and use. If degradation is suspected for polymer solutions, the viscosity can be measured several times over a few hours. A constant viscosity is reasonably good assurance of sample solution stability. This is because large molecules in a sample are, statistically, the ones most susceptible to chemical degradation, and it is these large molecules that will most influence the solution viscosity.
4. The solvent also must not corrode any of the components of the chromatograph.

Information for dissolving synthetic polymers is given in Section 7.2. An excellent reference text for solvent selection is the *Polymer Handbook* [8]. In this reference, the reader can find tables of solvents (including theta solvents) and nonsolvents for polymers, physical properties of polymers and solvents, specific refractive index increments and Mark–Houwink parameters, and tabulated values of second virial coefficients.

7.6.3 Effect on Column Packing

To be effective with organic-gel-type packings (e.g., cross-linked styrene/divinyl benzene-type packings), the solvent (mobile phase) must swell the packing. Thus, aqueous and certain organic solvents cannot be used with these organic packings. Table 7.7 lists, in order of increasing Hildebrand solubility parameter (increasing solvent polarity), the solvents most commonly used with styrene/divinylbenzene (S/DVB) column packings.

Most solvents can be used with silica packings because of the rigid, permanent nature of the particle and pore structure. Aqueous solvents with silica packings should be maintained at pH < 8, because at higher pH silica is slowly dissolved. A pH < 2 should also be avoided, if possible.

The solvent should have strong affinity for the packing to avoid sample partitioning or adsorption that will bias the size-exclusion mechanism. Unwanted solute retention can be recognized if the retention volume (V_R) of any portion of the sample exceeds the total permeating volume obtained for an inert single compound or monomer (see Section 7.12). Another way to check whether or not a non-size-exclusion mechanism is contributing to the separation is to determine K_{SEC} at

Table 7.7 Solvents commonly used with cross-linked S/DVB packings

Solvent	Solubility Parameter, δ [$(\text{cal/cm}^3)^{1/2}$, at 20°C]	Viscosity (cP, at 20°C)	Refractive Index, n_D^{20}	Boiling Point (°C)
Hexane	7.3	0.31	1.3749	68.7
Cyclohexane	8.2	1.0	1.4262	84.2
Toluene	8.9	0.59	1.4969	92.1
Tetrahydrofuran	9.1	0.55	1.4072	66
Ethyl acetate	9.1	0.45	1.3742	88.1
Chloroform	9.3	0.57	1.4458	61.2
Methyl ethyl ketone	9.3	0.43	1.3788	79.6
Dichloromethane	9.7	0.44	1.4241	39.8
Dichloroethane	9.8	0.79	1.4448	83.5
Acetone	9.9	0.36	1.3587	56.3
1,2,4-Trichlorobenzene	10.0	1.89 (25°C), 0.5–0.6 (145°C)	1.5717	213.5
o-Dichlorobenzene	10.0	1.32 (25°C)	1.5514	180.5
m-Cresol	10.2	12.8 (25°C)	1.5398	202.2
o-Chlorophenol	10.2	4.11	1.5473 (40°C)	175.6
Pyridine	10.7	0.95	1.5102	115.2
Dimethylacetamide	10.8	0.97	1.4384	166.1
N-Methylpyrrolidone	11.3	1.67 (25°C)	1.4700	202
Dimethyl sulfoxide	12.0	2.24	1.4783	189.0
Dimethylformamide	12.1	0.92	1.4305	153.0
1,1,1,3,3,3-Hexafluoroisopropanol	12.2	1.65	1.2750	58.2

the experimental temperature (Equation 2.8), then change the temperature (either up or down) and recalculate K_{SEC}. As the size-exclusion process is temperature-independent (see Section 2.4), K_{SEC} should show virtually no change with temperature (at most, a change of a few parts per hundred). If K_{SEC} changes with temperature, this is indicative of partitioning or solute absorption (enthalpic contribution to the separation) [34]. Results from this type of experiment are shown in Table 2.1.

7.6.4 Operation

With organic-gel column packings, the boiling point of the solvent should be 25 to 50°C above the column operating temperature and usually above the maximum temperature used to dissolve the sample. Operating the column close to the mobile-phase boiling point may cause outgassing (bubble formation), which in turn can upset column bed structures and interfere with detection. Outgassing normally does not interfere with the bed structure of well-packed porous silica columns.

The solvent should have a low viscosity (e.g., <1 cP) for maximum separation efficiency and minimum operating pressures. High pressure may lead to collapse or mechanical degradation of organic packings but has little effect on porous silica column packings. Therefore, it is more convenient to work with rigid material columns for some separations.

If the Mark–Houwink calibration concept is to be used (see Section 8.2.3), a solvent should be chosen for which literature values of K and a are known. Values of K and a for many polymer–solvent combinations can be found in Reference 8.

7.6.5 Safety

Generally, all solvents should be treated as dangerous and should be used only in well-ventilated areas. The use of gloves (e.g., Neoprene) and eye protection is highly recommended. Various acidic solvents such as phenols and cresols cause skin burns. *N*-Methylpyrrolidone, *N,N*-dimethylformamide, and dimethyl sulfoxide facilitate the transport of other chemicals through the skin, making them potentially more toxic than expected. Hexafluoroisopropanol is an extremely potent solvent for the cornea of the eye. Other solvents may be carcinogenic; for example, benzene, perchloroethylene, carbon tetrachloride, and chloroform have been implicated and should only be used in chemical hoods. Reference 35 or related publications may be consulted for additional information on solvent toxicity and handling. It is always prudent to review the material safety data sheet for a solvent before initiating work.

7.6.6 Solvent Purification and Modification

Solvents or mobile phases used should be of high purity. Use of HPLC-grade solvents is highly recommended. The reasons for using pure solvents are (1) to avoid suspended particulates that may abrade the solvent pumping system or cause plugging of small-particle columns, (2) to avoid impurities that may generate baseline noise, and (3) to avoid impurities that are concentrated by evaporation in preparative work. Solvents are purified by distillation, degassing, and filtration. At times it may be necessary to add certain oxidation or corrosion inhibitors to the solvent.

Distillation may be required for some solvents to eliminate impurities that cause baseline drift (e.g., with differential refractometers) and unwanted absorption with UV detectors. Distillation is required after solvent purification by silica adsorption because many solvents (or contaminating water) contain small amounts of dissolved silica which interfere in subsequent operations.

Degassing may be required to remove dissolved gases that can nucleate to form bubbles in the detector. Degassing is accomplished by warming the solvent under vacuum or by purging with helium. Ultrasonication will also degas the solvent, although this is a less desirable and temporary solution, as redissolution of air usually occurs.

If HPLC-grade solvents are not used, or if the solvent is not neat (e.g., electrolyte has been added to aid sample dissolution), filtration is essential to protect the high-performance column and pump. If the solvent contains a noticeable number of particles, it should be prefiltered through a coarse, sintered glass funnel. Solvents should be final-filtered under vacuum through a 0.5-μm or smaller membrane filter. Teflon (PTFE) filters are hydrophobic and not recommended for filtering aqueous solutions, but work quite well for most organic solvents. For aqueous solvents, nylon filters, which are hydrophilic, can be used. Filtered solvent should be stored under an inert gas in a container rinsed with the filtered solvent. The equipment should also be protected by a sintered-metal filter (0.5 μm) inserted in line between the solvent reservoir and the pump. If a light-scattering detector is part of the instrumental setup, an additional 0.2-μm membrane filter should also be inserted between the reservoir and the pump.

After obtaining a distilled, degassed, and filtered solvent, it is desirable in some cases to add certain inhibitors. With ethers such as tetrahydrofuran, a peroxide inhibitor such as BHT (butylated hydroxytoluene) should be added, especially for large-volume solvent use and storage. Whether an inhibitor is used or not depends on the purpose of the experiment. Usually, it is not desirable to have an inhibitor present in preparatory work, since it interferes with the characterization of the fractions. Examples where use of inhibitors may be desirable include (1) chloroform, where ethanol normally is used as a photolysis inhibitor; (2) 1,2,4-trichlorobenzene, where Santonox-R or another equivalent antioxidant is added to protect the sample, solvent, and packing from high-temperature oxidation effects; and (3) aqueous buffers, where for long-time use, antimicrobial agents such as 0.02% NaN_3 should often be used. Generally, columns should not be stored in phosphate buffers, as they encourage microbial growth.

7.7 SELECTION AND USE OF STANDARD REFERENCE MATERIALS

Standard reference materials are used both to evaluate system performance and to calibrate column retention in terms of specific molar masses. Usually, pure compounds, monomers, or narrow-molar-mass distribution polymer standards are used for evaluating column performance. These are available from a variety of commercial manufacturers. Certified standards for certain polymers are also available from the National Institute of Standards and Technology. For calculating accurate sample molar masses and molar mass distributions using calibration curves, it is necessary to calibrate SEC retention with known molar mass polymers of the same molecular type (chemistry and architecture) as the sample (Chapter 8). The number of useful standards is very limited relative to the large variety of molecular structures requiring analysis.

If standards of the sample type of interest are not available, one of several alternative approaches may be employed for molar mass calibration. The universal calibration and Mark–Houwink approaches use polystyrene or poly(methyl methacrylate) standards for organic solvents and dextrans, pullulans, sulfonated polystyrenes, or poly(ethylene glycol)s and poly(ethylene oxide)s for aqueous solvents. While often applicable, these calibration methods cannot be used with every system and for a Mark–Houwink calibration (Section 8.2.3), values for the constants K and a are required. In another approach, one or two of the sample materials of interest can be carefully characterized by light scattering and osmometry. The characterized sample then becomes the standard required for the single broad standard calibration method. By using the M_w and M_n values obtained by characterization of a single sample, calibration is possible with the Hamielec, GPCV2, and GPCV3 methods discussed in Chapter 8. A static light-scattering detector can also be used as an *in situ* molar mass detector, eliminating the need for calibration (Section 9.3).

7.8 DETECTOR SELECTION

A large number of detectors are available for use in SEC, in a variety of combinations. General characteristics of physical detectors are described in Chapter 9, whereas chemical detectors and their applications are discussed in Chapter 10. How physical detectors can combine to provide a wealth of information regarding polymer architecture and dilute solution thermodynamics is the subject of Chapter 11.

7.9 COLUMN SELECTION AND HANDLING

The general aspects of columns and column packings are discussed in detail in Chapter 6. Column selection in SEC is based largely on the required molar mass range of separation and the nature of the sample–solvent combination. Other considerations of convenience sometimes influence the selection: Are the needed columns readily available? Must several columns be coupled? Have the columns been in storage for some length of time, and could problems exist because of this?

7.9.1 Optimum Single Pore-Size Separations

The choice of column packing depends on the purpose of the separation. Pore size is normally the most significant parameter to be considered, since it dictates the

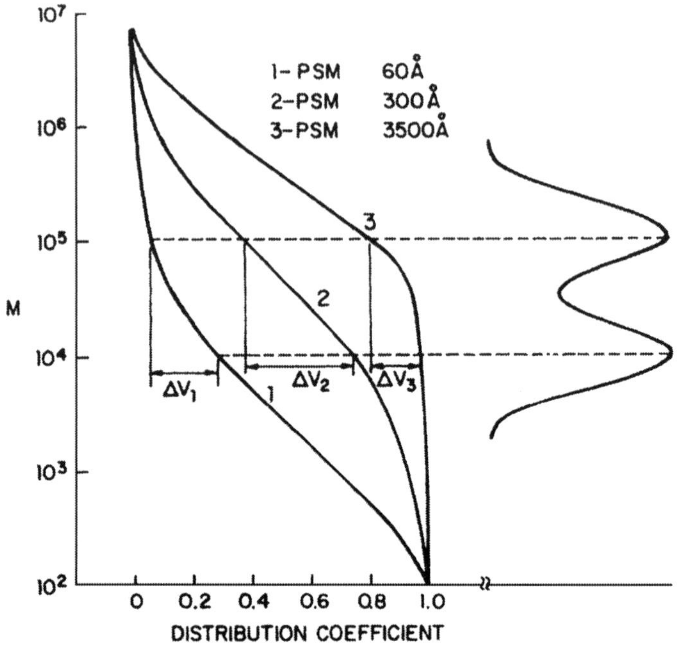

Figure 7.15 Selection of single-pore-size column packing for maximum resolution of two adjacent bands.

range of molar mass separation. Columns of the smallest single pore size (e.g., 40 to 60 Å) are desirable for separating mixtures of small molecules (<5000 g/mol) (Chapter 13). For separating individual components in samples of 5000 to 500,000 g/mol (e.g., proteins), columns of a single intermediate pore size (e.g., 200 to 500 Å) are normally used.

To select the proper pore size for optimizing a given separation, the individual column calibration curves (M versus K_{SEC} or V_R) are first obtained. The chromatographic peaks of two or more of the standards used in developing the calibration curves are then compared. The method is illustrated in Figure 7.15, where the calibration curves are real data for porous silica microsphere (PSM) columns, but the curves for the standards, although typical, are hypothetical. The distribution coefficient K_{SEC} is defined by Equations 2.7 and 2.8. For column packing 1 and the two standards in the example ($M = 10^4$ and 10^5, respectively), the separation is poor (small ΔV_1) and on a nonlinear portion of the calibration curve. The same is true for column packing 3 (nonlinear and small ΔV_3); the separation is greatest (largest ΔV) and on a linear portion of the calibration with column packing 2.

Once the pore size is selected, resolution can be improved by coupling together two or more columns of the same pore size. These individual columns should have very similar internal pore volumes, as discussed below. Figure 7.16 shows the effect of coupling two columns of the same pore size. Notice that while the range of

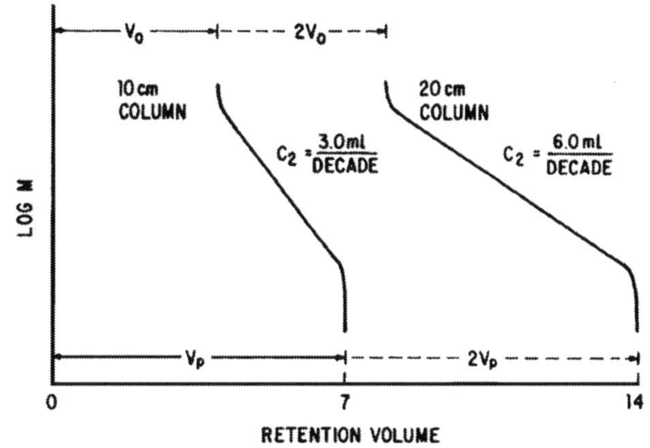

Figure 7.16 Pore volume determines the separation capacity (slope of M calibration curve).

separation remains the same (log M axis), the volume over which the separation is made (V_R axis) is doubled and the slope of the calibration curve C_2 is doubled by doubling the pore volume. The column coupling technique is discussed further in Section 7.10.

7.9.2 Bimodal Pore-Size Separations: Optimum Linearity and Range

For separating sample components that extend over more than two decades of molar mass (e.g., broad-MMD polymers), a set of bimodal pore sizes is optimum. Knowledge of the individual column calibration curves provides a useful guide for choosing which pair of pore sizes to use. For the broadest range of separation and maximum linearity, the individual column calibration curves should be adjacent but nonoverlapping. Also, the pore volumes of each size should be equal for best linearity. Optimized range and linearity yield the most accurate calculated molar mass values and permit the most accurate visual comparison of polymer chromatograms.

To better understand how bimodal coupling is easily accomplished in practice, Figure 7.17 should be consulted. Here, the linear portions of the calibration curves for two types of column packings in individual columns are presented. The vertical axis is the slope of the calibration, C_2, which is directly related to pore volume, and the horizontal axis is the log M range, which is separated by each pore size (in this case, for polystyrene standards in tetrahydrofuran). As mentioned above, maximum calibration linearity and fit are obtained by coupling columns of equal pore volume (equal C_2 values) and adjacent separation ranges. Recall also that pore volumes (C_2 values) are directly additive (Section 4.5). Using these concepts, we can now select a set of columns for wide linear M calibration. For example (Figure 7.17), to obtain a linear range of separation from about 500 to 1,000,000 g/mol with μ-Styragel, two columns of 500 Å (yielding a slope C_2 of $2 \times 1.6 = 3.2$) must be

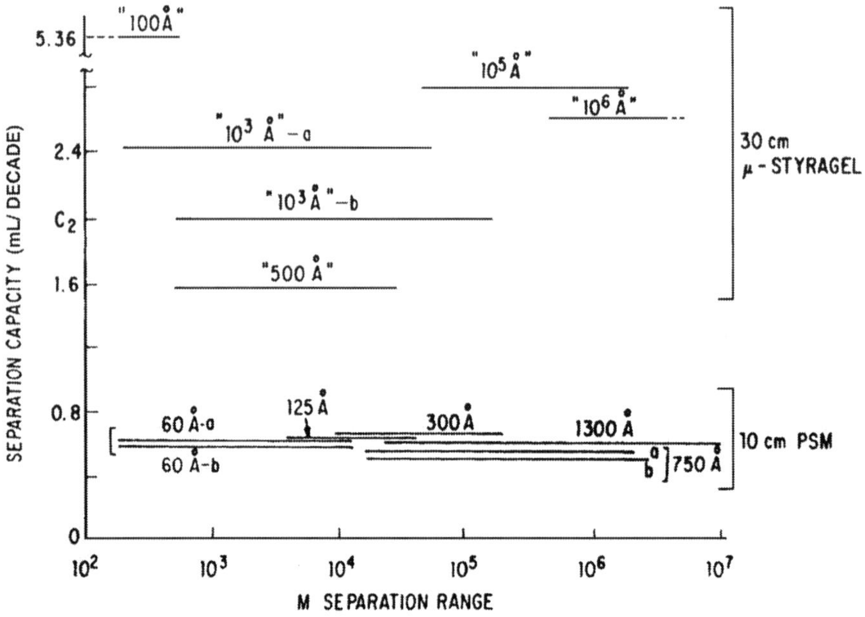

Figure 7.17 Selection of bimodal pore-size column packing for maximum range and linearity. Data for PSM (porous silica microsphere column packing), 10 × 0.78 cm: solvent, THF; temperature, 22°C; flow rate, 2.5 mL/min; detector, UV at 254 nm; sample, 25 μL. Data for μ-Styragel, 30 × 0.78 cm: solvent, THF; temperature, 23°C; flow rate, 1 mL/min; detector, DRI; sample, 100 μL. (Reprinted with permission from Ref. 26.)

coupled with one column of 10^5 Å ($C_2 \simeq 3$). To obtain a linear molar mass separation range of 1000 to 50,000, three columns of 500 Å ($C_2 = 3 \times 1.6 = 4.8$) should be coupled with one column of 100 Å ($C_2 = 5.4$). Although the linearity of the latter system is not perfect, it is the best available with the columns of this set.

Based on the data of Figure 7.17, it can be seen that arranging the porous silica microsphere (PSM) columns for bimodal operation is easier and more accurate because the internal pore volume (measured by C_2) and the separating range are much more nearly uniform for all pore sizes. For example, either 60-Å column (*a* or *b*) could be coupled with either 750-Å column to obtain a linear separation covering more than four decades (200 to 2,000,000 g/mol). A set of two columns each of 60 and 750 Å could conveniently be used to increase the resolution further.

7.9.3 Other Column Selection Guidelines

Sometimes column selection must be based on the solvent used for dissolving and separating the sample, because the solvent and column packing must be compatible. For example, organic-gel packings can only be used with organic solvents which swell them. A list of solvents compatible with styrene/divinyl benzene packings is given in Table 7.7. Rigid silica-based packings modified by grafting organic groups on the surface are available for organic and aqueous solvent use (Section 6.2.2).

An example involving column packing and solvent decisions in SEC is found in the characterization of poly(ethylene oxide). While water is an excellent solvent for this polymer, organic column packings such as styrene/divinyl benzene cannot be used because they are not swollen in water. On the other hand, with silica-based packings the poly(ethylene oxide) adsorbs from water onto the packing surface. Therefore, either a compatible organic solvent must be used with the organic-gel packing, or else the water must be modified with salt to eliminate adsorption on unmodified rigid silica packings. Alternatively, an appropriate, surface-modified porous silica packing may be used with either solvent system.

7.9.4 Column Handling

It is important to know the performance characteristics of new columns (Section 6.4). The separation range and performance of each new column should be checked by calibration under laboratory operating conditions. Particle and pore-size specifications provided by suppliers vary, and equipment variables (e.g., injection valve, connector tubing, and detector) all affect the values reported. Column performance should be tested at typical flow rates and pressures. A flow rate of about 1 mL/min (e.g., about 0.1 cm/s) is normally useful for operation. (As mentioned above, this depends on column dimensions, particle size, and so on.) The pressure at which the column has been packed and the vendor's specifications on flow rate should not be exceeded in the SEC experiment. If the column is found to be plugged during test, it should be returned to the vendor for exchange. High pressure should not be imposed on a column set suddenly, as this may disrupt the packing uniformity and may actually deform organic-gel packings. Especially with columns of organic gels, startup flow rates and pressures should be imposed gradually.

Columns should be stored at constant temperature in a solvent that is inert to the packing. Columns of silica packings are less susceptible to change on storage than are columns of organic-gel packings. Silica-based columns stored with aqueous solvents should be at pH 2 to 8 and include an antimicrobial agent (e.g., 0.02% sodium azide).

7.10 CHROMATOGRAPHIC DESIGN CONSIDERATIONS

Various chromatographic compromises are usually required for a given set of analyses, and these depend on the objectives of the SEC experiment. Efficiency and resolution, flow rate, pressure, time, temperature, column-packing particle size, cost, viscosity, sample concentration, and so on, must all be considered and suitably adjusted to meet the required conditions. For example, the plate number, N, decreases with an increase in flow rate or an increase in particle size. In addition, N increases linearly with L so that maximum efficiency usually is obtained with several small particle-size columns coupled together and used at low mobile-phase velocity.

Table 7.8 Connected columns for high-speed SEC

Column Design	Column Length (cm)	t_R, Toluene (min)	N	σ^2
PSM-50S	10	1.27	2970	5.43×10^{-4}
PSM-300S	15	1.92	3425	10.76×10^{-4}
PSM-800S	10	1.26	5400	2.94×10^{-4}
PSM-1500S	10	1.26	3130	5.07×10^{-4}
PSM-4000S	15	1.92	8145	4.53×10^{-4}
			$\sigma_t^2 = 28.73 \times 10^{-4}$	
			$N_t = 20{,}260$	
			$N_{\text{obs}} = 23{,}890$	

Source: Ref. 10.

Typically, flow rates of about 1 mL/min are used with 5- to 8-mm-i.d. columns packed with 5- to 20-μm particles.

Peak variances for individual columns are additive when the columns are coupled in a set:

$$\sigma_t^2 = \sigma_1^2 + \sigma_2^2 + \cdots + \sigma_n^2 \tag{7.10}$$

where σ_t^2 is the observed variance for the connected column set and σ_1^2, σ_2^2, and σ_n^2 are the variances for the individual columns 1 to n. If the test solute peaks are Gaussian and the retention times, t_R, are approximately equal for each column ($t_{R1} \simeq t_{Rn}$), then

$$N_t = \frac{n^2}{\sum(1/N_i)} \tag{7.11}$$

where N_t is the calculated total plate number, n the number of columns in series, and N_t the plate number for each column. As shown in Table 7.8, when connecting a set of columns of about 7-μm porous silica microspheres, a total plate number of 20,260 was calculated from Equation 7.11 for the test solute, toluene, compared to 23,890 plates actually measured at a flow rate of 2.5 mL/min.

It is important that matched columns always be used to assemble column sets. The variance relationship in Equation 7.10 predicts that if a low-efficiency column is connected to a high-efficiency column, the result is a total column set of poor efficiency. Therefore, a low-plate-number column never should be included with a high-plate-number column set. This includes use of a guard column, which, when included, will greatly reduce the efficiency of the column set. Generally, connecting columns of different diameter or of different packing types should be avoided. In coupling columns, there has been some debate over which column to place first in the series: the one with the large pores or the one with the small pores. No definitive experiments have been published for the bimodal pore systems recommended, although most column manufacturers recommend installing the columns in order of decreasing pore size (i.e., column with largest pore size first, closest to the injector, and column with smallest pore size last, closest to the first detector).

Increasing the column length by coupling increases the mobile-phase backpressure and the time required for analysis. At some point, adding more columns is not practical because the pressure required for useful mobile-phase flow rates will exceed the maximum safe operating pressure of the system (typically, 6000 psi or 400 bar).

Measurement of column efficiency by plate number indicates the extent of deleterious band spreading but does not provide information about the effectiveness of peak separation. The extent of separation of different molecular sizes is provided by the calibration curve (V_R versus molar volume for small molecules, or V_R versus M for macromolecules). Resolution accounts for both the extent of separation and column efficiency and is thus a measure of the useful separation (Section 4.2). Maximum resolution in SEC is usually obtained with a single pore-size column packing, which separates 1.5 to 2 decades of molar mass. If the sample encompasses a larger molar mass range than this, two pore-size distributions should be coupled to obtain maximum accuracy and range (approximately four to five decades of molar mass, Section 7.9).

To determine SEC column resolution, two very narrow MMD polymers, or two compounds of differing molar masses, are injected sequentially into the chromatograph and their peak positions (V_R) and baseline peak widths (W_b) evaluated for determining the resolution via

$$R_{\mathrm{sp}} = \frac{0.58}{D_2 \sigma} \tag{7.12}$$

Derivation of Equation 7.12 and theoretical concepts are covered in Section 4.2, as are typical performance values. If at any time the value of σD_2 becomes excessively large, either the columns have a low plate count (large σ values) or the column set is made up of improper pore sizes (large D_2) or both.

The hardware used for SEC which must be considered in experimental design has been discussed in Chapter 5 and will not be elaborated on further here. The equipment should be made ready for operation and a stable baseline obtained. The need for accurate, constant-flow-rate monitoring, and the effect of flow-rate variation on molar masses calculated from SEC experiments using calibration curves are discussed in Section 7.2.3.

For preparative work (Section 15.2), larger column capacities are needed and it is often more economical to operate with lower-efficiency, lower-cost, larger-diameter particles (30 to 50 μm). In terms of the column packing technology, it takes more sophisticated equipment and procedures to pack the smaller particles (3, 5, and 10 μm) into highly efficient columns than for the large particles. These procedures are discussed in Section 6.3.

7.11 MAKING THE SEPARATION

In this section we discuss preferred procedures for sample preparation, injection, and obtaining the chromatogram.

7.11.1 Dissolving the Sample and Standards

Before dissolution, it must be ascertained that the sample to be analyzed is accurately representative of the whole by grinding and thorough mixing of the larger bulk materials. The sample may also need to be dried (e.g., nylon picks up several weight percent of water at room temperature and 50% relative humidity), and for some synthetic polymers, melt quenching is desirable for facilitating dissolution by reducing the crystallinity.

The sample must be fully dissolved so that it is not fractionated by differential solubility based on molar mass, crystallinity, or composition. To eliminate spurious peaks and to provide optimum baseline stability, the solvent used for dissolving both the calibration standards (if applicable) and the sample should be the same mobile phase that is used for the separation. The dissolving solvent chosen should be based on the various criteria described in Section 7.6.

Most samples can be dissolved at room temperature by gentle agitation with a magnetic stirring bar or laboratory shaker. Ultrasonic devices should *not* be used for dissolution of macromolecules because flow-induced degradation may occur as a result of cavitational bubble collapse (Figures 7.8 and 11.11). Workers using ultrasonic devices to disrupt biological cells should take special note of this caution, because macromolecular materials such as high-molar-mass proteins can be degraded with this procedure. In sample dissolution, care should be taken not to let the sample adhere to the neck of the flask, where it may not be dissolved. For example, volumetric flasks should be filled to only one-third capacity during the initial dissolution step. With some synthetic polymers such as polyacrylonitrile, it is advantageous to add ice-cold solvent to the polymer sample and then to warm gradually with stirring to accomplish dissolution. This permits small pieces of polymer sample to dissolve before they swell and congeal into one large intractable mass. In other cases it is necessary to heat the mixture for complete dissolution. For example, there are no known solvents that will break up the crystalline bond forces of high-molar-mass polyethylene at room temperature.

To inspect the sample solution for complete dissolution, it is sometimes useful to direct a narrow beam of light through the solution flask in a dark box or darkroom to find undissolved or suspended particles. The viscosity of a macromolecular sample solution can be monitored as a function of time to indicate completeness of dissolution (or degradation), and the nuclear magnetic resonance (NMR) technique has also been used, by correlating proton band narrowing with the completeness of dissolution. For guidelines with regard to the total sample mass and concentration to be injected, the reader is referred to Section 7.4 and Table 7.9, which relates concentrations and molar mass for use with S/DVB columns.

Table 7.9 Molar mass versus sample concentration for S/DVB columns

Molar Mass Range	Maximum Sample Concentration (%)
Up to 20,000	0.25
34,000–200,000	0.10
400,000–2,000,000	0.05
2,000,000+	0.01

Source: Ref. 6.

7.11.2 Sample Solution Filtration

Although filtration may be necessary for samples from industrial streams which may contain contaminant particulate matter (e.g., carbon black), most samples do not need to be filtered. This is because filtration may remove or degrade high-molar-mass polymers in the sample. An example of degradation by filtration is seen in Figure 7.18, where a high-M ($M_w = 7.14 \times 10^6$ g/mol) polyacrylamide sample is observed to become increasingly degraded as it passes through filters of decreasing pore size. Evidence of degradation in this figure is given the shift of the polymer peak to larger elution volumes, as smaller samples elute after larger samples of the same polymer.

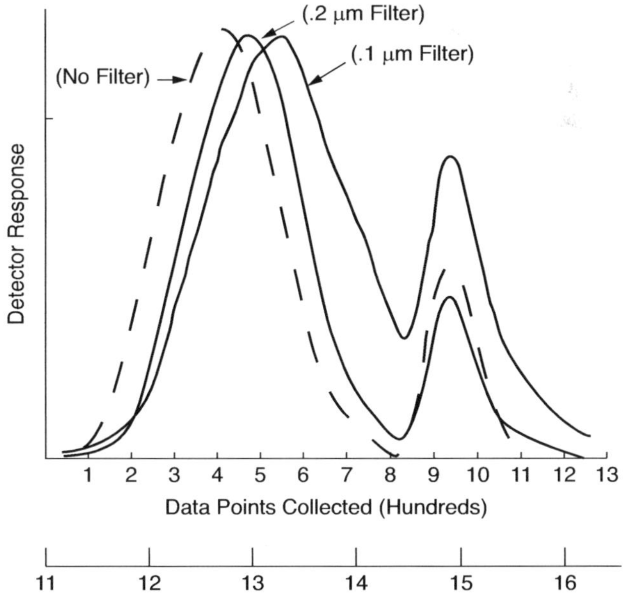

Figure 7.18 Degradation of high-molar-mass ($M_w = 7.14 \times 10^6$ g/mol) polyacrylamide sample after filtration through 0.1- and 0.2-μm filters. (Reprinted with permission from Ref. 36.)

Most samples are filtered for one of two reasons: (1) to improve the precision of the determination by generating a cleaner baseline and thus higher signal-to-noise ratios (S/N) for the analyte peak; or (2) because of a failure to identify the proper solvent–temperature conditions for complete sample dissolution; consequently, high-molar-mass portions of the sample remain undissolved. Erroneously, these portions are usually identified as "gels." With some care (see Sections 7.2 and 7.11.1), the proper conditions for full sample dissolution can be found. If solid contaminants or true gels are present after the main sample is dissolved, the resulting solution should be filtered gently through membrane filters with pore sizes of not less than about 0.4 μm and not greater than 1 μm. As with solvent filtration (Section 7.6.6), hydrophobic Teflon (PTFE) filters are not recommended for filtering aqueous solutions but work well for solutions using most organic solvents. For aqueous solutions, hydrophilic nylon filters can be used. Aluminum oxide membrane filters (e.g., Anotop) can be used with either aqueous or organic solvents. Regarding the desire for obtaining increased S/N through filtration, it should be noted that this is done at the possible expense of experimental accuracy, as shown in Figure 7.18.

With some polymer samples, the solvent type itself can have a striking effect on filtration. Depending on solvation effects, polymer chains may be expanded to a greater or lesser degree (i.e., their hydrodynamic volume may be larger or smaller). The solvent effect is particularly important in the case of elastomers, which frequently contain a small fraction of very high molar mass, soluble molecules as well as insoluble gel. If a good solvent is used (causing more expanded chain conformations), some of the very high molar mass, soluble material may be retained on the filter along with the true gel. On the other hand, if a less solvating but still adequate solvent is used, the polymer molecules uncoil to a lesser degree. Consequently, the larger molecules, which might have been filtered out when using the good solvent, will pass through the filter with the poorer solvent. Thus, the resulting chromatogram is more nearly representative of the total sample.

If any plugging of the filter pores by sample solution is indicated, chromatographic results obtained with this solution are of doubtful validity. All such observations should be considered as part of the SEC analysis. Even when the filtration is proceeding normally, the "edges" of the dissolution vessel should be checked for undissolved material. At this point the operator must decide whether or not to continue the analysis, depending on the degree of final sample dissolution. Again, except in situations where solid particulate matter or true gels are present, sample filtration is best avoided.

7.11.3 Sample Injection

Sample injection in SEC is usually performed using a multiport injector valve. The sample concentration should be low to facilitate filtration (where needed) and to avoid overloading the column. However, the solution must contain sufficient solute for detection, and the total weight of sample injected can be controlled by sample volume. Concentration limits for polymers analyzed on S/DVB columns are given in Table 7.9. In general, injection of 50- to 300-μL solutions of approximately

0.1% (w/v) sample is useful; however, the injection volume should be less than one-third the peak-width volume for a totally permeating peak (Section 7.4). For example, when using a set of two 300 × 7.5 mm columns packed with 5-μm particles to analyze a 0.1% (w/v) solution of a narrow polydispersity linear polymer of $M \approx 1 \times 10^5$ g/mol, an injection volume of 200 μL should produce a peak with good S/N, without column overloading or viscous fingering effects. Results will depend on the detection method used (see Chapters 9 and 10). When using a differential refractometer, for example, a good S/N value necessitates good RI contrast (specific refractive index increment, Section 9.2.1d) between the sample solution and the neat solvent. Preliminary runs may be required to optimize the sample concentrations. Highly accurate sample concentrations are generally not required for a determination of molar mass or molar mass distribution. However, concentration variations of no more than ±0.05% (w/v) should occur in a sample series. For preparative isolations (Section 15.2) it may be desirable to overload the columns and to use larger columns with macroparticle column packings to increase the mass of material that can be injected.

7.11.4 Baseline Stability

A stable detector baseline is required for a successful experiment, but little information has been published to specify the stability needed. For accurate, quantitative results it is necessary to have a flat, linear baseline prior to the emergence of the chromatographic peak and a return to that baseline after the peak has emerged. Therefore, the baseline desired may be specified in terms of acceptable percentage drift over the course of the calibration and sample analysis experiments and in terms of long- and short-term S/N ratios during the course of the experiment. Baselines that exhibit noise and drift of 0.5% or less of the height of the peak of interest will usually provide satisfactory data. Variations caused by temperature fluctuation, air bubbles, leaks, flow variation, solvent inhomogenities, bleed from an old adsorbed sample, and electronic noise may all contribute to baseline instability. The actual construction of the baseline is discussed in the next section.

7.11.5 Obtaining and Using a Chromatogram Baseline

When a macromolecular sample is chromatographed, an elution profile similar to that in Figure 7.19 occurs. Because this chromatogram is used to compute a sample M and MMD, it becomes especially important to accurately specify the start and end points of the curve envelope. In Figure 7.19, retention volumes V_a and V_b correspond to the beginning and end of the polymer chromatographic envelope, respectively. To establish the baseline, a straight line is drawn across the base of the chromatogram, as shown.

The definition of V_a, the low-retention-volume or high-molar-mass end of the chromatogram, is normally straightforward, since the baseline is usually flat and is not influenced by impurities at this point. Defining V_b frequently is more difficult and depends on the separation of the polymer peak from low-molar-mass materials and on the reestablishment of a stable flat baseline following elution of the peak of

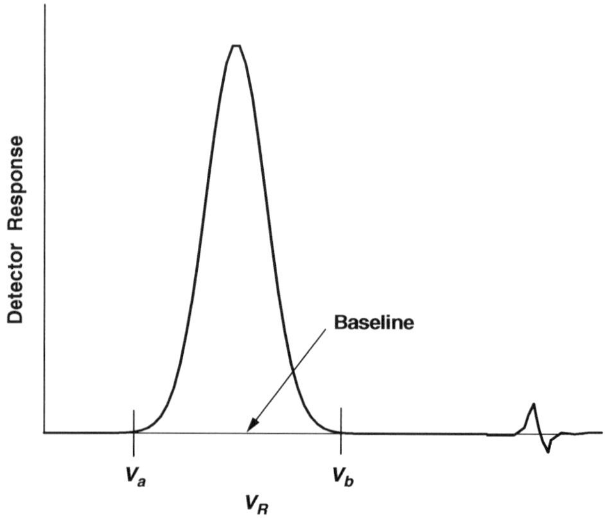

Figure 7.19 Typical chromatogram illustrating a good baseline.

interest. With baseline resolution of all peaks, the choice of V_b is obvious. For example, Figure 7.19 shows adequate separation between polymer and impurity peaks for establishing a good baseline. However, in Figure 7.20, the chromatogram for the main polymeric component of the sample overlaps with oligomer and impurity peaks. Here, there is a serious problem in establishing the correct baseline and the start- and end-point limits for the chromatogram. Sometimes (as in Figure 7.20),

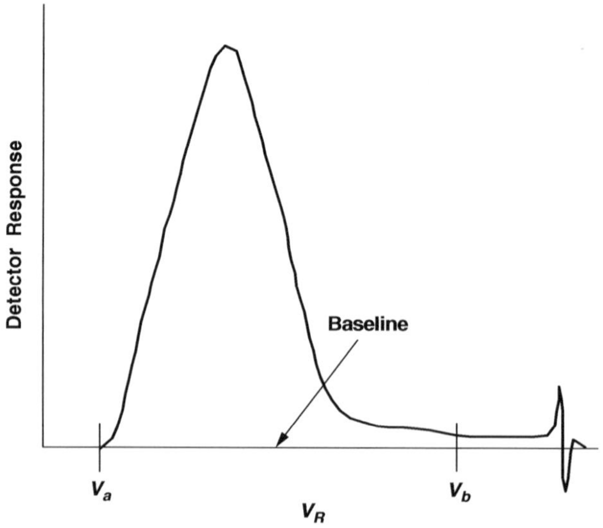

Figure 7.20 Typical chromatogram illustrating a poor baseline.

there is no choice in setting the limits except to be arbitrary in determining those components which adequately represent the sample. If a series of samples is being analyzed, all of which resemble one another with regard to baseline drift or the elution trace not returning to prepeak baseline level, the choice of V_a and V_b should be consistent within the sample set.

7.12 TROUBLESHOOTING

Experimental problems that may be encountered during the process of developing and making a separation and methods for overcoming some of these problems are discussed next.

7.12.1 Excessively High Pressure

If the pressure on the column is too high for acceptable operation (e.g., exceeds pressure specifications), one or more of several correction steps may be employed. If excessive backpressure is caused by high sample viscosity, the solution may be diluted, but detector sensitivity will have to be increased. Raising the temperature will reduce the viscosity of the mobile phase and, if applicable, may help to redissolve precipitated or adsorbed material in the columns or on the end fittings. Pressure can be reduced by removing some of the fractionating columns from the system, if a lower resolution can be tolerated. A sudden increase in column pressure is often indicative of partial plugging of the column inlet fitting assembly. Rigid particle columns that have been well prepared can sometimes be back-flushed to remove clogging particles from the column inlet, but this approach should only be used as a last resort. Back flushing of high-performance gel columns generally should not be attempted. Pressure increases may also be due to clogged inlet or in-line filters (items B and E, respectively, in Figure 5.1). Filter replacement or cleaning (e.g., by sonication or backflushing; see Section 5.3) will resolve the problem.

7.12.2 Column Plugging

If the flow of mobile phase stops completely, it is necessary to locate the source of the plug. Pump failure may be responsible for the stopped flow, and this possibility should be checked first. Plugged connector tubing or filters should be disconnected from the system and replaced. If the plugging has occurred in the columns or the column end fittings, eliminating the problem without loss of column efficiency should be attempted. Slowly raising the temperature of the columns may permit restoration of flow if a precipitated phase is clogging the inlet frit. If it becomes necessary to replace an inlet, the column should be equilibrated to room temperature and atmospheric pressure and mounted in a vise in a vertical position. The fitting should be removed carefully (do *not* touch the column, as the packing may be forced out by thermal expansion) and the end frit cleaned or replaced. The end fitting should be cleaned, checked for obstruction, and refitted onto the column. If some packing is unavoidably lost, additional packing can be added to the top of the column as a thick slurry, using a spatula. Such repairs should be used only as a last resort, and the efficiency of a repaired column should always be checked. Experience has shown that the success rate is low for reestablishing the original plate count of high-performance columns. However, if lower-performance columns can be tolerated, such repairs can be useful. Column repair accessories such as frit removal tools, end fittings, frits, and repair gel are available from several column manufacturers.

7.12.3 Air Bubbles and Leaks

Obvious solvent leaks should be repaired immediately. If air bubbles are observed in the detector, it is necessary to isolate parts of the system to find their source. Suspected fittings should be replaced with new fittings, since leaks may be caused by repeated tightening (i.e., by deformation; air can diffuse into the mobile phase even against high pressure). Small compression fittings are a particular problem in this regard. Bubbles from solvent outgassing are best prevented by degassing prior to use.

7.12.4 Poor Resolution

Poor initial column resolution can be improved by (1) reducing the flow rate of the mobile phase, (2) diluting the sample if the column is overloaded, (3) raising the column temperature, (4) adding column length (i.e., using a longer column or adding more columns), or (5) substituting more efficient columns. If the performance of an initially good column deteriorates significantly and cannot be restored as described above, it should be replaced with a good column. When using a set of several columns, replacement of the first column in the set often resolves the problem, as this is the column most likely to deteriorate first.

7.12.5 Low Solute Recovery

Low solute recovery means that the entire sample has not eluted from the column, usually because of adsorption of the sample on the column. This situation can be especially important with silica-based columns in the characterization of biomacromolecules. Suspect analyses can be checked by collecting and re-chromatographing the total sample. Peak areas should be equal for the original and rerun chromatograms. Alternatively, the sample peak can be collected, characterized by UV absorption or refractive index, and compared to known concentrations of the sample solution. If irreversible column adsorption has occurred, an alternative mobile phase or column packing may be required. Successful elution of the sample from the column may also be attempted by changing the temperature, ionic strength, pH, or solvent polarity.

7.12.6 Constancy of Separation

Periodic comparison of calibration runs is useful for column systems that have been used for a long time. If plate number N is slowly changing, the column packing or packing structure is probably deteriorating. If sudden changes in N are observed, checks should be made for plugging, leaks, pressure changes, and so on. If the changing of solvents and columns is common laboratory practice, one should ensure that the proper columns are in the instrument. The presence or absence of commonly used additives that could affect sample–substrate interactions should also be checked.

7.12.7 Peak Shape

If sudden or unexpected changes occur in sample peak shapes, the experiment should be suspect. Standard samples should be run and the peak shapes compared to those obtained previously. The user should always be on the lookout for effects that may add bias to the size-exclusion mechanism. Such effects can be noted by observing whether the retention volume for a particular solute has increased, or in extreme cases is greater than the retention volume for total permeation, or whether curve shapes occur other than those expected. These effects can often be eliminated by changing polarity, pH, or ionic strength of the solvent, or by changing the packing.

If negative peaks are obtained, the detector polarity can be reversed for visual convenience. If fused peaks are obtained for discrete species, more resolution is needed. As discussed in Chapter 8, it is not possible to obtain distinct peaks for each molecular species present in a high-molar-mass polymer.

REFERENCES

1. (a) H. C. Berg, *Random Walks in Biology*, expanded ed., Princeton University Press, Princeton, NJ, 1993, Chap. 2. (b) E. L. Cussler, *Diffusion: Mass Transfer in Fluid Systems*, 2nd ed., Cambridge University Press, Cambridge, UK, 1997, Chap. 5.
2. F. W. Billmeyer, Jr., *Textbook of Polymer Science*, 3rd ed., Wiley-Interscience, New York, 1984, Chap. 7.
3. H. Morawetz, *Macromolecules in Solution*, 2nd ed., Wiley, New York, 1975.
4. J. Hildebrand and R. Scott, *The Solubility of Non-electrolytes*, 3rd ed., Reinhold, New York, 1949.
5. J. C. Giddings, *Unified Separation Science*, Wiley-Interscience, New York, 1991.
6. D. W. Van Krevelen, *Properties of Polymers*, 3rd ed., Elsevier, Amsterdam, 1990, Chap. 7.
7. E. B. Bagley, T. P. Nelson, and J. M. Scigliano, *J. Paint Technol.*, **43**, 35 (1971).
8. J. Brandrup, E. H. Immergut, and E. A. Grulke, eds., *Polymer Handbook*, 4th ed., Wiley-Interscience, New York, 1999.
9. O. Fuchs, *Fortschr. Chem. Forsch.*, **11**, 74 (1968).
10. K. Krummen and R. W. Frei, *J. Chromatogr.*, **132**, 27 (1977).

11. J. Steinhardt and J. A. Reynolds, *Multiple Equilibria in Proteins*, Academic Press, New York, 1969, Chap. 4.
12. J. Wehrli, *J. Chromatogr.*, **149**, 199 (1978).
13. J. J. Kirkland, *J. Chromatogr.*, **125**, 231 (1976).
14. E. L. Slogowski, L. J. Fetters, and D. McIntyre, *Macromolecules*, **7**, 394 (1974).
15. A. M. Striegel, *J. Liq. Chromatogr. Rel. Technol.*, **31**, 3105 (2008).
16. D. J. Richard and A. M. Striegel, in preparation.
17. J. N. Little, J. L. Waters, K. J. Bombaugh, and W. J. Pauplis, *J. Polym. Sci. A-2*, **7**, 1775 (1969).
18. W. W. Yau, H. L. Suchan, and C. P. Malone, *J. Polym. Sci. A-2*, **6**, 1567 (1968).
19. S. Mori, *J. Appl. Polym. Sci.*, **21**, 1921 (1977).
20. J. H. Aubert and M. Tirrell, *Sep. Sci. Technol.*, **15**, 123 (1980).
21. A. M. Striegel, R. D. Plattner, and J. L. Willett, *Anal. Chem.*, **71**, 978 (1999).
22. G. Trenel, M. John, and H. Dellweg, *FEBS Lett.*, **2**, 74 (1968).
23. K. Unger and R. Kern, *J. Chromatogr.*, **122**, 345 (1976).
24. A. R. Cooper and A. R. Bruzzone, *J. Polym. Sci. A-2*, **11**, 1423 (1973).
25. J. J. Kirkland and W. W. Yau, U.S. patent 4,160,728 (1979).
26. W. W. Yau, C. R. Ginnard, and J. J. Kirkland, *J. Chromatogr.*, **149**, 465 (1978).
27. J. J. Kirkland and P. E. Antle, *J. Chromatogr. Sci.*, **15**, 137 (1977).
28. J. C. Sternberg, in *Advances in Chromatography*, Vol. 2, J. C. Giddings and R. A. Keller, eds., Marcel Dekker, New York, 1966.
29. J. J. Kirkland, W. W. Yau, H. J. Stoklosa, and C. H. Dilks, Jr., *J. Chromatogr. Sci.*, **15**, 303 (1977).
30. T. Bleha, D. Bakos, and D. Berek, *Polymer*, **18**, 897 (1977).
31. I. Teraoka, *Polymer Solutions*, Wiley-Interscience, New York, 2002, Chap. 1.
32. L. R. Snyder and J. J. Kirkland, *Introduction to Modern Liquid Chromatography*, Wiley, New York, 1974, Chap. 10.
33. K. J. Mayfield, R. A. Shalliker, H. J. Catchpoole, A. P. Sweeney, V. Wong, and G. Guiochon, *J. Chromatogr. A*, **1080**, 124 (2005).
34. M. A. Boone, H. Nymeyer, and A. M. Striegel, *Carbohydr. Res.*, **343**, 132 (2008).
35. D. J. De Renzo, *Solvents Safety Handbook*, Noyes Data Corp., Park Ridge, NJ, 1986.
36. M. A. Langhorst, F. W. Stanley, Jr., S. S. Cutié, J. H. Sugarman, L. R. Wilson, D. A. Hoagland, and R. K. Prud'homme, *Anal. Chem.*, **58**, 2242 (1986).
37. I. A. Haidar Ahmad, D. A. Striegel, A. M. Striegel, in preparation.
38. A. M. Striegel, S. L. Isenberg, G. L. Côté, in preparation.

8

CALIBRATION

8.1 INTRODUCTION

At the end of an SEC experiment, polymer molecules of different sizes are separated and their concentrations detected as a function of the retention volume V_R. In SEC, V_R increases with decreasing size of the eluting solute molecule. For SEC of biopolymers, solutes of distinctive sizes can appear as separate peaks in the chromatogram. On the other hand, broad chromatograms are often observed in SEC of most synthetic polymer samples, which tend to have broad molar mass distributions (MMDs). An SEC elution curve can be looked at as a profile of the molecular size distribution of the sample. Since many physical and chemical properties of polymers vary with molecular size, the raw-data sample elution curves are useful for relative sample comparisons. However, relative sample comparison is valid only for data obtained under the same experimental conditions, because the profile of an SEC elution curve is a function not only of the sample molecular-size distribution but also of the specific columns and instrumentation used in the experiment. Only when the elution curves are transformed into MMD curves for the polymer samples can the SEC data from different instruments be compared and treated quantitatively. Unlike elution curves, the MMD of a polymer sample is an intrinsic polymer property that determines the end-use properties of the polymer.

SEC elution curves contain the MMD information of the sample, and the task is to extract this MMD information from the elution curves with accuracy and precision. The objective is to remove from the SEC elution curves the influences that result

Modern Size-Exclusion Liquid Chromatography: Practice of Gel Permeation and Gel Filtration Chromatography, Second Edition
By André M. Striegel, Wallace W. Yau, Joseph J. Kirkland, and Donald D. Bly

from specific features of the particular experiment, but not the features of the intrinsic MMD of the sample. The first step in extracting molar mass information from SEC is to establish a calibration relating the retention volume V_R to the molar mass M of the polymer sample. The molar mass rather than the size of the polymer molecules is used to describe the calibration, because the actual size of polymer molecules in solution can change with temperature and solvent, but the molar mass of the polymer chain is directly proportional to its contour length, a more intrinsic property of the polymer. Like elution curves, SEC calibration curves should be reported in terms of retention volume V_R, not retention time t_R, to minimize the effect of flow-rate changes. In theory (Section 2.1), V_R is a more fundamental parameter than t_R for describing peak retention.

The molar mass calibration in SEC is an experimental approach that is valid only for the particular polymer–solvent system and the experiment in question. To describe the relationship between the polymer molar mass and the particular retention volume, one must have specific knowledge of the experimental conditions. SEC retention is determined by the relative sizes of the solute macromolecules and the sizes of the pores in the column packing. Different pore sizes of column packings can change the extent that macromolecules are excluded from the packing pore structures and thus can also affect retention. Therefore, the $M-V_R$ calibration relationship holds only for specified SEC columns as well as only for specified polymer–solvent systems.

SEC calibration should be repeated often to compensate for column deterioration. While small fluctuations in temperature or flow rate may not upset the calibration, changing columns, solvent, or the nature of the polymer samples will require recalibration of the SEC experiment. Frequently, the calibration curve is described as a property of the column sets. This is not true because the calibration also depends on the polymer–solvent system and the SEC instrumentation, including all external tubing, connectors, and detector volume.

The term *molecular size* is difficult to define quantitatively, as there are many size parameters that can be used to describe molecules or particles of different shapes and conformations. For a rigid spherical particle, the radius or diameter is the obvious size parameter to use. However, even for a simple shape such as a rigid cylindrical rod, there exist two characteristic geometric dimensions, the radius and the length of the rod. Neither of these two dimensions alone can uniquely define the size of a rod. For small organic molecules, the long axis of the molecules may still be used in the first approximation for rough size comparisons of solute molecules. An example of this is the calibration curve shown in Figure 8.1, where the length of the long axis of the molecules measured in angstroms is plotted against the SEC distribution coefficient K_{SEC}. For small molecules, the molar volume or the ratio of molar mass to density has also been used as the SEC calibration dimension [2].

For macromolecules there are conformations ranging from random coils to rigid spheres and rods (Section 2.5). To use a unique dimension to define the size of these molecules is not possible. Therefore, there cannot be a general way of relating molecular size to molar mass for macromolecules. Attempts have been made to use several of the polymeric radii (see Table 9.2) to relate elution behavior in SEC to a particular

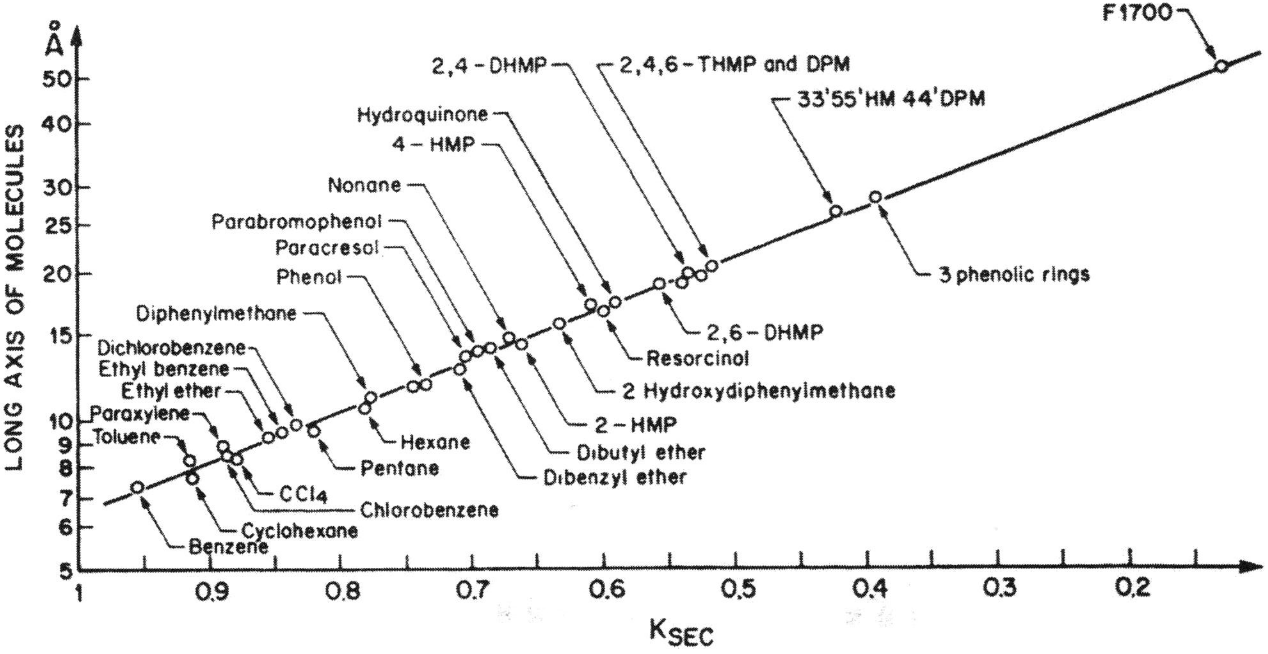

Figure 8.1 Calibration curve for small molecules. (Reprinted with permission from Ref. 1.)

size parameter in order to compare the retention characteristics of macromolecules of different shapes. Depending on the macromolecular architecture (e.g., random coil, rigid rod, star branching), varying levels of success have been found using the radius of gyration (R_G), the viscometric radius (R_η), and the hydrodynamic radius (R_H). No universal relationship between the retention behavior in SEC and any particular polymeric radius has emerged, however.

Since there are many polymer–solvent systems involved in SEC as well as many types of columns of different pore sizes, the best way to obtain accurate calibration is by experiment rather than by theoretical calculations. Errors in the molar mass calibration affect the accuracy in the measurement of molar mass averages and distributions by SEC. These errors are minimized by obtaining the experimental calibration curve under the same conditions as the samples.

In next two sections we describe various ways in which one can calibrate the SEC experiment so that a quantitative transformation of the SEC elution curve into a MMD curve can be made. Nowadays, the actual data transformation is almost always accomplished using commercially available software. Reference 3 provides a recent review of calibration methodologies in SEC.

8.2 CALIBRATION WITH NARROW-MMD STANDARDS

Narrow-MMD standards are available only for relatively few polymers: polystyrene, poly(methyl methacrylate), polyethylene, and poly(ethylene glycol)/poly(ethylene oxide). For these, standards are obtainable with $M_w/M_n < 1.1$ for molar masses extending from the oligomeric region to $>1 \times 10^6$ g/mol. For PS, somewhat broader standards, with a M_w/M_n ratio of 1.2 to 1.3, are available in ultrahigh molar masses of 7 to 20 $\times$ 10^6 g/mol. The polydispersity of other commercially available polyethylene standards is 1.2 to 1.3, of pullulan standards around 1.1 to 1.2, and for poly(acylic acid) the polydispersity of standards is usually in the range 1.3 to 1.8.

As we shall see, peak-position calibration cannot be used for molar mass analysis of polymers in general. The importance of this type of calibration is to check and optimize column resolution and separation conditions. As with peak-position calibration, the universal calibration is valuable for fundamental studies of SEC separation mechanisms and operating variables. The knowledge of these calibration concepts is important for a full appreciation of the fundamentals of SEC calibration.

8.2.1 Peak-Position (Calibrant-Relative) Calibration

Figure 8.2 shows how a calibrant-relative or peak-position calibration is performed and how this calibration is used to obtain the MMD of an unknown sample. In a relative or peak-position calibration procedure a series of usually linear, narrow polydispersity standards of known molar mass and chemistry is analyzed by SEC (Figure 8.2a), and the peak apexes of the chromatograms (which are usually obtained with a concentration-sensitive detector) are assigned molar mass values consistent with the peak-average molar mass (M_p) provided by the manufacturer. A calibration

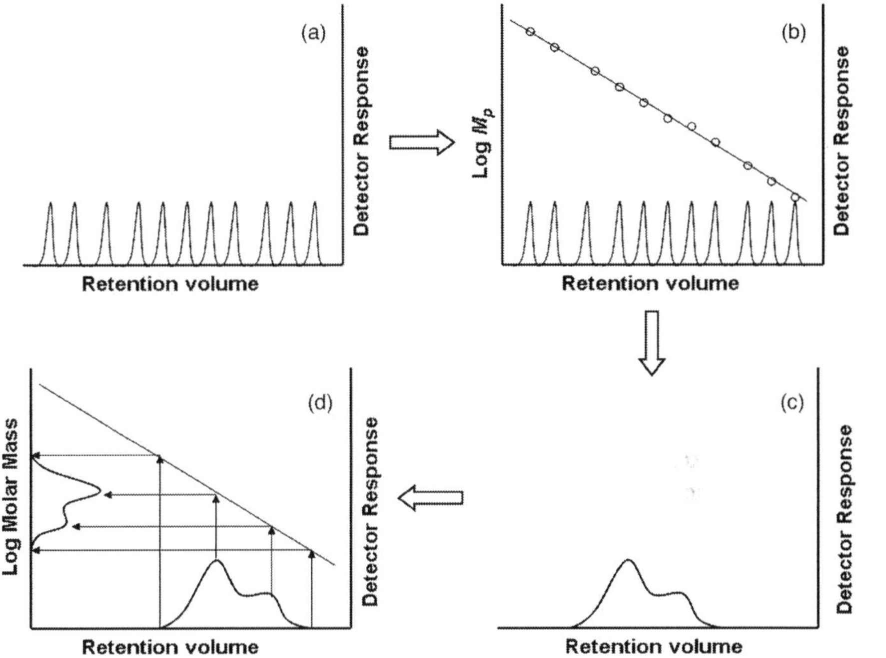

Figure 8.2 Constructing peak-position (calibrant-relative) calibration curve to obtain MMD of unknown sample: (a) narrow polydispersity standards of known molar mass are analyzed by SEC with a concentration-sensitive detector; (b) Peak apexes of chromatograms of standards are assigned molar mass values consistent with the peak-average molar mass, M_p, provided by the manufacturer. A calibration curve is constructed from the relation between retention volume and M_p of the standards. (c) Unknown sample is analyzed on the same system and under the same experimental conditions as were the calibration standards. (d) Elution profile of the unknown is "reflected" off the calibration curve onto the molar mass axis (logarithmic scale), to obtain the MMD of the unknown.

curve is constructed with log M_p as the ordinate and elution time or volume as the abscissa (Figure 8.2b). For extended molar mass ranges, these curves may be of order higher than 1 (e.g., second or third order). Fitting the calibration data to even higher-order (fourth or fifth) polynomials does give better fits (correlation coefficients closer to 1), but generates worse data than when lower-order fits are used. This is due to *overfitting*, in which not only the experimental data but also the noise and experimental error are being fit, thus reducing the predictive ability of the calibration curve [4]. Modern SEC columns of "linear calibration" are available from commercial vendors (Section 6.2). These columns can achieve linearity in SEC calibration by using mixed pore-size packings to provide broad MMD analytical capability. A first-order calibration fit is recommended when using these types of columns.

When an unknown sample is analyzed (Figure 8.2c), its elution profile is "reflected" off the calibration curve onto the molar mass axis: Each elution slice is converted into a molar mass slice. The height of the elution curve at each elution slice provides a measure of the amount of material in each slice. Thus, the combined retention volume and detector response information is used to obtain a molar mass distribution of the unknown (Figure 8.2d). Using Equations 1.1 to 1.3, the molar mass of each slice, M_i, and the height of each slice, h_i, are combined to obtain the number-, weight, and z-average molar masses of the unknown.

For columns with high plate number and low asymmetry ratio (Section 3.5.1), narrow standard calibration using peak maxima is acceptable. Under conditions where significant peak skew is observed in the narrow standards' peaks, more accurate calibration can be obtained by using the first moment of the peak retention volume about the peak center of gravity of the narrow standards to establish the calibration curve.

Relative calibration curves are quite precise, and their accuracy is excellent when the chemistry and architecture of the analyte are identical to those of the calibration standard. Unfortunately, the latter is not usually the case. For example, calibration curves constructed with linear, narrow polydispersity polystyrene standards are used to determine molar mass averages and distributions of both linear and branched analytes with a variety of chemistries. It thus becomes necessary to state the solvent and temperature conditions of the experiments, as well as the column set and method of detection, in order for results to be reproducible from lab to lab or instrument to instrument. "Calibrant-relative" molar mass averages are used to give some level of "quantitative feel" to the data, but the lack of accuracy of these calibrations should be a concern. An example of the error resulting from this type of calibration is given in Table 8.1, where results from SEC analysis using a PS-relative calibration curve are compared to those using SEC with online multiangle light-scattering (MALS) detection (Section 9.3.1) [5]. Large differences in M_w are observed for chemistries different from those of the calibrants (PMMA, three different PVAc samples). Moreover, when the chemistry of the analyte is identical to that of the calibrant but the

Table 8.1 Difference between M_w values obtained by SEC using a PS-relative calibration curve versus using an online MALS detector

Sample	SEC/PS-Relative[a]	SEC/MALS
PMMA	74,000	87,000
PVAc 1	220,000	367,000
PVAc 2	348,000	514,000
PVAc 3	504,000	525,000
PS 3-arm star	178,000	249,000
PS 8-arm star 1	58,000	77,000
PS 8-arm star 2	221,000	366,000

Source: Ref. 5.

[a] PMMA, poly(methyl methacrylate); PVAc, poly(vinyl acetate). Values obtained using a set of three PLgel 5-μm Mixed C columns operating in THF at 35°C and applying a third-order calibration curve. DRI vacuum wavelength of operation: 940 nm. Calibrants: narrow polydispersity linear PS standards in the range 162 to 1,130,000 g/mol.

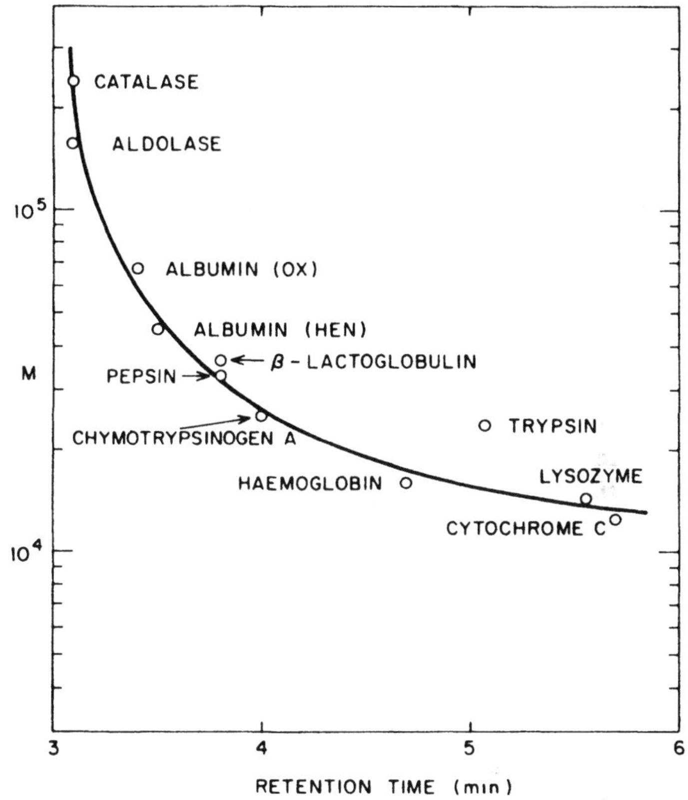

Figure 8.3 Protein peak-position calibration curve. (Data reproduced with permission from Ref. 6.)

architecture is not, as is the case with the three-arm star and the two eight-arm star polystyrenes, the M_w values obtained using a relative calibration curve also differ greatly from those obtained via light scattering.

Only for biopolymers are there truly monodisperse polymers, where all molecules in the sample are of the same molar mass. In fact, calibration curves in aqueous SEC are commonly obtained by the peak-position calibration method. An example of this type of peak-position calibration curve is shown in Figure 8.3 for a series of proteins of different molar mass. The accuracy of this calibration depends on whether the proteins all have a similar conformation and chemical structure, so that molecular size and molar mass are uniquely related.

As discussed in Chapter 1, a major application of SEC is to study the MMD of synthetic polymers. Before other, more practical calibration methods became available, SEC users learned about column characteristics only through the peak-position calibration curve, which was usually constructed using polystyrene standards. Therefore, a common practice was developed in interpreting the polymer MMD based on equivalent polystyrene molar mass units by means of the polystyrene calibration curve. However, for an unknown polymer, the molar mass values calculated by using the polystyrene calibration curve are just an arbitrary numerical ranking of sample molar mass. The absolute molar mass values of the unknown polymers can be quite different.

The problems associated with the equivalent polystyrene molar mass approach led to the development of more accurate calibration methods. The universal calibration curve concept discussed next evolved as a rigorous way to transform the polystyrene peak position calibration curve to a suitable calibration curve for characterizing many types of polymers.

8.2.2 Universal Calibration

The universal calibration method introduced by Grubisic et al. [7] utilizes the concept of the hydrodynamic volume of polymer molecules. From Einstein's viscosity law, the product $[\eta]M$ of a polymer chain in solution is directly proportional to the hydrodynamic volume V_h of an equivalent sphere [9]:

$$[\eta] \propto \frac{V_h}{M} \tag{8.1}$$

As such, the hydrodynamic volume can be expressed in terms of the product of the molar mass M and the intrinsic viscosity $[\eta]$ of the polymer sample (see Section 9.5.3 for a definition of $[\eta]$). In general, SEC calibration curves for polymers of different types merge into a single plot when the calibration data are plotted as log $[\eta]M$, as illustrated in Figure 8.4, instead of on the usual log M scale. Intrinsic viscosity $[\eta]$ is an experimental quantity derived from the measured viscosity of the polymer solution. The determination of $[\eta]$ when both a viscometer and a concentration-sensitive detector are used in SEC is described in Section 9.5.

As we will see in Chapter 9, there are a number of different macromolecular radii (see Table 9.2). Which of these adequately represents the radius of the hydrodynamic sphere occupied by a polymer in solution has been the subject of some debate in the literature. According to statistical theories of polymer solutions [8,9], V_h (or $[\eta]M$) is expected to be proportional to R_G to the third power, making the universal calibration concept consistent with Cassasa's SEC theory based on solute R_G considerations. Some authors have opted to use the viscometric radius R_η or the hydrodynamic (Stokes) radius R_H as the radius of the hydrodynamic sphere. Considerations using R_η have not met with great success [10]. More recently, calibration curves using R_H were found to be superior to those using R_G and comparable to results obtained using $[\eta]M$ [11]. The R_H results, however, did not apply to spheres and rigid rods.

The universal calibration concept has proven to be very useful: An SEC column set can be calibrated with a series of well-characterized narrow standards of a particular chemistry and architecture. Using this calibration curve, absolute (i.e., calibrant-independent) molar masses can be obtained for analytes of chemistry and/or architecture different from those of the calibrants. For example, universal calibration curves

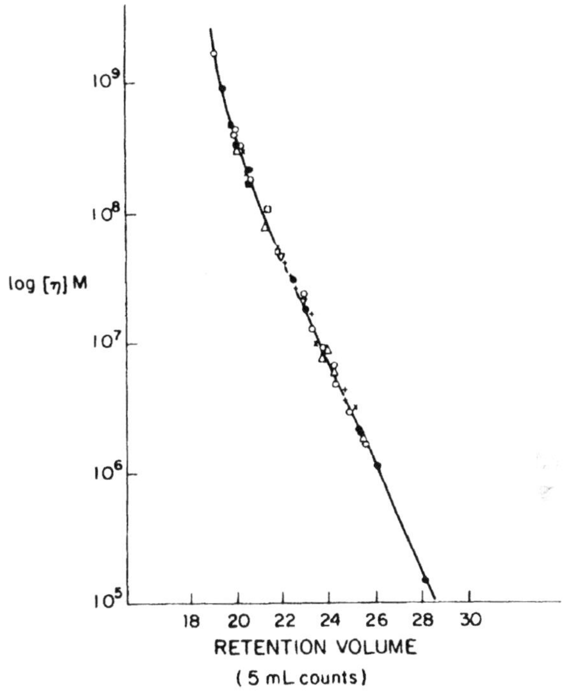

Figure 8.4 Universal calibration plot. Solvent, THF. ●, PS; ○, PS (comb.); +, PS (star); △, heterograft copolymer; ×, poly(methacrylate); ⬡, poly(vinyl chloride); ▽, graft copolymer PS/PMMA; ■, poly(phenyl siloxane); □, polybutadiene. (Reprinted with permission from Ref. 7.)

constructed with narrow polydispersity linear polystyrene standards have been used to obtain molar mass averages and distributions of linear and branched polysaccharides [12] and polyolefins [13]. Although exceptions to the universal calibration concept have been uncovered over the years [10,14], this type of calibration has proven to be quite robust: Universal calibration has been shown to be valid even in such architecturally "extreme" cases as dendrimers [15] and rodlike polymers [16].

A universal calibration curve is valid for a particular set of columns, in a particular solvent, at a particular temperature. If any of these parameters is changed, a new calibration must be performed. Universal calibration results have also been found to be extremely sensitive to measurement of interdetector delays (or to measurement of eluent split ratios if the viscometer and concentration-sensitive detector are connected in parallel rather than in series) and to determination of band-broadening parameters such as peak variance and skew. With the advent of online static light-scattering detectors, which provide absolute molar masses without the need to construct calibration curves of any kind (see Section 9.3), the use of universal calibration curves has begun to fade in analysis of homopolymers. These curves remain, however, an excellent alternative for analyzing copolymers and polymers of unknown chemical composition and branching structure.

Because $[\eta]$ is obtained from the ratio of the signals of the viscosity detector and the concentration-sensitive detector, universal calibration does not require a priori knowledge of the Mark–Houwink parameters K and a, relating intrinsic viscosity to molar mass according to Equation 8.2, of either the calibration standards or the analyte. This differentiates the universal calibration approach from Mark–Houwink calibrations, discussed next.

8.2.3 Mark–Houwink Calibration

The value of $[\eta]$ for a linear polymer in a specific solvent at a specific temperature is related to the polymer molar mass through the empirical Mark–Houwink equation,

$$[\eta] = KM^a \tag{8.2}$$

where values for the Mark–Houwink constants K and a vary with polymer type (chemistry and architecture), solvent, and temperature. They are tabulated for a wide variety of polymer–solvent–temperature combinations in Reference 17.

In the absence of an online viscometer, calibration curves of $[\eta]M$ (plotted on a logarithmic scale) versus retention volume can be constructed using narrow polydispersity standards of accurately determined M and of known K and a for the experimental conditions given. If the Mark–Houwink constants of the analyte are also known, absolute values of M can be determined for the analyte using the calibration curve. Letting the subscript 1 denote values for the standards and the subscript 2 denote values for the analyte, the molar mass of the analyte can be determined according to

$$[\eta]_1 M_1 = [\eta]_2 M_2 \tag{8.3a}$$

Thus

$$K_1 M_1^{1+a_1} = K_2 M_2^{1+a_2} \tag{8.3b}$$

Solving for M_2 gives

$$\log M_2 = \frac{1}{1+a_2}\left(\log \frac{K_1}{K_2}\right) + \frac{1+a_1}{1+a_2} \log M_1 \tag{8.3c}$$

Mark–Houwink constants for the standards and analyte can either be determined experimentally using a capillary viscometer (e.g., of the Ubbelhode or Ostwald–Fenske types) or may be sought in the literature. As with universal calibration, the analyte and standards need not have the same chemistry or architecture.

The main and very important differences between universal calibration and Mark–Houwink calibration are these:

1. An online viscometer is not needed to construct and apply a Mark–Houwink calibration, but it is needed for universal calibration. This reduces the hardware and software costs, analysis time, data processing, and training needed for the Mark–Houwink calibration compared to the universal calibration.
2. No a priori knowledge of Mark–Houwink constants of either standards or analyte is needed to construct and apply a universal calibration. Values of these constants can actually be obtained from applying a universal calibration curve to the analysis of a given analyte.
3. No assumptions about macromolecular architecture are made when performing a universal calibration. This is not true of Mark–Houwink calibrations.
4. The Mark–Houwink calibration is like conventional peak position calibration in that it does not account for the possibility of polymer branching. On the other hand, the universal calibration approach gives the true MMD, regardless of branching status.

Because great uncertainty exists in the published values of Mark–Houwink constants for most polymers, these values are best used for qualitative and semiquantitative comparisons, not for calculations where accuracy is important. Even if K and a values are known accurately for a given set of standards and for an analyte of a given chemistry, the architecture of the particular sample being analyzed may be different from that of the samples used to determine the Mark–Houwink constants. Let us say that the molar mass of a PMMA sample is desired. Utilizing an SEC system operating in THF at 35°C and with a single, concentration-sensitive detector, a calibration curve is constructed using narrow polydispersity PS standards of accurately characterized M, for which K and a have been determined at the given solvent–temperature conditions. The values of K and a for linear PMMA in THF at 35°C are found in the literature. Using the M values of the PS standards and the K and a values of PS and PMMA, a value for the molar mass of the PMMA sample is obtained by applying Equation 8.3c. This value will be correct only if the analyte is, indeed, PMMA and if the analyte is, indeed, linear. The former is easier to determine (e.g., using IR or NMR) than the latter. If the analyte turns out to be a branched PMMA, its Mark–Houwink constants K and a will be different from those found in the literature for linear PMMA, and the value of M calculated for this sample will be in error. Had an online viscometer been part of the SEC setup, this difference in the values of K and a between linear and branched PMMA would not have mattered: The intrinsic viscosity of the branched sample would have been measured directly (as would have been the intrinsic viscosities of the standards); that is, it would not have been calculated based on literature values of K and a and assumptions regarding the architecture of the analyte. Moreover, the K and a values of the particular PMMA sample (or of any given analyte) would have been determined as part of applying the universal calibration curve.

In most cases, Mark–Houwink calibrations are little better than peak-position, calibrant-relative calibrations. Mark–Houwink calibrations may actually be worse than their calibrant-relative counterparts, as Mark–Houwink calibrations provide a false sense of accuracy, through the generally unwarranted belief that the universal calibration principle has been applied. Caution should be exercised not only when applying Mark–Houwink calibrations but also when reading references that claim that molar mass averages and distributions have been determined using universal calibration. In many of these cases, it is not a universal calibration but a Mark–Houwink calibration that has been used and the accuracy of the results should be considered suspect if the sample is believed to be branched or its chemical identity is suspect.

8.3 CALIBRATION WITH BROAD-MMD STANDARDS

There are two different ways of using broad-MMD polymer standards for SEC calibration. The integral-MMD method utilizes the complete MMD curve of the polymer standard. The linear calibration methods use only the average M values of the polymer standard but assume a linear approximation of the SEC calibration curve. Both approaches are valid and useful at times, depending on SEC conditions. The linear calibration methods are more versatile for analyzing polymers of different types. Polymer standards are more readily available for these methods, because the average M values are more readily attainable than the complete polymer MMD.

The accuracy of the broad-standards calibration methods varies depending on the accuracy of the available MMD information on the standards and the accuracy of the linear-calibration approximation. When the separation columns are purposely selected to assure linearity in calibration (Section 8.6), the use of linear calibration methods is definitely recommended over integral-MMD methods.

Unlike those from narrow-standards methods, the calibration curves obtained by broad-standards methods are affected by instrumental peak broadening in the SEC experiment. Without corrections, this calibration error can cause errors in the molar mass analyses of polymer samples. Proper account of the effect of peak broadening on calibration is provided in GPCV2 and GPCV3 calibration methods. Recent developments in extending the GPCV2 approach to correct for the band-broadening effect over the entire MMD curve are discussed in Section 8.7.

8.3.1 Integral-MMD Method

The integral-MMD calibration method requires that the complete MMD of the broad polymer standard be known. For a known polymer MMD curve, there is a unique correspondence between molar mass and the weight fraction of the polymer below a given molar mass. Similarly, there is a unique correspondence between the retention volume V_R and the weight fraction or the fractional area under the observed SEC elution curve. The SEC calibration curve for the integral MMD method is obtained by matching those M and V_R values which correspond to the same value of sample weight fraction on the MMD and SEC elution curves.

Initially, complete MMD information for the broad polymer standards required for this calibration approach was obtained experimentally. A broad-MMD polymer sample intended for a calibration standard was fractionated by solvent extraction or column fractionation and the molar mass determined for each fraction. The MMD information obtained in this way for the whole polymer, now the calibration standard, was then used to establish the SEC calibration curve. This experimental polymer fractionation approach has not generally been followed in practice because of the tedious nature and the questionable molar mass precision of the experimental fractionation and the molar masses of the characterized fractions.

A practical alternative approach makes use of theoretical polymer MMD and average molar mass values to provide the needed MMD information for the integral-MMD calibration [18]. This approach has been applied to both water-soluble polymers [19] and organic-soluble polymers (e.g., nylon 6,6) [20]. We will now use the illustration in Reference 20 to explain the integral-MMD calibration method in more detail.

Many polymers follow predictable MMD curve shapes which depend on the type and condition of polymerization. For example, anionic polymerizations often follow a Poisson distribution; certain vinyl polymerizations (at low conversion and with termination via radical coupling) yield a Shultz–Zimm distribution (Figure 1.9); and condensation polymers give a Flory most probable distribution if prepared under equilibrium conditions (Chapter 1 and Reference 8). The Flory MMD, accepted as the idealized MMD function for nylon 6,6 polymers and for polyamides and polyesters in general, is now utilized to illustrate the use of the integral-MMD method. Consider the most probable distribution function,

$$W_X = (1 - p)^2(X)(p^{X-1}) \tag{8.4}$$

where W_X is the weight fraction of polymer with X repeat units of molar mass M_0 and p is the extent of reaction with $p = (M_n - M_0)/M_n$. As the value for p approaches unity (as is usually the case for high-molar-mass polymers), the value for polydispersity M_w/M_n approaches 2 for polymers of this theoretical MMD. Equation 8.4 is plotted in Figure 8.5, where the total area under the curve is unity. With a known M_n or M_w value and the M_0 value for the repeat-unit, the complete MMD of a particular sample of a condensation polymer can be predicted according to

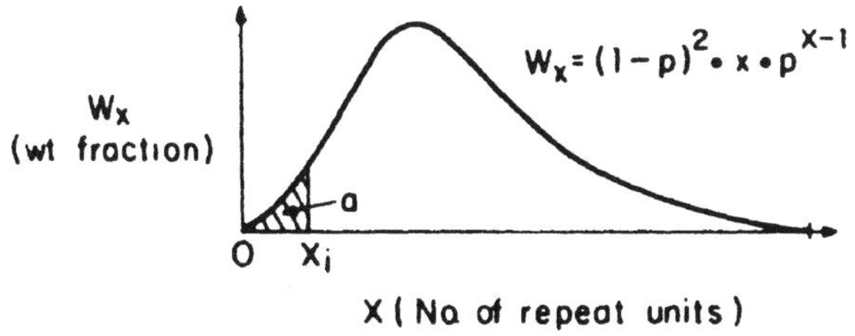

Figure 8.5 Flory most probable distribution function. (Reprinted with permission from Ref. 20.)

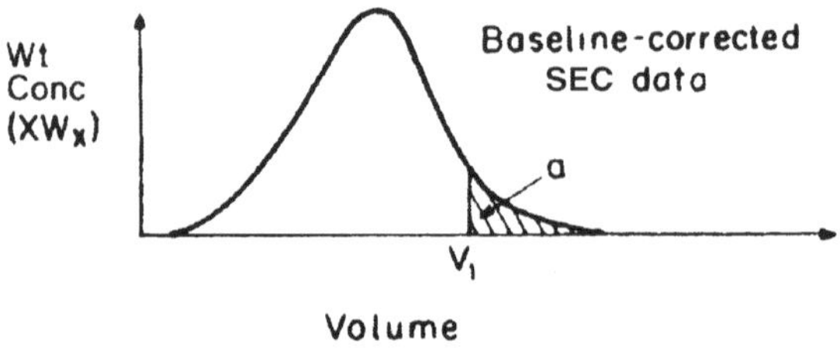

Figure 8.6 SEC elution-curve sketch. (Reprinted with permission from Ref. 20.)

Equation 8.4. This sample can now be used as the standard for the integral-MMD calibration. The shaded area a in Figure 8.5 represents the weight fraction of the molecules that have M values less than $X_i M_0$ in this polymer standard.

Now consider the SEC curve represented for a polymer standard, such as in Figure 8.6. The detected concentration response is normalized so that the area under the curve equals unity. Since small molecules elute last in SEC, the low-molar-mass fractions in the shaded area a of Figure 8.5 can be identified with the shaded area at the long retention time in Figure 8.6. For the same fractional area of a, a point V_i can be defined on the SEC elution curve. Without column dispersion, all molecules with molar mass less than $X_i M_0$ should elute after V_i. This gives a unique pairing between V_i and the M of $X_i M_0$. A series of these volume/M pairs can then be generated to produce the final SEC calibration curve.

The presumed advantage of the integral method is that it makes no assumptions regarding the calibration curve shape and thus permits an accurate search for nonlinear calibration curves. However, this advantage often does not occur in practice, owing to the rather poor precision of the method.

An example of a calibration curve obtained by this method using a nylon 6,6 standard is shown in Figure 8.7. The curve is compared with the effective linear calibration line from the Hamielec method, which is discussed in the next section. Differences between the calibration curves for these two methods occur at the extremes of very low ($<2\%$) or very high ($>98\%$) weight fractions of the calibration standard. At these extremes, the calibration points for the integral method are not very reliable, because they can be greatly affected by column dispersion, the choice of baseline, and the choice of the integration limits (i.e., the beginning and the end of the SEC elution-curve data points used in the calibration computations). Because the extrapolation of the calibration curve is questionable at the extreme molar masses, the useful calibration range of the integral method is limited to the narrow-molar-mass region, where the curve agrees with the linear calibration range. As a consequence, the integral calibration method is no more versatile or accurate than the linear calibration approaches.

To use the integral-MMD method of calibration, one must know that the polymer standard can be represented closely by the theoretical MMD function, meaning that there cannot be a mechanism or kinetic bias during the polymerization of the polymer standard. Unfortunately, this restriction can be prohibitive for many commonly occurring polymer MMD problems.

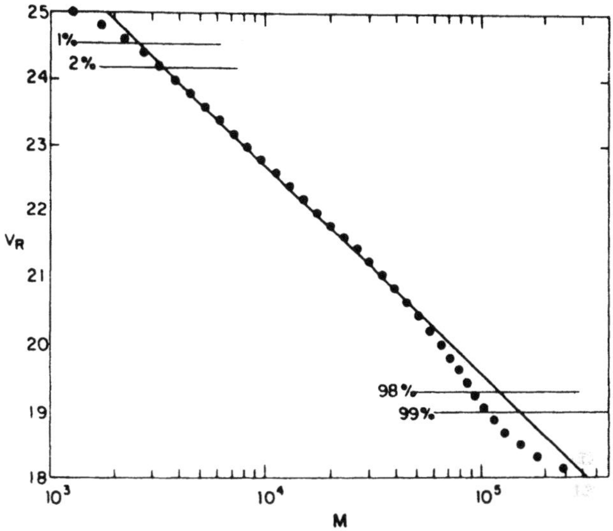

Figure 8.7 Broad-MMD calibration curves. Calibrations by most probable distribution method (circles) and by Hamielec's linear calibration method (solid curve) for a nylon 6,6 sample. The retention volumes that correspond to the cummulative elution peak areas of 1%, 2%, 98%, and 99% are marked in the figure. (Reprinted with permission from Ref. 20.)

8.3.2 Linear Calibration Methods

The linear calibration methods using broad polymer standards are discussed here beginning with the original form developed by Hamielec [21], followed by the discussions of the improved versions (GPCV2, GPCV3) of Yau et al. [22,26] and others [23–25].

The *Hamielec method* [21] consists of a search for an "effective" linear calibration $M_H(V)$ line having the form of Equation 8.5. The aim is to find the right values for the effective calibration constants D'_1 and D'_2 in Equation 8.5, so that the computed M_w and M_n values according to Equations 8.6 and 8.7 are in agreement with the known M values for the polymer standard.

$$M_H(V) = D'_1 e^{-D'_2 V} \tag{8.5}$$

$$M_w = \sum_V F(V) M_H(V) \tag{8.6}$$

$$M_n = \frac{1}{\sum_V F(V)/M_H(V)} \tag{8.7}$$

The SEC retention volume is represented by V, and the SEC elution curve is represented by $F(V)$. Since the experimental chromatogram $F(V)$ in Equations 8.6 and

8.7 is affected by instrumental peak broadening, the Hamielec method provides an effective calibration curve of the SEC experiment, not the true curve.

An alternative expression for the effective linear calibration line is

$$V = C_1' - C_2' \log M_H(V) \tag{8.8}$$

where the calibration constants C_1' and C_2' are the intercept and the slope of the calibration line as they appear in the usual logarithmic calibration plot. The two sets of calibration constants are related according to the following equalities:

$$C_1' = \frac{\ln D_1'}{D_2'}, \qquad C_2' = \frac{\ln 10}{D_2'} \tag{8.9}$$

or

$$D_1' = 10^{C_1'/C_2'}, \qquad D_2' = \frac{\ln 10}{C_2'} \tag{8.10}$$

The essential element of the computer program for the Hamielec method is a trial-and-error search routine that iteratively adjusts the values of C_1' and C_2' (or D_1' and D_2') until Equations 8.6 and 8.7 are satisfied by the known values of M_w and M_n of the standard. The desired calibration curve is defined by the final values of the calibration constants. The effective $M_H(V)$ calibration curve so obtained can then be used for calculating the molar mass averages for unknown samples from their experimental SEC elution curves.

The Hamielec method offers a truly practical way of obtaining SEC calibration curves that are specific to polymer type. The method needs only one broad-MMD standard of the same structure as the unknown samples. This can usually be provided either by commercial standards (Section 7.7) or by converting one of the samples into a working standard through independent determinations of its values of M_w and M_n using, for example, light-scattering and osmotic pressure techniques.

Although a single broad standard is often used, the Hamielec method can also be used with two different molar mass standards, as long as two average M values are known. Mathematically, two known M values in any combination of M_n and M_w are all that are needed to solve for the two unknown calibration constants. The precision of the method increases with the difference between the two M values used in calibration. For two standards, the two SEC elution curves, or $F(V)$ curves, are used in the search for the effective calibration constants D_1' and D_2' (or C_1' and C_2') from Equations 8.6 and 8.7. The calibration constants can be found by use of a trial-and-error computer algorithm, as in the case of a single broad standard.

In addition to dependence on the accuracy of the linear calibration approximation and on the experimental M_w and M_n values of the standard, the Hamielec method has two other weaknesses: (1) The physical significance of the effective calibration curve is not defined, and (2) the calculated M values are accurate only for samples having an MMD (or SEC elution) curve similar to that of the standard.

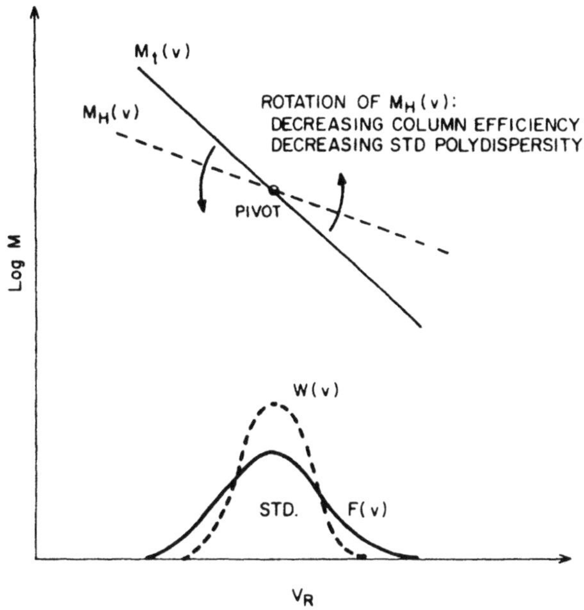

Figure 8.8 Effect of column dispersion and standard polydispersity on rotation of the Hamielec effective linear calibration line.

Unlike the true calibration curve, the effective calibration curve is not unique to the specific SEC column and polymer–solvent system, but varies as a function of the column efficiency and the M and MMD of the standard. The sketch in Figure 8.8 is helpful in explaining the properties of the calibration curves obtained by the Hamielec method. Besides the Hamielec calibration line $M_H(V)$ and the experimental SEC elution curve $F(V)$ of the broad standard, the sketch in Figure 8.8 also shows the true calibration curve $M_t(V)$ and the hypothetical dispersion-free elution curve $W(V)$. As discussed in Section 8.2, the true calibration curve $M_t(V)$ can be obtained experimentally by peak position calibration if there are narrow fractions of known M available for the polymer of interest. The linear approximation of the true calibration curve is described here by

$$M_t(V) = D_1 e^{-D_2 V} \tag{8.11}$$

where D_1 and D_2 are the true calibration constants in contrast to the effective calibration parameters D_1' and D_2' in Equation 8.5.

As illustrated in Figure 8.8, column band dispersion causes the experimental $F(V)$ to be broader than $W(V)$ and makes the apparent M_w value too high and the M_n value too low if the $M_t(V)$ calibration is used in the computation. To compensate for molar mass errors when forcing a fit to the known M_w and M_n of the standard, the required effective calibration curve $M_H(V)$ will have to be rotated away from $M_t(V)$ in a counterclockwise direction. This rotation of the $M_H(V)$ line increases with increasing column dispersion, or decreasing column efficiency, as the $F(V)$ curve becomes increasingly different from $W(V)$. Following the same reasoning, one can see that $M_H(V)$ is also sensitive to the polydispersity of the standard. For standards of narrower MMD, $M_H(V)$ will rotate more toward the horizontal. Of course, a horizontal effective calibration is the extreme case expected of a monodisperse molar mass standard. Also, the rotation pivot of the $M_H(V)$ line is expected to shift, depending on the M of the standard used in the calibration.

Knowing how the $M_H(V)$ line will rotate and shift, one can readily predict the trend of molar mass errors in the Hamielec method in isolated circumstances. If the sample is of higher M than the standard, the calculated molar mass values will be too low, and vice versa. If the sample is of broader MMD than the standard, the calculated sample polydispersity will be too small, and vice versa. Therefore, it is apparent that the limitation of the Hamielec method is the lack of proper compensation for column dispersion.

In an improved version of the Hamielec method, GPCV2, compensation is provided for the symmetrical peak broadening caused by column dispersion effects. In GPCV2, the instrumental peak broadening is approximated by a standard deviation (σ), which is assumed to be independent of retention volume [22]. The formulations that form the basis of GPCV2 are (see Section 4.3 and References 22 to 25 for derivations)

$$M_w = e^{-(1/2)(D_2\sigma)^2} \sum_V \left[F(V) D_1 e^{-D_2 V}\right] \tag{8.12}$$

$$M_n = \frac{e^{(1/2)(D_2\sigma)^2}}{\sum_V \left[F(V)/D_1 e^{-D_2 V}\right]} \tag{8.13}$$

These equations relate the average M values directly to the experimental SEC elution curve $F(V)$ and the true SEC calibration curve constants D_1 and D_2, including an exponential correction factor containing the column dispersion parameter σ. At $\sigma = 0$, Equations 8.12 and 8.13 reduce to Equations 8.6 and 8.7 of the original Hamielec method. To a first approximation, the peak standard deviation σ due to column dispersion is estimated as the minimum experimental value of peak σ for several narrow polystyrene standards (Section 3.5). The minimum value of σ is used to avoid overcorrection for column dispersion, since the experimental peak σ may include actual molar mass separation in addition to column dispersion for some polystyrene standards.

With known M_w, M_n, and σ values, Equations 8.12 and 8.13 can be solved in the same manner as Equations 8.6 and 8.7 of the original Hamielec method, except that the computer search now produces values of D_1 and D_2 and thus a more accurate calibration curve for the SEC experiment. Once the calibration curve is obtained, Equations 8.12 and 8.13 can be used again for calculating the M for unknown samples. Since the GPCV2 method maintains the integrity of the calibration constants D_1 and D_2 and provides correction for column dispersion, it is more accurate than the original Hamielec method for analyzing samples that have very different M or MMD from that of the standard. Actual SEC analyses have shown that the GPCV2 procedure can provide up to threefold improved M accuracy over the original Hamielec method (Reference 22; see also Section 8.4).

A more sophisticated version to further improve the Hamielec method, GPCV3 includes a consideration of the skewness of SEC column dispersion [26]. The expressions used in GPCV3 are

$$M_w = (1 + D_2\tau)e^{-(D_2^2\sigma^2/2)-D_2\tau} \sum_V \left[F(V)D_1 e^{-D_2 V}\right] \tag{8.14}$$

and

$$M_n = \frac{1}{1 - D_2\tau} \frac{e^{(D_2^2\sigma^2/2)-D_2\tau}}{\sum_V \left[F(V)/D_1 e^{-D_2 V}\right]} \tag{8.15}$$

where σ and τ are the two peak shape parameters of the assumed exponentially modified Gaussian peak shape model (Section 3.5 and Reference 27). The parameter τ is the time constant of the exponential modifier to the Gaussian component of standard deviation σ, and the τ/σ ratio relates to the peak skew. The procedure for extracting σ and τ from experimental SEC peaks is described in Section 3.5. The procedure for using GPCV3 is to insert experimental σ and τ values of column dispersion into Equations 8.14 and 8.15 and then solve for D_1 and D_2 to satisfy the known values of M_w and M_n for the standard. At $\tau = 0$, Equations 8.14 and 8.15 reduce to the GPCV2 Equations 8.12 and 8.13. Both GPCV2 and GPCV3 can use either one or two calibration standards, as does the original Hamielec method.

In contrast to the integral MMD method, linear calibration methods pose no restrictions on the MMD shape of the standard. In practice, linear calibration methods provide a much better compromise than does the integral MMD method, because the linearity of the SEC calibration curve can be improved by the proper selection of SEC columns, as described in Section 8.6. For SEC columns with nonlinear calibration curves, a modified broad standard method must be used for which the linear calibration approximation is not assumed. However, this approach can be used only when the universal calibration curve is known for the system [28].

8.4 ACCURACY OF CALIBRATION METHODS

The accuracy of the peak position, Hamielec, and GPCV2 calibration methods was tested in a specially designed experiment [22]. In the experiment, four polystyrene samples were prepared using blends of commercially available characterized standards. The expected values of M for the samples are listed as the calculated/reported values in Table 8.2. Four μ-Styragel and four Vit-X columns of different pore sizes were used in the experiment. These column sets were not optimized for range and linearity in M calibration by the bimodal approach discussed in Section 8.6. The

Table 8.2 Effect of SEC Calibration Method on Molar Mass Accuracy

	Sample 1[a]			Sample 2[a]			Sample 3[a]			Sample 4[a]		
Method	M_w[b]	M_n[b]	Average Error (%)	M_w[b]	M_n[b]	Average Error (%)	M_w[b]	M_n[b]	Average Error (%)	M_w[b]	M_n[b]	Average Error (%)
A. μ-Styragel Columns[c] ($N = 13{,}000$, *Toluene;* $\sigma = 0.70$ *mL for 4A*)												
Calculated/reported value	64	44	—	39	28	—	288	137	—	20	20	—
Peak position	74	34	19	47	22	21	454	107	40	23	17	15
Hamielec	64	44	—	46	32	16	210	107	25	30	26	40
GPCV2	64	44	—	42	30	7	314	129	7	23	23	15
B. Vit-X Columns ($N = 3500$, *Toluene,* $\sigma = 1.05$ *mL for 4A*)												
Calculated/reported value	64	44	—	39	28	—	288	137	—	20	20	—
Peak position	86	42	20	51	26	21	333	136	8	24	18	15
Hamielec	64	44	—	44	31	13	166	105	33	27	23	25
GPCV2	64	44	—	38	27	2	247	143	9	18	19	8

Source: Ref. 22.

[a]Sample 1 was a blend of 3 parts PS 4A, 2 parts 7B, and 1 part 2A. Sample 2 was a blend of 1 part 4A, 2 parts 7B, and 3 parts 2A. Sample 3 was NBS 706 polystyrene (National Institute of Standards and Technology, Washington, DC), and sample 4 was the narrow polystyrene standard 2A alone. Narrow standards 4A, 7B, and 2A (Pressure Chemical Co., Pittsburgh, PA) have reported polydispersity values of less than 1.06 and reported *M* values of 97,200, 37,000, and 19,800 g/mol, respectively.

[b]Values listed are $\times 10^3$ g/mol.

[c]Four 30 × 0.76 cm i.d. columns of 10^2-, 10^3-, 10^5-, and 10^6-Å μ-Styragel or 84-, 171-, 660-, and 1933-Å Vit-X were used in series.

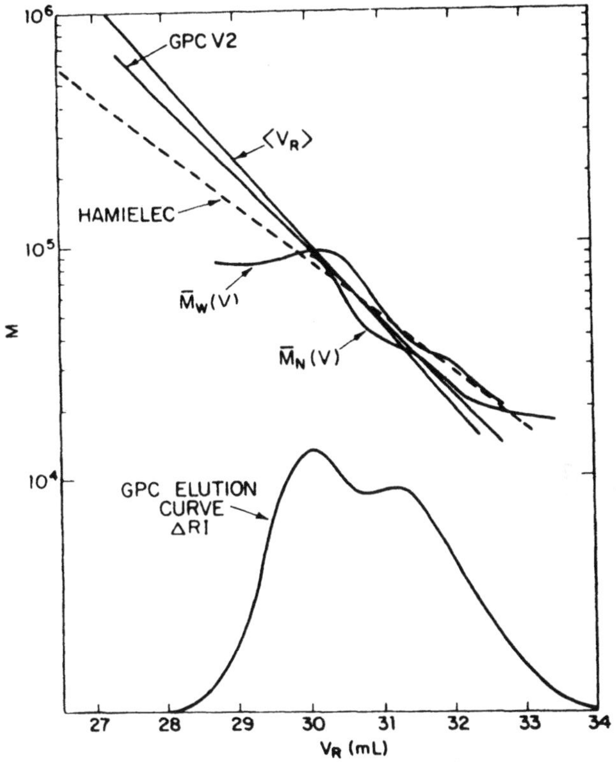

Figure 8.9 Actual *M* curves and calibration plots for polystyrene standard for sample 1 of Table 8.2; see the text for identification of labels. (Reprinted with permission from Ref. 22.)

experimental elution curves obtained on the μ-Styragel column series are shown in Figures 8.9 to 8.12.

In Figures 8.9 to 8.12 and Table 8.2 the peak position calibration was obtained by using the average retention volume $\langle V_R \rangle$ of the individual elution peaks of the narrow polystyrene standards. The Hamielec and the GPCV2 calibration lines (GPCV2 line shown in Figure 8.9 only) were calculated by using sample 1 (Table 8.2) as the calibration standard. The value for column dispersion σ was determined from the peak broadening of the polystyrene standard 4A ($M = 97{,}200$ g/mol). For the data in Table 8.2, the calculations of M_n and M_w were made using the appropriate calibration curve, the elution curve $F(V)$, and, where applicable, the values of σ.

The molar mass results calculated are shown in Table 8.2, in which the percent error in molar mass for various samples is listed under the headings of the three calibration methods. No entry of molar mass error is made for sample 1 under the Hamielec and GPCV2 methods, because this is the sample chosen as the calibrating standard for these two methods. For the other samples, the molar mass errors listed

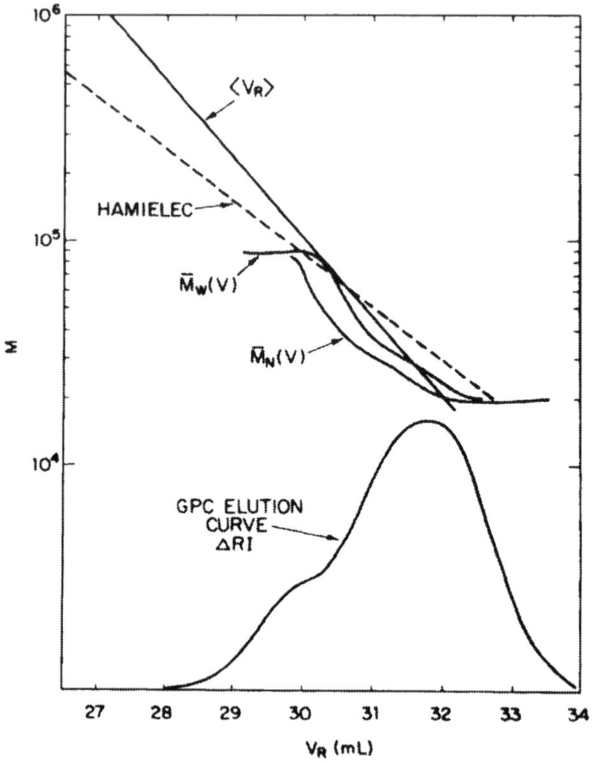

Figure 8.10 Actual *M* curves for unknown polystyrene sample and standard calibration plots for sample 2 of Table 8.2; see the text for identification of labels. (Reprinted with permission from Ref. 22.)

under GPCV2 are all more than a factor of 2 smaller than those under the Hamielec method, clearly showing the superior accuracy of the GPCV2 method. The molar mass results in Table 8.2 also show the limitation of the Hamielec method for correcting molar mass errors resulting from column dispersion for samples that do not have a MMD similar to that of the standard.

Although the data are affected by experimental uncertainties, GPCV2 still shows improved molar mass accuracy over the peak position method. Since the calibration curve obtained by the peak position method is not affected by instrumental peak broadening, the molar mass error of this method comes from the sample molar mass calculation rather than from the calibration step itself. The molar mass error in this case is simply caused by the instrumental peak broadening of the elution curves of the samples, not the standards. The residual molar mass error of the GPCV2 method is rather small considering the possible molar mass errors due to experimental uncertainties. The utility of the GPCV2 method is also demonstrated by the calibration

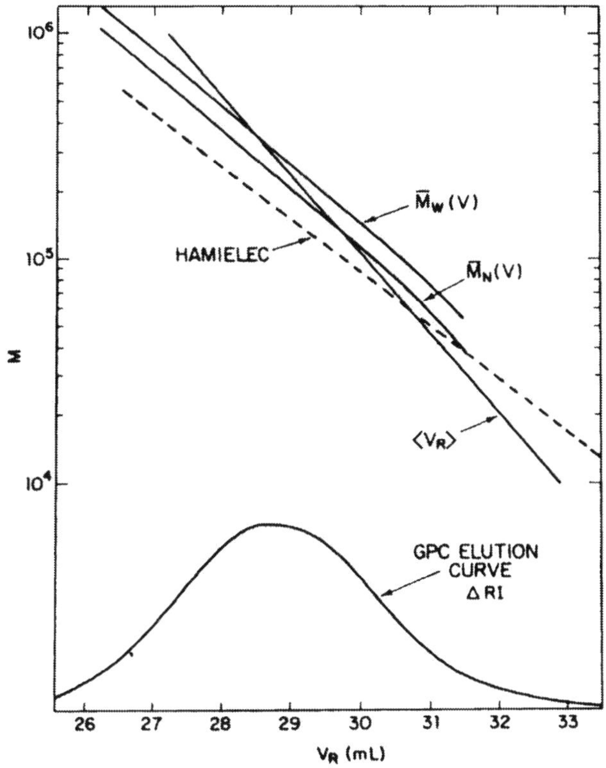

Figure 8.11 Comparison of actual molar mass curves and the calibration plots for sample 3 of Table 8.2; see the text for identification of labels. (Reprinted with permission from Ref. 22.)

lines of Figure 8.9, which show that the true calibration curve $\langle V_R \rangle$ is better approximated by GPCV2 than by the Hamielec calibration line.

8.5 ACTUAL MOLAR MASS ACROSS THE SEC ELUTION CURVE

The fundamental cause of the molar mass errors discussed in Section 8.4 is the inability of the various calibration curves to accurately describe the actual sample molar mass eluted at different retention volumes. Large discrepancies between actual and calibration molar mass values cause large errors in the calculated values of M_n and M_w for the sample. Therefore, a study of the actual molar mass across the sample elution curves is important for a full understanding of the SEC calibration problem. In the following we describe how to predict the actual molar mass and utilize it in studying SEC calibration methods.

In an SEC experiment for a broad-MMD sample, the species detected at any particular retention volume do not have a single specific molar mass. Neighboring

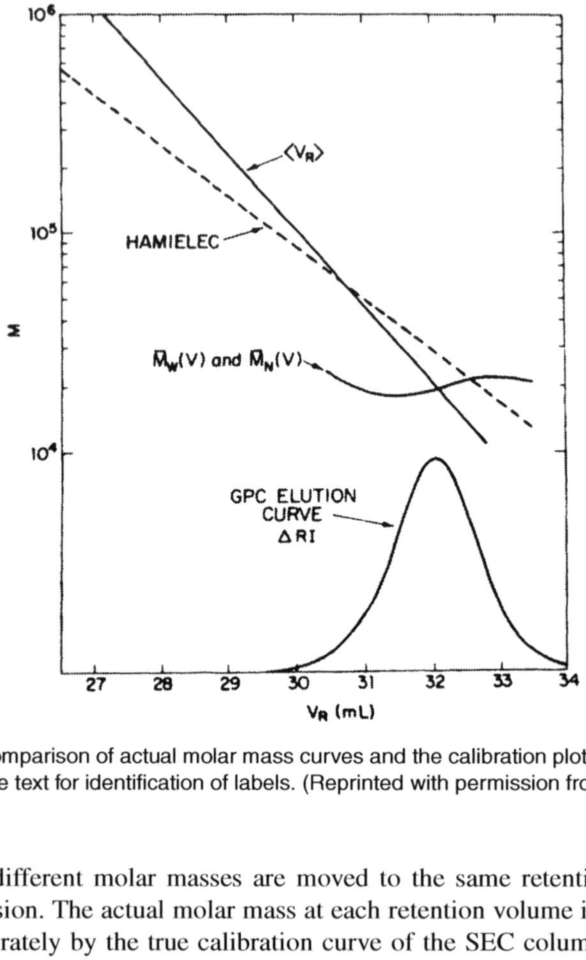

Figure 8.12 Comparison of actual molar mass curves and the calibration plots for sample 4 of Table 8.2; see the text for identification of labels. (Reprinted with permission from Ref. 22.)

molecules of different molar masses are moved to the same retention volume by column dispersion. The actual molar mass at each retention volume is therefore not described accurately by the true calibration curve of the SEC columns, except under the hypothetical condition of infinite column resolution. There are mixtures of molar masses within even infinitesimally small fractions at any retention volume (see Section 11.8). Expressions have been developed [22] to examine the actual change of molar mass with retention volume in the SEC experiment. The weight- and number-average molar mass at any retention volume V have the following functional dependence:

$$M_w(V) = \frac{F\left(V - D_2\sigma^2\right)}{F(V)} e^{(1/2)(D_2\sigma)^2} M_t(V) \tag{8.16}$$

and

$$M_n(V) = \frac{F(V)}{F\left(V + D_2\sigma^2\right)} e^{-(1/2)(D_2\sigma)^2} M_t(V) \tag{8.17}$$

where $M_t(V)$ is the M value at V defined by the true or peak position calibration curve; σ is the peak standard deviation of the assumed Gaussian column dispersion function; and $F(V)$ and $F(V \pm D_2\sigma^2)$ are the experimental SEC elution curve heights at the retention volume of V and $V \pm D_2\sigma^2$, respectively. It is interesting to note that the actual molar mass across the SEC elution curve is not defined uniquely but varies from sample to sample depending on the elution curve shape $F(V)$ and the values of σ.

With experimental values of $F(V)$, σ, and $M_t(V)$, Equations 8.16 and 8.17 can be used to predict the $M_w(V)$ and $M_n(V)$ curves. This was done for the four polystyrene samples chromatographed on a set of four μ-Styragel columns used in the experiment described in Section 8.4. The calculated curves for the actual molar mass variation $M_w(V)$ and $M_n(V)$ are plotted separately for each sample in Figures 8.9 to 8.12. Recall that sample 1 shown in Figure 8.9 was the single broad standard used for obtaining the Hamielec and GPCV2 calibration plot, which are shown in the same figure with the peak position line $\langle V_R \rangle$. The same Hamielec and $\langle V_R \rangle$ calibration lines are reproduced in Figures 8.10 to 8.12. The features of the peak position and the Hamielec calibrations are compared to the actual molar mass variations $M_w(V)$ and $M_n(V)$. In Figure 8.9 the Hamielec line encompasses the actual molar mass curves over most of the molar mass range. This is understandable since the sample in this case is the Hamielec calibration standard itself. In this case the molar mass calculation for sample 1 using either GPCV2 or the Hamielec method will, by definition, give accurate results.

However, the actual molar masses $M_w(V)$ and $M_n(V)$ begin to differ from the Hamielec and peak position lines in Figure 8.10. This is the case in which the sample has an MMD different from that of the standard. As the sample differs more and more from the standard, the Hamielec line becomes an increasingly poorer estimate of the actual molar mass elution behavior. For example, in Figure 8.11 neither the $M_w(V)$ nor the $M_n(V)$ curve ever intersects with the Hamielec line; and in Figure 8.12 the respective curves are nearly at right angles. Since Figure 8.12 is the chromatogram of a very narrow MMD sample, most of its elution-curve profile is caused by peak broadening, and therefore the $M_w(V)$ and $M_n(V)$ curves calculated merge and do not vary much with V_R.

The fact that the $M_w(V)$ and $M_n(V)$ curves vary as a function of column dispersion and shape of the sample MMD indicates that the SEC elution curve profiles obtained by an online viscometer or light-scattering detector [29] would be similarly affected by column dispersion and sample MMD. These effects should be taken into consideration in interpreting results from these molar-mass-specific SEC detectors to achieve absolute molar mass calibration. A discussion of the band-broadening methods for triple-detector SEC is given in Section 8.7.4.

The preceding results also show that the actual molar mass variation across an SEC elution curve is a very complex function of the combined effect of column dispersion and elution-curve profile. The $M_w(V)$ and $M_n(V)$ curves can vary drastically from sample to sample, and can in no way be fitted well by a single linear calibration curve. Generally, tilting and shifting the calibration by way of the Hamielec method cannot properly compensate for column dispersion. The dispersion problem is accounted for, however, in GPCV2 or GPCV3 by including the σ correction in the calibration computation (Equations 8.12 to 8.15).

GPCV2 or GPCV3 should be helpful in the study of the universal calibration concept [7]. With these calibration methods one can obtain nearly the true calibration curves in various polymer–solvent systems for evaluating the accuracy and the precision of the universal calibration method. A discussion of the band-broadening method for universal calibration is given in Section 8.7.2.

8.6 LINEAR CALIBRATION RANGES

The discussions in the earlier sections have pointed out that the broad standard linear calibration method is the choice for quantitative SEC-MMD analyses, since this approach provides the best compromise between practical convenience and molar mass accuracy. For many commercial polymers, linear calibration may be the only workable method that can provide the desired calibration curve for the specific polymer–solvent system. Since this calibration method works best for SEC columns of wide and linear molar mass separation range, it is important that the SEC columns used in the experiment are optimized for range and linearity. To handle most commercial polymers adequately, a linear molar mass range of four decades covering molar mass values of a few hundred to a few million is needed.

A popular empirical guideline for linear calibration recommends the use of columns of each pore size that have finite fractionation capacity in the molar mass range of interest [31,32]. The calibration in Figure 8.13 illustrates this empirical approach of connecting columns with different pore sizes to obtain fractionation over a wide molar mass range. Note that this approach produces relatively large deviations from a linear fit with a linearity range of less than three decades. The empirical approach is not as effective as the bimodal-pore-size distribution method [30], the basic theory of which was discussed in Section 4.5. The rules for optimizing bimodal column sets using available SEC columns are described below (see also Section 7.9 for experimental practice).

The bimodal concept involves coupling SEC columns containing only two discrete pore sizes and approximately equal pore volumes for the two pore sizes. This column selection rule was developed so that the linear portion of the calibration plot for the individual pore-size columns is substantially parallel but nonoverlapping. The bimodal approach to column selection relies on the proper recognition of two concepts: (1) Pores of only one size can fractionate polymers over nearly two decades of molar mass range; and (2) the separation capacities of individual columns are additive in a column set.

The bimodal SEC theory (Sections 4.5 and 7.9.2) predicts that wide-range linear calibrations are possible for bimodal column sets with pore sizes differing by about one decade or more. However, there is considerable leeway in selecting the pore-size separation around the optimum bimodal arrangement. Therefore, a detailed knowledge of pore size or pore-size distribution is not necessarily required to assemble a column set with reasonable molar mass range and calibration linearity. A very

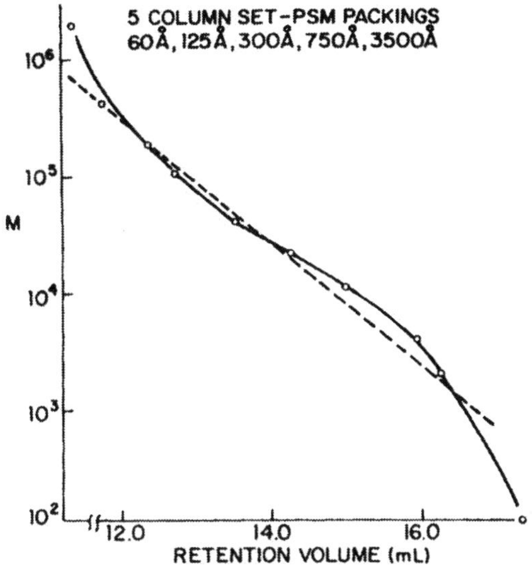

Figure 8.13 Calibration plot from the empirical approach of connecting columns of different pore sizes. Columns, 10 × 0.78 cm each porous silica microsphere; mobile phase, THF; temperature, 22°C; flow rate, 2.5 mL/min; detector, UV at 254 nm; sample, 25-μL solutions of polystyrene standards. (Reprinted with permission from Ref. 30.)

wide linear molar mass calibration range can be obtained at the expense of only slightly poorer linearity by increasing the pore-size separations of the bimodal approach. Figure 8.14 demonstrates the molar mass calibration of such a column set, which exhibits almost five decades of molar mass linearity. (It should be noted that at around $M < 10^3$ g/mol, the random-coil model for describing polymer conformation becomes less definitive. Thus, linear extrapolation of the calibration data points to molar masses of a few hundred should be interpreted with caution [33]). The increase in molar mass error caused by the poorer linearity is modest, and in many cases the increased range of linear calibration is more valuable in view of the added versatility and convenience of linear columns of wide range.

A bimodal column set designed for optimum performance for one polymer–solvent system should function equally well in other systems, although the actual calibration curve may shift somewhat.

8.7 RECENT DEVELOPMENTS AND RECOMMENDATIONS ON BAND-BROADENING CORRECTION

An extension of the GPCV2 method has been made recently [34]. Beyond correcting molar mass averages (M_n, M_w, M_z, . . .), the extended method corrects and removes band-broadening effects on the entire MMD curve.

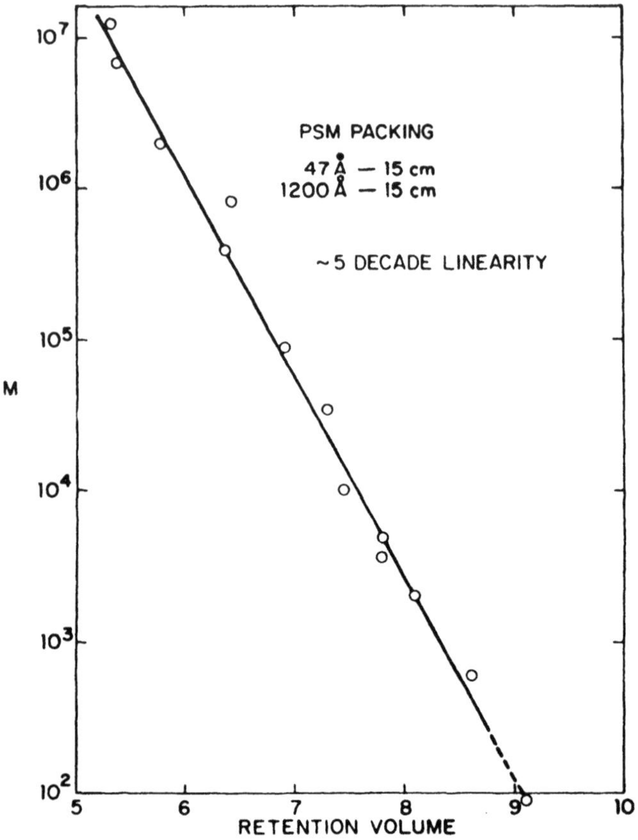

Figure 8.14 Very wide linear molar mass range calibration curve with bimodal column set. Conditions same as for Figure 8.13. (Reprinted with permission from Ref. 30.)

8.7.1 Algorithm for BBC in Conventional SEC Analysis with Only a Concentration-Sensitive Detector

Key concepts are shown in Figure 8.15, while Reference 35 provides a detailed derivation of the band-broadening correction (BBC) formulations.

Column Molar Mass and Band-Broadening Calibration Using a Broad MMD Standard

1. Establish a narrow-MMD-standard Mark–Houwink calibration curve for the polymer of interest [i.e., $M_t(V)$ in Equation 8.11].
   a. Start with a polystyrene narrow-MMD standard calibration using the peak apex retention volume and a first-order linear calibration fit.

Band-Broadening Correction for Conventional GPC

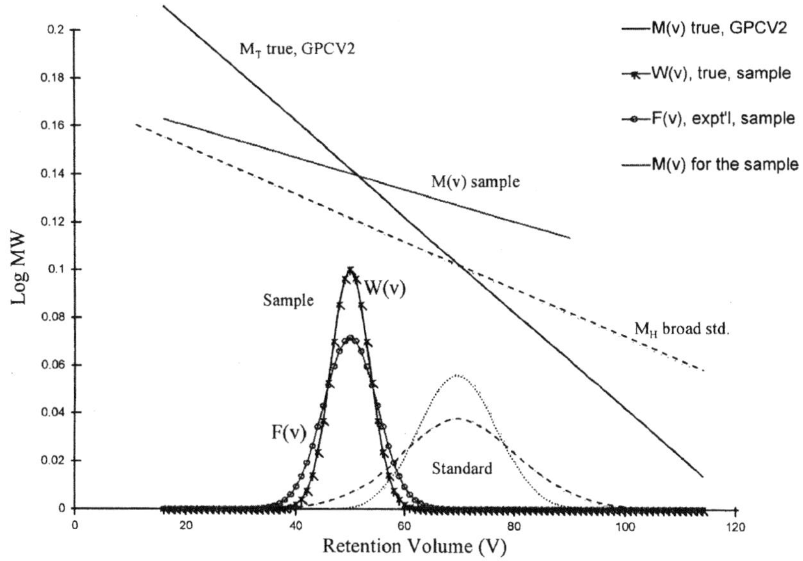

Calibration Method	M_W , M_N , Pd Corrections	MWD-Overlay Precision	Remarks
Method-1	Yes	Better	Corrections apply to bell-shaped MWD and linear columns

Algorithms:

1. Use narrow-standard calibration if available, or, obtain GPCV2 calibration line from broad standard.
2. Calculate $M_{W,\ expt'l}$ and $M_{N,\ expt'l}$ from $F(v)_{\text{sample}}$.
3. Calculate $M_{W,\ true}$ and $M_{N,\ true}$ of the sample from:

   $M_{W,\ true} = e^{-(D_2\sigma)^2/2}\ M_{W,\ expt'l}$, and, $M_{N,\ true} = e^{+(D_2\sigma)^2/2}\ M_{N,\ expt'l}$.
4. Search the sample M(v) function [every sample has its own unique $M(v) = D_1'' e^{-D_2'' V}$] :

   $M_W = D_1'' \Sigma F(v) e^{-D_2'' V}$, and, $M_N = D_1'' / \Sigma F(v) e^{+D_2'' V}$.
5. Calculate sample MWD curve from F(v) and M(v) of the sample.

Figure 8.15 BBC of conventional SEC calibration.

b. Convert the polystyrene-relative calibration curve to a curve for the polymer type of interest for Equation 8.11 by using Equation 8.3c.

c. By using Equation 8.11 and the elution curve of the broad standard, calculate $(M_w)_p$, $(M_n)_p$, and $(M_g)_p$, where

$$(M_g)_p \equiv \sqrt{(M_w)_p \times (M_n)_p} \tag{8.18}$$

M_g is the geometric-average M, and the subscript p is used to indicate the use of peak-position calibration of Equation 8.11 in calculating these M averages.

2. Search for an effective broad-MMD-standard calibration (Hamielec) line $M_H(V)$ of Equation 8.5 from known values of M_w and M_n for the broad-MMD linear standard (see Section 8.3.2). Then, using Equation 8.5 and the elution curve of the broad-MMD standard, calculate $(M_w)_H$, $(M_n)_H$, and $(M_g)_H$, where

$$(M_g)_H \equiv \sqrt{(M_w)_H \times (M_n)_H} \tag{8.19}$$

The subscript H is used to indicate the use of a Hamielec calibration of Equation 8.5 in calculating these M averages.

3. Estimate the column performance σ and τ from a broad standard:
   a. Calculate τ:

$$\tau = \frac{\ln(M_g)_H - \ln\left(M_g\right)_p}{D_2} \tag{8.20}$$

   b. Calculate σ:

$$\sigma = \frac{\left\{\ln\left[(M_w/M_n)_p\right] - \ln\left[(M_w/M_n)_H\right]\right\}^{0.5}}{D_2} \tag{8.21}$$

   Steps 1 to 3 complete the calibration algorithm for obtaining the system $M_t(V)$, σ, and τ, which is applicable to unknown samples around the retention volume range of the broad-MMD standard.

Processing Sample M-Averages and MMD Using BBC

4. MMD calculation of unknown sample
   a. Calculate the band-broadening-corrected (BBC) M_w and M_n for a sample using the $M_t(V)$ calibration of Equation 8.11 and σ and τ from Equations 8.20 and 8.21, with the sample elution chromatogram $F(V)$, in accord with the following equations:

$$(M_w)_{\text{BBC}} = e^{D_2\tau-(1/2)(D_2\sigma)^2}\sum_V\left[F(V)D_1e^{-D_2V}\right] \tag{8.22}$$

$$(M_n)_{\text{BBC}} = \frac{e^{D_2\tau+(1/2)(D_2\sigma)^2}}{\sum_V\left[F(V)/D_1e^{-D_2V}\right]} \tag{8.23}$$

b. Use the $(M_w)_{\text{BBC}}$ and $(M_n)_{\text{BBC}}$ values of the samples from step 4a to obtain an effective Hamielec calibration line for the sample:

$$M_{H,\text{sample}}(V) = D_1'' e^{-D_2'' V} \tag{8.24}$$

where D_1'' and D_2'' are the *sample-specific* Hamielec-like M-to-retention-volume "effective calibration" constants. *Note: Every sample has its own Hamielec-like "effective* M *calibration" curve!*

c. Finally, calculate the sample *conventional chain-backbone* MMD from the sample $F(V)$ elution curve and its own $M_H(V)$ calibration curve (Equation 8.24) obtained from step 4b.

8.7.2 Algorithm for BBC in Dual-Detector SEC Using an Online Static Light-Scattering Detector

1. Align the static light-scattering (LS) and concentration detectors with the geometric volume offset so that the M_w values across each peak of the narrow-MMD polystyrene standards calculated by LS tend to approach a constant plateau. This is because the molar mass polydispersity of the narrow standards is low, and most of the peak width of commercial narrow standards is caused by instrumental band broadening.

   It is highly recommended that the two detectors be connected in a series configuration. Do not connect these detectors in a parallel configuration with split streams. The quality of dual-detector SEC analyses is highly sensitive to the integrity of maintaining a constant-volume offset between detectors (i.e., a constant interdetector delay). A constant-volume offset can be maintained more accurately when the detectors are connected in series than when connected in parallel.

   Due to differences in detector cell volumes and in the lengths of connecting tubing, the band broadening observed by the LS and concentration detectors may be different. To optimize the BBC approach, it is useful to match the detector band broadening first, preceding all the BBC data-processing steps. Matching the bend broadening of detectors involves digitally convoluting the right amount of σ and τ of one detector to make its band broadening larger in order to match that of the other detector.

2. MMD calculation of an unknown sample which can be linear or branched:

   a. Calculate the *experimental* $M_w(V)$ by dividing the sample LS signal $LS(V)$ by the concentration detector signal $F(V)$ across the sample elution curve, using the appropriate LS detector calibration constant K_{LS}:

$$M_{w,\text{LS}}(V)_{\text{exp}} = K_{\text{LS}} \frac{\text{LS}(V)}{\text{Conc}(V)} \tag{8.25}$$

b. Calculate the BBC using the following equation:

$$M_{w,\mathrm{LS}}(V)_{\mathrm{BBC}} = M_{w,\mathrm{LS}}(V)_{\mathrm{exp}} \times \mathrm{BBC}_{\mathrm{LS}}(V) \tag{8.26}$$

where the LS correction factor as a function of retention volume is given in the following equation:

$$\mathrm{BBC}_{\mathrm{LS}}(V) = \frac{D_1'' e^{-D_2'' V}}{\left[F(V - D_2\sigma^2)/F(V)\right] e^{D_2\tau + (1/2)(D_2\sigma)^2} D_1 e^{-D_2 V}} \tag{8.27}$$

Equation 8.27 is derived from a combination of the formulations in Equations 8.11, 8.20, 8.21, and 8.24.

c. Finally, calculate the weight-average MMD by LS from the sample concentration profile $F(V)$ and the $M_{w,\mathrm{LS}}(V)_{\mathrm{BBC}}$ formulation in Equation 8.26.

8.7.3 Algorithm for BBC in Universal Calibration Using an Online Viscosity Detector

This is the universal calibration using an online viscometer (UC-VISC) method of MMD calculation. Key concepts are shown in Figure 8.16. A detailed derivation of the BBC formulations is given in Reference 36.

1. Align the concentration and viscosity detectors (DP for differential pressure viscometer) with the geometric volume offset. The IV values calculated across each peak of the narrow-MMD polystyrene standards should approach a constant plateau. Again, the molar mass polydispersity of the narrow standards is low, and most of the peak width of the commercial narrow standards is caused by instrumental band broadening.

   Again, in order to maintain a constant offset between detectors, which improves the quality of dual-detector SEC analyses, it is highly recommended that the detectors be connected in series rather than in parallel.

   Due to differences in detector cell volumes and in the lengths of connecting tubing, the band broadening observed by the viscometer is usually higher than that in the concentration detector. To optimize the BBC approach, one may find it useful to match the detector band broadening first, preceding all the BBC data-processing steps. In this case, matching the band broadening of detectors usually means digitally convoluting the right amount of σ and τ of the concentration detector to make its band broadening larger in order to match that of the viscometer.

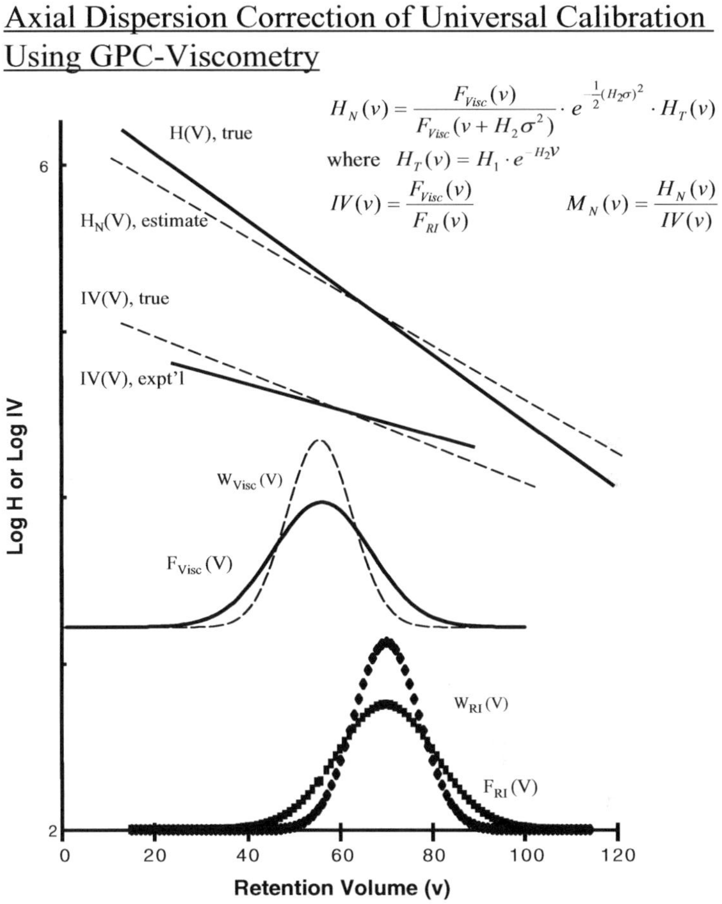

Figure 8.16 BBC of universal calibration using SEC-viscometry.

2. Establish the narrow-MMD-standard universal calibration curve for the SEC column system [i.e., $H_t(V)$ in Equation 8.28]:

$$H_t(V) \equiv M_t(V)\,\mathrm{IV}_t(V) = \left\lfloor D_1 e^{-D_2 V} \right\rfloor \left\lfloor E_1 e^{-E_2 V} \right\rfloor \equiv H_1 e^{-H_2 V} \qquad (8.28)$$

where, $H(V)$ stands for the hydrodynamic volume calibration curve used in universal calibration, and D_1, D_2, E_1, E_2, H_1, and H_2 are the calibration constants for molar mass M, intrinsic viscosity IV, and hydrodynamic-volume calibration, respectively.

Start with a narrow-MMD polystyrene standard universal calibration, from the IV $\times$ M values, using the peak apex retention volume and a first-order linear calibration fit. The IV values used in this step can either be determined experimentally using an online viscometer, or calculated from

M using the Mark–Houwink constants for polystyrene at the particular solvent–temperature conditions of the SEC experiment.

Processing Sample M and MMD with BBC

3. Obtaining Hamielec-like effective universal calibration $H_H(V)$ for unknown samples:
   a. Calculate the band-broadening-corrected (BBC) H_w and H_n for the sample using the $H_t(V)$ calibration of Equation 8.28, σ and τ from Equations 8.20 and 8.21, with the sample elution chromatogram $F(V)$ in accordance with the following equations:

$$(H_w)_{\text{BBC}} = e^{H_2\tau-(1/2)(H_2\sigma)^2} \sum_V \left[F(V)H_1e^{-H_2V}\right] \tag{8.29}$$

$$(H_n)_{\text{BBC}} = \frac{e^{H_2\tau+(1/2)(H_2\sigma)^2}}{\sum_V \left[F(V)/(H_1e^{-H_2V})\right]} \tag{8.30}$$

   b. Use the $(H_w)_{\text{BBC}}$ and $(H_n)_{\text{BBC}}$ values of the samples from step 3a to obtain an effective Hamielec calibration line for the sample:

$$H_{H,\text{sample}}(V) = H_1''e^{-H_2''V} \tag{8.31}$$

   where H_1'' and H_2'' are the *sample-specific* Hamielec-like hydrodynamic-volume-to-retention-volume "calibration" constants. *Note: Every sample has its own Hamielec-like "H calibration" curve!*
4. Intrinsic viscosity calculation of an unknown sample that can be linear or branched:
   a. Calculate the *experimental* IV(V) by dividing the sample viscosity detector signal DP(V) by the concentration detector signal $F(V)$ across the sample elution curve, using the appropriate viscosity detector calibration constant K_{DP}:

$$\text{IV}_{w,\text{DP}}(V)_{\text{exp}} = K_{\text{DP}}\frac{\text{DP}(V)}{F(V)} \tag{8.32}$$

   b. Calculate the BBC using the following equation:

$$\text{IV}_{w,\text{DP}}(V)_{\text{BBC}} = \text{IV}_{w,\text{DP}}(V)_{\text{exp}} \times \text{BBC}_{\text{IV}}(V) \tag{8.33}$$

where the IV correction factor as a function retention volume is given in the following equation:

$$\mathrm{BBC}_{\mathrm{IV}}(V) = \frac{E_1'' e^{-E_2'' V}}{\left[F(V - E_2\sigma^2)/F(V)\right] e^{E_2\tau + (1/2)(E_2\sigma)^2} E_1 e^{-E_2 V}} \tag{8.34}$$

Equation 8.34 for IV BB correction is derived in a fashion analogous to the derivation of Equation 8.27 for the LS-M_w BB correction factor in Section 8.7.2.

c. Calculate the BB-corrected intrinsic viscosity distribution, measured with an online viscometer from the sample concentration profile $F(V)$ and the $\mathrm{IV}_{w,\mathrm{DP}}(V)_{\mathrm{BBC}}$ formulation in Equation 8.33.
d. Calculate the BB-corrected $M_n(V)$ "calibration curve" by UC-VISC using the H_H and IV_w results from Equations 8.31 and 8.33:

$$M_{n,\mathrm{UC-IV}}(V)_{\mathrm{BBC}} = \frac{H_{H,\mathrm{Sample}}(V)}{\mathrm{IV}_{w,\mathrm{DP}}(V)_{\mathrm{BBC}}} \tag{8.35}$$

e. Finally, calculate the number-average MMD by using Equation 8.35 and the sample elution curve $F(V)$.

8.7.4 Algorithm for BBC in Triple-Detector SEC Using Online Static Light Scattering, Viscosity, and Concentration Detectors

For triple-detector SEC, BBC is not different from its individual parts as outlined in Sections 8.7.1 to 8.7.3. The result of BBC of MMD gives three representations of the MMD: (1) *chain-backbone MMD* by conventional SEC calculation, (2) *absolute M_w-based MMD* by LS and (3) *absolute M_n-based MMD* by universal calibration. For *linear polymers without BBC*, these three MMD profiles have different widths due to differences in instrumental band broadening. Without BBC, the widths of MMD profiles display the following trend: MMD-universal-calibration > MMD-conventional > MMD-LS. *With BBC*, one should expect all three MMD curves to converge and approach an MMD curve that lies between the MMD-conventional and MMD-LS curves.

The interpretation of the three MMD profiles for branched polymers is different. For polymers *with long-chain branching* (LCB, Section 11.2), all three MMD are not expected to be the same even after BB correction. This is good. The absolute MMD curve obtained by either LS or by UC-VISC contains information on the branched structures across the MMD, whereas the chain-backbone MMD by conventional SEC does not. LS provides the weight-average MMD. UC-VISC provides the number-average MMD. The difference between these two MMD curves provides important information on the local polydispersity of LCB across the SEC elution curve and polymer MMD. Local polydispersity is discussed in Section 11.8.

REFERENCES

1. M. Duval, B. Block, and S. Kohn, *J. Appl. Polym Sci.*, **16**, 1585 (1972).
2. J. Cazes and D. Gaskill, *Sep. Sci.*, **2**, 421 (1967).
3. L. K. Konstanski, D. M. Keller, and A. E. Hamielec, *J. Biochem. Biophys. Methods*, **58**, 159 (2004).
4. Y. V. Heyden, S. T. Popovici, and P. J. Schoenmakers, *J. Chromatogr. A*, **957**, 127 (2002).
5. A. M. Striegel, unpublished results.
6. K. K. Unger and N. P. Becker, Pittsburgh Conference on Analytical Chemistry and Applied Spectroscopy, Cleveland, OH, 1977, paper 171.
7. Z. Grubisic, R. Rempp, and H. Benoit, *J. Polym. Sci. B*, **5**, 753 (1967).
8. P. J. Flory, *Principles of Polymer Chemistry*, Cornell University Press, Ithaca, NY, 1953.
9. O. B. Ptitsyn and Yu. E. Éizner, *Sov. Phys. Tech. Phys.*, **4**, 1020 (1960).
10. P. L. Dubin and J. M. Principi, *Macromolecules*, **22**, 1891 (1989).
11. I. Teraoka, *Macromolecules*, **37**, 6632 (2004).
12. A. M. Striegel and J. D. Timpa, *Carbohydr. Res.*, **267**, 271 (1995).
13. A. M. Striegel and M. R. Krejsa, *J. Polym. Sci. B*, **38**, 3120 (2000).
14. J. Pannell, *Polymer*, **13**, 277 (1972).
15. A. M. Striegel, R. D. Plattner, and J. L. Willett, *Anal. Chem.*, **71**, 978 (1999).
16. E. Temyanko, P. S. Russo, and H. Ricks, *Macromolecules*, **34**, 582 (2001).
17. M. Kurata and Y. Tsunashima, in *Polymer Handbook*, 4th ed., J. Brandup, E. H. Immergut, and E. A. Grulke, eds., Wiley-Interscience, New York, 1999.
18. A. R. Weiss and E. Cohn-Ginsberg, *J. Polym. Sci. A-2*, **8**, 148 (1970).
19. A. H. Abdel-Alim and A. E. Hamielec, *J. Appl. Polym. Sci.*, **18**, 297 (1974).
20. T. D. Swartz, D. D. Bly, and A. S. Edwards, *J. Appl. Polym. Sci.*, **16**, 3353 (1972).
21. S. T. Balke, A. E. Hamielec, B. P. LeClair, and S. L. Pearce, *Ind. Eng. Chem. Prod. Res. Dev.*, **8**, 54 (1969).
22. W. W. Yau, H. J. Stoklosa, and D. D. Bly, *J. Appl. Polym. Sci.*, **21**, 1911 (1977).
23. S. T. Balke and A. E. Hamielec, *J. Appl. Polym. Sci.*, **13**, 1381 (1969).
24. A. E. Hamielec, *J. Appl. Polym. Sci.*, **14**, 1519 (1970).
25. T. Provder and E. M. Rosen, *Sep. Sci.*, **5**, 437 (1970).
26. W. W. Yau, H. J. Stoklosa, C. R. Ginnard, and D. D. Bly, *12th Middle Atlantic Regional Meeting, American Chemical Society*, Apr. 5–7, 1978, paper PO13.
27. E. Grushka, *Anal. Chem.*, **44**, 1733 (1972).
28. T. Provder, J. C. Woodbrey, J. H. Clark, and E. E. Drott, *Adv. Chem. Ser.*, **125**, 117 (1973).
29. A. C. Ouano, D. L. Horne, and A. R. Gregges, *J. Polym. Sci. A-1*, **12**, 307 (1974); A. C. Ouano and W. Kaye, ibid., **12**, 1151 (1974).
30. W. W. Yau, C. R. Ginnard, and J. J. Kirkland, *J. Chromatogr.*, **149**, 465 (1978).
31. Waters Associates, *Plastics and Polymers*, 1992.

32. M. R. Ambler, L. J. Fetters, and Y. Kesten, *J. Appl. Polym. Sci.*, **21**, 2439 (1977).

33. R. R. Chance, S. P. Baniukiewicz, D. Mintz, G. ver Strate, and N. Hadjichristidis, *Int. J. Polym. Anal. Charac.*, **1**, 3 (1995).

34. W. W. Yau, D. Gillespie, R. Brown, J. Langston, R. Cong, T. Huang, L. Hazlitt, and W. deGroot, *Macromol. Symp.*, in press.

35. W. W. Yau, in *Chromatography of Polymers: Hyphenated and Multidimensional Techniques*, ACS Symp. Ser. 731, T. Provder, ed., American Chemical Society, Washington, DC, 1999, p. 35.

36. W. W. Yau, in *Chromatography of Polymers: Hyphenated and Multidimensional Techniques*, ACS Symp. Ser. 731, T. Provder, ed., American Chemical Society, Washington, DC, 1999, p. 44.

9

PHYSICAL DETECTORS

9.1 INTRODUCTION

As can be seen in the last column of Table 1.1, a multiplicity of detection methods is necessary for accurate characterization of the many different types of macromolecular distributions that may be present in complex polymers [1,2]. For the purposes of our discussion, detection methods will be divided into two classes, chemical and physical. The information obtained from "chemical" detectors, such as ultraviolet (UV)/visible (when not being used as a concentration-sensitive detector), infrared (IR), nuclear magnetic resonance (NMR), mass spectroscopy (MS), and so on usually combines in additive fashion. "Physical" detectors such as the viscometer and the light-scattering photometer [3], are defined as those which generally combine in synergistic fashion.

The values of the properties measured using different physical detectors (e.g., the specific viscosity, translational diffusion coefficient, or angular distribution of scattered radiation) may themselves not be particularly important. Combining these measurements into molar mass and size data, for example, and then observing how polymer size scales with molar mass can tell us much about polymer architecture and dilute solution thermodynamics (the subjects of Chapter 11). This knowledge would be inaccessible if physical detectors did not combine synergistically.

In this chapter and Chapter 10 we shall begin to explore the type of information provided by the two different classes of detectors, chemical and physical. First, we focus on physical detectors, next on their chemical counterparts. We defer until

Modern Size-Exclusion Liquid Chromatography: Practice of Gel Permeation and Gel Filtration Chromatography, Second Edition
By André M. Striegel, Wallace W. Yau, Joseph J. Kirkland, and Donald D. Bly

Chapter 11 discussion of the architectural and thermodynamic information obtained by employing a multiplicity of detectors. Unless otherwise specified, all detectors are able to operate in both aqueous and organic environments. Due to the large number of detectors being covered, it is not possible to include here an explanation of the fundamental principles of operation of each instrument, and a certain familiarity with the chemical detectors will be assumed in Chapter 10.

9.2 CONCENTRATION-SENSITIVE DETECTORS

This type of detector, which measures the concentration of analyte at each slice eluting from the SEC column, is exemplified by the differential refractive index (DRI), UV/visible, IR, and evaporative-type detectors. Concentration-sensitive detectors are by far the most widely used in SEC, as they meet the minimum detection requirement for calculation of molar mass averages and distributions using peak-position and Mark–Houwink calibration curves (Sections 8.2.1 and 8.2.3, respectively). Concentration-sensitive detectors are also needed for molar mass and related calculations when using static light scattering (SLS) or viscometry (VISC) as detection methods, as will be seen in Sections 9.3 and 9.5.

9.2.1 Differential Refractometers

The concentration, c, of analyte dissolved in a solution can be expressed in terms of the refractive indices of the solution, n, of the neat solvent, n_0, and of the analyte, n_p, by

$$c \propto \frac{n - n_0}{n_p - n_0} \tag{9.1}$$

Reorganizing Equation 9.1 it is seen how the refractive index of a solution changes with concentration of dissolved solute:

$$n \propto n_0 + (n_p - n_0)c \tag{9.2a}$$

or in terms of molecular polarizability, α:

$$n^2 = n_0^2 + 4\pi\alpha\frac{cN_A}{M} \tag{9.2b}$$

for $cN_A/M \ll 1/\alpha$.

9.2.1.1 Deflection-Type Differential Refractometers. The most common differential refractive index (DRI) detector is a *deflection-type system* employing the principle of refraction, shown schematically in Figure 9.1. In this instrument, the light emitted from a source (pulsed light-emitting diode or tungsten lamp) is lensed

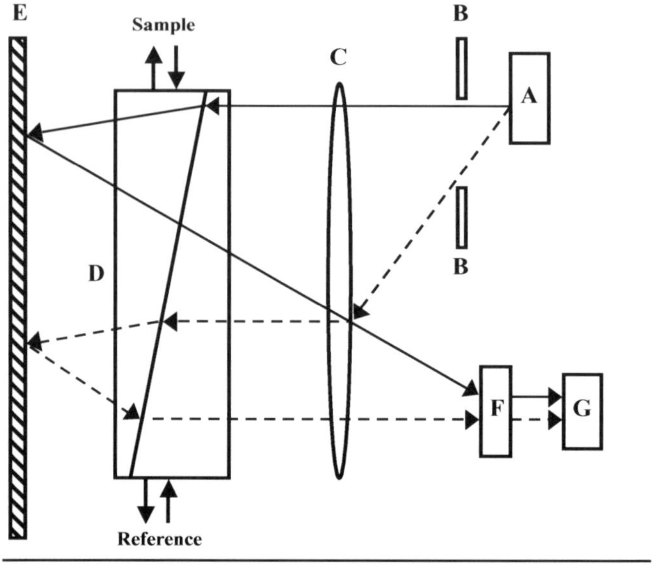

Figure 9.1 Deflection differential refractive index detector. A, light source; B, mask; C, lens; D, refractometer cell, showing sample and reference sides; E, mirror; F, zero adjust; G, photosensor.

through the sample and reference sides of a flow cell. The light then strikes a mirror and reflects back through the cell and lens to the detector, which consists of either two photodiodes mounted on a single chip or, in the newest generation of instruments, of a photodiode array. If the liquid in the sample side of the cell has the same refractive index (as determined using Snell's law) as that in the reference side of the cell, the photodiodes will produce equal signals. If the liquids in the two sides of the cell have different refractive indices, a voltage difference results between the photodiodes. This difference produces a signal, given in refractive index units (RIUs) or in volts, proportional to the concentration of the analyte in solution in the particular slice that has eluted from the SEC column and is currently passing through the detector.

9.2.1.2 Interferometric Differential Refractometers. Another type of DRI detector operates based on the principles of *interferometric refractometry.* A schematic of an interferometric DRI is shown in Figure 9.2. Light from a tungsten lamp or light-emitting diode (LED) is first masked and collimated, then passes through a polarizer oriented at 45° with respect to the horizontal. The linearly polarized beam (referred to as the original beam) strikes a Wollaston prism, creating two orthogonally polarized beams of equal intensity. These beams are then focused by a lens, such that one beam passes through a reference cell and the other through a sample cell. The light passing through the reference cell is vertically polarized; that passing through the sample cell is horizontally polarized. As the beams enter

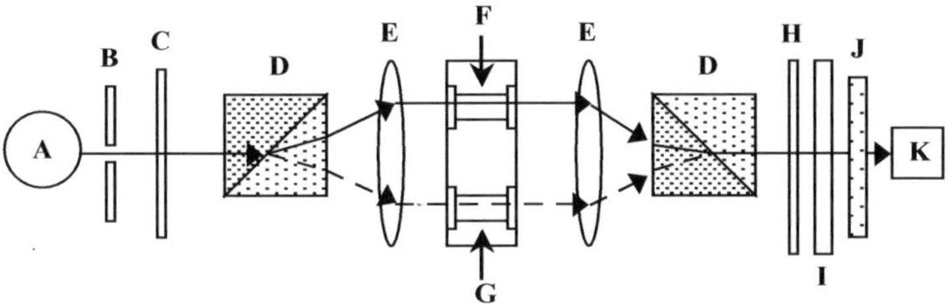

Figure 9.2 Interferometric differential refractive index detector. A, light source; B, mask; C, polarizer; D, Wollaston prism; E, lens; F, sample cell; G, reference cell; H, quarter-wave plate; I, analyzer; J, interference filter; K, photosensor.

the cells they are in phase with one another. The wavelength of light in a particular medium (λ) is proportional to the refractive index of the medium (n), as given by

$$\lambda \equiv \frac{\lambda_0}{n} \tag{9.3}$$

where λ_0 is the wavelength of light in vacuum. The refractive index of a solution changes in direct proportion to the concentration of dissolved solute (this is, of course, the same fundamental principle by which the deflection-type instrument operates), as seen in Equation 9.2. As the light beams emerge from the sample and reference cells, they will again have the same wavelength, but they will differ in phase by an amount proportional to the concentration of dissolved solute. The phase angle (in radians), ϕ, between the two waves is

$$\phi = \frac{2\pi L\, \Delta n}{\lambda_0} \tag{9.4}$$

with

$$\Delta n = n_s - n_r \tag{9.5}$$

where n_s is the refractive index of the fluid in the sample cell, n_r the refractive index in the reference cell, and L the length of the cells. The emerging beams are then focused by another lens onto a second Wollaston prism, which recombines the two beams. Due to the relative phase shift, the light emerging from the prism is elliptically polarized. A quarter-wave plate (with its fast axis at 45° to the horizontal) converts the elliptically polarized beam into a horizontally polarized beam, rotated by an angle $\phi/2$ with respect to the original beam. After emerging from the quarter-wave plate, the beam strikes a plane polarizer (the analyzer in the optical train) placed at an angle $90° - \beta$ with respect to the axis of the incident plane polarizer. The angle θ between the combined beam and the analyzer is

$$\theta = 90° - \beta - \frac{\phi}{2} \tag{9.6}$$

The intensity of the beam (I) relative to the incident intensity of the combined beam (I_0) is given by:

$$\frac{I}{I_0} = \sin^2\left(\beta + \frac{\phi}{2}\right) \tag{9.7}$$

By measuring the ratio I/I_0 and the phase angle $\phi/2$, the refractive index difference Δn can be deduced. This difference is directly proportional to the concentration of dissolved solute.

Modern DRI detectors (both deflection and interferometric) possess flow cells with minimal total volume (<8 μL), operate in the range 4 to 80°C, and operate up to 150°C and 210°C in high-temperature SEC instruments (see Chapter 16). Advanced detectors have the ability for off-line measurement of the specific refractive index increment ($\partial n/\partial c$; see Section 9.2.1.3) and for online measurement of the absolute refractive index of the solution. Generally, these last two measurements can be done at virtually the same operating wavelength (i.e., <5 nm difference) as that of the laser in an accompanying light-scattering photometer. This is done by either selectively filtering the light from the tungsten lamp light source in the refractometer or by choosing an LED of the appropriate wavelength to serve as the light source.

9.2.1.3 Specific Refractive Index Increment. Even though the DRI detector is a concentration-sensitive detector, the signal produced by a given concentration of polymer A in solution will generally be different from that produced by the same concentration of polymer B, all other experimental conditions being equal. The reason for this discrepancy is that the response of the DRI detector is directly proportional to the concentration of dissolved analyte as well as to a factor known as the specific refractive index increment of the solution, or $\partial n/\partial c$, as defined by

$$\frac{\partial n}{\partial c} \equiv \lim_{c \to 0} \frac{n - n_0}{c} \tag{9.8}$$

where n and n_0 are the refractive indices of the solution and solvent, respectively, and c is the concentration of the analyte in solution (i.e., DRI $\propto c \times \partial n/\partial c$). Specific refractive index increment values, which may be positive or negative, depend on the chemistry of the analyte, the solvent–temperature conditions of the experiment, and the vacuum wavelength of the incident radiation. Figure 9.3 shows the wavelength dependence of $\partial n/\partial c$ for both low- and high-density polyethylene.

The $\partial n/\partial c$ is generally constant across the MMD of a polymer at a given set of experimental conditions. $\partial n/\partial c$ values can, however, show a steady increase or decrease between consecutive members of a homologous series in the oligomeric region due to end-group effects. Examples of this type of behavior are shown in Figures 9.4 and 13.2.

For a given polymer, values of $\partial n/\partial c$ are obtained by injecting a series of concentrations into the DRI detector off-line (i.e., decoupled from the SEC columns), at a low (0.1 to 0.2 mL/min) flow rate, and observing the resulting change in DRI that

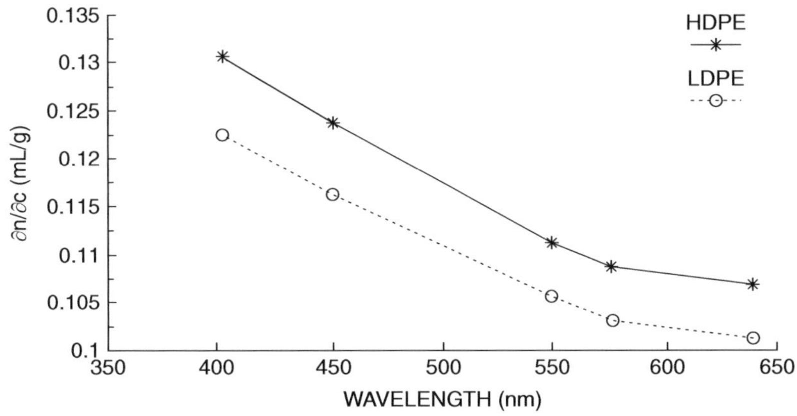

Figure 9.3 Wavelength dependence of the specific refractive index increment, $\partial n/\partial c$. HDPE, high-density polyethylene; LDPE, low-density polyethylene. See Ref. 4 for experimental details. (Reprinted with permission from Ref. 4.)

occurs as a function of concentration. Figure 9.5a shows a typical data collection plot for measuring the $\partial n/\partial c$ of a solution of 1.86×10^5 g/mol PS sample in THF. The eight plateaus in this plot correspond to eight different concentrations of the polymer in solution after solvent background subtraction; increasingly concentrated solutions produce increasingly higher DRI plateaus. The resulting differential refractive index versus concentration plot for this polymer, the slope of which corresponds to the $\partial n/\partial c$, is shown in Figure 9.5b. A $\partial n/\partial c$ value of 0.192 ± 0.002 mL/g was obtained.

For random, block, and alternating copolymers, the specific refractive index increment can generally be expressed in terms of the weight fractions (w) and specific refractive increments of each species, A and B, as [7]:

$$(\partial n/\partial c)_{\mathrm{AB}} = w_{\mathrm{A}}(\partial n/\partial c)_{\mathrm{A}} + w_{\mathrm{B}}(\partial n/\partial c)_{\mathrm{B}} \tag{9.9}$$

This relationship extends to terpolymers, and so on. An extensive table of $\partial n/\partial c$ values may be found in Reference 8.

9.2.2 UV/Visible Detectors

Detectors based on ultraviolet absorption are widely used in modern LC as well as in SEC with aqueous solvents, but have more restricted application when organic solvents are used in SEC. Often, a UV-transmitting solvent cannot be found that meets the solubility requirements of polymers. Obviously, solutes must absorb in the ultraviolet to be detected (e.g., UV detection of polyethylene dissolved in 1,2,4-trichlorobenzene would be impossible).

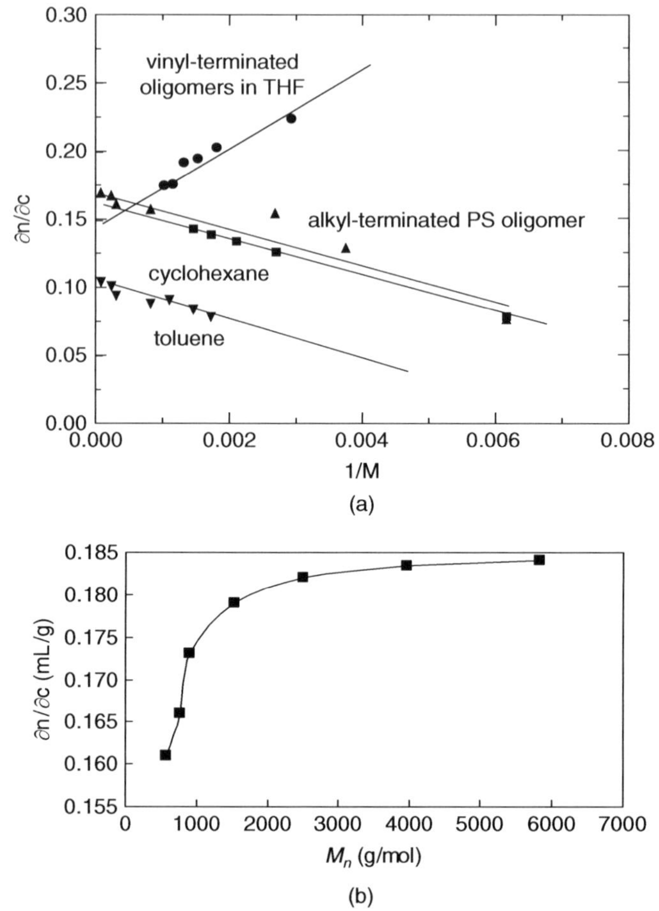

Figure 9.4 Dependence of $\partial n/\partial c$ on molar mass in the oligomeric region: (a) specific refractive index increment versus 1/(molar mass) for various oligomers; (b) dependence of $\partial n/\partial c$ on M_n for oligostyrenes. (Reprinted with permission from Refs. 5 and 6.)

The relation between absorbance, A, and concentration, c, is given by Beer's law:

$$A = \log \frac{I_o}{I} = \varepsilon bc \tag{9.10}$$

where I_o and I are the intensities of the incident and transmitted radiation, respectively, b is the path length of the detector cell, and ε is the molar absorptivity. The molar absorptivity may be considered the UV-detector equivalent of the specific refractive index increment, $\partial n/\partial c$, for differential refractometry detection. As is the case for the $\partial n/\partial c$, values of ε are specific to a given polymer in a given solvent at a

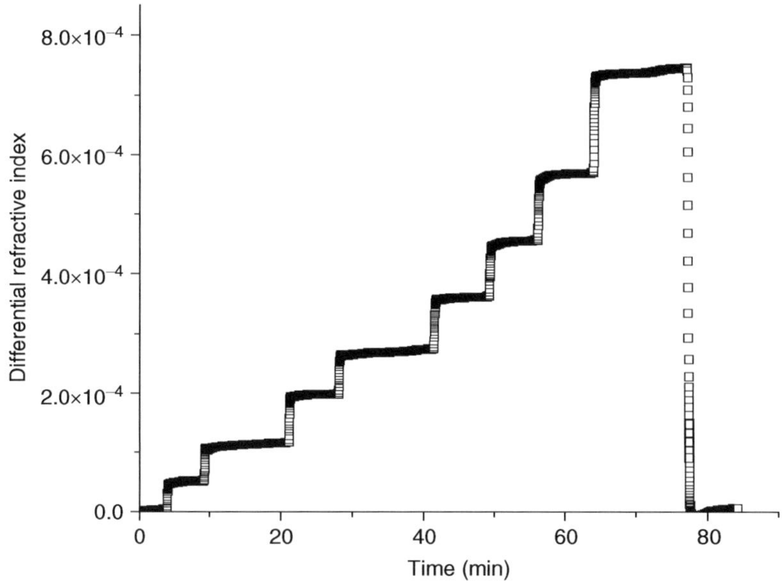

(a)

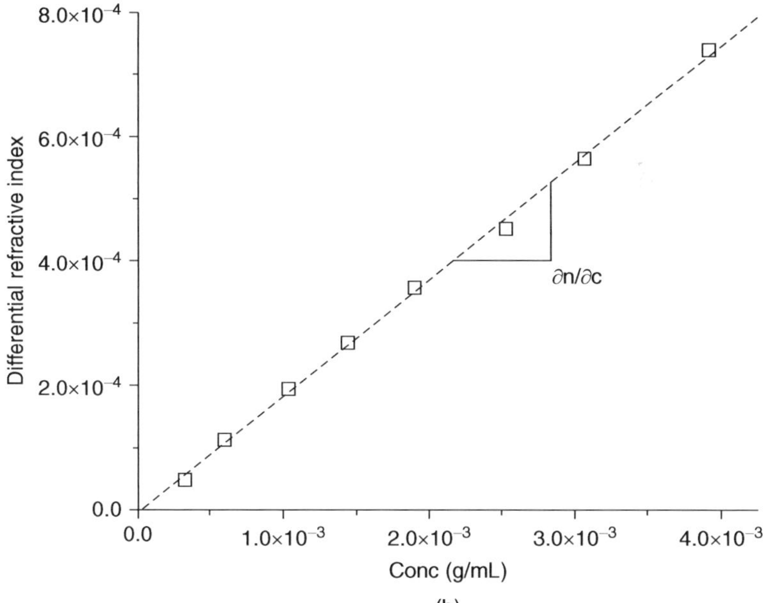

(b)

Figure 9.5 Measuring the specific refractive index increment $\partial n/\partial c$ of a polymer. Sample, 1.86×10^5 g/mol PS; solvent, THF; temperature, 25°C, flow rate, 0.1 mL/min; $\lambda_0 = 690$ nm; concentration range, 0.2 to 7 mg/mL. (a) DRI (after solvent baseline subtraction) of eight PS solutions of increasing concentration as a function of flow rate; (b) DRI (after solvent baseline subtraction) as a function of PS solution concentration. Slope of plot corresponds to the $\partial n/\partial c$ value of the polymer solution, which was found to be 0.192 ± 0.002 mL/g. (Reprinted with permission from Ref. 7.)

given temperature at a given wavelength, and are generally constant across the MMD of a polymer, except in the oligomeric region (see Figure 13.2).

Very sensitive photometers operating over a wide wavelength range provide utility in terms of both applicability and sensitivity. These devices require only a single double bond or similar chromophore per molecule for solute detection, and in favorable cases a few nanograms of a solute can be sensed.

The high versatility of UV and visible light-absorption detectors is due largely to the availability of many solvents of widely varying solvent power which have high transmittance at the wavelengths required for detecting many solutes. Lists of UV cutoff wavelengths for a number of these useful solvents are readily found in the literature (e.g., Reference 9).

Monochromatic photometric detectors are more often used when aqueous solvents are employed. Many of the single-wavelength commercial UV detectors exploit the high-energy 254-nm emission from low-pressure mercury lamps, which permits excellent detectability with low-volume cells. Many biologically important compounds (e.g., proteins, enzymes, aromatic amino acids, nucleic acids) absorb strongly at 254 nm, but because of high inherent sensitivity, the modern photometric detector is useful even when the absorption maximum is not at 254 nm.

UV filter photometers operating at wavelengths other than 254 nm are also available. Light from a highly regulated source passes through sample and reference cells onto silicon photodiodes. The amplified output of these sensors yields a signal that is linear with solute concentration. With a medium-pressure mercury vapor lamp, wavelengths of 254, 280, 312, 316, 436, and 546 nm can be selected simply by changing a filter. However, signal-to-noise specifications are still somewhat inferior to those of photometers operating with a low-pressure mercury source at 254 nm, because of the higher energy associated with this source. Relative detector response at two different wavelengths (e.g., 254 and 280 nm) is sometimes used to confirm the identity of a suspected solute.

Variable-wavelength spectrophotometric detectors are also very useful for SEC. Specially designed equipment, now available for use over a very wide wavelength range (e.g., 190 to 700 nm), employs a monochromator and continuous-spectrum energy source (usually, dual deuterium and tungsten lamps) for maximum versatility and convenience. Spectrophotometric detectors may be set to the absorption maximum of a solute for maximum detection sensitivity, or alternatively, at a wavelength that provides maximum freedom from possible interferences. The example of the latter in Figure 9.6 shows that detection of a yellow impurity in a polymer extract (separated on an oligomeric SEC column set; see Chapter 13) was enhanced by working at 445 nm rather than to 254 nm.

Multiwavelength detectors are also available, employing a 1024-element diode array. These detectors cover a wide spectral range (e.g., 190 to 950 nm) and can acquire many compound-specific wavelengths simultaneously. In some cases, programmable change of wavelength is possible.

UV detectors of all types have high sensitivity (but samples must have some UV absorptivity) and good linearity and can be made with very small cell volumes. They are relatively insensitive to mobile-phase flow rate and temperature changes, and they

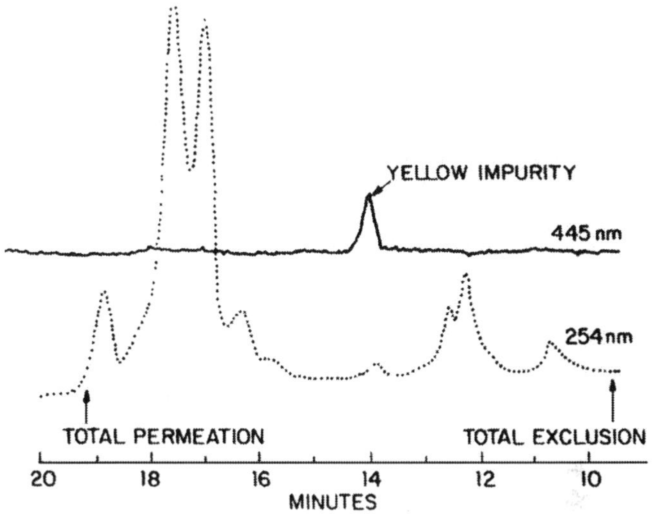

Figure 9.6 Selective detection with UV spectrophotometer. Columns, four 25 × 0.62 cm PSM-50S (75 Å, silanized); mobile phase, methanol; flow rate, 1.25 mL/min; temperature, 27 °C; detectors, UV photometer, 0.02 AUFS at 254 nm, UV spectrophotometer, 0 02 AUFS at 445 nm; sample, 50 μL methanol extract. (Reprinted with permission from Ref. 10.)

are very reliable, easy to operate, and attractive for routine operation with relatively inexperienced personnel. UV detectors for SEC typically exhibit the following properties: cell volume, 1 to 20 μL; noise, $<1 \times 10^{-5}$ absorbance unit full-scale (AUFS) at 254 nm; linear range, 10^4 to 10^5; drift, <0.0005 AU/h; sensitivity to favorable sample, 5×10^{-10} g/mL; absorbance range, 0.005 to 2.54 for full-scale deflection.

9.2.3 Evaporative-Type Detectors

Variously referred to as evaporative light-scattering detectors (ELSD) or evaporative mass detectors (EMDs), in evaporative-type detectors the column effluent is nebulized to form an aerosol. This aerosol then enters a heated drift tube (evaporator), where the mobile phase evaporates and leaves behind particles of analyte (the analyte must be less volatile than the solvent). The analyte particles then enter the optical cell of the detector, where they cause a light beam from a collimated light source to be scattered. The scattered radiation is detected by a detector (photomultiplier tube or photodiode), providing the output signal. A schematic of an ELSD is shown in Figure 9.7.

Although having the disadvantage of being a destructive detector, the ELSD contributes negligible backpressure to an SEC system and can be added in serial fashion at the end of most multidetector setups. If connecting subsequent to a DRI, though, backpressure restrictions for the refractometer should be obtained from the manufacturer in order not to risk damaging the refractometer cell. Other advantages of the ELSD are its ability to handle solvent gradients, its compatibility with a number of

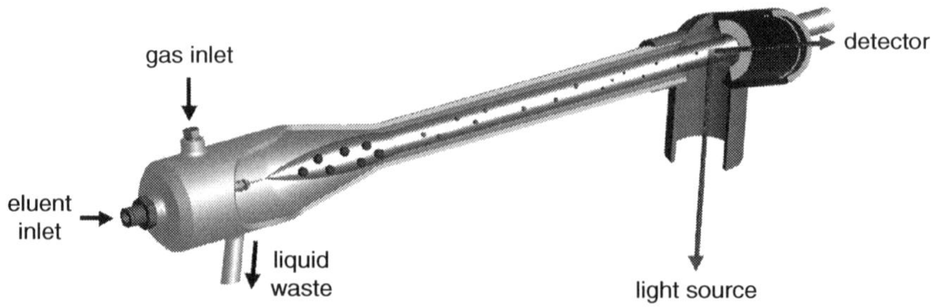

Figure 9.7 Evaporative light-scattering detector. (Courtesy of Varian/Polymer Laboratories.)

solvent modifiers, its relative insensitivity to pump pulsations, the fact that it does not need a chromophoric group on the analyte or analyte derivatization in order to add a chromophore, and its ability to handle strongly absorbing solvents [11–13].

A large droplet size after sample evaporation is desired in order to scatter the largest amount of light. The dependence of droplet size (d_o) on instrumental parameters is described empirically as [11]

$$d_o = \frac{585\sqrt{\sigma_L}}{(u_G - u_L)\sqrt{\rho_L}} + 597\left(\frac{\eta_L}{\sqrt{\sigma_L \rho_L}}\right)^{0.45}\left(\frac{1000 Q_L}{Q_G}\right)^{1.5} \qquad (9.11)$$

where σ_L, ρ_L, and η_L are the surface tension, density, and viscosity of the eluent, respectively; u_G and u_L are the linear velocities of the gas and the liquid stream, respectively; and Q_G and Q_L are the volumetric flow rates of the gas and the liquid stream, respectively.

Higher flow rates of nebulizer gas (air, N_2, He, Ar) result in the formation of smaller droplets, which scatter less light. Because of this, signal response decreases with increasing gas flow. This effect should be counterbalanced by the fact that at very low gas flow rates, nebulization of eluent will be inefficient and baseline noise will increase. Gas flow rates of several liters per minute are normally used, although newer instruments can handle less than that. High-purity gas is not necessary, but the gas should be free of oil and particulate matter. When using house air, filtering is recommended.

A maximum in particle diameter (i.e., signal response) is usually observed with temperature. The temperature must be high enough to evaporate the solvent being used but without thermally degrading the analyte. Temperature also has an effect (although less pronounced) on surface tension, density, and viscosity. Too low a temperature leads to inefficient evaporation of the solvent, resulting in increased baseline noise. For method development, it is recommended that the user start at a high temperature (relative to the boiling point of the solvent) and work downward from there to find the optimal operating temperature. The dependence of particle size on temperature is generally not as strong as the dependence on air pressure. Modern instruments can operate in the range 30 to 220°C for nebulizer temperature and

30 to 300°C for evaporator temperature. Subambient detectors, with integrated cooled evaporators, can operate as low as 10°C. These subambient systems are particularly useful when analyzing thermally labile compounds.

Droplet size will decrease with decreasing solvent flow rate, as a lower flow rate provides more time for the solvent to evaporate. The range over which flow rate can be varied will depend on the choice of SEC column and is usually specified by the column manufacturer. The lowest possible (and practical) flow rate should be employed. In optimizing this parameter it is best to go from low to high, as a nonoptimal plateau can be reached at higher flow rates (e.g., 2 to 3 mL/min). The influence of eluent velocity on detector response is less significant than the influence of temperature or gas flow rate, due to the smaller range over which the eluent velocity can be altered. The current generation of evaporative instruments can handle solvent flow rates in the range 0.1 to 5 mL/min for both aqueous and organic solvents.

Droplet size will also increase with decreasing analyte concentration. ELSD response usually follows a log-linear relationship with concentration given by $A = bm^x$, where A is the peak area observed, m the analyte mass, and b and x are coefficients that depend on both analyte and operating conditions. A plot of log A versus log m should give a straight line with slope x and intercept log b. Curves of order higher than 1 may be necessary to achieve a good fit. It is best to begin with the lowest concentration deemed achievable and to work upward from that.

Care should be taken when working with small molecules. While ELSD response above the oligomeric region (see Section 13.2) is independent of M, oligomers run the risk of being swept away with the nebulizing gas.

9.3 STATIC LIGHT-SCATTERING DETECTION

Static light scattering (SLC; sometimes referred to as *total-intensity light scattering*) has emerged in the last two decades as one of the most powerful detection methods in SEC [14,15]. SLS has the ability to provide information about the molar mass and size of a polymer, dilute solution thermodynamics of the polymer solution, long-chain branching, aggregation, conformation, fractal dimension, and more [16]. Discussion of most of these properties is deferred to Chapter 11. In this chapter we provide an overview of the technique itself. We do this also with quasielastic light scattering (Section 9.4) and viscometry (Section 9.5). In discussing both static and quasielastic LS, the absence of absorption effects by the sample in solution is assumed. In all cases, except for off-line experiments (Section 9.3.3), a concentration-sensitive detector is assumed to be part of the chromatographic setup in addition to the light-scattering photometer.

9.3.1 Multiangle Light Scattering

The amount of light scattered by a dilute polymer solution at a given angle, θ, in excess of that scattered by the neat solvent at the same angle, corrected for distance dependence and incident light intensity, is referred to as the *excess Rayleigh ratio*, $R(\theta)$:

$$R(\theta) = \frac{I_\theta r^2}{I_0} \tag{9.12}$$

where I_θ is the scattered intensity per unit scattering volume, I_0 the intensity of the incident radiation, and r the distance between the scattering volume element and the photodetector of the light scattering unit. In Equation 9.12 it is assumed that background scattering by the neat solvent has been subtracted from the scattering by the polymer solution.

The basic equation of light scattering, known as the *Rayleigh–Gans–Debye* (RGD) *approximation*, is

$$\frac{K^* c}{R(\theta)} = \frac{1}{P(\theta)}\left(\frac{1}{M_w} + 2A_2 c + 3A_3 c^2 + \cdots\right) \tag{9.13}$$

where

$$K^* = \frac{4\pi^2 n_0^2 (\partial n/\partial c)^2}{\lambda_0^4 N_A} \tag{9.14}$$

and

$$\frac{1}{P(\theta)} = 1 + \frac{16\pi^2}{3\lambda^2}\left\langle r^2 \right\rangle_z \sin^2\frac{\theta}{2} + \cdots \tag{9.15}$$

where N_A is Avogadro's number, n_0 the refractive index of the neat solvent, λ_0 the vacuum wavelength of the incident radiation [$\lambda \equiv \lambda_0/n_0$ (Equation 9.3)], c the concentration of polymer in solution, and M_w the weight-average molar mass (Equation 9.16, and Equation 1.2). The specific refractive index increment of the solution, $\partial n/\partial c$, was discussed in Section 9.2.1.3.

The second virial coefficient, A_2, provides a measure of the excess chemical potential (excess Gibbs free energy of dilution) between polymer and solvent molecules in solution (see Section 7.2.1) [17]. $A_2 = 0$ defines theta (Θ) solvent–temperature conditions, a thermodynamically pseudoideal state analogous to the Boyle temperature for gases (this θ should not be confused with the scattering angle, to which it is unrelated). $A_2 > 0$ signifies that the polymer is dissolved in a thermodynamically "good" solvent at the given temperature, while $A_2 < 0$ signifies that the solvent–temperature conditions are thermodynamically "poor." At good conditions, the chemical potential between analyte and solvent is minimized, the molecule is more extensively solvated, and the chain assumes a more extended configuration, due to the buttressing effect of the solvent. The opposite of this is true at poor conditions. The value of A_2 can depend on the method of determination [i.e., if determined by SLS (as outlined below) or by colligative methods (e.g., osmometry)]. It should be noted that terms dependent on higher virial coefficients ($A_3, \ldots$) can also be added to Equation 9.13. These terms, and even the A_2 term, tend to vanish at the near-infinitely dilute concentrations employed in SEC experiments, which are enhanced by the dilutory effects of the separations mechanism itself.

The particle scattering factor, $P(\theta)$, describes the angular dependence of the scattered light. The root-mean-square (rms) radius, $\langle r^2\rangle^{1/2}$, often referred to as the radius of gyration, R_G, is defined as the root-mean-square distance of an array or group of atoms from their common center of mass (i.e., the center of mass of the molecule). R_G can be related directly to the sum of the squares of the distances between every pair of particles, which defines this radius only in terms of the internal coordinates of the polymer, independent of the center of mass. Angular brackets, $\langle\ \rangle$, are used in this book to signify an average over all configurations. $P(\theta)$ was shown by Debye to always be represented by a power series in $\sin^2(\theta/2)$. A plot of $R(\theta)/(K * c)$ versus $\sin^2(\theta/2)$ (appropriately known as a Debye plot) yields a straight line with slope equal to $\langle r^2\rangle$ and y-intercept equal to M_w^{-1}. The angular dependence of the scattered light thus emerges as a necessary measurement to determine the size (actually, *a size*; see Table 9.2) of the polymer in solution, although not as a necessary condition for determining the molar mass.

In an SEC-MALS experiment the scattered light is detected, at a multiplicity of angles θ, for each slice i eluting from the SEC column. M_w is a fundamental parameter measured by SLS (multi- or single-angle) at each slice, and its value is "absolute," that is, independent of such experimental conditions as solvent and temperature. As shown in Equation 9.13, the excess Rayleigh ratio is proportional to the product of the concentration and the molar mass: $R(\theta) \propto c \times M_w$. For this reason, the light-scattering photometer is referred to as a molar-mass-sensitive detector [3]. This makes SEC-SLS an ideal method for the detection and study of solution aggregates and aggregation mechanisms, as even a small amount of high-molar-mass aggregates will scatter a substantial amount of light.

From Equation 9.13 we see that $M_w \propto R(\theta)/c$. This means that for each individual chromatographic slice i, the ratio of the signals from the SLS photometer $R(\theta)_i$ and from the concentration-sensitive detector c_i is proportional to $M_{w,i}$, the weight-average molar mass of that slice. The slices eluting from an SEC column are sufficiently narrow that they are assumed to be monodisperse. For any given slice i, $M_{z,i} \approx M_{w,i} \approx M_{n,i} \approx M_i$. Using a combination of concentration-sensitive and SLS detection, the c_i and M_i at each elution slice are combined to yield the MMD and associated M-averages of the polymer via

$$M_\beta = \frac{\sum c_i M_i^x}{\sum c_i M_i^{x-1}} \qquad x = 0, \beta = n; x = 1, \beta = w; x = 2, \beta = z \tag{9.16}$$

This calculation requires no a priori knowledge of the concentration of the polymer solution (i.e., it is independent of whether or not 100% of the polymer actually dissolves in the solvent). In the case of partial dissolution, the MMD and M-averages calculated are those of the dissolved portion of the analyte.

$\langle r^2 \rangle$ is also a fundamental parameter measured by MALS (although not by single-angle SLS) at each slice. The value of $\langle r^2 \rangle$ is considered "relative" in that the radius will change as solvent, temperature, and so on, are varied. A distribution of the mean-square radii is also obtained from an SEC-MALS experiment, with the measured statistical moments defined as [15]

$$\langle r^2 \rangle_\beta = \frac{\sum c_i M_i^x r^2}{\sum c_i M_i^x} \qquad x = -1, \beta = n; x = 0, \beta = w; x = 1, \beta = z \qquad (9.17)$$

A useful rule of thumb is that the minimum value of $\langle r^2 \rangle^{1/2}$ measurable by MALS is approximately $\lambda/20$ ($= \lambda_0/20n_0$). This value depends on the solvent–temperature conditions of the experiment, which enter into the equation through the refractive index of the solvent, n_0. Molecules smaller than $\lambda/20$ generally do not possess sufficient angular dissymmetry to enable accurate measurements of the root-mean-square radius. As seen in Chapter 11, the molar mass dependence of $\langle r^2 \rangle^{1/2}$ is of particular importance (e.g., for determining long-chain branching or fractal dimension). Henceforth, we use R_G to abbreviate the root-mean-square radius (i. e., $R_G \equiv \langle r^2 \rangle^{1/2}$).

9.3.1.1 MALS Instrumentation. MALS instruments are currently available in two-angle, three-angle, seven-angle, and eighteen-angle configurations, all using linearly polarized lasers. Product specifications of the various commercially available instruments are summarized in Table 9.1. There are two dual-angle instruments on the market, although one of these, with photodetectors at 7° and 90°, operates in either low- or right-angle mode but does not use both angles simultaneously. Low-angle light scattering is discussed in Section 9.3.2. The combination of right-angle light scattering with viscometry and refractometry, termed SEC^3, is discussed in Section 9.6.

The second dual-angle instrument uses photodetectors located at 15° and 90°. This detector can be housed in the same unit as a refractometer and a viscometer in addition to a quasielastic light-scattering (QELS; see Section 9.4) photometer, which minimizes interdetector band broadening. A schematic of this detector is shown in Figure 9.8.

The three-angle MALS unit is a room-temperature system with available high-temperature option for operation up to 150°C. Photodiodes are placed at 45°, 90°, and 135°, the first and last being nominal angles (i.e., dependent on solvent–temperature conditions). Embedding of a QELS unit is also possible. A schematic of the cell in this detector is shown in Figure 9.9; the flow cell assembly is shown in Figure 9.11b.

The seven-angle instrument, the cell of which is shown in Figure 9.10, has photodetectors [charge-coupled devices (CCDs)] mounted at angles ranging from 35° to 145°. This is a self-venting cylindrical cell with solvent flow from bottom to top, to reduce bubbles. The CCDs are attached directly in the flow cell, with no glass–glass reflectance that must be factored in to obtain the true angle.

The 18-angle MALS instrument has photodiodes placed at nominal angles ranging from 10° to 160°, although in flow-through mode (e.g., when coupled to an SEC

Table 9.1 Product specifications of commercial light-scattering detectors

	LALS	Two-Angle	Three-Angle	Seven-Angle	18-Angle
Manufacturer	Malvern/Viscotek	Varian/Precision Detectors	Wyatt Technology Corp.	PSS Polymer Standards Service	Wyatt Technology Corp.
Angular placement of photodiodes (deg)	7	15, 90	45, 90, 135	35, 50, 75, 90, 105, 130, 145	22, 28, 32, 38, 44, 50, 57, 64, 72, 81, 90, 99, 108, 117, 126, 134, 141, 147
Scattering volume (μL)		0.01	0.02	0.02	0.07
Laser wavelength (nm)	670	680 or 809	658	635	658
Laser source (mW)		35 or 100	60	35	120
M range (g/mol)	10^3 to $>10^6$	10^3 to $>10^7$	10^3 to 10^6	10^3 to 10^8	10^3 to 10^9
R_G range (nm)	10 to 150	10 to 200	10 to 50	10 to 150	10 to 500
Temperature range (°C)	30 to 80	4 to 50 or 4 to 90	Ambient	30 to 80	−15 to 80
Limit for high temperature (°C)	175	250	175		150 or 210
Options/features	90° measurement for SEC^3	QELS	QELS at 90°	Self-cleaning CCDs attached with no glass–glass reflectance	Variable-angle QELS; polarization option; two flow cells, depending on solvent refractive index; integrated ultrasonic cell cleaner

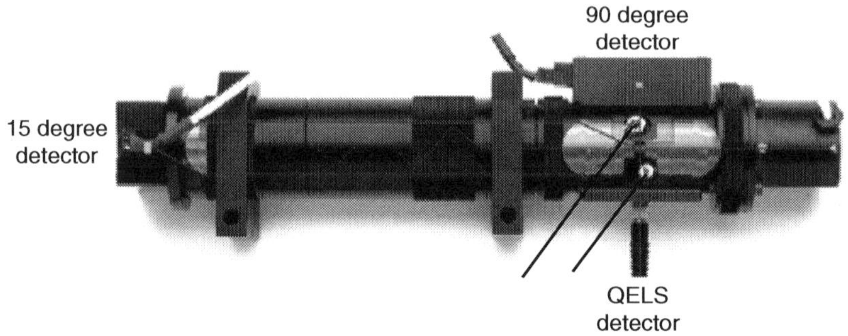

Figure 9.8 Two-angle static light-scattering detector, including QELS detector. (Courtesy of Precision Detectors Inc.)

system) the lowest angle is not accessible. Flow path is collinear with the laser. Two types of flow cell are available, depending on whether the refractive index of the solvent is above or below 1.50. The system comes equipped with an integrated ultrasonic cell cleaner and with a cell retainer for optional addition of polarization filters (see Section 9.3.4 for a discussion of depolarized MALS). This system can be housed in the same unit as a QELS system (Section 9.4). Schematics of the read head and flow cell assembly for this detector are shown in Figure 9.11. The flow cell for this detector is the same as that for the three-angle MALS system described above.

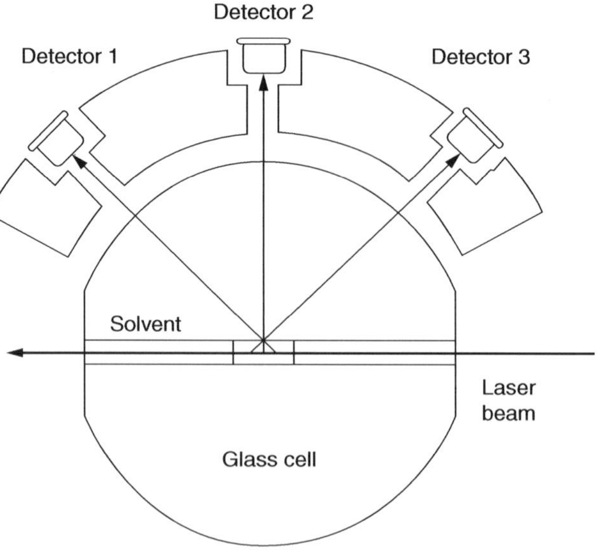

Figure 9.9 Cell of a three-angle MALS system. (Courtesy of Wyatt Technology Corp.)

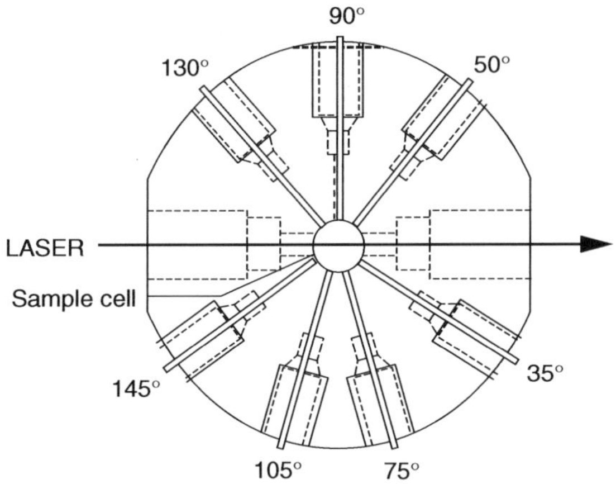

Figure 9.10 Cell from a seven-angle MALS detector. (Courtesy of PSS Polymer Standards Service.)

9.3.2 Low-Angle Light Scattering

In a LALS experiment the particle scattering factor, $P(\theta)$, may be ignored with a high degree of confidence. At $\theta = 0°$ (an highly impractical measurement to perform, as the unscattered laser light would probably overload the photodiode of an SLS unit), $P(\theta) = 1$. At $\theta = 7°$, $P(\theta) = 0.98$ for macromolecules with $R_G \leq 150$ nm. From Equations 9.13 to 9.15, we see that measuring the intensity of scattered light at a multiplicity of angles is only necessary to determine the mean-square radius of the polymer. Determination of M_w can be performed using a single angle and, indeed, is performed most accurately by low-angle measurements.

The current generation of LALS detectors measures light scattered at 7°, with an option to measure at 90° instead [18]. The detector is available with either a differential viscometer, a differential refractometer, a UV detector, or a combinations of these. Product specifications are given in Table 9.1.

9.3.3 Off-Line, Batch-Mode MALS

As the title suggests, this is an off-line experiment (i.e., without the separation capabilities of SEC). It is included because it provides one key size parameter that will be used synergistically with size parameters determined by SEC with various online detection methods.

The basic off-line, batch-mode MALS experiment consists of measuring the scattering, at a multiplicity of angles, for a series of solutions of different concentrations [19]. The results for each concentration at each angle are plotted together in what is known as a *Zimm plot*, with $K^*c/R(\theta)$ as the ordinate and $\sin^2(\theta/2) + kc$ as the abscissa. k is a number, with dimensions of reciprocal concentration (i.e., volume/mass), chosen to give a good visual distribution of data points on the graph. The

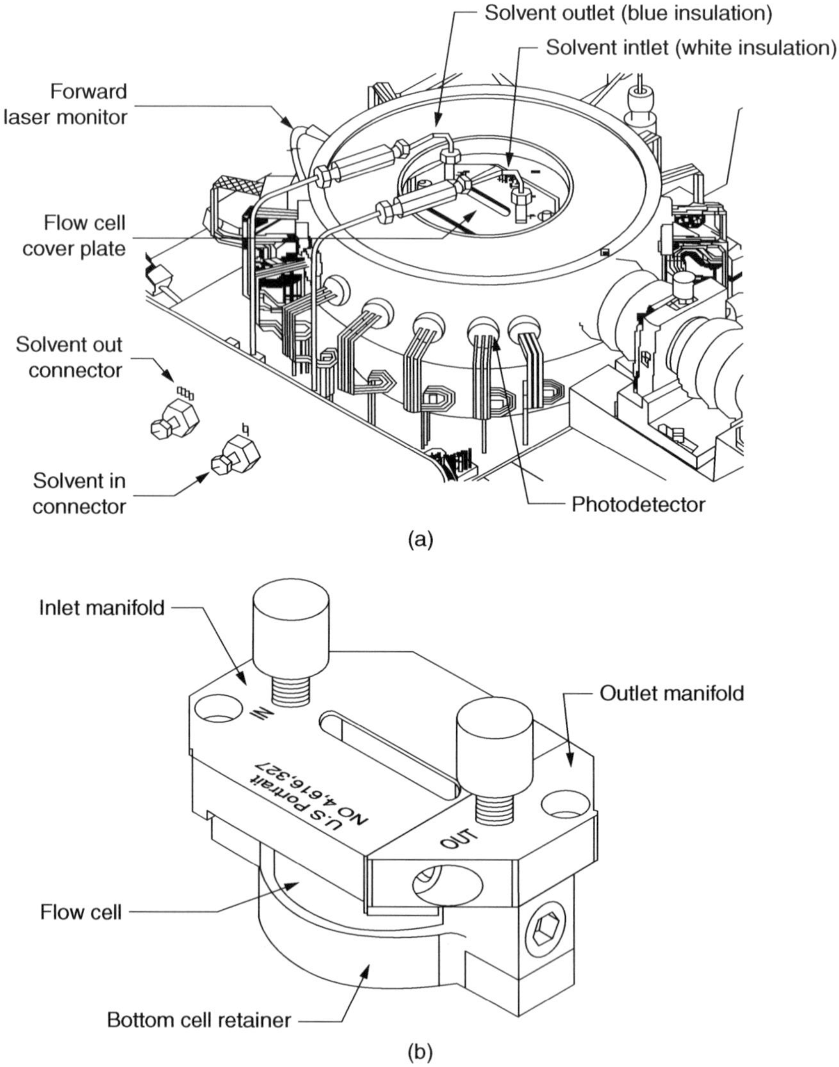

Figure 9.11 (a): Read head of an 18-angle MALS unit; (b) flow cell assembly of three- and 18-angle MALS units. (Courtesy of Wyatt Technology Corp.)

slope of the line constructed from the angular data extrapolated to zero concentration is proportional to $R_{G,Z}$, the z-average radius of gyration. When the concentration data are extrapolated to zero angle, the slope of this extrapolated line is proportional to A_2, the second virial coefficient. The common y-intercept of the two extrapolated lines equals $1/M_w$.

A variation of the Zimm plot is the *Berry plot*, where the ordinate is the square root of $K^*c/R(\theta)$ and the abscissa remains $\sin^2(\theta/2) + kc$. For branched polymers or for polymer solutions at good solvent–temperature conditions, the Berry plot often

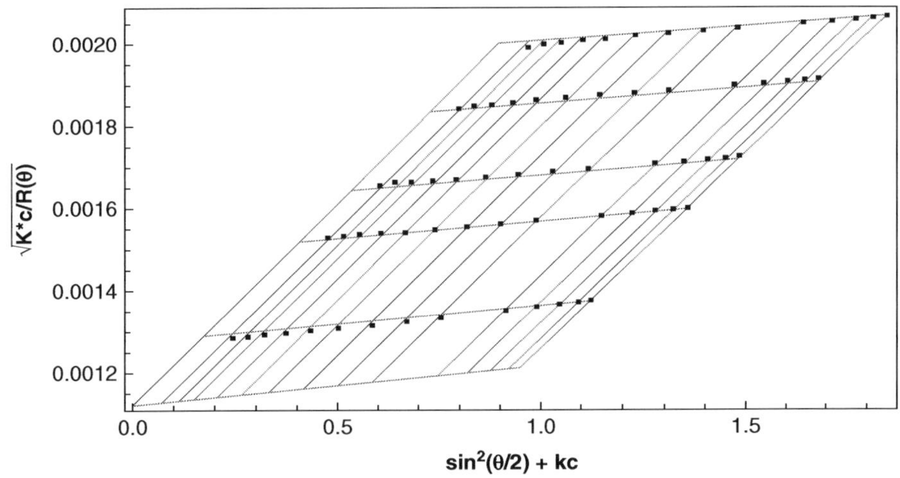

Figure 9.12 Berry plot for off-line, batch-mode MALS analysis. Sample, PMMA; solvent, THF; temperature, 20 °C. Scattering from five concentrations, ranging from 1–5 $\times$ 10^{-3} mg/mL, was measured at fourteen different angles, ranging from 31–155°. Results: $M_w = 7.90\ (\pm 0.06) \times 10^5$ g/mol, $R_G = 28\ (\pm 1)$ nm, $A_2 = 1.87\ (\pm 0.02) \times 10^{-4}$ mol mL/g^2. (Reprinted with permission from Ref. 20.)

provides a better fit to the data than does the Zimm plot. An example of a Berry plot, for a 7.90×10^5 g/mol PMMA sample, is shown in Figure 9.12 [20].

The seven- and eighteen-angle instruments described above can be used for this type of analysis, which can also be done using a variable-angle instrument. (Variable-angle instruments are not used as SEC detectors, however, so they are not discussed here.) A caveat to batch-mode MALS experiments is that they measure average values of the molar mass or radius of the sample but provide no information about the distribution of these values. Consequently, a small amount of aggregation may skew M_w and R_G to much higher values than the true value for the unaggregated sample. Whether or not aggregation is present should become obvious by coupling the light-scattering detector to an SEC system: The larger aggregates will be separated from the smaller, unaggregated polymers. The size and molar mass of the aggregated and unaggregated components of the sample can then be measured, individually.

Reasons to perform an off-line, batch-mode MALS experiment include:

1. A separation system is not available.
2. The sample availability is limited.
3. The sample is known to be monodisperse (or virtually monodisperse) and/or not to aggregate at the solvent–temperature conditions of the experiment.
4. Only average values of M or R_G are needed for a particular study.
5. To avoid on-column, flow-induced degradation during an SEC experiment or to determine whether the sample has degraded during its passage through an SEC column (Section 7.2.3) [21]. If degradation has occurred, the size and molar mass of the sample will be larger when measured in off-line mode, as long as aggregation is not present.

6. To determine the value of the second virial coefficient, in order to ascertain whether the polymer solution is at good, poor, or theta conditions.
7. To obtain the value of the thermodynamic radius, R_T, defined as [17]

$$R_T \equiv \left(\frac{3A_2 M_w^2}{16\pi N_A} \right)^{1/3} \tag{9.18}$$

This radius can be thought of as the radius of a hard sphere with the same excluded volume as the polymer. How R_T is used to determine, for example, the coil interpenetration function of polymers, is discussed in Section 11.4.3.

9.3.4 Depolarized MALS

Various features of polymers and polymer solutions are associated with the ability to depolarize incident radiation: chain stiffness, aggregation, sample contamination, and so on. By coupling depolarized static light scattering to SEC, the depolarization characteristics of a sample can be measured across the MMD [22–24]. In this section we describe the online coupling of depolarized MALS (D-MALS) to SEC and give examples of the type of information obtained from SEC/D-MALS. As with other types of SLS detection, a concentration-sensitive detector is needed as part of the experimental setup if one wishes to relate depolarization information to molar mass across the elution profile of the sample.

In an SEC/D-MALS experiment, two optical filters, each consisting of a strip of Polaroid film, are placed around the sides of the flow cell of the MALS unit. One strip is polarized vertically and the other is polarized horizontally. The incident light from the laser in the light-scattering unit is also polarized, usually vertically. Scattering data from a polymer solution are initially acquired with the vertically polarized filter facing the odd-numbered photodiodes in a MALS photometer and the horizontally polarized filter facing the even-numbered photometers. The placement of the filters is then reversed and data are reacquired for the same sample dissolution as examined using the original placement of the filters.

Isotropic components of the sample should not depolarize the incident radiation. Therefore, when vertically polarized incident radiation interacts with an isotropic molecule in the flow cell of the MALS photometer, none of the scattered radiation should reach the photodiodes facing the horizontal polarization filter and all of the scattered radiation should reach the photodiodes facing the vertical polarization filter. (In actuality, there is some absorption of radiation by the filters. A method of correcting for this absorption is described in References 22 and 23.) In this case, the depolarization ratio equals zero. The depolarization ratio is given the symbol ρ_θ^v, where the superscript v denotes the polarization state of the incident radiation and the subscript θ denotes the scattering angle. The depolarization ratio is defined as

$$\rho_\theta^v = \left(\frac{I_h^v}{I_v^v} \right)_\theta \tag{9.19}$$

where I_h^v and I_v^v are the intensities of the horizontally and vertically polarized components of the scattered radiation, respectively.

If the scattering particle is not completely isotropic, fluctuations in the orientation of the particle offer the possibility of additional scattering, as the induced dipole moment is generally no longer parallel to the electric vector of the incident light. The excess scattering due to anisotropy is related to the depolarization ratio. For a system of anisotropic particles, the excess Rayleigh ratio $R(\theta)^{v,\mathrm{tot}}$ is

$$R(\theta)^{v,\mathrm{tot}} = R(\theta)^{v,\mathrm{iso}} \frac{3 + 3\rho_\theta^v}{3 - (4 + 7\cos^2\theta)\rho_\theta^v} \tag{9.20}$$

where $R(\theta)^{v,\mathrm{iso}}$ is the excess Rayleigh ratio for isotropic particles. For a given scattering angle θ,

$$R(\theta)^{v,\mathrm{tot}} = R(\theta)^{v,\mathrm{iso}} C^v(R(\theta)) \tag{9.21}$$

The correction term $C^v(R(\theta))$, known as the *Cabannes factor*, is defined as

$$C^v(R(\theta)) = \frac{3 + 3\rho_\theta^v}{3 - (4 + 7\cos^2\theta)\rho_\theta^v} \tag{9.22}$$

Striegel has used SEC/D-MALS to plot the distribution of $C^v(R(\theta))$ as a function of M and, simultaneously, as a function of θ. As shown in Figure 9.13 for a sample of PSBr dissolved in DMAc/LiCl, the depolarization behavior of dilute polymer solutions can possess both molar mass dependence and angular dependence.

SEC/D-MALS can also be used to detect sample contamination if the sample and contaminant are of different size and if the contaminant depolarizes light differently than does the sample. An example of this is shown in Figure 9.14 for an industrial sample (Sample A) containing a small amount of impurity [22]. In Figure 9.14, the smooth solid line corresponds to the SEC/MALS signal (i.e., when no polarization filters were installed around the MALS flow cell), the dashed line is the SEC/D-MALS signal with the filters in the vertical position (i.e., in aligned-polarization mode, with the transmission axis of the optical filters parallel to the polarization axis of the laser), and the jagged line is the SEC/D-MALS signal with the filters in the horizontal position (i.e., in cross-polarization mode, with the transmission axis of the filters perpendicular to the polarization axis of the laser). The retention time of Sample A is about 25 minutes; the retention time of the impurity is about 35 minutes. Without optical filters and in aligned-polarization mode, S/N for the impurity is about 2 : 1 or less, and the impurity is barely distinguishable from the chromatographic baseline. In SEC/D-MALS operating in cross-polarization mode, S/N for the peak at about 35 minutes is above 60 : 1 and the presence of the impurity becomes evident.

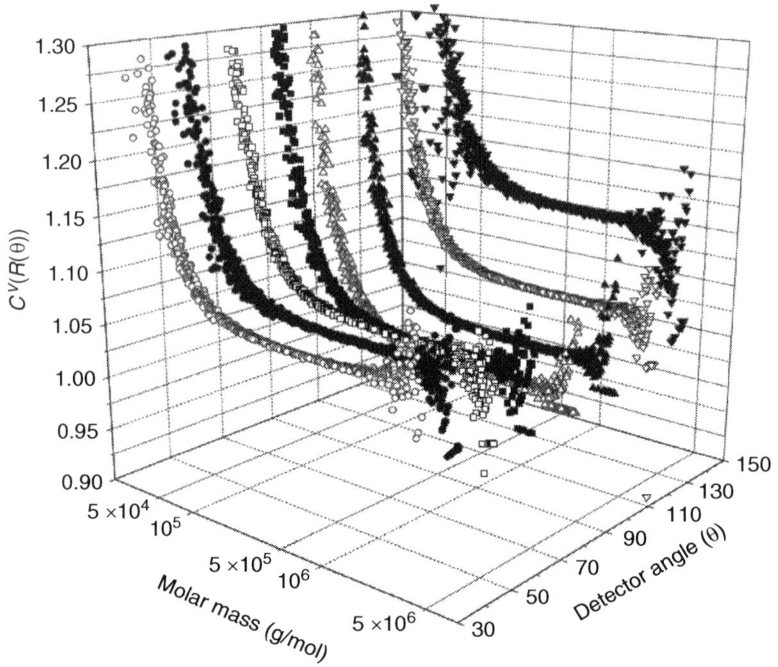

Figure 9.13 SEC/D-MALS to study depolarization of dilute polymer solutions: distribution of the Cabannes factor $C^v(R(\theta))$ as a simultaneous function of M and of scattering angle θ. Sample, PSBr; solvent, DMAc/0.5% LiCl; temperature, 35°C; flow rate, 1 mL/min; columns, set of three GRALlinear and one GRAL1000 10-μm columns, preceded by a guard column; detectors, DRI and 18-angle MALS. Scattering angles plotted in figure: 30°, 43°, 56°, 72°, 90°, 108°, 127°, 142°. (Reprinted with permission from Ref. 23.)

9.4 QUASIELASTIC LIGHT-SCATTERING DETECTION

Quasielastic light scattering, also referred to as *dynamic light scattering* (DLS) and *photon correlation spectroscopy* (PCS), is used extensively for particle-size analysis of systems such as latexes, lubricants, and additives. In recent years, QELS has become available as an online detection technique for SEC.

In a dilute solution or suspension, the result of the dissolved particles colliding continuously with solvent molecules is random thermal motion, better known as *Brownian motion*. This motion will cause the intensity of scattered light reaching a photodetector to fluctuate with time about some average value $\langle I \rangle$, as shown in Figure 9.15a. The time scale of the fluctuations is related to the time scale of the particle motions, as characterized by the translational diffusion coefficient D_T of the particles. Large particles possess small diffusion coefficients, and the intensity of the light scattered by these particles will fluctuate slowly. The opposite is true of small particles. The Brownian motion produces a Doppler-type effect in which the scattered light possesses a range of frequencies very slightly shifted from the frequency of the

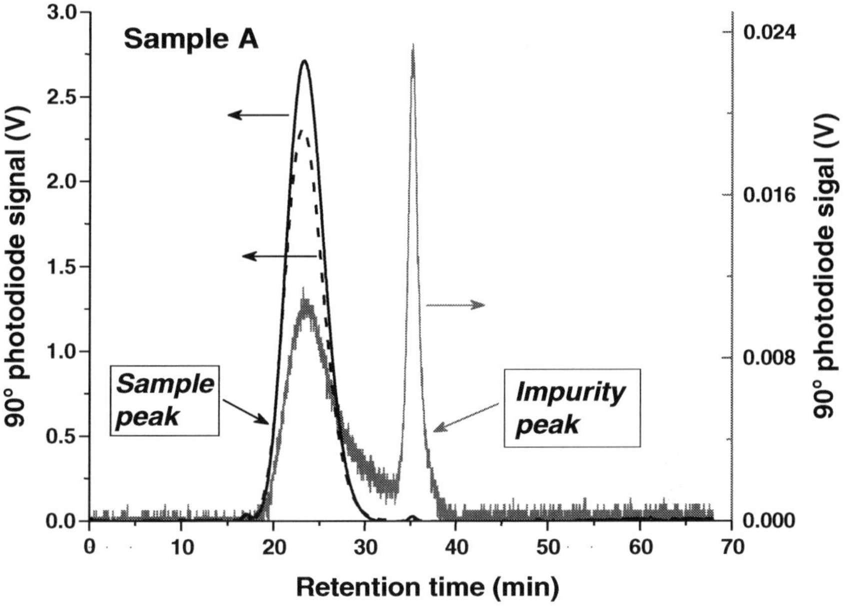

Figure 9.14 Using SEC/D-MALS to detect sample contamination: signals from the 90° photodiode of the MALS detector for industrial Sample A. Smooth solid line, SEC/MALS; dashed line, SEC/D-MALS in aligned-polarization mode; jagged line, SEC/D-MALS in cross-polarization mode. Experimental conditions same as for Figure 9.13. (Reprinted with permission from Ref. 22.)

incident light. This type of scattering is referred to as quasielastic. In static (elastic) light scattering, the time-averaged fluctuations of the scattered light are measured, whereas in QELS it is the time-dependent fluctuations that are of interest.

The average of the scattering intensities at two times, $I(t)$ and $I(t+\tau)$, separated by a period (delay time) τ, is known as the *autocorrelation function* and is written as $\langle I(t)I(t+\tau)\rangle$. Over a long period, T_A, the autocorrelation function is given by [25]

$$\langle I(t)I(t+\tau)\rangle = \lim_{T_A\to\infty}\left[\frac{1}{T_A}\int_0^{T_A} I(t)I(t+\tau)\,dt\right] \tag{9.23}$$

As shown in Figure 9.15b, when $\tau = 0$, $\langle I(t)I(t+\tau)\rangle = \langle I^2\rangle$. As τ increases, the value of the autocorrelation function asymptotically approaches the baseline $\langle I\rangle^2$. The difference between these two values is $\langle \Delta I^2\rangle$; that is,

$$\langle \Delta I^2\rangle = \langle I^2\rangle - \langle I\rangle^2 \tag{9.24}$$

The autocorrelation function can be written as the sum of $\langle I\rangle^2$, the baseline, and $\langle \Delta I(t)\,\Delta I(t+\tau)\rangle$, the average of the fluctuating components of the scattering at the

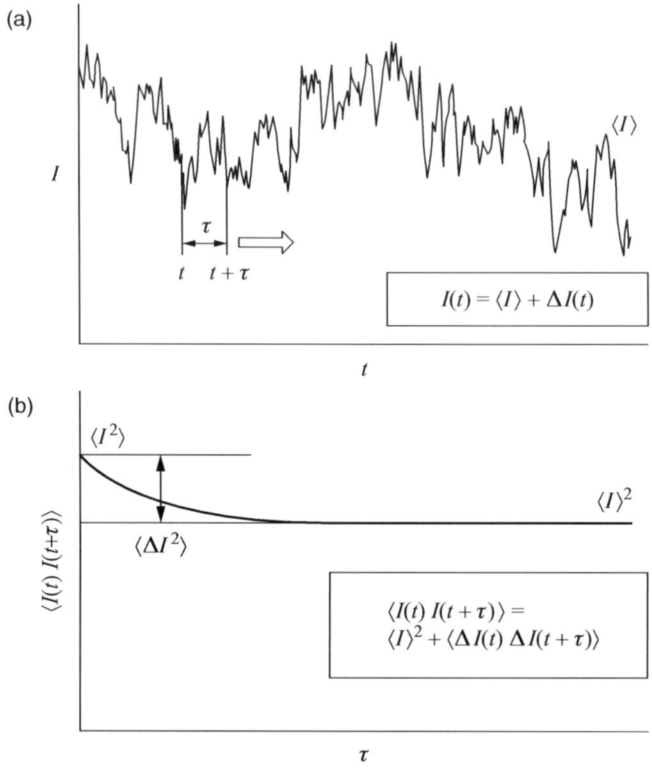

Figure 9.15 Light-scattering intensity fluctuations and autocorrelation function in QELS: (a) fluctuations in light-scattering intensity, $I(t)$, about the mean, $\langle I \rangle$; (b) decay in autocorrelation function over time. (Reprinted with permission from Ref. 25.)

two scattering times t and $t + \tau$:

$$\langle I(t)I(t + \tau)\rangle = \langle I\rangle^2 + \langle \Delta I(t)\, \Delta I(t + \tau)\rangle \tag{9.25}$$

After normalization and baseline subtraction, the following relationship is obtained:

$$\frac{\langle I(t)I(t + \tau)\rangle}{\langle I\rangle^2} = 1 + f_{\text{coh}} g_2(\tau) \tag{9.26}$$

The coherence factor, f_{coh}, depends on the coherence of light incident on the photodetector and is defined as

$$f_{\text{coh}} \equiv \frac{\langle \Delta I^2\rangle}{\langle I\rangle^2} \tag{9.27}$$

In general, $0 < f_{coh} < 1$. The factor $g_2(\tau)$ in Equation 9.26 is the baseline-subtracted, normalized intensity autocorrelation function, defined as

$$g_2(\tau) \equiv \frac{\langle \Delta I(t)\, \Delta I(t+\tau) \rangle}{\langle \Delta I^2 \rangle} \tag{9.28}$$

The relationship between $g_2(\tau)$ and the autocorrelation function of the electric field of the scattered light, $g_1(\tau)$, is given by

$$g_2(\tau) = |g_1(\tau)|^2 \tag{9.29}$$

A decay in $|g_1(\tau)|$ is observed with increasing τ. This decay is directly related to the rate of movement of the analyte. For a monodisperse species (as each slice eluting from the SEC columns is assumed to be), the decay curve is that of a single exponential where

$$|g_1(\tau)| = \exp(-\Gamma\tau) \tag{9.30}$$

Here, Γ is the decay rate, related to the translational diffusion coefficient by

$$\Gamma = D_T q^2 \tag{9.31}$$

where q is the scattering vector, defined as

$$q = \frac{4\pi}{\lambda} \sin(\theta/2) \tag{9.32}$$

From the translational diffusion coefficient D_T, a third size parameter is obtained, the hydrodynamic or Stokes radius (R_H):

$$R_H = \frac{k_B T}{6\pi \eta_0 D_T} \tag{9.33}$$

where k_B is Boltzmann's constant, T the absolute temperature, and η_0 the viscosity of the solvent. R_H can be considered the radius of an equivalent hard sphere that feels the same force due to flow as does the macromolecule. Measurement of R_H is possible down to a few nanometers.

Figure 9.16 shows the QELS-obtained intensity correlation function of a 1.86×10^5 g/mol narrow polydispersity PS sample, at theta conditions. From these data, a $D_{T,z}$ value of $3.02 \pm 0.13 \times 10^{-7}$ cm^2/s was derived. Applying Equation 9.33, an $R_{H,z}$ of 7.5 ± 0.3 nm was obtained [17].

The translational diffusion coefficient measured by QELS is a *mutual* diffusion coefficient, driven by a concentration gradient of solute molecules. This differs from *tracer diffusion*, where the migration of a labeled solute molecule through a medium of uniform solute concentration is followed. This difference probably explains many

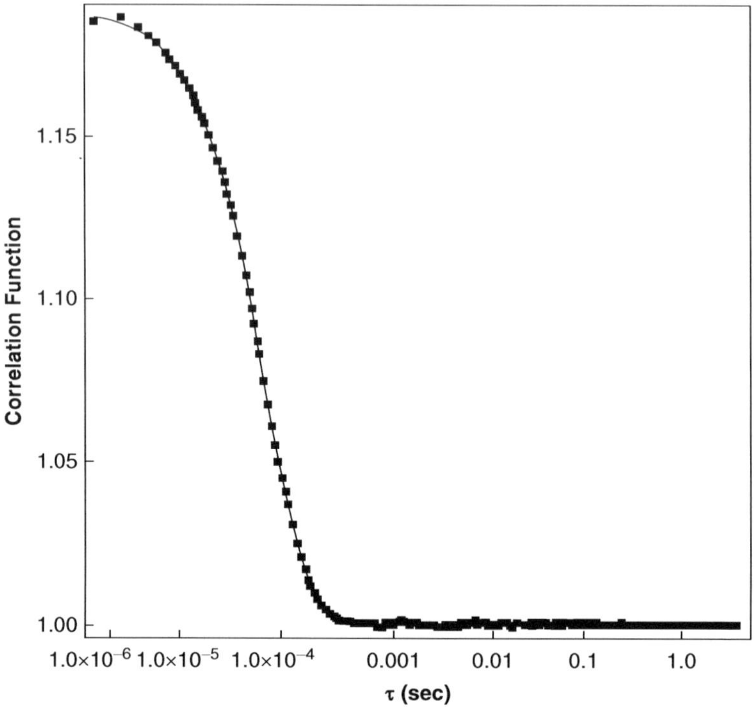

Figure 9.16 Correlation function from QELS. Sample, 1.86×10^5 g/mol PS, with $M_w/M_n \leq 1.02$; solvent, cyclohexane; temperature, 34°C (theta-state conditions); scattering angle, 108°. (Adapted from results in Ref. 17.)

of the discrepancies between diffusion coefficients measured by QELS versus those measured by membrane methods, for example.

9.4.1 QELS Instrumentation

We discuss here only QELS instruments available as online detectors for SEC, not detectors that can be used exclusively in batch mode. The QELS detectors are all available for use with the specific MALS detectors discussed in Section 9.3.1.1, at the same operating conditions mentioned (temperature range, vacuum wavelength of incident laser), and all use avalanche photodiodes. Because the QELS photodiodes are part of the same hardware unit as the MALS systems, there is no need to measure interdetector delays between the MALS and QELS.

The QELS unit in the two-angle SLS system (Figure 9.8) samples a light-scattering volume of 0.01 μL, and the manufacturer claims the ability to measure R_H from 1 to 1000 nm. The QELS in the three- and 18-angle MALS systems (Figures 9.9 and 9.11) examines a measured volume of 0.2 nL. The manufacturer claims an R_H range of 1 to 30 nm in flow-through mode and up to 500 nm in batch mode.

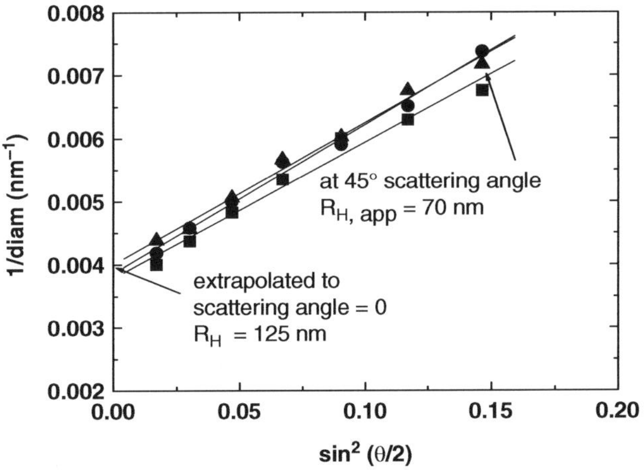

Figure 9.17 Angular dependence of quasielastic light scattering. Sample, 1.3×10^7 g/mol PS. (Reprinted with permission from Ref. 26.)

In the two- and three-angle systems, the QELS performs a 90° measurement. In the eighteen-angle unit, it is possible to place the QELS photodiode and any of the eighteen angular positions except 90°.

The need for variable detector placement is demonstrated in Figure 9.17 for an ultrahigh molar mass PS of 1.3×10^7 g/mol. For samples that are not hard spheres, the center-of-mass translational diffusion coefficient is obtained only in the limit of $q \to 0$. As shown in Figure 9.17, at a scattering angle θ of 45°, an apparent value of R_H of 70 nm is determined, and the apparent R_H is seen to be angle dependent. Measurement at a variety of angles and extrapolation to $\theta = 0°$ gives the actual R_H of the sample, which is 125 nm.

9.5 VISCOMETRIC DETECTION

For laminar flow between two plates or within a cylinder, driven by a constant pressure gradient, flow is steady and the time derivatives can be eliminated from the Navier–Stokes equations [27]. An exact solution of these equations is obtained for what is known as *Poiseuille flow*. This solution relates the pressure drop ΔP across a capillary to the length L and radius r of the tube as well as to the viscosity η of the solution flowing through the tube and to the volumetric flow rate Q through the tube. The relation is known as the *Poiseuille* (or *Hagen–Poiseuille*) *equation* [1,14]:

$$\Delta P = \frac{8L\eta Q}{\pi r^4} \tag{9.34}$$

When L, r, and Q are known quantities, η can be directly related to ΔP. This is the case for most experimental setups, which employ capillaries of known length and radius and where the flow rate through the tube is carefully controlled. In addition to

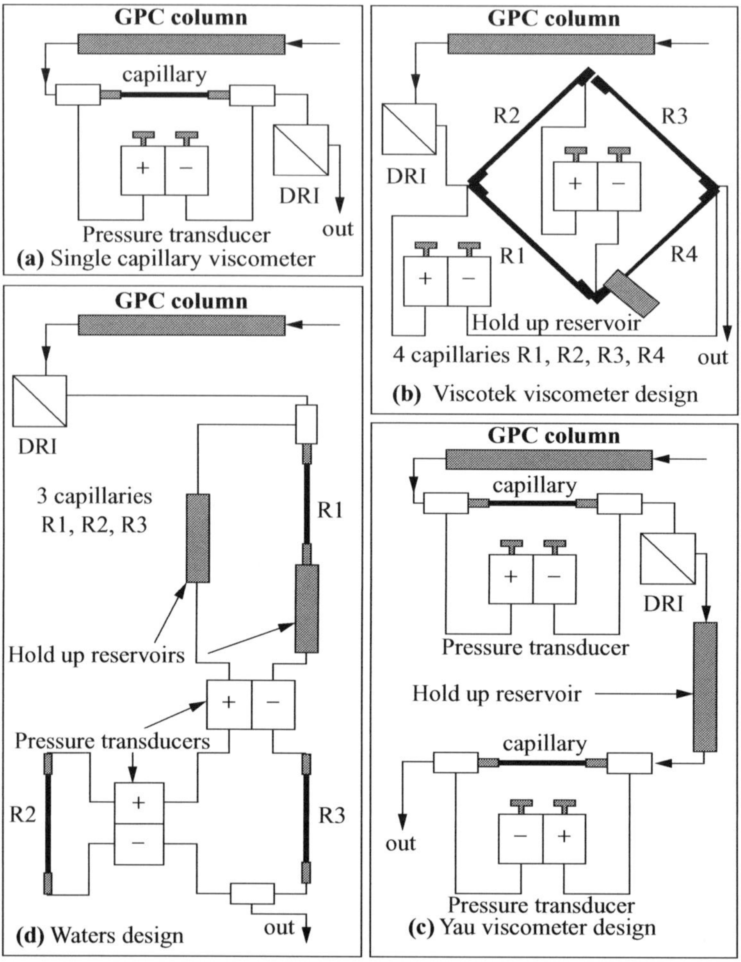

Figure 9.18 Design of various viscometric detectors. See Sections 9.5.1 and 9.5.2 for discussions of each type of viscometer. (Reprinted with permission from Ref. 28.)

laminarity, the Poiseuille equation assumes a no-slip boundary condition at the walls of the capillary and also assumes that the capillary possesses a uniform cross section.

Figure 9.18 shows the schematics of a single-capillary viscometer and of three different types of differential viscometers [28], all discussed in the following sections.

9.5.1 Single-Capillary Viscometers

In a single-capillary system (Figure 9.18a), a differential pressure transducer measures the pressure drop across the tube. For a constant flow rate, the viscosity of each slice i eluting from the SEC columns is calculated using Poiseuille's law (Equation

9.34). The specific viscosity of the solution, η_{sp}, is then calculated via [3]

$$\eta_{sp} = \frac{\eta - \eta_0}{\eta_0} = \frac{\eta}{\eta_0} - 1 = \frac{P_i - P_0}{P_0} = \frac{V_i - V_0}{V_0} \tag{9.35}$$

where η and η_0 are the viscosities of the polymer solution and of the neat solvent, respectively; P_i and V_i represent, respectively, the pressure drop along the capillary and the voltage signal corresponding to that pressure drop, for slice i; and P_0 and V_0 correspond to the average solvent baseline values of the same quantities. Because only one transducer is present, no calibration is necessary. The extreme sensitivity of single-capillary viscometers to even minor flow-rate variations was one reason behind the development of the differential viscometer.

9.5.2 Differential Viscometers

In differential viscometry (Figure 9.18b–d), the polymer solution flows through one part of the viscometric detector and pure solvent through another part, creating a pressure imbalance. For this type of arrangement to work, half the sample is delayed in a hold-up reservoir. Due to emptying of this hold-up reservoir, the presence of a large, negative hold-up peak in the SEC-viscometer trace must be taken into account when setting acquisition times in SEC, in order not to have this peak interfere with the next injection.

The most common type of differential viscometer is the commercial Viscotek design, which is therefore discussed here more extensively than are the other types of viscometers [29]. As shown in Figure 9.18b, this detector is a fluid flow analog of the classic Wheatstone bridge electrical circuit, where the electrical resistors have been replaced with capillaries that measure the flow impedance (i.e., the viscosity, η) by means of Equation 9.34. The bridge inlet pressure (IP) is measured with respect to the outlet, and the differential pressure (ΔP) is measured at any point in time. For the arrangement as shown in Figure 9.18b, flow through the bottom (R1R4 side) of the bridge is given by

$$Q_{\text{bottom}} = \frac{\text{IP} - P_1}{\eta R_1} = \frac{P_1}{\eta_0 R_4} \tag{9.36}$$

where P_1 corresponds to the pressure drop across R4. The R_1 term corresponds to the flow impedance through capillary R1, of length L_1 and radius r_1, according to

$$R_1 = \frac{8L_1}{\pi r_1^4} \tag{9.37}$$

In what follows, R_2, R_3, and R_4 correspond to the flow impedances through capillaries R2, R3, and R4, respectively.

Flow through the top side (R2R3 side) of the bridge is given by

$$Q_{\text{top}} = \frac{\text{IP} - P_2}{\eta R_2} = \frac{P_2}{\eta R_3} \tag{9.38}$$

where P_2 corresponds to the pressure drop across R3. Solving Equations 9.37 and 9.38 for IP gives

$$\frac{\text{IP}}{P_1} = \frac{\eta R_1}{\eta_0 R_4} + 1 \tag{9.39}$$

$$\frac{\text{IP}}{P_2} = \frac{R_2}{R_3} + 1 \tag{9.40}$$

The ratio of the differential pressure across the bridge to the inlet pressure is

$$\frac{\Delta P}{\text{IP}} = \frac{P_2 - P_1}{\text{IP}} = \frac{1}{(R_2/R_3) + 1} - \frac{1}{(\eta R_1/\eta_0 R_4) + 1} \tag{9.41}$$

If the flow impedances of all four capillaries are equal (i.e., if $R_1 = R_2 = R_3 = R_4$) results are independent of bridge capillary impedance and Equation 9.41 can be solved for the specific viscosity:

$$\eta_{\text{sp}} = \frac{\eta}{\eta_0} - 1 = \frac{4\Delta P}{\text{IP} - 2\Delta P} \tag{9.42}$$

This result is also obtained if both $R_1 = R_4$ and $R_2 = R_3$, or if the ratio of the impedances on the left side of the bridge balances the ratio of the impedances on the right side (i.e., if $R_1/R_2 = R_4/R_3$).

Figure 9.18c shows the Yau viscometer, which is composed of two single-capillary viscometers in series, a hold-up reservoir preceding the second capillary [30]. The first viscometer in the series measures the viscosity of the polymer solution; the second capillary measures the viscosity of the neat solvent. The difference between the signals from the two single-capillary viscometers corresponds to the polymer contribution to the viscosity of the solution.

Figure 9.18b is a triple-capillary differential viscometer used by Waters Corp. in their high-temperature instrument. In this design, the differential pressure transducers are connected in flow-through mode, reducing the need for frequent purging of the transducers. This detector has the advantage of not requiring perfect matching of the capillaries (i.e., the capillaries need not have equal impedances), as in the Viscotek design.

9.5.3 Intrinsic Viscosity and the Viscometric Radius

Of main interest in SEC with both concentration-sensitive and viscometric detection (SEC/DRI/VISC) is the determination of the intrinsic viscosity $[\eta]$, defined as

$$[\eta] \equiv \lim_{c \to 0} \frac{\eta_{\text{sp}}}{c} \tag{9.43}$$

where c is the concentration of the polymer solution. We recognize $[\eta]$ as the ratio of the signals from the viscometer (which measures η_{sp}) and the concentration-sensitive detector (which measures c), for the same data slice subsequent to correction for interdetector delay and interdetector band broadening. The units of $[\eta]$ are volume/mass (usually, either mL/g or dL/g). Because of this, the intrinsic viscosity may be thought of as an "inverse density" of the polymer in solution, that is, the more extended (less dense) a polymer is in solution, the higher its intrinsic viscosity will be, and vice versa.

In practice, $[\eta]$ is usually determined by a procedure that solves simultaneously for the reduced and inherent viscosities (η_{red} and η_{ihn}, respectively) in the limit of near-infinite dilution [31]. These viscosities are defined as follows:

$$\eta_{red} = \frac{\eta_{sp}}{c} \tag{9.44}$$

$$\eta_{inh} = \frac{\ln(\eta_{rel})}{c} \tag{9.45}$$

where

$$\eta_{rel} = \frac{\eta}{\eta_0} \tag{9.46}$$

For dilute solutions, the intrinsic viscosity is thus given by

$$[\eta] = \frac{\sqrt{2[\eta_{sp} - \ln(\eta_{rel})]}}{c} \tag{9.47}$$

The measurement of intrinsic viscosity may be combined with that of molar mass to define a fourth size parameter, the viscometric radius of a macromolecule [17]:

$$R_\eta \equiv \left(\frac{3[\eta]M}{10\pi N_A}\right)^{1/3} \tag{9.48}$$

The viscometric radius can be considered the radius of a solid sphere that can dissipate (as heat) the same amount of the flowing solvent's kinetic energy as does the polymer, or the radius of a solid sphere that increases the fluid viscosity by the same amount as does the polymer. As with R_H, measurement of R_η down to a few nanometers is usually possible. The four main macromolecular radii are summarized in Table 9.2.

9.5.4 Viscometry Instrumentation

Because single-capillary viscometers are no longer manufactured commercially, discussion here is limited to differential viscometers. As mentioned above, the overwhelming majority of available differential viscometers are based on the Wheatstone bridge design shown in Figure 9.18b and discussed in detail in Section 9.5.2.

Table 9.2 Definitions of the four main macromolecular radii

Radius	Equation	Definition	Method of Determination
Root-mean-square radius (radius of gyration, R_G)	$R_G = \left[\frac{1}{n+1}\sum_i (r_i - R_{\rm cm})^2\right]^{1/2}$	RMS distance of array of atoms from their common center of mass	MALS, other scattering techniques
Thermodynamic radius	$R_T \equiv \left(\frac{3A_2M_w^2}{16\pi N_A}\right)^{1/3}$	Radius of hard sphere with same excluded volume as the macromolecule	MALS
Hydrodynamic (Stokes) radius	$R_H = \frac{k_BT}{6\pi\eta_0 D_T}$	Radius of equivalent hard sphere that feels the same force due to flow as does the macromolecule	QELS, other methods for determining D_T
Viscometric radius	$R_\eta \equiv \left(\frac{3[\eta]M}{10\pi N_A}\right)^{1/3}$	Radius of solid sphere that increases the fluid viscosity by the same amount as does the macromolecule	MALS + VISC

Source: Ref. 17.

Depending on the manufacturer, flow rates of up to 3 mL/min may be used without damaging the instrument. Operating temperatures of most systems range from room temperature to either 60 or 80°C, with high-temperature options extending the upper temperature limit to 150, 175, or 220°C. Some systems are capable of sub-ambient operation if dry gas is used to cool the system. Temperature can usually be controlled to within ±0.01°C or better. Most bridge viscometers have fixed-volume hold-up reservoirs, although in a newer model the user can choose between 4-, 8-, and 15-mL hold-up volumes, or any combination of these volumes.

9.6 SEC[3]

The type of information (i.e., structural, thermodynamic, etc.) obtained from SEC analysis employing physical detectors such as MALS, QELS, VISC, and DRI, as well as "chemical" detection methods, is explored in Chapter 11. We conclude this chapter by mentioning the combination of right-angle light scattering (RALS), which measures scattered light at a single 90° angle, with differential viscometry and differential refractometry as SEC detection methods. This combination has been termed SEC^3 [32].

In an SEC^3 experiment, after determining interdetector delays and band-broadening parameters, an estimate of the weight-average molar mass ($M_{w,\mathrm{est}}$) of the sample is obtained from the Rayleigh–Gans–Debye (RGD) Equation 9.13. This is done by assuming that the form factor $P(\theta) = 1$ and that all virial coefficients equal zero (i.e., $A_2 = A_3 = \cdots = 0$), with all symbols retaining their previous meanings:

$$M_{w,\mathrm{est}} = \frac{R(\theta = 90^\circ)}{K^* c} \tag{9.49}$$

$P(\theta)$ close to unity is encountered in biopolymers such as globular proteins and certain highly branched systems. The intrinsic viscosity is then calculated from the combined viscometer and refractometer signals, as above, and inserted into the Flory–Fox equation [33] to obtain an estimate of the root-mean-square radius, $R_{G,\mathrm{est}}$:

$$R_{G,\mathrm{est}} = \frac{1}{\sqrt{6}} \left(\frac{[\eta] M_{w,\mathrm{est}}}{\Phi} \right)^{1/3} \tag{9.50}$$

where Φ is the value of Flory's universal constant, Φ_0, after undergoing a correction for non-theta solvent–temperature conditions according to the theory of Ptisyn and Éizner [34]:

$$\Phi = \Phi_0(1 - 2.63\varepsilon + 2.86\varepsilon^2) \tag{9.51}$$

At theta conditions $\varepsilon = 0$, whereas at non-theta conditions it is related to the exponent a of the Mark–Houwink Equation 8.2:

$$\varepsilon = \frac{2a - 1}{3} \tag{9.52}$$

$R_{G,\mathrm{est}}$ is then inserted into the Debye Equation 9.15 to obtain a better estimate of $P(\theta)$:

$$P(\theta)_{\mathrm{est}} = \frac{2}{X^2} \left(e^{-X} + X - 1 \right) \tag{9.53}$$

where

$$X^{1/2} = \left(\frac{4\pi}{\sqrt{3}\lambda} \right) R_{G,\mathrm{est}} \sin(\theta/2) \tag{9.54}$$

This newly estimated $P(\theta)$ is then reinserted into the RGD equation to obtain an improved estimate of the weight-average molar mass, $M_{w,\text{est2}}$:

$$M_{w,\text{est2}} = \frac{M_{w,\text{est}}}{P(\theta)_{\text{est}}} \tag{9.55}$$

The loop is iterated until the values of M_w, R_G, and $P(\theta)$ converge, which typically takes less than five iterations.

A few comments about this approach:

1. It should be noted that at least a dozen different values for Flory's constant, Φ_0, have been reported in the literature. A commonly used value is 2.87×10^{23} mol^{-1}.
2. The value of Φ for a particular polymer under a particular set of conditions is not necessarily independent of M [35].
3. It is assumed that a single factor, ε, can adequately correct for non-theta solvent–temperature effects.
4. The procedure begins its assumptions near a local minimum [i.e., by assuming that $P(\theta) = 1$]. Eventual convergence to a value near this local minimum does not necessarily signify that said value is also the global minimum of the system.
5. Despite many of the foregoing points, the SEC^3 approach appears to have been applied successfully to a number of polymer architectures and chemistries.

Just as the Flory–Fox/Ptitsyn–Eizner relations allow calculation of R_G from viscometric data, they also permit calculation of the intrinsic viscosity from MALS-generated R_G and M_w data. Given the caveats mentioned above, the most accurate route toward M_w is through LALS/DRI, to R_G through MALS, and to $[\eta]$ through VISC/DRI.

REFERENCES

1. A. M. Striegel, *Anal. Chem.*, **77**, 104A (2005).
2. A. M. Striegel, ed., *Multiple Detection in Size-Exclusion Chromatography*, ACS Symp. Ser. 893, American Chemical Society, Washington, DC, 2005.
3. T. H. Mourey, *Int. J. Polym. Anal. Charact.*, **9**, 97 (2004).
4. J. Horska, J. Stejskal, and P. Kratochvíl, *J. Appl. Polym. Sci.*, **28**, 3873 (1983).
5. A. A. Gridnev, P. M. Cotts, C. Roe, and H. G. Barth, *J. Polym. Sci. A*, **39**, 1099 (2001).
6. S. Podzimek, in Ref. 2, Chap. 5.
7. I. A. Haidar Ahmad, Master's thesis, Florida State University, 2008.
8. S. Michielsen, in *Polymer Handbook*, 4th ed., J. Brandup, E. H. Immergut, and E. A. Grulke, eds., Wiley-Interscience, New York, 1999, p. VII/547.
9. *HPLC Solvent Reference Manual*, J.T. Baker Chemical Co., Phillipsburg, NJ, 1985.

10. J. J. Kirkland and P. E. Antle, *J. Chromatogr. Sci.*, **15**, 137 (1977).
11. R. Schulz and H. Engelhardt, *Chromatographia*, **29**, 517 (1990).
12. C. S. Young and J. W. Dolan, *LCGC*, **21**, 120 (2003).
13. C. S. Young and J. W. Dolan, *LCGC*, **22**, 244 (2004).
14. W. F. Reed, in Ref. 2, Chap. 2.
15. P. J. Wyatt, *Anal. Chim. Acta*, **272**, 1 (1993).
16. M. Rubinstein and R. H. Colby, *Polymer Physics*, Oxford University Press, New York, 2003.
17. M. J. Smith, I. A. Haidar, and A. M. Striegel, *Analyst*, **132**, 455 (2007).
18. M. A. Haney, *Lab. Equip.*, **39**, S-12 (2003).
19. B. H. Zimm, *J. Chem. Phys.*, **16**, 1099 (1948).
20. I. A. Haidar Ahmad and A. M. Striegel, *Instrum. Sci. Technol.*, in press.
21. A. M. Striegel, *J. Liq. Chromatogr. Rel. Technol.*, **31**, 3105 (2008).
22. A. M. Striegel, in Ref. 2, Chap. 4.
23. A. M. Striegel, *Polym. Int.*, **52**, 1863 (2003).
24. A. M. Striegel, *Anal. Chem.*, **74**, 3013 (2002).
25. I. Teraoka, *Polymer Solutions: An Introduction to Physical Properties*, Wiley-Interscience, New York, 2002.
26. P. M. Cotts, in Ref. 2, Chapt. 3.
27. R. F. Probstein, *Physicochemical Hydrodynamics*, Butterworth Boston, 1989.
28. J. Lesec, in *Encyclopedia of Chromatography*, 2nd ed., J. Cazes, ed., Marcel Dekker, New York, 2005, p. 1767.
29. M. A. Haney, *J. Appl. Polym. Sci.*, **30**, 3037 (1985).
30. W. W. Yau, S. D. Abbott, G. A. Smith, and M. Y. Keating, in *Detection and Data Analysis in Size Exclusion Chromatography*, ACS Symp. Ser. 352, T. Provder, ed., American Chemical Society, Washington, DC, 1987, Chap. 5.
31. O. F. Solomon and I. Z. Ciută, *J. Appl. Polym. Sci.*, **6**, 683 (1962).
32. S. V. Greene, in *Encyclopedia of Chromatography*, 2nd ed., J. Cazes, ed., Marcel Dekker, New York, 2005, p. 1516.
33. P. J. Flory, *Principles of Polymer Chemistry*, Cornell University Press, Ithaca, 1953, Chap. XIV.
34. O. B. Ptitsyn and Yu. E. Éizner, *Sov. Phys. Tech. Phys.*, **4**, 1020 (1960).
35. B. S. Farmer, K. Terao, and J. W. Mays, *Int. J. Polym. Anal. Charact.*, **11**, 3 (2006).

10

CHEMICAL DETECTORS

10.1 INTRODUCTION

In this chapter we deal with the use of so-called "chemical" techniques as detection methods in SEC. Primary topics are spectrometric and spectroscopic techniques, although other methods are also included. Some of the detection methods in this chapter may appear also to be physical techniques and, admittedly, the distinction between the two is somewhat artificial and to a large extent, organizational. We repeat here our rationale for distinguishing between the two classes of detection methods, outlined originally in Chapter 9. Physical detectors are those that combine synergistically; chemical detectors combine additivively. Ultraviolet (UV), infrared (IR), or nuclear magnetic resonance (NMR) are thus considered chemical detectors. While these particular methods can provide either chemical or physical information about the polymer (e.g., either chemical composition or short-chain branching data), this physical information combines in additive fashion with that provided by the physical detectors described in Chapter 9. The advantages gained from multidetector synergisms are explored in Chapter 11.

Contrary to the case with physical detectors (Chapter 9), here we assume a familiarity by the reader with most of the chemical methods of detection. This stems from the fact that the principles of operation and the instrumentation for chemical methods of detection are generally taught in undergraduate instrumental analysis courses. We instead provide a survey of the many different chemical detectors currently available to the SEC practitioner, with examples of the types of information these generate.

Modern Size-Exclusion Liquid Chromatography: Practice of Gel Permeation and Gel Filtration Chromatography, Second Edition
By André M. Striegel, Wallace W. Yau, Joseph J. Kirkland, and Donald D. Bly

Some of the methods have been the subject of recent review articles and books [1,2]. We focus on methods that can be used either online or in continuous off-line fashion. The latter term refers to the continuous, or nearly continuous, deposition of eluate onto a plate or disk for subsequent off-line analysis. This type of setup is seen in certain SEC/FTIR and SEC/MALDI-TOF-MS analyses. Only in select cases do we mention the use of off-line detectors where a small number of discrete SEC elution fractions ("heart cuts") are collected for off-line analysis by, for example, MS or NMR. These off-line approaches yield a paucity of data, and although quite useful in many cases, MS, NMR, and so on, may in this context be thought of more appropriately as methods of analysis than as detection techniques.

SEC with chemical detection is often part of a two-dimensional liquid chromatographic (2D-LC) setup. The role of SEC in 2D-LC analysis is the topic of Chapter 14.

10.2 MASS SPECTROMETRY

Many different mass spectrometric methods have been used in polymer analysis [3]. Of these, matrix-assisted laser desorption/ionization time-of-flight (MALDI-TOF), electrospray ionization (ESI), and inductively coupled plasma (ICP) mass spectrometry have shown the greatest promise as SEC detectors. SEC coupled to either MALDI-TOF-MS or to ESI-MS is used mostly in the analysis of synthetic polymers; SEC/ICP-MS is used chiefly for elemental speciation in environmental and biological samples.

10.2.1 Electrospray Ionization Mass Spectrometry

The ion source of an electrospray ionization mass spectrometer (ESI-MS), an atmospheric pressure ionization technique, is shown schematically in Figure 10.1. The formation of charged droplets, from which gas-phase ions are obtained, initiates at the electrospray capillary tip (needle), which is held at a potential of a few kilovolts relative to a counterelectrode (nozzle) located several centimeters away. In response to an applied electric field, charges accumulate at the surface of the liquid at the capillary tip. Eventually, destabilization of this surface occurs and charges are drawn toward the counterelectrode but are unable to escape the liquid. The surface is drawn out and a liquid cone (a Taylor cone) forms. Charged-droplet emission from the Taylor cone occurs at an onset potential V_{on} that depends on the surface tension of the solvent, the radius of the capillary, and the distance between the capillary tip and the nozzle. Once a charged droplet has become severed from the Taylor cone, droplet shrinkage begins. Shrinkage occurs by evaporation of the solvent molecules until repulsive Coulombic forces between surface charges overcome the cohesive surface tension force holding the droplet together. The droplet fissions in an uneven fashion, yielding a tail of smaller offspring droplets which carry off only about 2% or so of the mass of the parent droplet but about 15% of the parent droplet's charge. The offspring droplets are monodisperse and their radius is approximately one-tenth that of

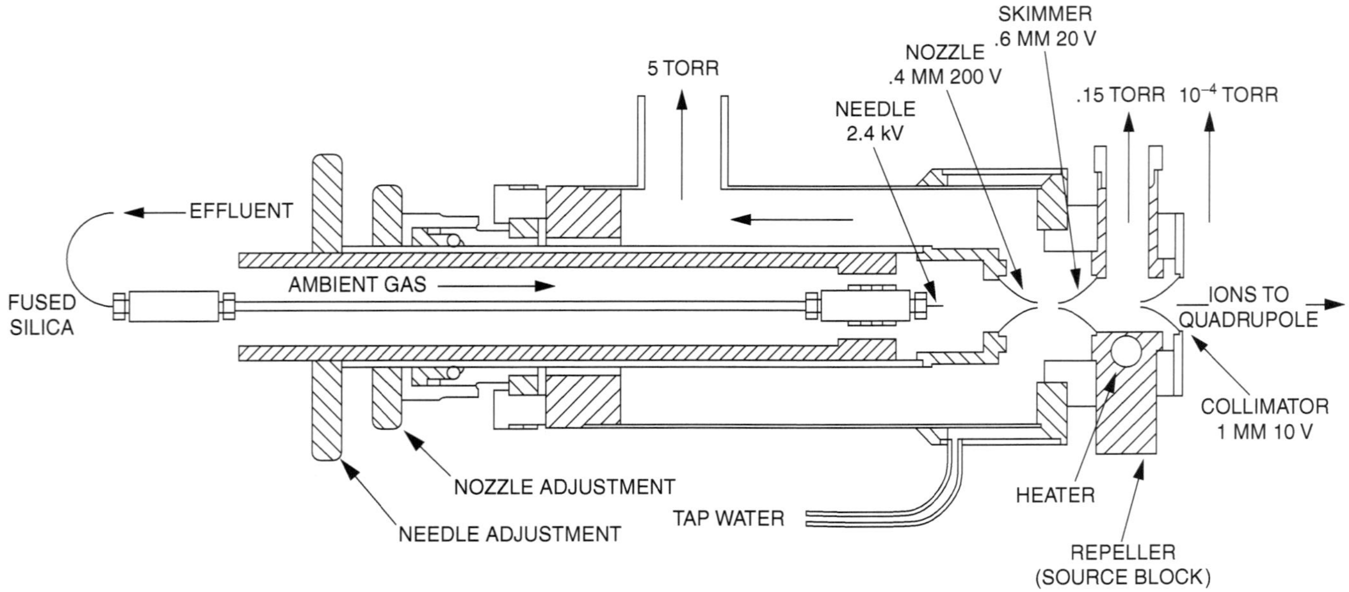

Figure 10.1 ESI-MS ion source. (Reprinted with permission from Ref. 4.)

the parent droplet. It is these offspring droplets that are ultimately expected to lead to gas-phase ions. For macromolecules, this last step appears to follow the *charged residue model*, according to which solvent continues to evaporate from the offspring droplets and the excess charges situate themselves on analyte sites that will result in the most stable gas-phase analyte ion. An excellent treatment of most aspects of ESI-MS may be found in Reference 5.

ESI is the softest ionization method currently available in MS. Unlike most other MS techniques (e.g., electron or chemical ionization), the "soft" ionization that occurs in the electrospray process does not generally lead to fragmentation. The presence of a mass-to-charge (m/z) envelope of multiply charged peaks is a distinguishing feature of many ESI mass spectra. This combination of soft ionization capability and multiple charging has made ESI-MS an ideal tool for the study of perfectly monodisperse biopolymers such as proteins and peptides and for accurate determination of their molar masses. For most synthetic polymer, however, the combination of molar mass polydispersity and multiple charging that occurs in ESI-MS has the potential to generate a fairly incomprehensible mass spectrum. This is because each degree of polymerization can produce its own m/z envelope. For a given polydisperse macromolecule, all the individual envelopes are then superimposed upon each other in the ESI mass spectrum of the sample. The application of ESI-MS to polymer applications is reviewed in References 6 and 7.

Only about 1% of the SEC effluent is needed for ESI-MS, so other detectors can also be used (and usually are). Microcolumn SEC is often coupled to ESI-MS, because low flow rates of a few μL/min work best with ESI-MS. As an online detection method, ESI-MS has been used in the study of humic and fulvic acids [8]; to obtain chemical composition distribution information on copolymers, provided that the individual monomers differ in molar mass [9]; and to study polysaccharides such as dextrans and arabinoxylan digests [10]. SEC/ESI-MS^n ($n = 2, 3$) has been used to determine glycosidic linkage information of permethylated arabinogalactan oligomers [10]. By analyzing PMMA samples by SEC coupled to ESI-FTMS (FT: Fourier transform), the presence of a secondary distribution due to the formation of cyclic reaction products was discovered atop the MMD of PMMA [9]. The setup for this type of experiment is shown in Figure 10.2.

It is assumed that each slice eluting from the SEC columns is virtually monodisperse with respect to molar mass. Therefore, a hyphenated SEC/ESI-MS experiment allows for calculation of extremely accurate molar masses at each slice. This, in turn, allows for construction of an accurate SEC calibration curve, based on the analyte itself, from which the molar mass distribution may be computed. For example, using SEC/ESI-MS, more accurate SEC calibrations than those obtained by traditional approaches were generated for linear, cyclic, and hyperbranched polyester [9].

SEC/ESI-MS has been employed in structural investigations of the reversible addition fragmentation chain transfer (RAFT) polymerization of methyl acrylate [12]. Experimental results suggested that either the RAFT intermediate disproportionation product or the actual RAFT intermediate radical was present in the ESI mass spectrum. Examples of experimental results are shown in Figure 10.3. In Figure 10.3a are the size-exclusion chromatograms of polymeric material generated after varying

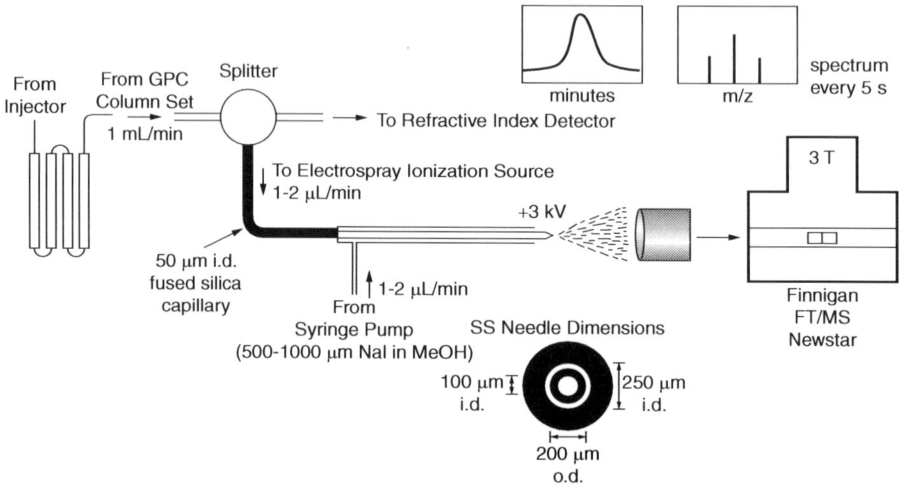

Figure 10.2 Experimental setup for SEC/ESI-FTMS. (Reprinted with permission from Ref. 11.)

reaction times: 10 minutes (solid line), 15 minutes (dashed line), and 20 minutes (dotted line). The solid vertical lines in Figure 10.3a indicate the positions in the molar mass distribution of the ESI mass spectra shown in Figure 10.3b (18.40 minutes) and Figure 10.3c (20.04 minutes).

10.2.2 Matrix-Assisted Laser Desorption/Ionization Time-of-Flight Mass Spectrometry

In matrix-assisted laser desorption/ionization (MALDI), analyte is embedded in a solid or liquid matrix consisting of a small, highly absorbing species. Typical analyte/matrix ratios are on the order of $1:10^4$. Energy from a pulsed laser is deposited in the matrix, and a small portion of this energy is transferred to the analyte in an efficient and controlled manner, sparing the analyte from excessive energy deposition that can result in decomposition. Diluting the analyte in the matrix also serves to prevent analyte clustering, which can lead to high-mass complexes that are too large for desorption and analysis. The matrix thus serves two purposes: It absorbs energy from the laser and it isolates the analyte molecules from each other. The fraction of energy from the laser that is transferred to the analyte via the matrix is enough to desorb the analyte as a dense cloud that expands supersonically (at about Mach 3) into the vacuum. Ionization of the analyte is believed to occur in the expanding plume resulting from collisions between analyte neutrals, excited matrix ions, and charge carriers (e.g., protons, sodium cations). The analyte ions are next routed into the mass analyzer, most often a time-of-flight (TOF) mass analyzer. The principles of operation of a TOF mass analyzer are described in detail in Reference 13; the

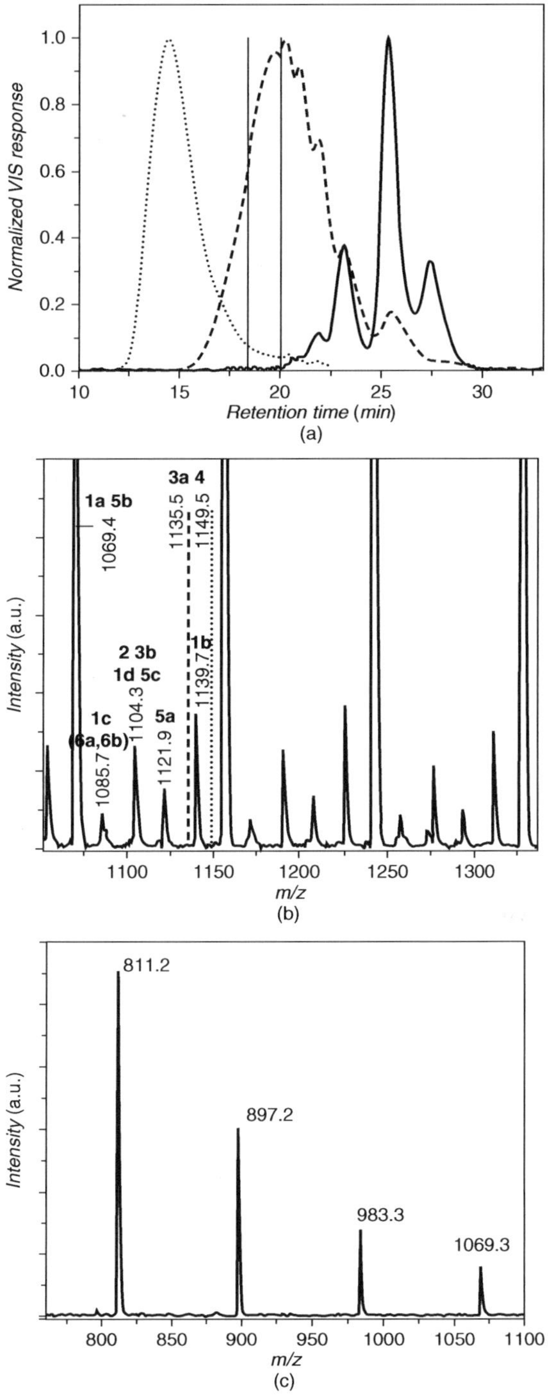

Figure 10.3 Monitoring RAFT polymerization of methyl acrylate by SEC/ESI-MS: (a) SEC traces of polymerization products at 10 min (solid line), 15 min (dashed line), and 20 min (dotted line). Solid vertical lines indicate positions in the molar mass distribution of ESI mass spectra shown in (b) (18.40 min) and (c) (20.04 min.) (Reprinted with permission from Ref. 12.)

application of MALDI-TOF-MS to the analysis of synthetic polymers is covered in Reference 14.

The choice of MALDI matrices is about as extensive as is the variety of polymers that have been analyzed by this technique. Caveats in choosing a matrix are sample incompatibility, background effects, mass discrimination, and contamination of the ion source by the matrix. An extensive tabulation of matrices for UV- and IR-MALDI of synthetic polymers may be found in Reference 15.

Multiple charging is not usually observed in MALDI-MS. This renders interpretation of the information obtained, and its use in molar mass calculations (especially of synthetic polymers), more straightforward than in ESI-MS.

Most SEC/MALDI-TOF-MS experiments employ a continuous off-line approach using a commercially available interface (newer interface models can also be used for SEC/FTIR; see Section 10.3). The SEC eluent is sprayed directly through a heated capillary nozzle onto a moving MALDI target precoated with an appropriate matrix. For polydisperse macromolecules, a continuous track of sample is deposited onto the matrix surface of the target. The target is then introduced into the mass spectrometer, and spectra are obtained from different positions of the track. This approach was used for the study of PS 32500 and PMMA 10900, as well as to obtain both molar mass and copolymer composition of a diblock copolymer of *n*-butyl methacrylate and PMMA [16]. A three-dimensional plot of the results of the SEC/MALDI-TOF-MS analysis of the PMMA sample is shown in Figure 10.4, where one axis ("mass/charge" or *m*/*z*) corresponds to the mass range, another axis ("pulses")

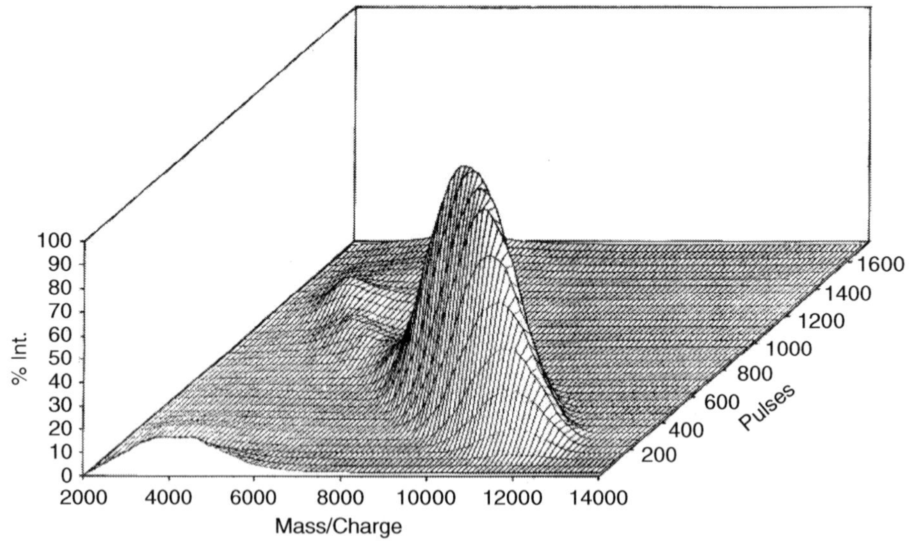

Figure 10.4 Three-dimensional plot of a 10,900-g/mol PMMA, obtained from SEC/MALDI-TOF-MS analysis. (Reprinted with permission from Ref. 16.)

corresponds to the position of the MALDI-TOF target (equivalent to the SEC elution volume), and the third axis ("% int.") plots the mass peak intensity.

Continuous off-line SEC/MALDI-TOF-MS and online SEC/ESI-TOF-MS have been compared using poly(dimethyl siloxane) oligomers. The electrospray technique was found to be more effective at reporting low-M oligomers; the hyphenated MALDI technique was better at characterizing high-M oligomers [17].

10.2.2.1 Comparison Between SEC and MALDI-TOF-MS. Recently, SEC with both DRI and multiwavelength UV detection was compared to MALDI-TOF-MS in both linear and reflectron mode, using a series of narrow polydispersity PMMA standards and one broad polydispersity sample [18]. Results of this comparison showed that the breadth of the distribution correlated well with the mass range observed by the mass spectrometric method. However, this was the case only when said breadth was defined as the standard deviation of the number distribution, σ_n, calculated from both the polydispersity M_w/M_n and M_n using Equation 10.1, but not when the breadth was defined simply as the polydispersity M_w/M_n:

$$\sigma_n = M_n\sqrt{\frac{M_w}{M_n} - 1} \tag{10.1}$$

The molar mass distributions obtained for a broad polydispersity PMMA sample by both MALDI and SEC are shown in Figure 10.5. SEC and MALDI-TOF-MS were determined to be complementary methods. However, large differences between the molar mass averages as determined by both methods were also found. On the part of

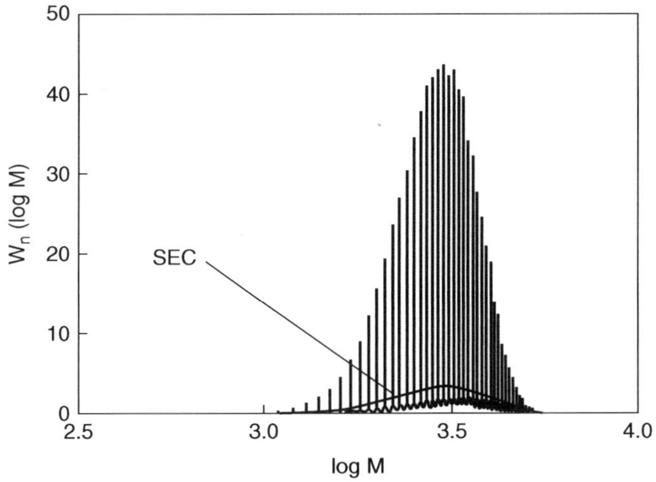

Figure 10.5 Differential molar mass distributions of a 2990-g/mol PMMA, determined by MALDI-TOF-MS and SEC. (Reprinted with permission from Ref. 18.)

MALDI-TOF-MS, these differences were attributed to mass discrimination effects. On the part of SEC, the differences were due to inaccurate calibration curves, probably resulting from inaccurate M_p values for the calibration standards. Cumulative MMDs, rich in information content, were easily obtained using both methods. The notion that current MALDI technology yields spectra that are representative of the MMD of polymers with polydispersity less than 1.2 was shown to be false. Errors exceeding 11% were found in M averages obtained by MALDI for polymers with $M_w/M_n \leq 1.1$. This limitation of the MALDI technique was found to be due to the breadth or mass range that the distribution covers, not to the polydispersity of the sample. For all narrow standards examined, which ranged from 2.26×10^3 to 5.58×10^4 g/mol, $M_w/M_n \leq 1.1$. However, over this same mass range σ_n was found to increase with mass range with a proportionality constant of 8.69.

10.2.3 Inductively Coupled Plasma Mass Spectrometry

The online coupling of SEC to ICP-MS has proven quite popular for the quantitation of metals across the elution profile of natural and synthetic polymers, biological fluids, and environmental samples. A setup similar to that shown in Figure 10.6 is usually employed. As the principles of ICP and ICP-MS can be found in virtually

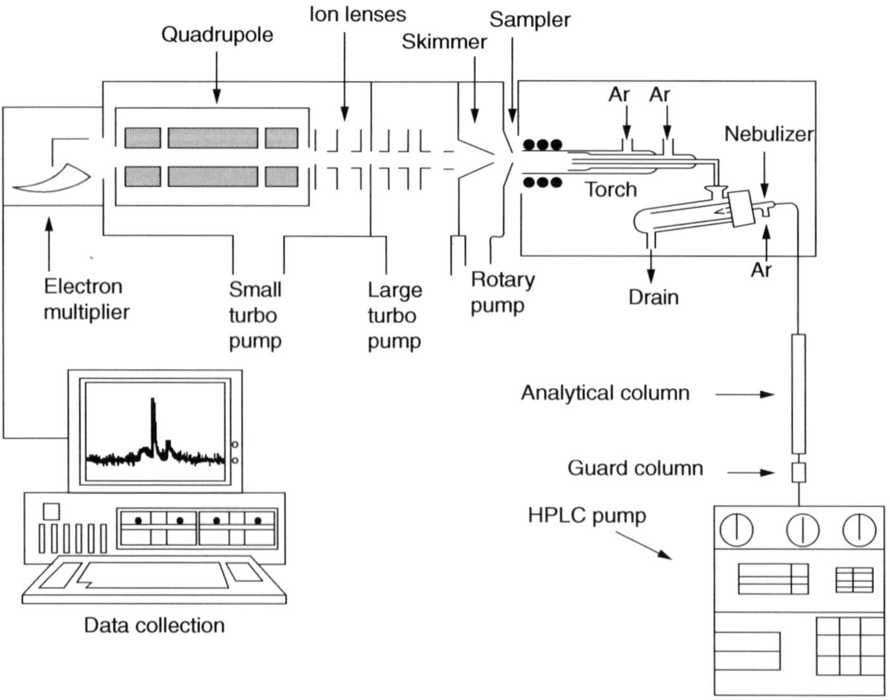

Figure 10.6 Coupling of SEC to ICP-MS. (Reprinted with permission from Ref. 20.)

any instrumental analysis book, a description of the methods and instrumentation is not given here. A good source of information on the subject is Reference 19.

In humic substance analysis, SEC/ICP-MS has been used to determine the distribution of Co, Cu, Ni, Pb, and Zn in humic substances derived from municipal compost [21]. Results of these experiments are shown in Figure 10.7. Analysis of the elution profiles obtained with and without ethylenediaminetetraacetic acid (EDTA) in the mobile phase suggested that both complexation and chelation by individual molecules, and bridging between small molecules, are involved in the binding of metal ions to humic substances.

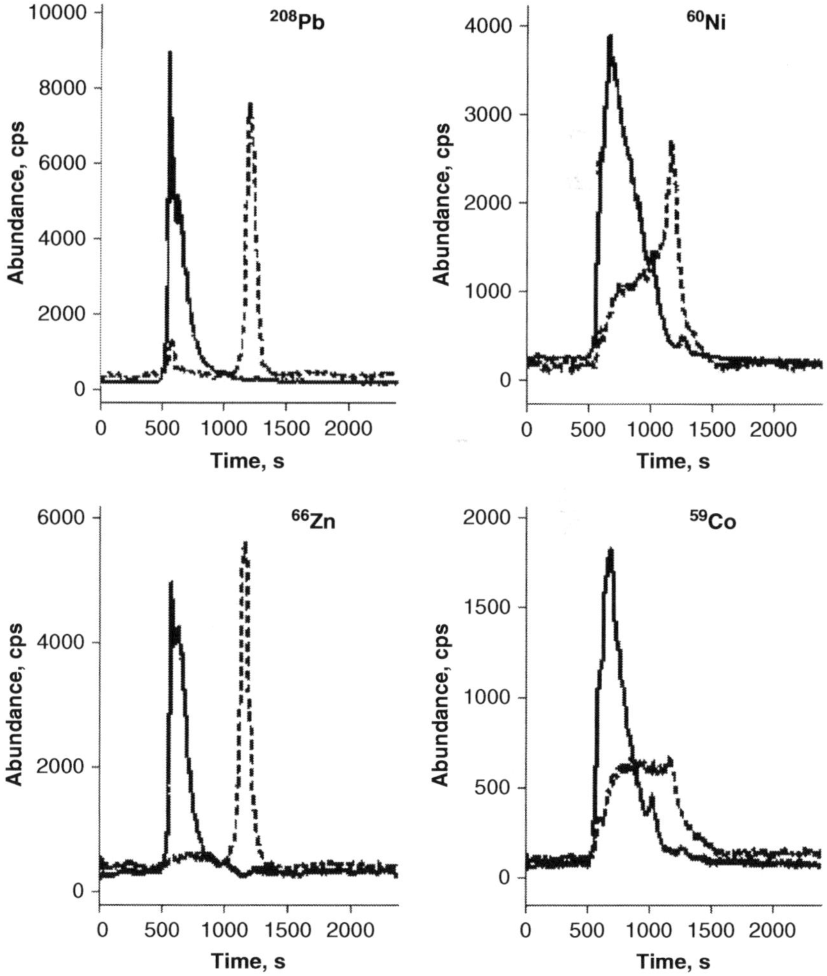

Figure 10.7 SEC/ICP-MS analysis of compost extract. Clockwise from top left: detection for lead (*m*/*z* 208), nickel (*m*/*z* 60), cobalt (*m*/*z* 59), and zinc (*m*/*z* 66). Solid lines indicate analysis in the presence of EDTA in the mobile phase, dashed lines in the absence of EDTA. (Reprinted with permission from Ref. 21.)

Precautions that must be taken when coupling SEC to ICP-MS include [22]:

1. Use organic solvents minimally, because of their adverse effect on plasma stability.
2. Pay careful attention to the ionic strength of mobile-phase buffers because although buffers can help reduce non-size-exclusion behavior, they may also denature the organometallic complexes of interest.
3. The salt content of the mobile phase should be kept to a minimum to avoid clogging the nebulizer and to reduce wear on the sampler and skimmer cones.
4. Careful sample preparation is required to ensure that the species of interest are extracted from the matrix without altering the nature of these species.

For glycopolymer analysis, SEC/ICP-MS has been used along with a DRI detector [23]. The concentration-sensitive detector served to characterize the elution profiles of water-soluble and of enzymatically digested polysaccharides. The mass spectrometer provided the distribution patterns of Mg, Cu, and Zn in the elution profiles of the low-M, noncarbohydrate fractions of apple and carrot samples, and the distribution patterns of Pb, Ba, Sr, Ce, and B in the high-M polysaccharide fraction. SEC/ICP-MS was also used to determine that the metal-binding polysaccharide ligand in the apple samples appears to be a dimer of rhamnogalacturonan II.

SEC/ICP-MS has also been applied to investigating the kinetics of metallodrug–protein interactions [24]. The effects of different SEC column lengths and packings were assessed for the analysis of two platinum-based and three ruthenium-based drugs. The online chromatographic–mass spectrometric coupling was seen to offer considerable advantages in terms of speed, simplicity, precision, and selectivity over traditional methods of determining binding kinetics based on ultrafiltration followed by off-line metal determination.

10.3 FOURIER TRANSFORM INFRARED SPECTROSCOPY

As mentioned in Section 10.1, the infrared detector may in some cases seem to act as a physical detector (pseudophysical detector), providing physical, not chemical, information in a nonsynergistic fashion. Infrared detection may be used in online or in continuous off-line modes, and a comparison of these is presented in Section 10.3.3. In off-line mode the same commercially available hardware used for SEC/MALDI can often be used for SEC/FTIR, with only minor modifications. Infrared detection as applied to polymer analysis is reviewed in Reference 25.

10.3.1 FTIR as a Pseudophysical Detector: Short-Chain Branching Distribution of Polyolefins

The uses of FTIR as a pseudophysical detector for SEC have thus far been limited to the characterization of the short-chain branching distribution (SCBD) in polyolefins.

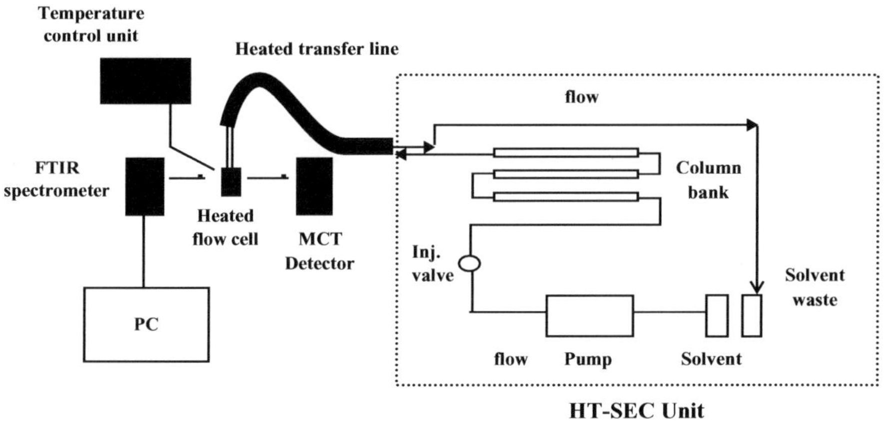

Figure 10.8 Typical on-line SEC/FTIR setup. The arrangement shown is standard for high-temperature SEC (see Section 16.2), with the addition of a heated external transfer line, heated flow cell, and mercury cadmium telluride (MCT) FTIR cell. (Reprinted with permission from Ref. 27.)

Characterizing the SCBD is of great interest, because it has been found to influence end-use properties such as environmental stress crack resistance [26]. The online SEC/FTIR arrangement shown in Figure 10.8, with the type of flow cell shown in Figure 10.9, was used to quantitate the ethyl and/or butyl content of ethylene 1-olefin copolymers as a function of molar mass and to detect trends resulting from catalysis and process changes [27]. Spectra from individual time slices were analyzed, using chemometrics, for co-monomer branch levels in terms of total methyl content and then converted to short-chain branching (SCB) by subtracting the molecular chain end contributions. SCB levels were accurately predicted to within ±0.5/1000 total carbons (TC) (about 0.1 mol%). Figure 10.10 shows an example of this type of characterization of the SCBD. The FTIR was also used as a concentration-sensitive detector, to obtain calibrant-relative M-averages and distributions of the samples analyzed.

10.3.2 FTIR as a Chemical Detector

The large majority of applications of FTIR as a chemical detector involve continuous off-line detection in which the SEC eluent is directed to a heated nozzle for evaporation of the solvent and deposition of the analyte onto a rotating germanium disk. The disk can be analyzed off-line by FTIR to obtain spectra from any location on it. If the rate of rotation of the disk and the chromatographic flow rate are both known and coordinated to each other, spectra from any given spot on the disk can be correlated to particular slices of the SEC elution profile and, consequently, to particular slices of the MMD of a sample. A schematic of the nozzle arrangement for a commercially available continuous off-line SEC/FTIR interface is shown in

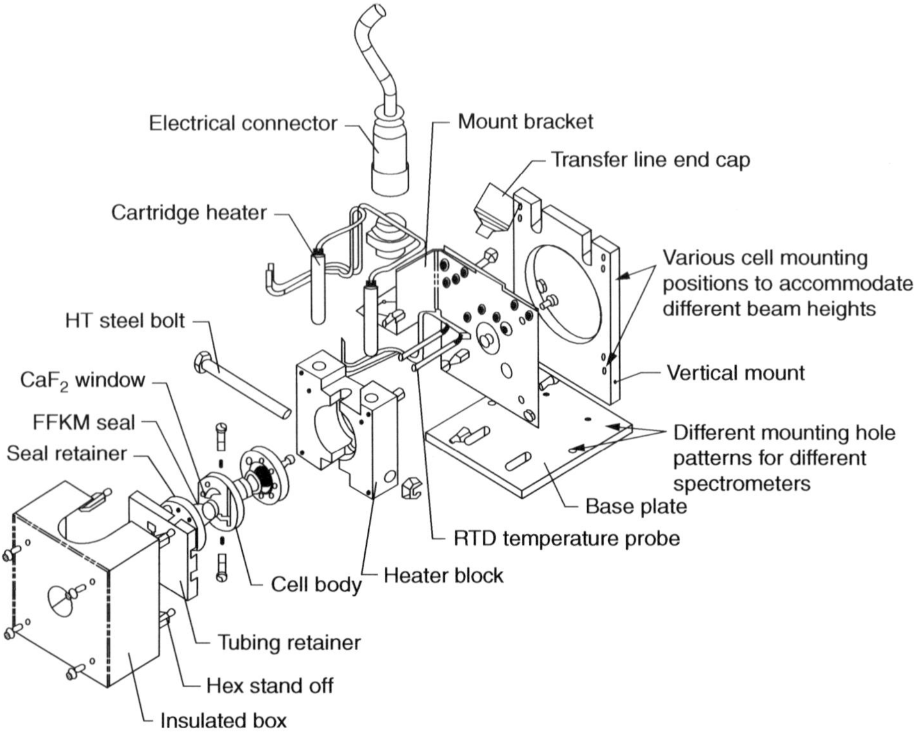

Figure 10.9 Exploded view of heated FTIR flow cell for on-line coupling to SEC. (Reprinted with permission from Ref. 27.)

Figure 10.11. A comparison of on- and off-line modes of FTIR detection in SEC is given in Section 10.3.3.

In continuous off-line mode, the majority of SEC/FTIR uses have been to monitor comonomer content as a function of M of polyethylene copolymers. Figure 10.12 shows how SEC/FTIR was used to determine the mole percent of styrene across the MMD of ethylene/styrene copolymers [29]. For this figure it should be noted that the stated styrene contents of 51.2 mol% was determined by ^{1}H-NMR. Differences in reference samples used for ^{1}H-NMR and for FTIR are responsible for the different styrene contents determined by each technique.

In online mode, SEC/IR was employed, in combination with DRI detection, to determine the dependence of the vinyl alcohol content on the molar mass of poly(vinyl butyral-*co*-vinyl alcohol-*co*-vinyl acetate) terpolymer [30]. Both the OH stretch of the hydroxyl group and the butyral ring vibration were monitored by IR, and the normalized IR chromatograms were found to be superimposable on the DRI chromatograms. This superimposition confirmed that the vinyl alcohol content in the terpolymer samples was independent of M.

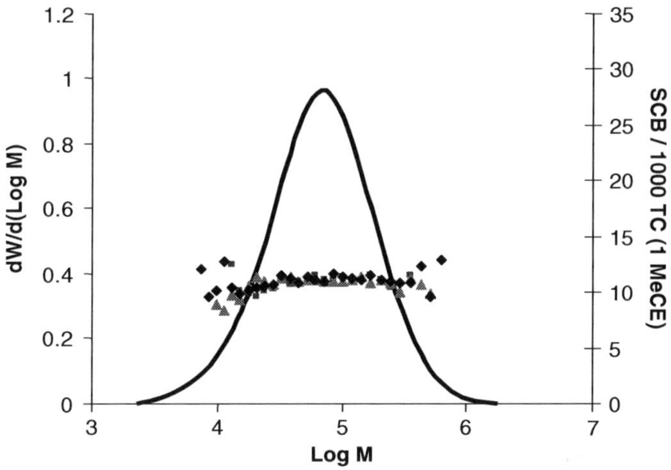

Figure 10.10 On-line SEC/FTIR determination of the SCBD in polyolefins: distribution of the number of short-chain branches with one methyl chain end (1 MeCE) per 1000 total carbons (TCs) across the molar mass distribution (solid line) of an ethylene 1-olefin copolymer. Each type of symbol represents one run through the column. (Adapted from Ref. 27, courtesy of Paul J. DesLauriers.)

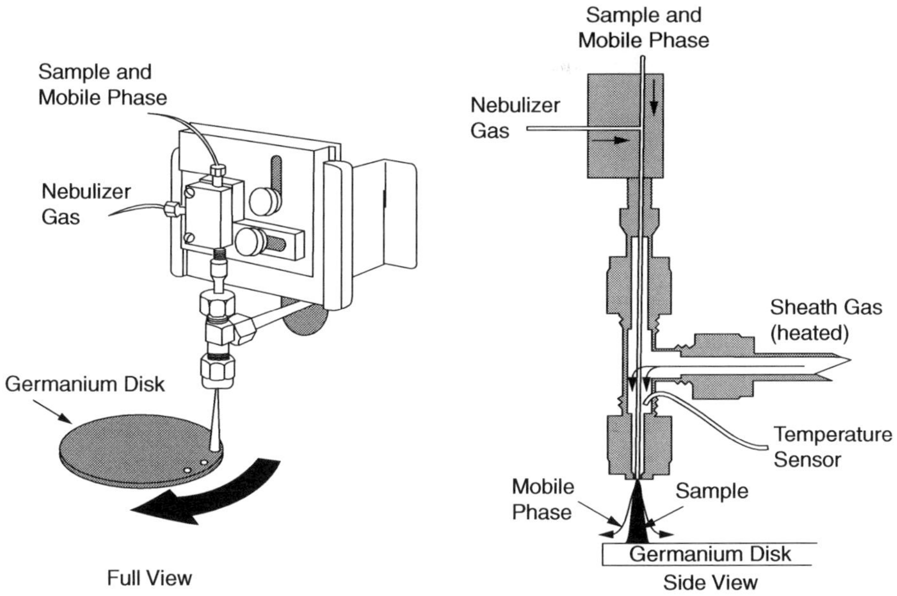

Figure 10.11 Nozzle design for a commercial SEC/FTIR interface. (Reprinted with permission from Ref. 28.)

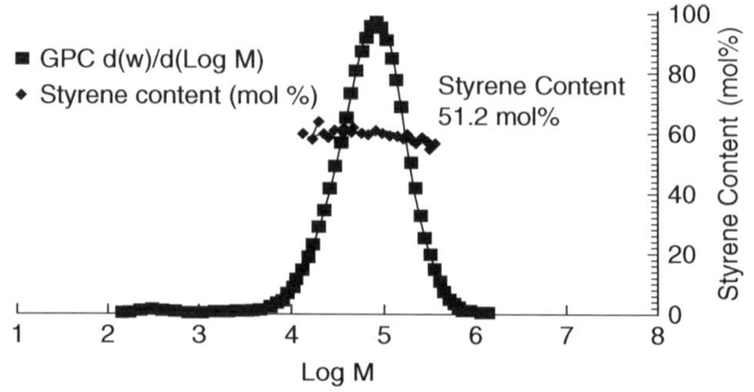

Figure 10.12 SEC/FTIR analysis of poly(ethylene-*co*-styrene)s. SEC/FTIR provides the styrene content across the MMD, which in this case is PS-relative. Styrene content was determined by ^{1}H-NMR using different reference samples than were used for FTIR; this difference accounts for the apparent discrepancy in results. (Reprinted with permission from Ref. 29.)

10.3.3 Comparison of Online and Continuous Off-Line SEC/FTIR

Online flow-cell SEC/FTIR has been compared to the continuous off-line technique employing a solvent-elimination interface of the type shown in Figure 10.11 [31]. The probe species were a polycarbonate/aliphatic polyester blend and a poly(carbonate-*co*-dimethyl siloxane) copolymer. In select cases, both forms of SEC/FTIR were also compared to SEC with UV detection. The findings from this study can be summarized as follows:

1. A comparable decrease in chromatographic resolution was found to result from the relatively large volume of the flow cell and from the inherent deposition characteristics of the solvent-elimination interface.
2. Compared to SEC/UV, separation of oligomers was diminished in both forms of SEC/FTIR.
3. Peak asymmetry was not affected significantly by either interface.
4. For both online and continuous off-line interfacing, a linear relationship was obtained for FTIR response versus injected sample concentration.
5. Solvent-elimination interfacing was found to have higher sensitivity and better limits of detection than those of online flow cell interfacing.
6. Reproducibility of the online flow cell results was considerably better than that of its off-line counterpart. This was ascribed to the well-defined optical path length of the sample in the flow cell.
7. Using spectral subtraction, it was possible to obtain qualitative functional group identification by both techniques.

Typical characteristics of the interfaces are also compared in Table 10.1.

Table 10.1 Typical characteristics of SEC/FTIR online flow-cell and off-line solvent-elimination interfaces

Condition	Flow-Cell Interface	Solvent-Elimination Interface
Gradient separations	No	Yes
Qualitative information	Limited, depends on eluent	Yes
Quantitative information	Excellent	Limited
Sensitivity	Moderate	Excellent
Limit of detection	Low, depends on eluent	High
Spectral signal-to-noise ratio	Moderate, spectra collection on the fly	High, extended post-run scanning possible
Ease of operation	User-friendly	Time-consuming optimization
Application area	SEC	SEC, gradient HPLC

Source: Ref. 31.

10.4 NUCLEAR MAGNETIC RESONANCE SPECTROSCOPY

While it has been stated that "One of the most impressive advantages of continuous-flow ^{1}H-NMR spectroscopy is the direct monitoring of the change in chemical composition of polymers and copolymers during gel permeation [size-exclusion] chromatography" [32], there is still a paucity of publications in this regard. To date, all SEC/NMR reports appear to use ^{1}H-NMR, where the mobile phase is either a deuterated solvent or a nondeuterated solvent containing a known amount of deuterated solvent. Figure 10.13 shows the design of a continuous-flow NMR probe for SEC/NMR coupling. The application of NMR spectroscopy to polymer analysis is covered in Reference 34.

The use of NMR as a pseudophysical detector has been rather limited and appears to be restricted to measuring the tacticity distribution in mixtures of isotactic and syndiotactic PMMAs [35,36]. As an on-line chemical detector for SEC, NMR can be used to determine the MMD of polymers and chemical heterogeneity of copolymers. As seen in Figure 10.14, SEC/NMR was used to measure the mol% of ethyl acrylate across the elution profile of a series of poly(styrene-*co*-ethyl acrylate) copolymers [37].

10.5 OTHER CHEMICAL DETECTORS

10.5.1 Ultraviolet Detection

In addition to its widespread use as a concentration-sensitive detector (Section 9.2.2), UV detection has also been used to determine the chemical heterogeneity of copolymers. In studying poly(styrene-*co*-methyl methacrylate) copolymers by SEC/DRI/UV/MALS, the DRI served as the concentration-sensitive detector. The UV detector monitored $\lambda = 262$ nm, a wavelength where styrene presents a strong

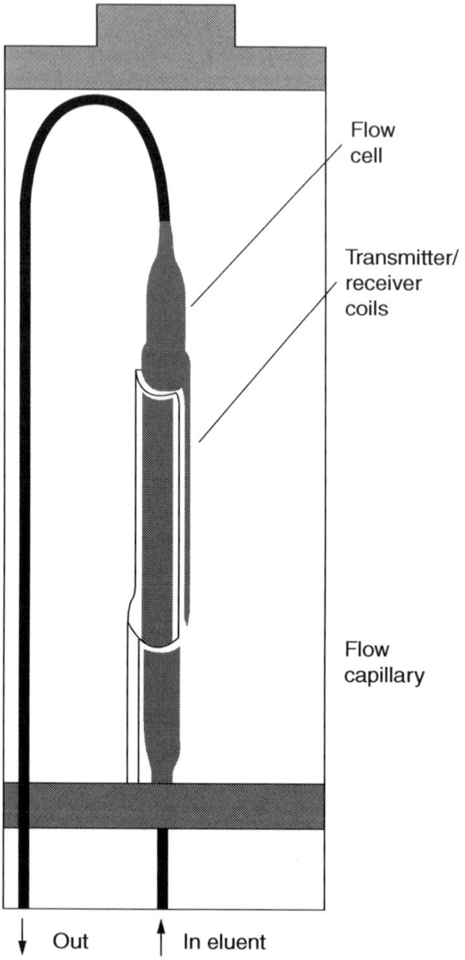

Figure 10.13 Design of a continuous-flow NMR probe for on-line SEC/NMR coupling. (Reprinted with permission from Ref. 33.)

absorption band but where MMA does not absorb. This selective absorption by the different components of the copolymer permitted quantitation of the weight fraction of styrene across the copolymer MMD [38].

SEC/UV/ELSD, monitoring $\lambda = 254$ nm with the UV detector, has been used to analyze the product of a grafting reaction of PMMA and EPDM (ethylene–propylene–diene monomer) rubber [39]. The result is shown in Figure 10.15. Because PMMA has a higher UV response at 254 nm than EPDM does, it was concluded that the higher-M (smaller elution volume) portion of the sample was richer in PMMA than the lower-M (larger elution volume) portion.

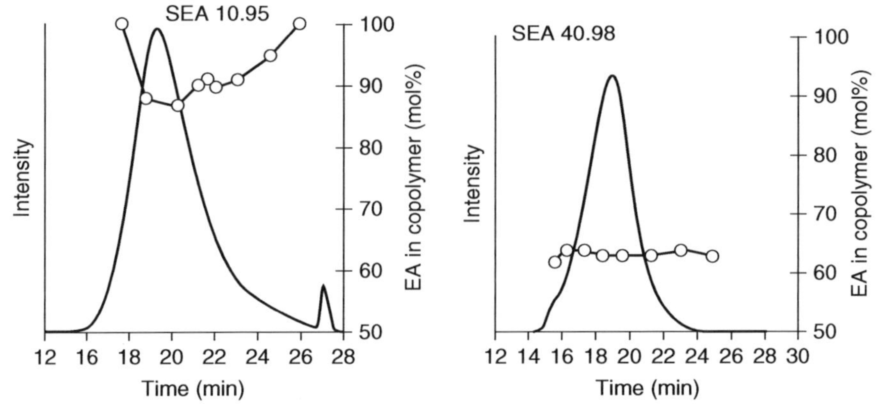

Figure 10.14 On-line SEC/NMR for chemical heterogeneity analysis. Solid lines, elution profiles of styrene/ethyl acrylate copolymers, obtained by SEC/ELSD; lines with circles, distribution of ethyl acrylate content (as mol%) across the elution profiles of the copolymers, obtained by SEC/NMR. (Reprinted with permission from Ref. 37.)

10.5.2 Fluorescence

A niche coupling has been that of SEC and fluorescence. Together, these techniques have been employed to measure polymer–polymer chain end reaction rate constants [40], to separate and identify airborne organic molecules from bitumen [41], and to study dissolved organic matter from Baltic Sea water [42]. SEC/DRI/fluorescence has been used to monitor the coupling of fluorescence-labeled anhydride-functional

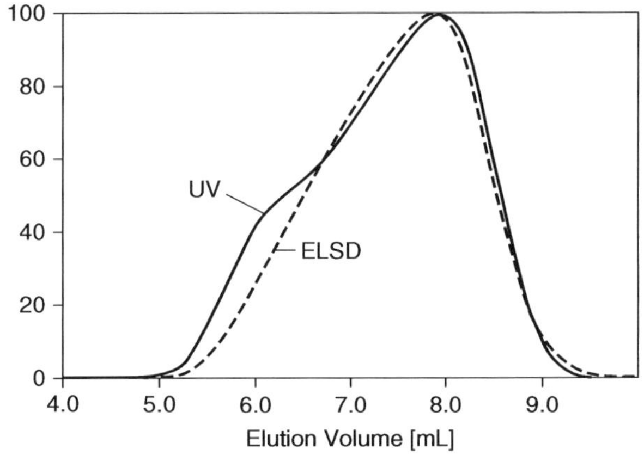

Figure 10.15 SEC/UV/ELSD chromatogram of the graft product of PMMA and EPDM copolymer. (Reprinted with permission from Ref. 39.)

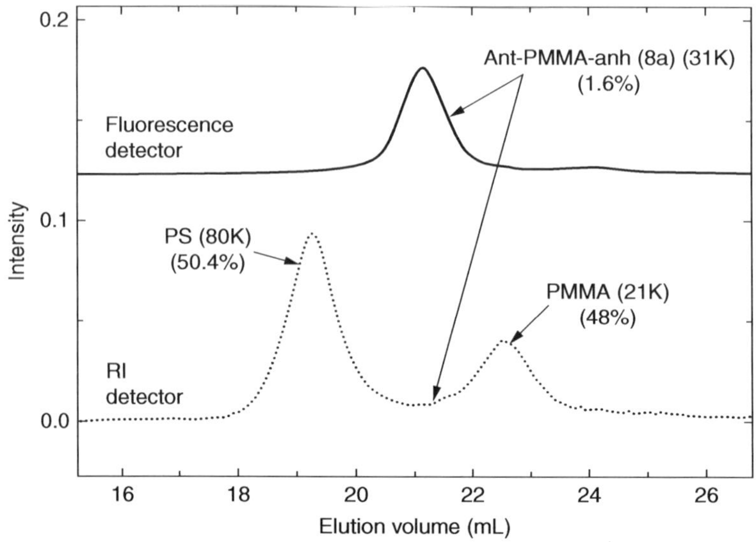

Figure 10.16 SEC/DRI/fluorescence analysis of a mixture of 1.6 wt% fluorescent polymer, 50.4 wt% PS, and 48 wt% PMMA. For fluorescence detection, $\lambda_{excitation} = 358$ nm, $\lambda_{emission} = 405$ nm. (Reprinted with permission from Ref. 43.)

PS and PMMA in dilute polymer blends [43]. Figure 10.16 shows the DRI and fluorescence detector SEC traces for a mixture of 1.6 wt% of a fluorescent polymer, 50.4 wt% PS, and 48 wt% PMMA. The fluorescent polymer is not observable by refractometry. The unlabeled PS and PMMA are not visible in the SEC/fluorescence chromatogram, even though these two compounds are present in approximately 30 times the weight percent of the fluorescent polymer.

10.5.3 Conductivity

The use of conductivity detection in SEC has been rather sparse, although it has great potential in the analysis of polyelectrolytes. SEC/DRI/VISC/conductivity has been used to measure the charge distribution, as a function of degree of substitution, of sodium carboxymethylcelluloses [44]. SEC with conductivity detection was also used to estimate the molar mass of per-*O*-sulfonated polysaccharides, including glycosaminoglycans (GAGs) and hyaluronan [45]. Figure 10.17 shows the conductivity detector traces for a series of unmodified and *O*-sulfonated GAGs. Because the conductivity detector response depends on a number of parameters, such as analyte concentration, charge, and molar mass, identical sample sizes of different analytes give different peak areas.

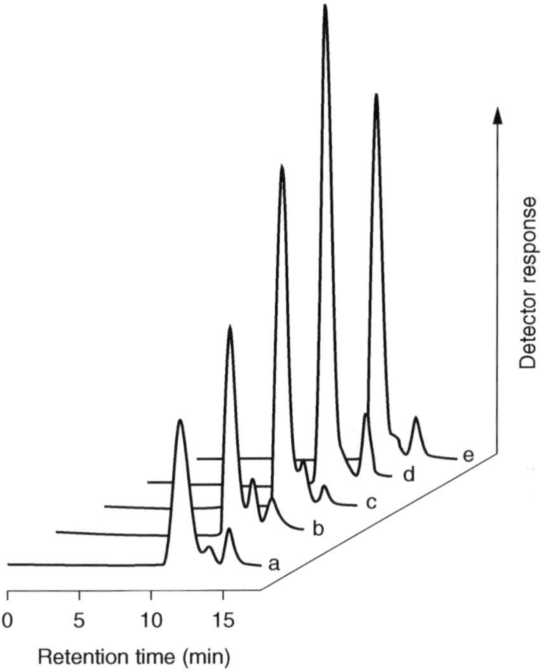

Figure 10.17 SEC/conductivity analysis of unmodified and *O*-sulfonated GAGs: (a) Hyaluronan; (b) dermatan sulfate; (c) chondroitin sulfate; (d) heparin; (e) heparin sulfate. All analytes are per-*O*-sulfonated. (Reprinted with permission from Ref. 45.)

10.5.4 Dynamic Surface Tension Detection

The schematic of an SEC system with DSTD [46] is shown in Figure 10.18; Figure 10.19a and b show the pressure sensor in this detector and the capillary tip during drop growth. Compared to the surface tension of the neat solvent, the lower surface tension of a polymer solution causes a relative displacement of a pressure sensor membrane and, consequently, a recorded pressure drop. The surface tension at the air–liquid interface is related to the time-dependent modified Young–Laplace equation:

$$P(t) = \frac{2\gamma(t)}{r(t)} + P_c \tag{10.2}$$

where $P(t)$ is the differential pressure growth across the drop interface throughout drop growth, relative to atmospheric pressure, $\gamma(t)$ the surface tension at the interface, and $r(t)$ the radius of the drop as a function of time during drop growth. P_c, which represents the offset pressure and viscous losses in the tubing, is generally time independent in most SEC applications.

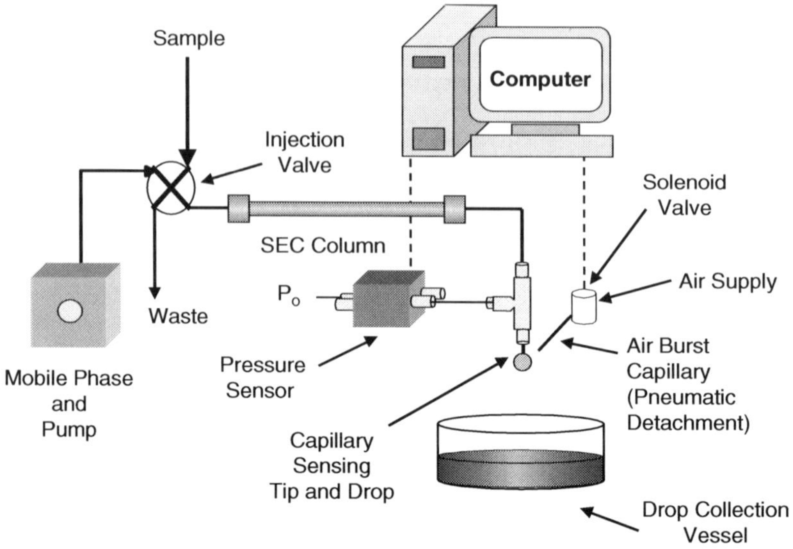

Figure 10.18 SEC system with on-line dynamic surface tension detection (DSTD). (Reprinted with permission from Ref. 46.)

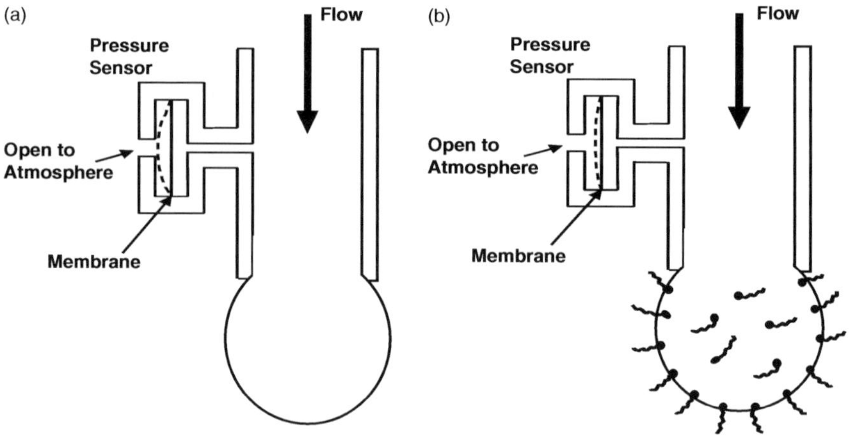

Figure 10.19 Dynamic surface tension detector pressure sensor and capillary tip during growth: (a) Displacement of the sensor membrane during formation of a mobile-phase droplet; (b) displacement of the sensor membrane during formation of a droplet containing surface-active analyte. (Reprinted with permission from Ref. 46.)

SEC/DSTD has been used to separate mixtures of polydisperse poly(ethylene glycol)s (PEGs) and to measure the polydispersity of the individual PEGs, to distinguish between star polymers with the same chemistry and molar mass but varying number of arms, and to separate and characterize protein mixtures.

10.5.5 Microscale Molecular Mass Sensor

The microscale mass sensor (μ-MMS) is a novel detection method that relies on the simultaneous measurement of the refractive index gradient (RIG) between adjacent laminar flows at two different positions along a microchannel [47]. A flow stream containing only solvent and a flow stream containing analyte solution merge on a microchip and flow parallel to each other down the microchannel. The analyte concentration gradient between, and analyte diffusion across, the two streams creates an RIG. The RIG is measured at a position close to the merging point (upstream signal) and simultaneously at a distance farther down the microchannel (downstream signal), providing real-time data about analyte diffusion. These data can be correlated to analyte molar mass by taking the ratio of the downstream-to-upstream signals. To the extent correlated to molar mass, the diffusion coefficient may also be determined by SEC/μ-MMS. The online technique has been applied successfully to the study of PEGs over a molar mass range of 106 to 22,800 g/mol, and off-line to both PEGS and to small sugars such as mono- and oligosaccharides. A schematic of a μ-MMS assembly is shown in Figure 10.20. Results of an SEC/μ-MMS experiment using PEG samples are shown in Figure 10.21.

10.6 COUPLING OF CHEMICAL DETECTORS

Because of their synergistic nature, discussion of the advantages of coupling of physical detectors is deferred to Chapter 11. Here, we mention a few representative instances of recent couplings of chemical detectors.

SEC with online UV (254 nm), ^{1}H-NMR, and ESI-MS detection, and off-line continuous FTIR detection, has been used for the analysis of polymer additives [49]. The set-up, shown in Figure 10.22 using deuterated chloroform as eluent, was applied to the analysis of BHT (2,6-di-*tert*-butyl-4-methoxyphenol), the antioxidant Irganox 1076 [octadecyl-3-(3,5-di-*tert*-4-hydroxyphenyl)propionate], and the plasticizer diisooctyl phthalate.

SEC with online diode array UV, MALS, DRI, and fluorescence detection (in that order) has been used in the characterization of soluble glucan polymers [50]. The light-scattering and concentration-sensitive detectors allowed calculation of molar mass and size averages. A derivatization reactor was placed between the DRI and fluorescence detectors. When multiple components were present in the sample, derivatization of the glucans in this reactor, using a fluorescent complex (Calcofluor White), enabled positive identification of the peak of interest.

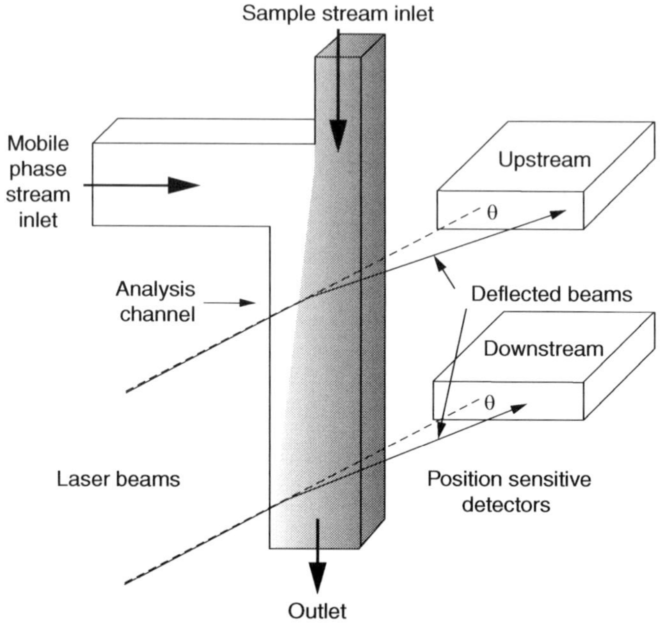

Figure 10.20 Dual-beam μ-MMS for on-line SEC detection. (Reprinted with permission from Ref. 48.)

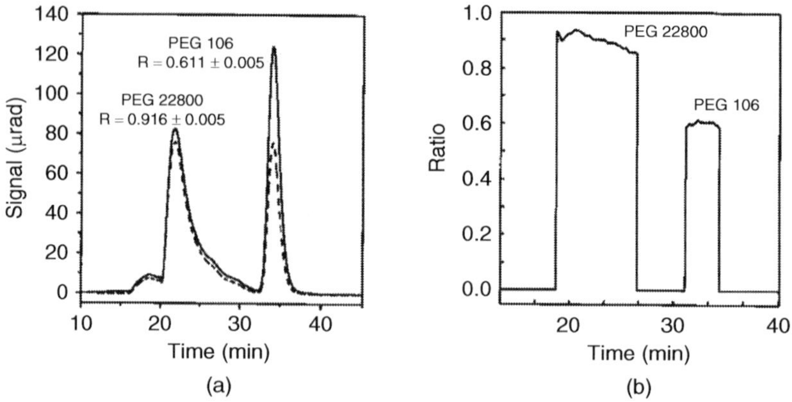

Figure 10.21 SEC/μ-MMS analysis of a two-component PEG mixture: (a) SEC chromatogram collected with on-line μ-MMS. Solid line, upstream data; dashed line, downstream data. Flow rate, 20 μL/min; sample concentration, 2%; injection volume, 5 μL; time delay between detection positions, 5.3 s. (b) Ratiogram of the chromatograms in (a). (Reprinted with permission from Ref. 48.)

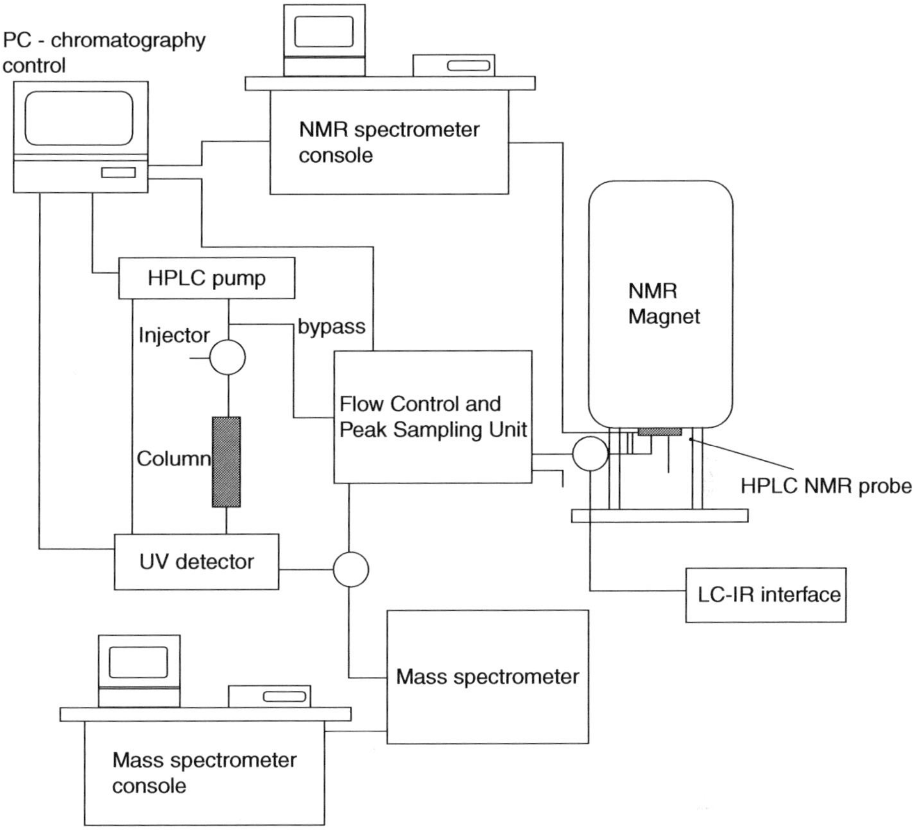

Figure 10.22 SEC/UV/^{1}H-NMR/ESI-MS/FTIR. UV, NMR, and ESI-MS are on-line detectors, FTIR is a continuous off-line detector. (Reprinted with permission from Ref. 49.)

REFERENCES

1. A. M. Striegel, *Anal. Chem.*, **77**, 104A (2005).
2. A. M. Striegel, ed., *Multiple Detection in Size-Exclusion Chromatography*, ACS Symp. Ser. 893, American Chemical Society, Washington, DC, 2005.
3. M. Montaudo and R. P. Lattimer, *Mass Spectrometry of Polymers*, CRC Press, Boca Raton, FL, 2002.
4. M. H. Allen and M. L. Vestal, *J. Am. Soc. Mass Spectrom.*, **3**, 18 (1992).
5. R. B. Cole, ed., *Electrospray Ionization Mass Spectrometry*, Wiley, New York, 1997.
6. R. Saf, C. Mirtl, and K. Hummel, *Acta Polym.*, **48**, 513 (1997).
7. L. Prokai, *Int. J. Polym. Anal. Charact.*, **6**, 379 (2001).
8. A. These and T. Reemtsma, *Anal. Chem.*, **75**, 6275 (2003).
9. L. Prokai, S. M. Stevens, Jr., and W. J. Simonsick, Jr., in Ref. 2, Chap. 12.

10. M. J. Deery, E. Stimson, and C. G. Chappell, *Rapid Commun. Mass Spectrom.*, **15**, 2273 (2001).
11. D. J. Aaserud, L. Prokai, and W. J. Simonsick, Jr., *Anal. Chem.*, **71**, 4793 (1999).
12. A. A. Toy, P. Vana, T. P. Davis, and C. Barner-Kowollik, *Macromolecules*, **37**, 744 (2004).
13. R. J. Cotter, *Anal. Chem.*, **64**, 1027A (1992).
14. G. H. Theissen, W. Schrepp, and H. Pasch, *MALDI-TOF Mass Spectrometry of Synthetic Polymers*, Springer-Verlag, Berlin, 2003.
15. M. W. F. Nielen, *Mass Spectrom. Rev.*, **18**, 309 (1999).
16. E. Esser, C. Keil, D. Braun, P. Montag, and H. Pasch, *Polymer*, **41**, 4039 (2000).
17. X. M. Liu, E. P. Maziarz, D. J. Heiler, and G. L. Grobe, *J. Am. Soc. Mass Spectrom.*, **14**, 195 (2003).
18. T. H. Mourey, A. J. Hoteling, S. T. Balke, and K. G. Owens, *J. Appl. Polym. Sci.*, **97**, 627 (2005).
19. S. J. Hill, ed., *Inductively Coupled Plasma Spectrometry and Its Applications*, 2nd ed., Blackwell Publishing, Oxford, UK, 2007.
20. G. K. Zoorob, J. W. McKiernan, and J. A. Caruso, *Mikrochim. Acta*, **128**, 145 (1998).
21. K. Wrobel, B. B. M. Sadi, K. Wrobel, J. R. Castillo, and J. A. Caruso, *Anal. Chem.*, **75**, 761 (2003).
22. B. B. M. Sadi, A. P. Vonderheide, J. S. Becker, and J. A. Caruso, in Ref. 2, Chap.10.
23. J. Szpunar, P. Pellerin, A. Makarov, T. Doco, P. Williams, and R. Łobiński, *J. Anal. At. Spectrom.*, **14**, 639 (1999).
24. J. Szpunar, A. Makarov, T. Pieper, B. K. Keppler, and R. Łobiński, *Anal. Chim. Acta*, **387**, 135 (1999).
25. J. L. Koening, *Rapra Rev. Rep.*, **12**, 1 (2001).
26. J. B. P. Soares, R. F. Abbott, and K. S. Kim, *J. Polym. Sci. B*, **38**, 1267 (2000).
27. P. J. DesLauriers, in Ref. 2, Chap. 13.
28. J. N. Willis and L. Wheeler, in *Chromatographic Characterization of Polymers: Hyphenated and Multidimensional Techniques,* Adv. Chem. Ser. 247, T. Provder, H. G. Barth, and M. W. Urban, eds., American Chemical Society, Washington, DC, 1995, Chap. 19.
29. K. Nomura, H. Okumura, T. Komatsu, and N. Naga, *Macromolecules*, **35**, 5388 (2002).
30. P. M. Cotts and A. C. Ouano, in *Microdomains in Polymer Solutions,* P. Dubin, ed., Plenum Press, New York, 1985, Chap. 7.
31. S. J. Kok, C. A. Wold, Th. Hankemeier, and P. J. Schoenmakers, *J. Chromatogr. A*, **1017**, 83 (2003).
32. K. Albert and E. Bayer, *Anal. Methods Instrum.*, **2**, 302 (1995).
33. K. Albert, M. Dachtler, T. Glaser, H. Händel, T. Lacker, G. Schlotterbeck, L.-H. Tseng, and U. Braumann, *J. High Resolut. Chromatogr.*, **22**, 135 (1999).
34. P. A. Mirau, *A Practical Guide to Understanding NMR of Polymers*, Wiley-Interscience, New York, 2004.
35. K. Hatada, *J. Polym. Sci. A*, **37**, 245 (1999).
36. K. Ute, R. Niimi, M. Matsunaga, K. Hatada, and T. Kitayama, *Macromol. Chem. Phys.*, **202**, 3081 (2001).
37. I. Krämer, H. Pasch, H. Händel, and K. Albert, *Macromol. Chem. Phys.*, **200**, 1734 (1999).

38. R. Medrano, M. T. R. Laguna, E. Saiz, and M. P. Tarazona, *Phys. Chem. Chem. Phys.*, **5**, 151 (2003).

39. H. Pasch, in Ref. 2, Chap.14.

40. A. J. Maliakal, B. O'Shaughnessy, and N. J. Turro, in Ref. 2, Chap.6.

41. P. Fauser, J. C. Tjell, H. Mosbaek, and K. Pilegaard, *Pet. Sci. Technol.*, **18**, 989 (2000).

42. V. Lepane, *Oil Shale*, **18**, 239 (2001).

43. B. Moon, T. R. Hoye, and C. W. Macosko, *J. Polym. Sci. A*, **38**, 2177 (2000).

44. M. Rinaudo, J. Danhelka, and M. Milas, *Carbohydr. Polym.*, **21**, 1 (1993).

45. A. Chaidedgumjorn, A. Suzuki, H. Toyoda, T. Toida, T. Imanari, and R. J. Linhardt, *J. Chromatogr. A*, **959**, 95 (2002).

46. R. E. Synovec, B. A. Staggemeier, E. Bramanti, W. W. C. Quigley, and B. J. Pranzen, in Ref. 2, Chap. 16.

47. E. M. Humston, A. D. McBrady, M. Valero, and R. E. Synovec, *Talanta*, **73**, 287 (2007).

48. C. D. Costin, R. K. Olund, B. A. Staggemeier, A. K. Torgerson, and R. E. Synovec, *J. Chromatogr. A*, **1013**, 77 (2003).

49. M. Ludlow, D. Louden, A. Handley, S. Taylor, B. Wright, and I. D. Wilson, *J. Chromatogr. A*, **857**, 89 (1999).

50. W. H. Yokoyama and B. E. Knuckles, in Ref. 2, Chap. 8.

11

POLYMER ARCHITECTURE AND DILUTE SOLUTION THERMODYNAMICS

11.1 INTRODUCTION

The difference between physical and chemical detectors was discussed in Chapters 9 and 10. Specifically, this difference was explained as one based on detector synergies, with the physical detectors having a greater capability in this regard than their chemical counterparts. A generic example may help clarify the point.

Take the case of a polydisperse, branched, random AB copolymer. The molar mass averages and distribution are determined, most conveniently, by SEC/SLS/DRI. Should short-chain branching be present, this may be determined by addition of FTIR detection, thus providing the distribution of short-chain branches as a function of molar mass. The chemical heterogeneity (distribution of the percent of component A or B as a function of copolymer molar mass) may also be measured by adding FTIR, NMR, or UV detection, as appropriate. As can be seen, each of these detectors (IR, NMR, UV) provides a piece, or sometimes several pieces, of information, and adding more chemical detectors serves to provide an additive amount of information.

Looking at this same generic AB copolymer using SEC with physical detection techniques such as MALS, QELS, and VISC (in addition to the perfunctory concentration-sensitive detector, e.g., DRI), the molar mass distribution can be determined, and also the distribution of long-chain branches (LCBs) and the branching frequency. This combination of physical detectors can also provide the fractal dimension of the polymer and the change in this dimension as a function of molar mass. Each of the physical detectors listed measures a distinct polymeric radius (see

Modern Size-Exclusion Liquid Chromatography: Practice of Gel Permeation and Gel Filtration Chromatography, Second Edition
By André M. Striegel, Wallace W. Yau, Joseph J. Kirkland, and Donald D. Bly

Table 9.2). The distribution of the hydrodynamic radius as a function of molar mass, determined using QELS detection, may be combined with the distribution of the radius of gyration, determined using MALS, to show whether the polymer is linear or branched. In the latter case, this approach will also indicate if branching is random, starlike (with arms that are either of uniform length or polydisperse), dendritic, and so on. Off-line MALS analysis can help define the thermodynamic state of the solution (i.e., good, poor, or theta solvent–temperature conditions) via the second virial coefficient, and also provides the thermodynamic radius of the polymer. Comparison of this radius with R_G or R_η provides information on coil interpenetration and polymer draining, respectively. Combining these two new pieces of information, additional insight into polymer branching can be gleaned. Combining viscometric data with results from MALS detection may also yield information about persistence length and local polydispersity that either detector individually would not have provided. Again, this last point, the synergistic nature of physical detectors, constitutes our basis for detector classification.

With the notable exception of the type of short-chain branching information obtained using SEC/FTIR, which was discussed in Section 10.3.1, the architectural and thermodynamic knowledge discussed in this chapter will be arrived at through multidetector SEC with physical detection methods. In the case of copolymers, some type of chemical detection will usually be necessary if the chemical properties of the polymer are to dictate end-use performance, or for more accurate quantitation of the physical detector data than would be possible without the chemical methodologies. For example, if constituents A and B of a generic AB copolymer have different specific refractive index increments ($\partial n/\partial c$; see Section 9.2.1.3) than each other, the chemical heterogeneity data can combine with the MALS molar mass data to increase the accuracy of the latter. A similar case may be made for polyelectrolytes.

The generic random AB copolymer may also possess a chemical composition distribution (CCD), which would be determined by a non-SEC separation method (e.g., GPEC). Determining the distribution of the CCD as a function of the MMD falls within the purview of two-dimensional liquid chromatography (e.g., SEC $\times$ GPEC). The role of SEC in 2D-LC experiments is discussed in Chapter 14.

Reference 1 provides a review of the applications of physical detectors, including a discussion of the detection principles of each technique. Additional discussion and applications may be found in References 2 and 3.

11.2 LONG-CHAIN BRANCHING

Long-chain branches are generally perceived as having lengths comparable to, or a substantial fraction of, the length of the main macromolecular backbone [4,5]. As shown in Table 1.1, LCBs can influence a variety of chemical, physical, processing, and end-use properties of polymers and polymer solutions. The changes that LCBs bring about in space-filling physical properties affect the viscosity and elasticity of melts as well as the viscosity, sedimentation behavior, and angular distribution of scattered radiation of dilute polymer solutions. Additionally, the multiplicity of endgroups created as a result of an increase in branches can affect the chemical reactivity of the molecule, as in the case of dendrimers and hyperbranched polymers. Consider the words of Walter Stockmayer and Marshall Fixman, written in 1953 [6]: "In an earlier era of high polymer chemistry, branching not infrequently served as a whipping boy to explain deviations from an expected physical behavior often without any independent or clear-cut evidence for the occurrence of chemical reactions leading to branched molecular structures. . . . A sounder and more quantitative attack on the question of branching is therefore mandatory on both experimental and theoretical fronts." The methods described in this section aim to provide the type of sound, quantitative reasoning that the problem of polymer LCB demands.

In an SEC experiment, a branched macromolecule will elute from the column later than a linear macromolecule of the same chemistry and equal molar mass as a result of the smaller hydrodynamic (or solvodynamic) volume occupied by the branched species with respect to the linear analyte. In this section we describe the analysis of LCBs in macromolecules using SEC with multiple detection. Although the qualitative information obtained from this type of analysis is mentioned, particular attention is paid to the requirements necessary for accurate, quantitative determination of LCB, of the long-chain branching distribution (LCBD), and of the fractal dimension (d_f) of macromolecules. An excellent treatment of the solution properties of branched macromolecules may be found in Reference 4.

11.2.1 Quantitating the Long-Chain Branching Distribution by SEC/MALS

In 1949, Zimm and Stockmayer published a now-classic paper that laid out the requirements for quantitating branching in trifunctional ($f = 3$) and tetrafunctional ($f = 4$) polymers [7]. Zimm and Stockmayer's derivations are based on the contraction factor, g, which they defined as the ratio of the mean-square radius, R_G^2, of a branched polymer (subscript B) to the mean-square radius of its linear counterpart (subscript L):

$$g = g_M = \left[\frac{(R_G)_B^2}{(R_G)_L^2}\right]_M \tag{11.1}$$

The subscript M refers to values obtained for the same molar mass. ($R_G \equiv \langle r^2 \rangle^{1/2}$, i.e., the radius of gyration, is defined as the root-mean-square (RMS) radius of a polymer; see Section 9.3.1.) Strictly speaking, it is the z-average radii that are being compared, although in the case of an SEC separation we follow the accepted assumption that each slice eluting from the column is monodisperse within the limits of experimental accuracy (this assumption is questioned in Section 11.8).

Zimm and Stockmayer recognized at the time the requirements needed for accurate branching calculations:

1. A linear standard is needed in order to obtain the contraction ratio g.
2. Only materials of equal chemistry should be compared.

3. The molar mass distributions (MMDs) of the branched and linear polymers should overlap in the region of interest.
4. The type (functionality f) of the branch points should remain uniform across the MMD.

Results from the LCB calculations were shown to be invariant to changes in the length of the branches throughout the MMD. The number-average number of branch points, B_n, per molecule for a monodisperse system is given by Equation 11.2 for the case of trifunctional ($f = 3$) branching and by Equation 11.3 for the case of tetrafunctional ($f = 4$) branching. The weight-average number of branch points, B_w, for a trifuctionally branched polydisperse species is given by Equation 11.4, while B_w for a tetrafunctionally branched polydisperse species is given by Equation 11.5:

$$g = \left[\left(1 + \frac{B_{3n}}{7}\right)^{1/2} + \frac{4B_{3n}}{9\pi}\right]^{-1/2} \tag{11.2}$$

$$g = \left[\left(1 + \frac{B_{4n}}{6}\right)^{1/2} + \frac{4B_{4n}}{3\pi}\right]^{-1/2} \tag{11.3}$$

$$g = \frac{6}{B_{3w}}\left\{\frac{1}{2}\left(\frac{2 + B_{3w}}{B_{3w}}\right)^{1/2} \ln\left[\frac{(2 + B_{3w})^{1/2} + B_{3w}^{1/2}}{(2 + B_{3w})^{1/2} - B_{3w}^{1/2}}\right] - 1\right\} \tag{11.4}$$

$$g = \frac{\ln(1 + B_{4w})}{B_{4w}} \tag{11.5}$$

The symbols used here are those that have found favor in the literature over the years; occasionally, they may differ from the symbols used in the original publication. It is important to realize that, excluding the backbone, one branch emanates from each branch point when $f = 3$, whereas two branches emanate from each branch point when $f = 4$.

The plot of size versus molar mass, with each axis on a logarithmic scale (i.e., the plot of log R_G versus log M), is commonly referred to as a *conformation plot*. Comparing the conformation plot of a branched species to that of a linear standard allows determination of g at each molar mass slice and, from there, calculation of the number of branches as a continuous function of the molar mass of the branched macromolecule, the LCBD. An additional parameter that can be calculated using the branching number, B, is the branching frequency, λ, defined as the number of long-chain branches per 1000 repeat units:

$$\lambda = \frac{1000BR}{M} \tag{11.6}$$

where R is the molar mass of the repeat unit. Figure 11.1 shows the application of Zimm–Stockmayer theory to SEC/MALS data to calculate LCB in a branched

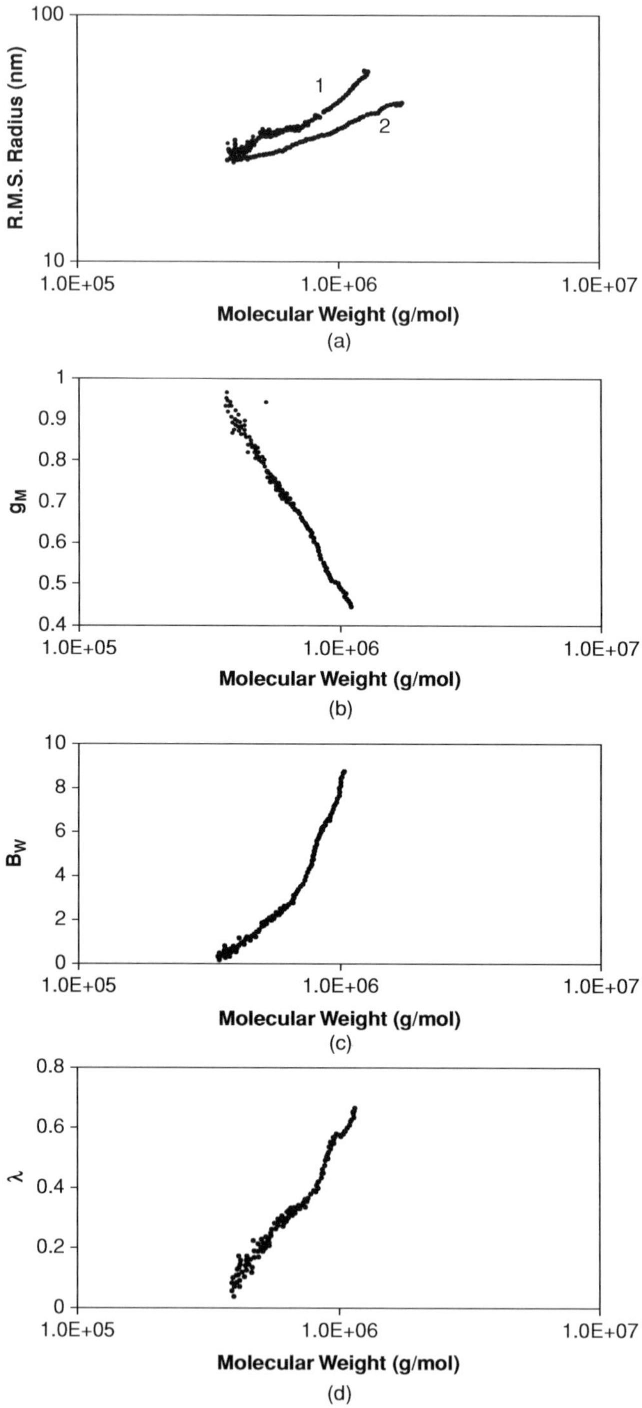

Figure 11.1 Measuring the LCBD of PVAc by SEC/MALS: (a) conformation plot of linear (sample 1) and branched (sample 2) PVAc; (b) ratio, *g*, of the mean-square radii of branched and linear samples; (c) weight-average number of branches, B_w, in sample 2; (d) branching frequency, λ, of sample 2. (Reprinted with permission from Ref. 8.)

poly(vinyl acetate) (PVAc) sample by comparison with a linear PVAc, assuming trifunctional branching and using Equation 11.4 [8].

11.2.2 Qualitative and Semiquantitative Descriptions of the Long-Chain Branching Distribution by SEC/VISC

If an online viscometer (VISC) is used instead of a MALS detector in SEC (again in conjunction with a concentration-sensitive detector), the ratio g' of the intrinsic viscosities of the branched molecule $[\eta]_B$ and of the linear standard $[\eta]_L$, at the same molar mass M, has been used for the branching calculations:

$$g' = \left(\frac{[\eta]_B}{[\eta]_L}\right)_M \neq \left(\frac{[\eta]_B}{[\eta]_L}\right)_{v,\mathrm{SEC}} \tag{11.7}$$

As noted in Equation 11.7, the ratio of the intrinsic viscosities of linear and branched polymers with the same molar mass (subscript M) is *not* equal to the ratio of the intrinsic viscosities of linear and branched samples eluting at the same retention volume in an SEC experiment (subscript v, SEC).

The relationship between g and g' is given by Equation 11.8, where ε (also referred to in the literature by the symbols b and e) is known as the *viscosity shielding ratio*:

$$g' = g^{\varepsilon} \tag{11.8}$$

The *viscosity shielding ratio* is defined as the distance within the hydrodynamic sphere occupied by the molecule in solution over which solvent flow decreases by a factor $1/e$ of that in the free solution (corresponding to an approximately 37% reduction in flow), divided by the radius of said hydrodynamic sphere. The value of ε is dependent on a number of factors, including solvent, temperature, and branching. ε has been found to generally fall in the range 0.5 to 1.5. A table of ε values may be found in Reference 9.

The viscosity shielding ratio can be determined, as a continuous function of molar mass, through a single SEC/DRI/MALS/VISC experiment, where ε is the slope of a plot of log g' versus log g. This type of measurement has shown ε to be constant for certain randomly branched poly(methyl methacrylate)s, as shown in Figure 11.2. The viscosity shielding ratio has also been found to be nonconstant across the MMD of certain low-density polyethylenes, as shown in Figure 11.3. As such, quantitation of the LCBD by SEC/VISC should be approached with caution.

One may alternatively calculate g by what is known as the *mass method* [12], by comparing the molar masses of the branched molecule (M_B) and of the linear standard (M_L) at the same elution volume (V):

$$g = \left(\frac{M_L}{M_B}\right)_V^{(a+1)/\varepsilon} \tag{11.9}$$

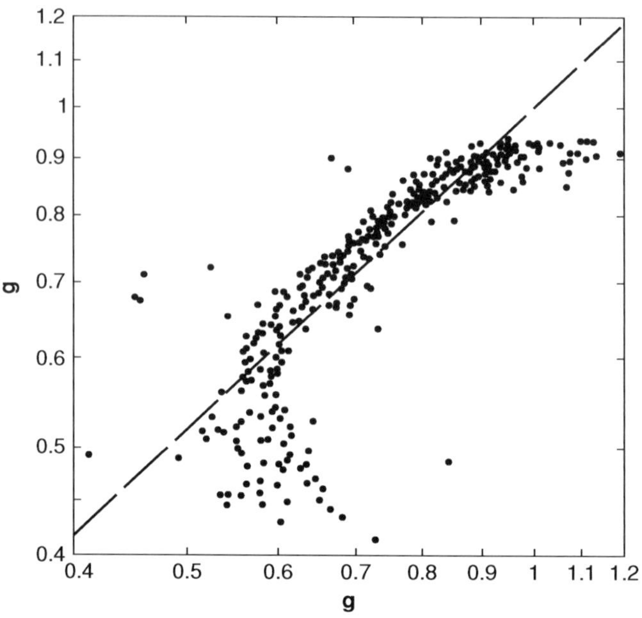

Figure 11.2 Determining the viscosity shielding ratio, ε, of branched PMMA using SEC/DRI/MALS/VISC. Dots represent experimental points from three different PMMA samples, giving ε values in the range 0.8 to 1.0. The dashed line is a least-squares fit to the data and has a slope of 0.9. (Reprinted with permission from Ref. 10.)

Here a is the exponent in the Mark–Houwink equation, $[\eta] = KM^a$ (Equation 8.2). Because results from the mass method are also dependent on the value of the viscosity shielding ratio, ε, quantitation of the LCBD by the mass method should also be approached with caution.

Errors can result if LCB calculations are performed using a standard that is believed to be linear but in fact is not; by applying the equations for tetrafunctional branching to calculate the branching in trifunctionally branched polymers (or vice versa); and by using incorrect values for the viscosity shielding ratio ε [13]. Individually, any of these cases can result in gross misestimations of polymer LCB.

While generally less accurate than SEC/MALS measurements, SEC/VISC measurements are usually both more precise and extend to lower M. The increased precision stems from the higher signal-to-noise ratio in viscometric measurements of the intrinsic viscosity than in most corresponding light-scattering measurements of R_G. Because the response of LS detectors is proportional to M, the precision of LS measurements increases with increasing M.

In summary, SEC determination of the LCBD of polymers is performed most accurately using a MALS detector. In the absence of the latter, SEC with a viscometric detector may also provide the LCBD. While the latter method offers the advantages over SEC/MALS of generally increased precision and of being able to access smaller

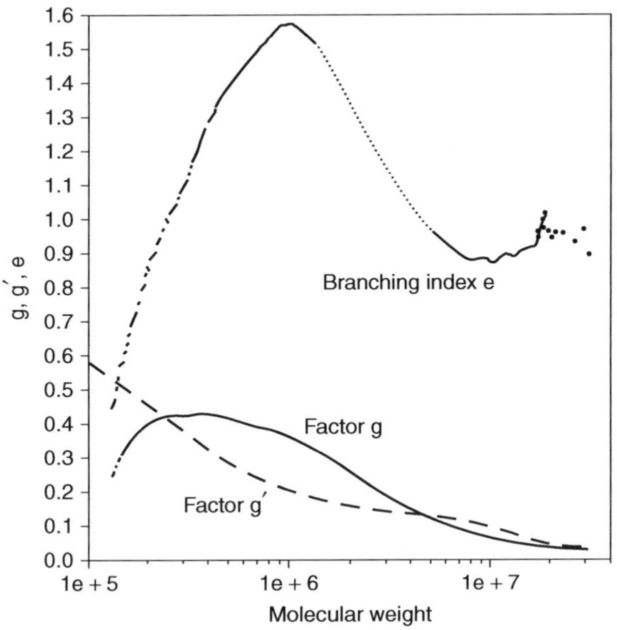

Figure 11.3 Variation in viscosity shielding ratio, ε, as a function of molar mass. SEC/DRI/MALS/VISC analysis of low-density polyethylene shows variability in ε (referred to in figure as "e") across the MMD. (Reprinted with permission from Ref. 11.)

sizes, it suffers from the need to know the viscosity shielding ratio, ε, which has been shown to be M-dependent in certain polymers.

11.2.3 Average Molar Mass Between Long-Chain Branches

For randomly branched polymers, the molar mass of the point of intersection of the power laws describing the linear and branched regions of the Mark–Houwink plot, or of the conformation plot, corresponds to the average molar mass between long-chain branches. This conclusion was arrived at through two different routes: first, by applying percolation theory to results of SEC/DRI/MALS/VISC analysis of randomly branched polyesters [14]; and second, by using Zimm–Stockmayer theory (Section 11.2.1) combined with frictional arguments to interpret SEC/DRI/MALS/VISC data of long-chain branched random terpolymers [15].

As seen in Figure 11.4 for a poly(vinyl butyral) terpolymer, similar results are obtained using either the Mark–Houwink or conformation plot [18]. In this figure, the sample termed PVBX contains both native branching and cross-link-induced branching. The power laws describing the linear and cross-link-induced branching regions of PVBX intersect at $M \approx 2 \times 10^5$ g/mol (denoted as $M_{\text{seg}}^{\text{ave}}$ in Figure 11.4). This value of M corresponds to the average molar mass between cross-link-induced branches in PVBX.

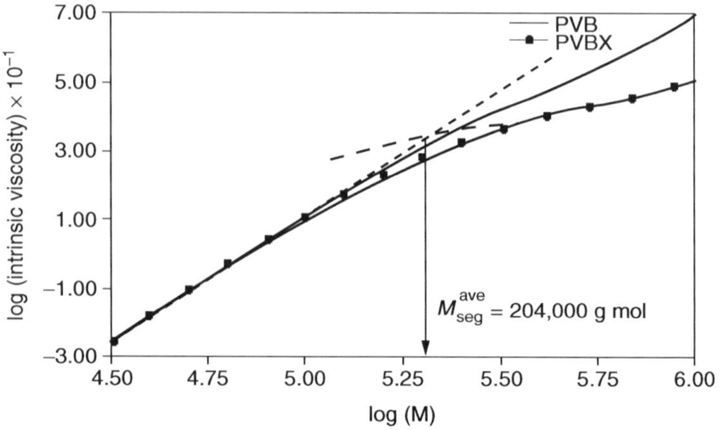

(a)

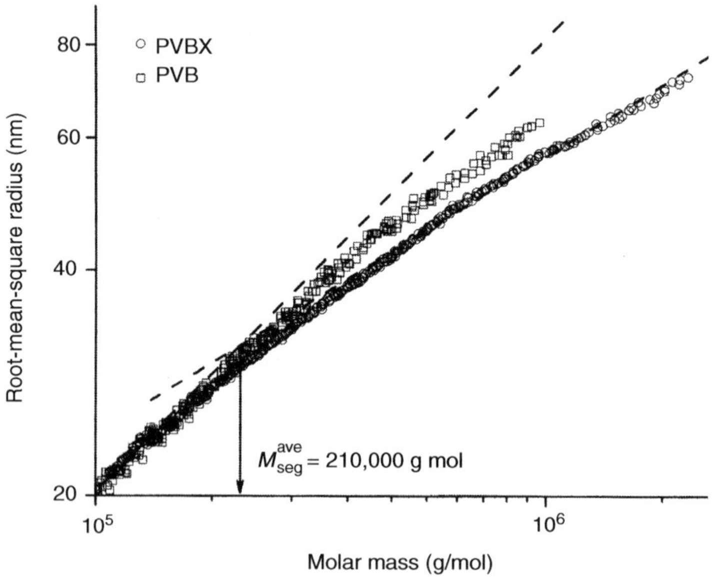

(b)

Figure 11.4 Average molar mass between branches, from SEC/DRI/MALS/VISC: (a) Mark–Houwink plot overlay; (b) conformation plot overlay of PVB and PVBX. PVB is a poly(vinyl butyral) sample with native long-chain branching (LCB). PVBX is a poly(vinyl butyral) sample with both native and cross-link-induced LCB. Dashed lines correspond to power laws describing the linear and cross-link-induced LCB regions of PVBX. In each figure, the point of intersection of the power laws corresponds to the average M between cross-link-induced branches in PVBX (denoted as M_{seg}^{ave}). (Reprinted with permission from Ref. 15.)

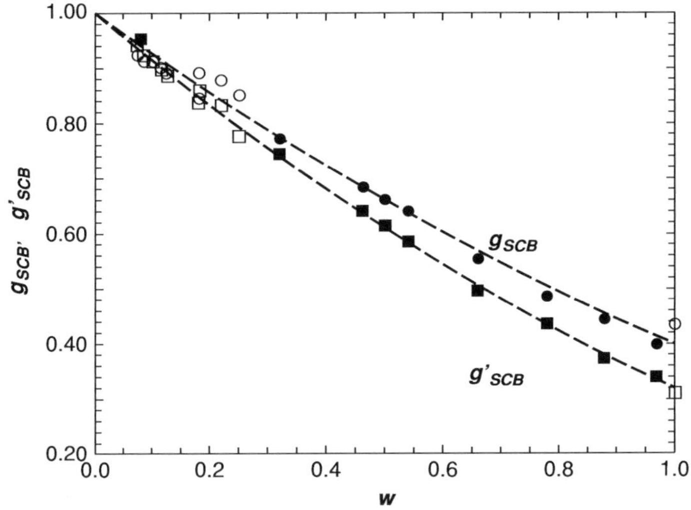

Figure 11.5 SCBD content of polyolefins by SEC/DRI/MALS/VISC. Open symbols indicate values for ethylene–butene copolymers, filled symbols for model materials; circles correspond to values of g_{SCB}, squares to values of g'_{SCB}; w is comonomer weight fraction. Dashed lines are placed to guide the eye. Data obtained in 1,2,4-trichlorobenzene at 135°C. (Reprinted with permission from Ref. 16.)

11.3 DETERMINING THE SHORT-CHAIN BRANCHING DISTRIBUTION

Determination of the short-chain branching distribution (i.e., of the short-chain branch content of a polymer as a continuous function of M), by means of SEC/FTIR was described in Section 10.3.1 for the case of polyolefins. For samples devoid of LCB, Equations 11.1 (for g) and 11.7 (for g') can be employed in conjunction with SEC/MALS or SEC/VISC, respectively, to determine the SCB content and the SCBD of a polymer, through [16]

$$g'_{\mathrm{SCB}} = (g_{\mathrm{SCB}})^{3/2} \tag{11.10}$$

Results of SCB calculations for ethylene–butene copolymers are shown in Figure 11.5.

In the case of polymers with both LCB and SCB, the following relation was proposed [17]:

$$g' = g'_{\mathrm{LCB}} g'_{\mathrm{SCB}} \tag{11.11}$$

More recently, however, SEC/MALS analysis in combination with SEC/FTIR and off-line NMR was used to show that in polyolefin samples with both LCB and SCB, correction for the short-chain branch content is necessary for accurate determination of the LCBD [18]. To effect this correction, two relationships are needed, that between the branching-index correction factor and the SCB content, and that between the SCB content and M. These two relationships may be obtained from SEC/MALS and SEC/FTIR analysis of polyethylene standards with known SCB content (as determined by NMR or FTIR), which is constant as a function of M (as determined by SEC/FTIR). Mathematically, the product of the two relationships gives the branching index correction factor as a function of molar mass.

In summary, determining the SCBD in polymers otherwise devoid of LCB appears relatively straightforward and can be done using a number of SEC-based methods, including those used in determining the LCBD. In polymers with both LCB and SCB, correcting for LCB when determining SCB is a nontrivial matter in that a number of analytical techniques are needed as well as standards where the SCB content is both known and constant across the MMD.

11.4 POLYMER ARCHITECTURE: CONFORMATION AND TOPOLOGY

The term *conformation* refers to the spatial structure of a macromolecule in dilute solution where, depending on solvent–temperature conditions, the polymer may adopt random coil, compacted sphere, or highly extended conformations. It becomes obvious that polymer conformation depends on the thermodynamics of the dilute polymer solution. Conversely, *topology* refers to the branching status of the macromolecule, which may be star, comb, dendritic, hyperbranched, random, or nonbranched (i.e., linear), and which also includes ring structures. Unlike conformation, however, polymer topology is thermodynamically invariant. Often, the terms *topology* and *architecture* are used interchangeably; here, we use the term *architecture* to encompass both conformational and topological phenomena.

The thermodynamic status of a dilute polymer solution is determined most accurately by means of a Zimm or related plot, obtained from an off-line, batch-mode MALS or variable-angle light-scattering experiment (Section 9.3.3). For example, should the fractal dimension (Section 11.4.1) or the dimensionless radii ratios (Section 11.4.2) indicate that the polymer is either a linear random coil at theta conditions, or a randomly branched polymer at good solvent–temperature conditions, experimental determination of the second virial coefficient (A_2) should help eliminate one of these choices.

11.4.1 Determining the Fractal Dimension

As depicted in Figure 11.6, the mass of a rigid rod, a one-dimensional object with $d = 1$, scales with the first power of the radius of a circle circumscribing the rod. The mass of a thin, flat disk, a two-dimensional object with $d = 2$, scales with the second power of the radius of the disk. The mass of a homogeneous sphere of constant density, a three-dimensional object with $d = 3$, scales with the third power of the radius of the sphere. For these simple one-, two-, and three-dimensional objects, the relationship between size and mass is given by the *topological* or *Euclidean dimension*, d_T.

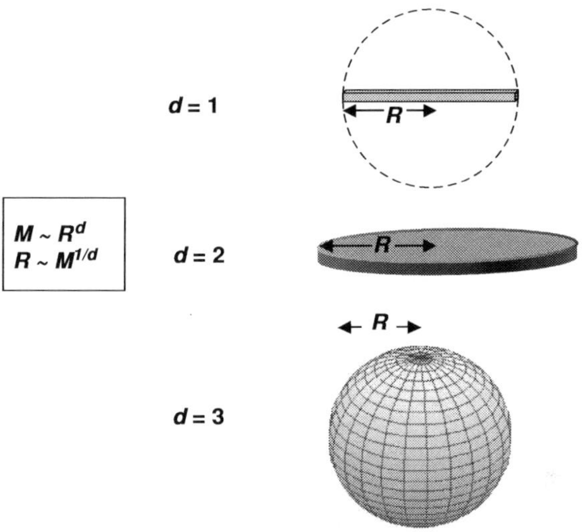

Figure 11.6 Relationship between the mass, size, and dimensionality of an object: top, thin, rigid rod of length $L = 2R$; middle, thin, flat disk of radius R; bottom, homogeneous sphere of constant density with radius R. (Courtesy of Deborah Striegel.)

From the study of fractal geometry, percolation theory, critical phenomena, and so on, it is known that self-similar objects must follow power-law behavior [19,20]. For macromolecules, *self-similarity* can be interpreted to mean that regardless of what scale is used to measure the root-mean-square (RMS) radius (R_G), whether this measurement is performed using the bond length, the length of a Kuhn statistical segment, or some other arbitrarily chosen scale, R_G should scale with molar mass in a unique fashion. Should this scaling relationship change, said change is indicative of a fundamental architectural change in the polymer (Figure 11.7c and d), or of a fundamental thermodynamic change in the polymer solution.

The relationship between radius and molar mass for macromolecules is given by the conformation plot, the plot of R_G versus M with each axis plotted logarithmically. The slope of this plot, α, is defined as the inverse of the fractal dimension, d_f, of the polymer (Figure 11.7a):

$$\alpha \equiv \frac{1}{d_f} \tag{11.12}$$

The fractal dimension has also been defined in terms of a, the slope of the Mark–Houwink plot of intrinsic viscosity versus molar mass (Figure 11.7b):

$$a = \frac{3}{d_f} - 1 \tag{11.13}$$

The viscometric definition relies on the assumption that the polymer draining (Φ) and coil interpenetration (Ψ^*) functions (see Section 11.4.3) do not change as the

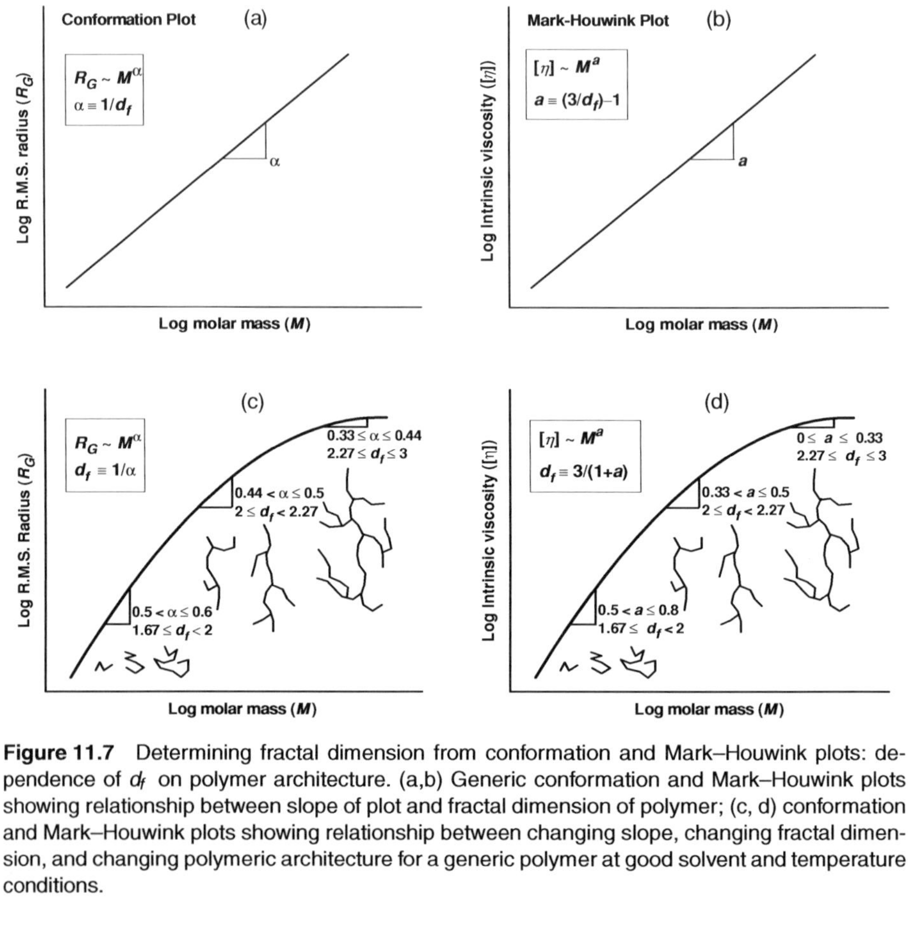

Figure 11.7 Determining fractal dimension from conformation and Mark–Houwink plots: dependence of d_f on polymer architecture. (a,b) Generic conformation and Mark–Houwink plots showing relationship between slope of plot and fractal dimension of polymer; (c, d) conformation and Mark–Houwink plots showing relationship between changing slope, changing fractal dimension, and changing polymeric architecture for a generic polymer at good solvent and temperature conditions.

molar mass and/or number of branches in the molecule increase. Although the viscometric definition is rarely fulfilled in its strictest sense, it still provides for useful approximations. These approximations become increasingly accurate as the polymer approaches being a linear rigid rod.

In the study of polymer architecture, the fractal dimension proves to be more useful than the topological dimension. For example, a rigid rod and a random coil both belong to the same topological form class (i.e., they are homeomorphs) and possess the same topological dimension: namely, $d_T = 1$ in both cases. However, a rigid rod and a random coil are quite different from each other. Solutions of a rigid rod, of a random coil at good solvent–temperature conditions, and of a random coil at the theta state are not distinguished through topology. For a linear random coil, $d_f = 2$ at theta condition and $1.67 \leq d_f < 2$ at good solvent–temperature conditions. For a randomly branched polymer, $d_f = 2.27$ at theta conditions, and $2 \leq d_f < 2.27$ at good solvent–temperature conditions. As expected, for a rigid rod, $d_f = 1$. These results are summarized in Table 11.1.

Table 11.1 Relationship between fractal dimension and polymeric architecture

Architecture	d_f
Rigid rod[a]	1
Linear random coil (good)[b]	$1.67 \leq d_f < 2$
Linear random coil (θ)[b]	2
Random branching (good)[b]	$2 \leq d_f \leq 2.27$
Random branching (θ)[b]	2.27
Hard sphere[a]	3

[a] Technically, these are not fractal objects.
[b] "Good" and "θ" correspond to values at good and theta solvent–temperature conditions, respectively.

Overlaid in Figure 11.8 are the conformation plots of polystyrene (PS), poly(γ-benzyl-L-glutamate) or PBLG, and two poly(vinyl butyral) samples, one with native long-chain branching (PVB) and one with both native and cross-link-induced long-chain branching (PVBX) [2,15,21]. All plots were determined by SEC/MALS. Additional off-line, batch-mode MALS analysis showed that the experiments were performed at good solvent–temperature conditions, as indicated by high positive values of the second virial coefficient A_2. The random coil nature of PS is evidenced by the fact that $d_f = 1.85$, as expected for a linear random coil at good solvent/temperature

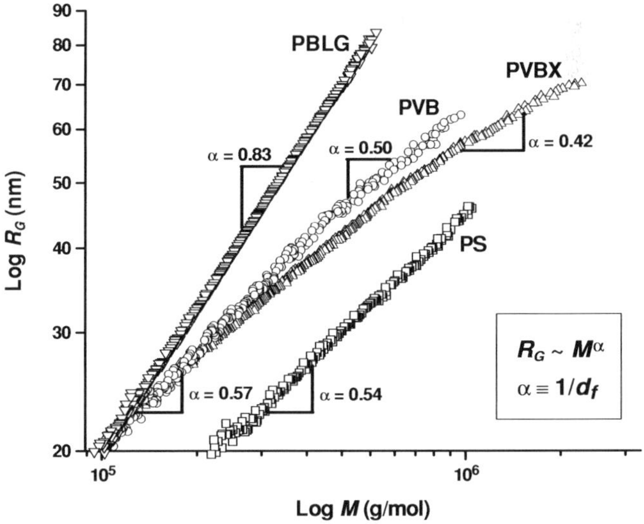

Figure 11.8 Dependence of d_f on macromolecular architecture for several polymers. PBLG, poly(γ-benzyl-L-glutamate); PS, polystyrene; PVB, poly(vinyl butyral) with native branching; PVBX, poly(vinyl butyral) with both native and cross-link-induced branching. Data from SEC/MALS analysis in DMAc/0.5% LiCl, 35°C. (Adapted from Refs. 2, 15, and 21.)

conditions (see Table 11.1). PBLG, which is known to adopt a highly extended conformation in certain solvents, resembles a rigid rod more than it does a random coil [24]. PBLG is often referred to as a *semiflexible polymer*, as evidenced by its fractal dimension $d_f = 1.20$. For the two poly(vinyl butyral) samples, the low-M region where the conformation plots of PVB and PVBX overlap appears to be a region where both polymers are linear. The fractal dimension for this region of the conformation plot ($d_f = 1.75$) is in the range expected for linear random coils at good solvent–temperature conditions. The fractal dimension of the branched region of PVB ($d_f \approx 2$) is as expected for a solution of a randomly branched polymer at good thermodynamic conditions. The relatively high fractal dimension of the branched region of PVBX ($d_f = 2.4$) gives an indication of the heterogeneity of branching present. This polymer contains both native and cross-link-induced branching, which may also include branch-on-branch structures. Figure 11.8 provides clear evidence of the usefulness of the fractal dimension, as determined by SEC/MALS experiments. This approach characterizes polymeric architecture and its changes as a continuous function of M.

It should be noted that polymers are "random" or "statistical" fractals, quite different from such well-known "regular" fractals as Sierpiński carpets, Koch curves, and Cantor dusts (see Figure 11.9). The constructs in Figure 11.9 show how seemingly

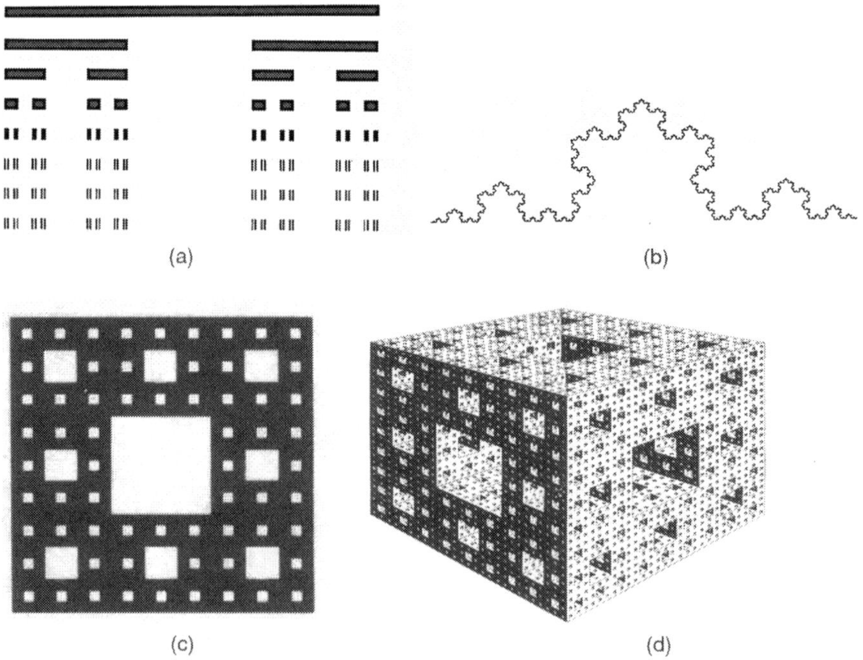

Figure 11.9 Examples of regular fractals: (a) Cantor dust, $d_T = 0$, $d_f \approx 0.63$; (b) triadic Koch curve, $d_T = 1$, $d_f \approx 1.2618$; (c) Sierpiński carpet, $d_T = 2$, $d_f \approx 1.89$; (d) Menger sponge, $d_T = 3$, $d_f \approx 2.73$. ((a–c) Courtesy of Deborah Striegel; (d) reprinted with permission from Ref. 20.)

zero-, one-, two-, and three-dimensional structures actually occupy intermediate dimensions by virtue of their fractality. For macromolecules, the statistical relationship comes from the fact that root-mean-square polymeric radii are derived from averaging over the different conformations polymer chains may adopt in solution.

11.4.2 Dimensionless Radii Ratios

Table 9.2 summarized the four main macromolecular radii: R_G, R_H, R_η, and R_T. These have been combined in the form of dimensionless ratios which can define polymeric architecture. Determining A_2 using off-line MALS (Section 9.3.3) provides information on the thermodynamic state of the dilute polymer solution, narrowing the choices between the good and theta regimes. Coupling of SLS to SEC (Section 9.3) provides the polydispersity of the sample, narrowing the choices between those for mono- and polydisperse species. Section 11.5 shows how to determine the arm number in select types of star polymers.

11.4.2.1 $\boldsymbol{R_{G,z}/R_{H,z}}$**.** To date, the best known and most employed dimensionless ratio has been ρ, the ratio of the z-averages of R_G and R_H [4]:

$$\rho \equiv \frac{R_{G,z}}{R_{H,z}} \tag{11.14}$$

Values of ρ have been calculated for a number of polymeric architectures, some examples of which are given in Table 11.2.

Table 11.2 Relation between the dimensionless ratio ρ and polymer architecture

Architecture	ρ
Homogeneous (hard) sphere	0.778
Gaussian "soft sphere"	0.977
(e.g., dendrimers generation $\geq$ 10)	
Monodisperse linear random coil	
θ-Conditions	1.504
Good conditions	1.78
Polydisperse linear random coil	
θ-Conditions	1.73
Good conditions	2.05
Regular stars with uniform arm length	
θ-Conditions, $f = 4$	1.333
θ-Conditions, $f \gg 1$	1.079
Regular stars with polydisperse arm length	
θ-Conditions, $f = 4$	1.534
θ-Conditions, $f \gg 1$	1.225

Source: Ref. 4.

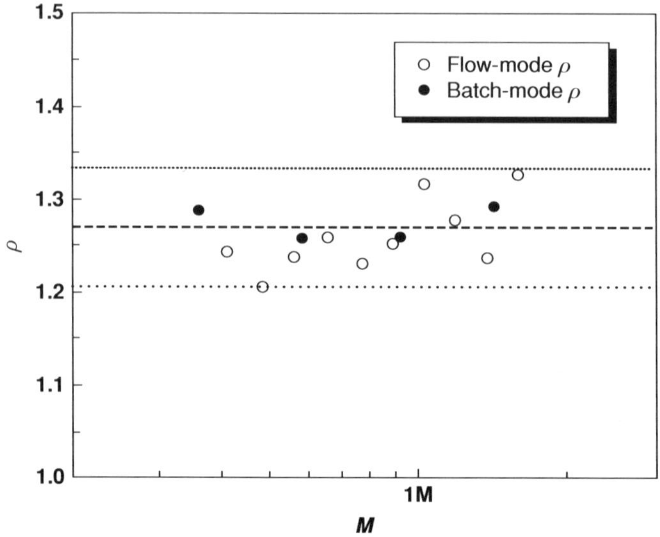

Figure 11.10 Dimensionless parameter ρ as a function of molar mass of PDMS: SEC/MALS/QELS analysis of virtually monodisperse linear PDMS in TCB at 150°C. ρ is constant as a function of *M*, indicating architectural invariance with varying molar mass. (Reprinted with permission from Ref. 22.)

The current availability of combined MALS/QELS detectors for SEC (Sections 9.3 and 9.4) permits determination of R_G and R_H across the molar mass distribution of polymers in a single experiment. An example of this is shown in Figure 11.10 for a virtually monodisperse linear random coil poly(dimethylsiloxane) sample [22]. The parameter ρ remains relatively constant across the distribution, indicating architectural invariance across the MMD. The relatively low value of ρ, which was approximately 1.3, was interpreted as the polymer solution being at poor solvent–temperature conditions, causing the chain to adopt a compact conformation in solution. The conformation plot for this sample, however, yielded a fractal dimension of about 2, consistent with that of a linear random coil at θ-conditions. This type of discrepancy, and the agreement between experimental and theoretical values of the dimensionless ratios, is discussed more extensively in Section 11.4.4.

11.4.2.2 $R_{\eta,w}/R_{G,z}$. Other dimensionless radii ratios have also been employed, although they have not been studied as thoroughly as the ρ parameter and have thus found more limited use. The ratio of the weight-average viscometric radius $R_{\eta,w}$ to the z-average radius of gyration $R_{G,z}$ has found most of its application in the study of star polymers [23]. The range for $R_{\eta,w}/R_{G,z}$ spans from about 0.36 for rigid rods to about 1.29 for homogeneous hard spheres [24]. For linear random coils, the value of $R_{\eta,w}/R_{G,z}$ is about 0.73 at good conditions, and approaches the theta-state value of about 0.87 with decreasing goodness of solution thermodynamics. At good solvent–temperature conditions, there is a weak dependence of $R_{\eta,w}/R_{G,z}$ on M.

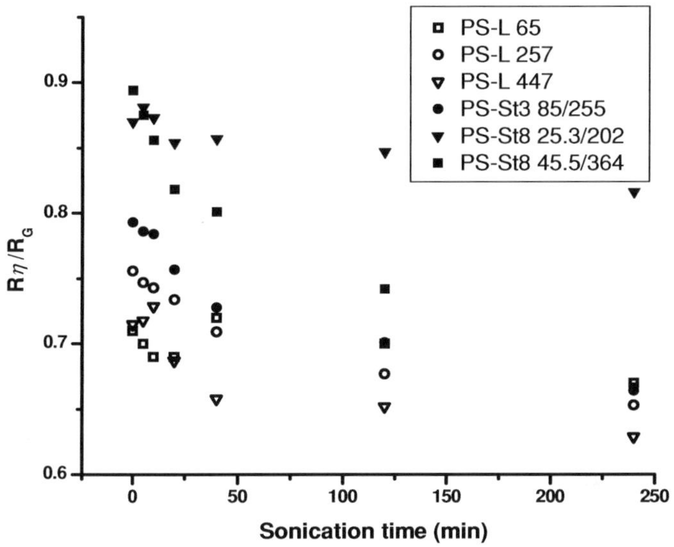

Figure 11.11 Change in polymer architecture during ultrasonic degradation, as determined by the dimensionless radii ratio $R_{\eta,w}/R_{G,z}$. PS-L W refers to linear polystyrenes where W = M_w/1000. PS-StX Y/Z refers to star polystyrenes with X number of arms, where Y = (M_n of arm)/1000 and Z = XY. Radii determined via SEC/DRI/MALS/VISC experiments in DMAc/0.5% LiCl at 35°C. Ultrasonic degradation conditions: 47 kHz, 185 W, 20 to 25°C. (Adapted from Ref. 26.)

All the values given are for narrow polydispersity homopolymers. For copolymers, other effects, such as chemical heterogeneity or sequence length distribution, can contribute to a nonconstancy in the value of $R_{\eta,w}/R_{G,z}$ across the MMD of random copolymers [25].

An example of the utility of the ratio $R_{\eta,w}/R_{G,z}$, obtained from SEC/DRI/MALS/VISC analysis, is the ability to show that the architecture of linear polystyrenes remains constant when subjected to ultrasonic degradation. The same type of degradation can substantially alter the architecture of three- and eight-arm star polystyrenes [26]. This is shown in Figure 11.11, where the decrease in the value of $R_{\eta,w}/R_{G,z}$ for the linear polymers as a function of sonication time is due to the concomitant increase in polydispersity. For narrow molar mass distribution (small polydispersity) polymers, $R_{\eta,w}$ and $R_{G,z}$ lie close to each other on the MMD, while for broad molar mass distribution (high polydispersity) polymers, $R_{\eta,w}$ characterizes a point near the middle of the distribution, while $R_{G,z}$ characterizes a point at the high end of the MMD. Thus, the denominator in the radii ratio grows faster than does the numerator with increasing polydispersity. Regardless, it can be seen in Figure 11.11 that the three-arm star and one of the eight-arm stars (PS-St8 45.5/364), after sufficient exposure to ultrasonic irradiation, resemble their linear counterparts. The three-arm star does this more quickly than the eight-arm star, because loss of one arm in the three-arm species results in two linear species, whereas loss of one arm in the eight-arm star results in one linear species and one seven-arm species (and so on with increasing degradation). The eight-arm star with shorter arms (PS-St8 25.3/202), which did not show appreciable evidence of degradation, also shows negligible change in R_η/R_G. Results were explained on the basis of a "path theory" of transient elongational flow degradation, which relates the ability of a polymer to degrade to its persistence length L_p (see Section 11.6 for a discussion of L_p) [24,26,27].

11.4.2.3 Other Ratios. Very limited data and theories exist relating the viscometric, hydrodynamic, and thermodynamic radii to each other in the form of dimensionless radii ratios [4,28]. In the hard-sphere limit, the viscometric and hydrodynamic radii should be identical to each other (i.e., $R_\eta/R_H = 1$). Similar behavior is expected at theta conditions. At the hard-sphere limit, $R_T/R_G = R_\eta/R_G = (5/3)^{0.5} \approx 1.29$. While the ratio R_T/R_H can be considered a measure of the relative thermodynamic and hydrodynamic interactions of polymer chains in dilute solution, limited observations with both linear and branched species indicate that this parameter remains close to unity.

11.4.3 Dimensionless Functions

Several of the polymeric radii may also combine in slightly more complicated ways. This is the case with the polymer draining (Φ) and coil interpenetration (Ψ^*) functions [4,28]:

$$\Phi = \frac{10\pi N_A}{3} \frac{R_\eta^3}{R_G^3} \tag{11.15}$$

$$\Psi^* = \frac{4}{3\sqrt{\pi}} \frac{R_T^3}{R_G^3} \tag{11.16}$$

The polymer draining function (technically, not a dimensionless function, as it has units of mol^{-1}) is meant to describe how deeply a polymer in solution is drained by the solvent. Deep draining corresponds to a reduction in effective hydrodynamic volume, and consequent reduction in R_η, and shallow draining corresponds to an increase in the hydrodynamic volume and viscometric radius. Φ is determined by the resistance to solvent penetration by the clouds of connected segments of the macromolecule, whereas Ψ^* is determined by the resistance to interpenetration by the clouds of two connected segments. While both functions depend on segment density, small species such as solvent molecules can penetrate the cloud occupied by a chain segment more easily than two such clouds (corresponding to two different section of the macromolecule) can penetrate each other. This leads to the last dimensionless function, $V_{A_2\eta}$, which relates Φ and Ψ^* and provides a relationship between the thermodynamic and viscometric radii [29]:

$$V_{A_2\eta} \equiv \frac{A_2 M_w}{[\eta]} = 4\pi^{3/2} N_A \frac{\Psi^*}{\Phi} = \frac{8}{5} \frac{R_T^3}{R_\eta^3} \tag{11.17}$$

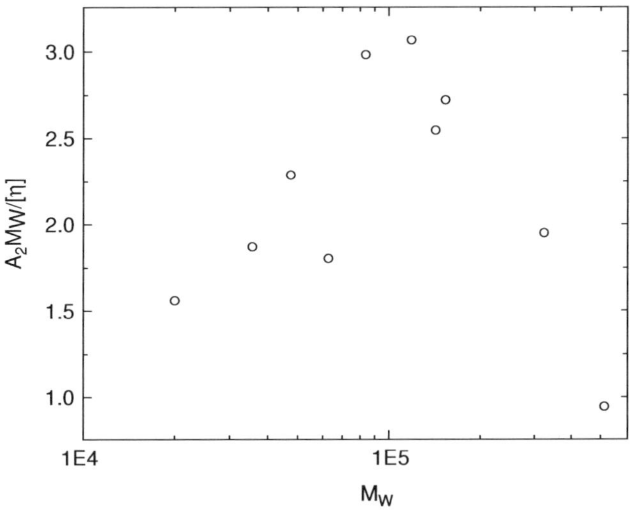

Figure 11.12 Values of the dimensionless function $V_{A_2\eta}$ for a hyperbranched polyester: $A_2M_w/[\eta] = V_{A_2\eta}$, as defined in Equation 11.17. Most values in figure are seen to lie above the hard-sphere value of 1.60, in accordance with independent observations for hyperbranched polymers. (Reprinted with permission from Ref. 30.)

Again, limited data exist for $V_{A_2\eta}$. An asymptotic limit of 1.07 ± 0.03 appears to have been identified for linear species, while for hard spheres a value of 1.60 was calculated. The latter value should be looked at with caution, as highly branched and hyperbranched polymers and stars with more than 5 to 10 arms will approach the hard-sphere value from above while linear species, lightly branched polymers, and stars with fewer than 5 to 10 arms approach the hard-sphere limit from below [4]. For linear and lightly branched polymers, the increase in $V_{A_2\eta}$ with increase in branching seems related to the fact that interpenetration of two polymer segments will be more strongly inhibited by long-chain branching than will be the penetration of small solvent molecules into the polymeric coil when the segment density is increased as a result of branching. The $V_{A_2\eta}$ parameter has been used in the study of star polymers [4], in the comparison of polymers with native branching to those with both native and cross-link-induced branching [15], and in the characterization of hyperbranched polyesters [30]. In the latter case, Figure 11.12 shows how values of $V_{A_2\eta}$ for the polyesters were higher than the hard-sphere value, in accordance with the observations mentioned previously for hyperbranched species.

11.4.4 Caveats Regarding Dimensionless Parameters

By using SEC with a multiplicity of physical detectors it is, in principle, possible to measure any of the ratios or functions discussed in Sections 11.4.2 and 11.4.3 across the MMD of polymers. (A function such as Ψ^* cannot be measured across the MMD, because Ψ^* depends on R_T, which is obtained from off-line, not online

MALS analysis). Such measurements provide valuable architectural or thermodynamic information about the analytes.

In practice, it may not be possible to measure R_G with acceptable accuracy and precision for relatively small-sized species, even though the molar mass of the analyte may be considerable. For example, a monodisperse linear PS of molar mass 150,000 g/mol dissolved in tetrahydrofuran (THF) at room temperature has an R_G of around 12 nm, barely measurable by MALS using 690-nm incident radiation. Solvents with a high Rayleigh ratio may also not permit measurement of the R_H distribution using SEC/QELS. However, for small polymers, discrepancies between experimental and theoretical radii of only a few nanometers may result in values of ρ, Φ, and so on, that differ greatly from the theoretical values for the type of polymer analyzed at the thermodynamic conditions of the experiment. This effect becomes even more pronounced at or near the theta point, where polymers occupy a relatively collapsed volume and therefore have relatively small radii [31].

As a generic example, take a polymer with $R_G = 21$ nm and $R_H = 12$ nm. From these measurements, a ρ value of 1.75 is obtained. Examination of the conformation plot shows a constant slope, indicating that the polymer is most likely linear. The SEC/MALS-obtained molar mass averages and distribution indicate that the polymer is polydisperse ($M_w/M_n > 1$). From this discussion, and using the information in Table 11.2, it is concluded that the polymer is a polydisperse linear random coil at theta conditions. Another group performs the same measurements, for the same polymer under the same experimental conditions, but their radii differ from the initial measurements by only one nanometer (i.e., $R_G = 22$ nm and $R_H = 11$ nm). Conformation plot and polydispersity data are the same as above, a value of $\rho = 2$ is obtained from ratioing the radii, and the second group concludes that the polymer solution is at good solvent–temperature conditions. Which group is correct? One way to answer this question is to perform an off-line MALS experiment to determine A_2 from a Zimm or related plot. A_2 will be equal to (or very near) zero at theta conditions and will have a high, positive value at good solvent–temperature conditions. Zimm plots are somewhat laborious to make, however, sample requirements are greater, and it will be necessary to decouple the MALS detector from the SEC system to perform this measurement. None of these possibilities are particularly appealing [32].

A second, more straightforward approach is to reexamine the conformation plot (and also the Mark–Houwink plot, if available). Initial examination of this plot showed a constant slope, from which it was concluded that the polymer is linear. A second look reveals that the slope, α, of the plot is 0.58, corresponding to a fractal dimension d_f of 1.72, in accordance with a linear random coil at good solvent temperature conditions (see Table 11.1 and Figure 11.7). This example illustrates that if determining the thermodynamic state of the solution is critical, using the first suggestion of an off-line batch-mode MALS experiment is recommended.

In summary, no individual parameter, be it the slope of a plot or a dimensionless ratio, should be the sole criterion by which one arrives at conclusions regarding polymer architecture. Because multidetector SEC has the ability to provide a number of such parameters, this approach allows us to achieve a higher degree of confidence in understanding macromolecular structure.

11.5 STAR POLYMERS

Star polymers present a special case of branching, in which at least three long-chain branches extend from a common core [23]. A review of recent work on star polymers using SEC with molar-mass-sensitive detectors is given in Reference 1. Based on the contraction factor g, defined in Equation 11.1, the following equation was derived to calculate the number of arms, f, in star polymers where the arm length is polydisperse, that is, when all arms are not of equal length [7]:

$$g = \frac{6f}{(f+1)(f+2)} \tag{11.18}$$

Equation 11.18 is sometimes referred to as the *random distribution equation* for stars. An equation for stars with monodisperse arm length (all arms of equal length), referred to as the *regular distribution* or *regular star equation*, is [6]

$$g = \frac{3f-2}{f^2} \tag{11.19}$$

Equation 11.19 is strictly valid only under theta conditions, for stars having random walk arms. In both of the cases above, it is assumed that the star polymer is being compared to a linear polymer of the same chemistry and molar mass. The arms of the star polymer are considered to be of a single and equal type of chemistry, that is, each arm is a homopolymer and all arms are chemically equal. There will probably be a chemical inequality between the star and the linear standard, due to the chemical makeup of the core of the star, but because in most cases $M_{\text{arm}} \gg M_{\text{core}}$, this inequality can usually be ignored [26].

Because of the more compact structure of stars, as compared to their linear analogs, it is often difficult the measure R_G for stars with acceptable accuracy and/or precision. While the contraction factor g' is easier to measure, its relationship to arm number f is less certain. Originally, Stockmayer and Fixman defined h as the ratio of the hydrodynamic radius of a regular star to that of a linear polymer of the same chemistry and molar mass [6]. Now, the symbol h is used to denote the more general ratio of the hydrodynamic radii of branched (B) and linear (L) polymers of the same chemistry and molar mass (subscript M):

$$h = \left(\frac{R_{H,B}}{R_{H,L}}\right)_M \tag{11.20}$$

For the case of stars at theta conditions and having random walk arms (with arms "obeying random-flight statistics"), the following expression relates the ratio h to the number of arms f [6]:

$$h = \frac{\sqrt{f}}{2-f+\sqrt{2}(f-1)} \tag{11.21}$$

The use of multidetector SEC to study arm loss during ultrasonic irradiation of star polymers, and the behavior of stars and their linear analogs under ultrasonic conditions, was described earlier (see Section 11.4.2 and Figure 11.11) [26]. Recently, SEC/DRI/MALS/VISC/UV was used to attempt to determine the number of arms across the MMD of star polymers [33]. Initial attempts using the contraction factor g' were defeated by axial dispersion and branching homogeneity. A method of "component chromatograms" or deconvolution was developed in which the chromatogram from each detector is fitted as the sum of component chromatograms. For a five-arm star PMMA, Figure 11.13 shows how the DRI chromatogram could be fitted as the sum of four Gaussian components. The viscometer and light-scattering photometer, both more sensitive to earlier-eluting species than the DRI, necessitated five Gaussian components for an adequate fit.

The component method seems to provide reasonable values of the arm number of stars over a wide range of f without the need for axial dispersion correction or determination of interdetector delays. A significant limitation of this method is that a particular component chromatogram needs to correspond to the same molecules in each detector's chromatogram. This is complicated by the fact that both the shape and number of component chromatograms can be different for DRI, VISC, and SLS detectors, due to differences in detector sensitivities and responses.

11.6 DETERMINING THE PERSISTENCE LENGTH

As implied by its name, the persistence length, L_p, measures the length over which the chain "persists" in the same direction as the first bond. It is a measure of chain stiffness, with higher values of L_p denoting stiffer chains. The persistence length is defined as the average projection of the end-to-end distance vector $\mathbf{r}$ onto the first bond of the chain $\hat{\mathbf{I}}_1$, in the limit of infinite chain length (i.e., as degree of polymerization, n, goes to infinity) [34]:

$$L_p = \left\langle \frac{\hat{\mathbf{I}}_1}{l_1} \sum_{i=1}^{n} \hat{\mathbf{I}}_1 \right\rangle \qquad \text{as } n \to \infty \tag{11.22}$$

where $\hat{\mathbf{I}}_1 / l_1$ is a unit vector in the direction of $\hat{\mathbf{I}}_1$. The persistence length is shown schematically in Figure 11.14 for a particular conformation. The value of L_p is obtained by averaging over all conformations.

For the wormlike chain model (see Appendix G in Reference 35), the following relationship between the persistence length, the radius of gyration, and the contour length, L_c, of macromolecules was derived:

$$R_G^2 = \frac{L_p L_c}{3} - L_p^2 + \frac{2L_p^3}{L_c}\left\{1 - \frac{L_p}{L_c}\left[1 - \exp\left(\frac{-L_c}{L_p}\right)\right]\right\} \tag{11.23}$$

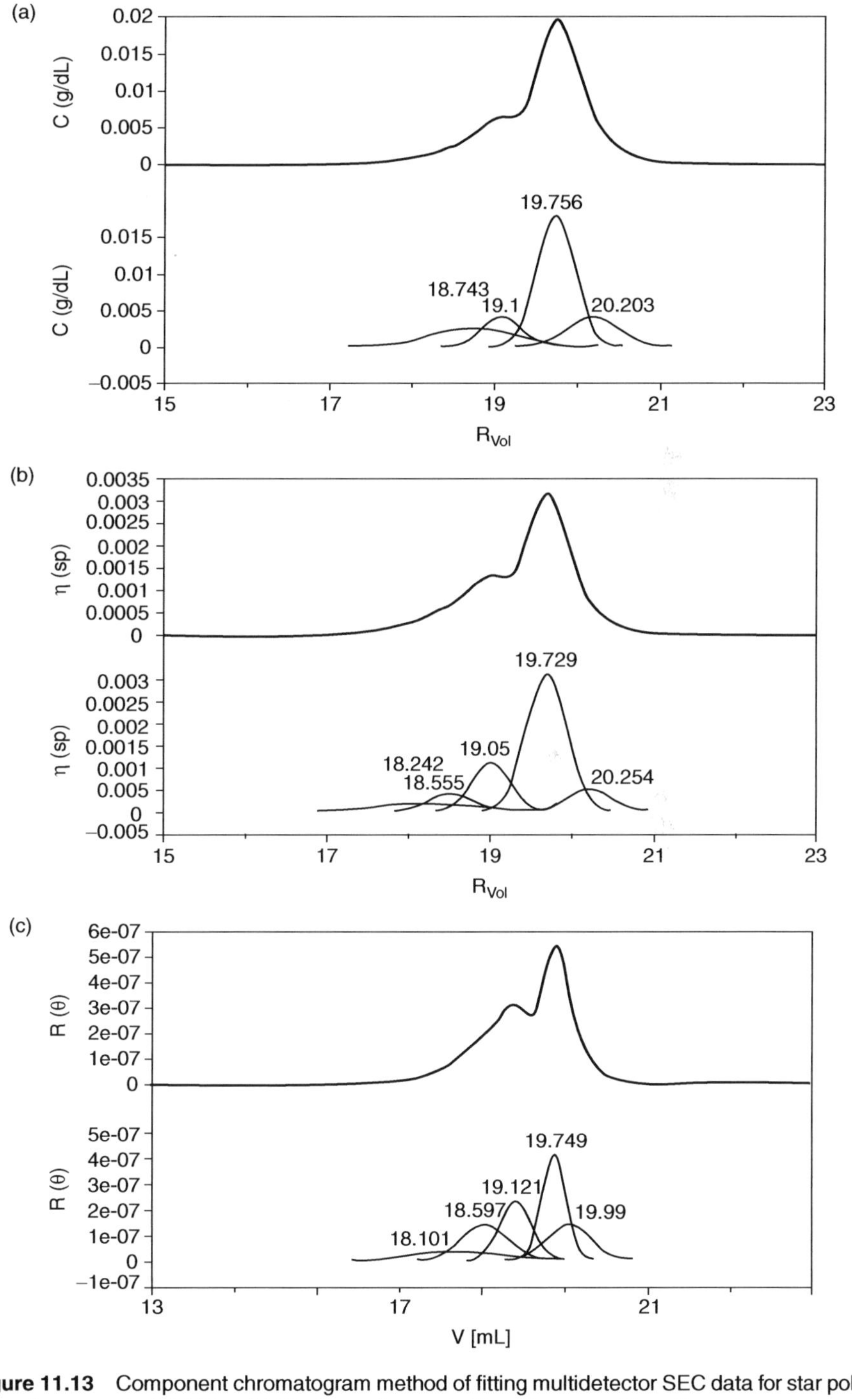

Figure 11.13 Component chromatogram method of fitting multidetector SEC data for star polymers: five-arm star PMMA. (a) DRI chromatogram is fit as the sum of four Gaussian component chromatograms; (b) VISC and (c) 90° SLS chromatograms are each fit as a sum of five Gaussian component chromatograms. (Reprinted with permission from Ref. 33.)

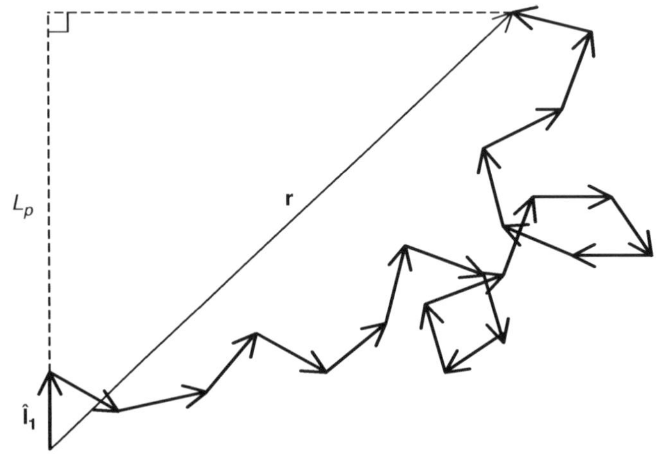

Figure 11.14 Persistence length, L_p. L_p is the projection of **r** onto $\hat{l}_1$ in the limit of $n \rightarrow \infty$, averaged over all conformations.

Although this is a theta-state relationship, most MALS and SEC/MALS measurements of R_G are performed at nontheta conditions. As such, most measurements of L_p are actually of an "apparent" L_p, occasionally denoted L'_p [36].

Writing Equation 11.23 in terms of M_L, the molar mass per unit contour length, the resultant expression can be approximated, with errors of less than 1%, as

$$\left(\frac{M^2}{12R_G^2}\right)^{2/3} = M_L^{4/3} + \frac{2M_L^{1/3}M}{15L_p} \qquad \text{for } \frac{M}{2L_pM_L} < 2 \tag{11.24}$$

and as

$$\left(\frac{M}{R_G^2}\right)^{1/2} = \left(\frac{3M_L}{L_p}\right)^{1/2}\left(1 + \frac{3L_pM_L}{2M}\right) \qquad \text{for } \frac{M}{2L_pM_L} > 2 \tag{11.25}$$

Equation 11.25 was used to determine the persistence length of poly(n-hexyl isocyanate) (PHIC) [37] and, more recently, of cellulose, amylose, and poly(γ-benzyl-L-glutamate) [24,27]. L_p and M_L were determined from the intercept, $(3M_L/L_p)^{1/2}$, and slope, $(3/2)M_L(3L_pM_L)^{1/2}$, of a plot of $(M/R_G^2)^{1/2}$ versus $1/M$. Results of the PHIC experiments, in both dichloromethane and THF, are shown in Figure 11.15. While the dichloromethane data were unsuitable for least-squares linear regression, fitting of the THF data gave values of $L_p = 40$ nm and $M_L = 730$ nm^{-1}.

The persistence length may also be derived from molar mass and intrinsic viscosity data. This can prove advantageous over the use of light scattering, for polymers with low values of $\partial n/\partial c$ and/or low M, for which obtaining adequate R_G data is difficult. The data for PHIC in THF from Figure 11.15 can be replotted in the form of $(M^2/[\eta])^{1/3}$ versus $M^{1/2}$. This *Bohdanecký plot*, shown in Figure 11.16, is

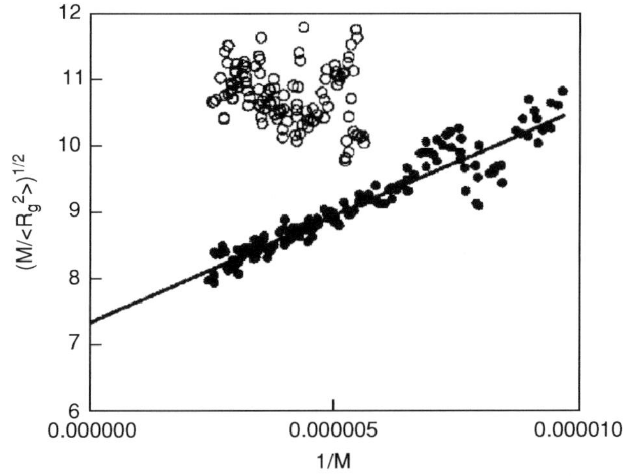

Figure 11.15 Determining persistence length L_p and molar mass per unit contour length M_L from SEC/MALS data: plotting radius of gyration and molar mass data for PHIC as per Equation 11.25. Filled symbols correspond to data in THF, open symbols to data in dichloromethane. Solid line is unweighted least-squares regression linear fit of data in THF. L_p and M_L are evaluated from slope and intercept of line. See the text for details. (Reprinted with permission from Ref. 37.)

considerably more linear and less noisy than the corresponding plot in Figure 11.15. To derive L_p from a Bohdanecký plot, the ordinate and abscissa are related via [37]

$$\left(\frac{M^2}{[\eta]}\right)^{1/3} = A_\eta + B_\eta M^{1/2} \tag{11.26}$$

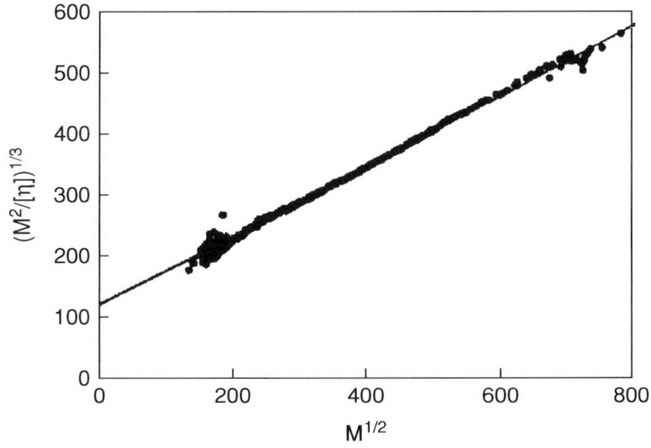

Figure 11.16 Bohdanecký plot for determining persistence length L_p from SEC/DRI/MALS/VISC data: replotting of data for PHIC in THF, from Figure 11.15. See the text for details. (Reprinted with permission from Ref. 37.)

where the intercept is

$$A_\eta = \frac{A_0 M_L}{\Phi_\infty^{1/3}} \tag{11.27}$$

and the slope is

$$B_\eta = \frac{B_0}{\Phi_\infty^{1/3}} \left(\frac{2L_p}{M_L} \right)^{-1/2} \tag{11.28}$$

Φ_∞ is Flory's constant (see Section 9.6) for nondraining coils in the limit of infinite molar mass, $A_0 = 0.46$ to $0.53d_r$, $B_0 = 1.00$ to $0.0367(\log d_r)$, and d_r is the reduced hydrodynamic diameter ($d_r = d/2L_p$).

11.7 DETERMINING THE CHARACTERISTIC RATIO

The characteristic ratio, C_n, is defined as the ratio of the unperturbed dimensions of a polymer to the dimensions of a freely jointed chain [34]:

$$C_n = \frac{\langle r^2 \rangle_\theta}{nl^2} \tag{11.29}$$

where $\langle r^2 \rangle_\theta$ is the root-mean-square end-to-end distance of a polymer molecule, n the number of bonds in the polymer, l the bond length, and the subscript θ denotes the unperturbed dimensions of the polymer.

For long, flexible chains with skeletal bond angle ϕ, the characteristic ratio is related to the persistence length via

$$C_n = \frac{2L_p}{l} \sin(\phi/2) \tag{11.30}$$

In accordance with the random walk model, C_n is independent of n and the rate of convergence of C_n to its asymptotic value, C_∞, is a measure of the stiffness of the chain [34]. This asymptotic value is defined in terms of the persistence length as

$$C_\infty = \frac{2L_p}{l} - 1 \tag{11.31}$$

A number of groups have used values of L_p, calculated by one of the methods described in Section 11.6, to determine C_n. For example, the wormlike chain model was applied to SEC/MALS data in studying poly(*N*-vinylcarbazole) (PVCz) in THF at 25°C and poly(vinylpyrrolidone) (PVP) in H_2O/0.1 *M* $NaNO_3$ at the same temperature [38]. Measurements of A_2 showed that these solvent–temperature conditions

Table 11.3 Characteristic ratio, persistence length, and contour length of PVCz and PVP, obtained from SEC/MALS

Parameter	PVCz	PVP
C_n	15.9	14
L_c/n (nm)	0.128	0.128
L_p (nm)	1.48	1.30

Source: Ref. 38.
n is the number of skeletal bonds. All the valence angles in the chain were assumed to be 112° for calculation of L_c.

were close to the theta state, and C_n was calculated as

$$C_n = \frac{6M_o R^2_{G,\theta}}{2l^2 M} \tag{11.32}$$

where M_o is the molar mass of the repeat unit of each polymer, and $R_{G,\theta}$ and M were obtained from SEC/MALS analysis. Results of the measurements and calculations of C_n, L_p, and L_c/n are shown in Table 11.3. Figure 11.17 shows the calculation of the characteristic ratio as a function of the number of skeletal bonds, n, according to the wormlike chain model and the rotational isomeric state model (where C_n is plotted as a function of the degree of polymerization, X). While both the wormlike chain and RIS models mimic the behavior of PVCz and PVP well, the RIS model somewhat underestimates the value of C_n for the PVP sample. A discussion of both models can be found in Reference 35.

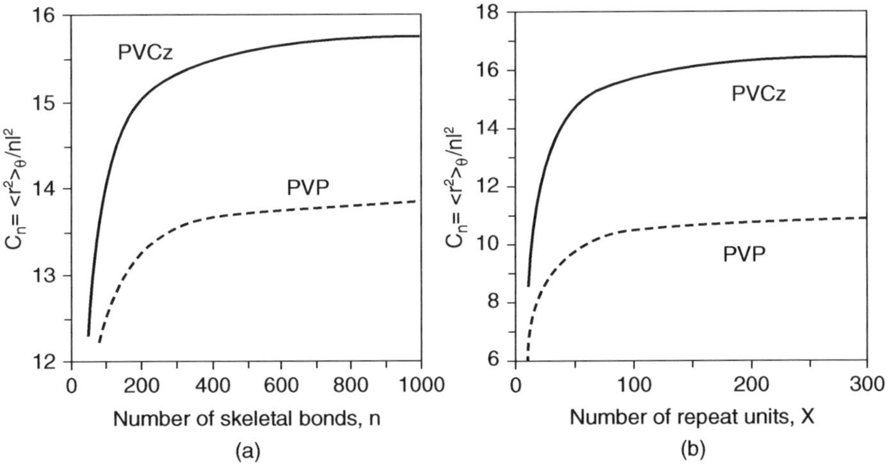

Figure 11.17 Characteristic ratio of PVCz and PVP according to (a) the wormlike chain model and (b) the rotational isomeric state model. See Table 11.3 for comparison to results based on SEC/MALS data. (Reprinted with permission from Ref. 38.)

11.8 LOCAL POLYDISPERSITY

Each elution slice in an SEC chromatogram is considered to be monodisperse with respect to molar mass, architecture, and chemical composition. The term *local polydispersity* denotes the heterogeneity of molecules present in the same SEC retention volume. For linear homopolymers, this is generally caused by band broadening, and the local polydispersity is thus a polydispersity of molar mass within a chromatographic slice. When analyzing copolymers and branched polymers, local polydispersity arises due to a coincidence in the hydrodynamic volumes of different species. For example, in a polymer sample that contains chains of a single type of chemistry but is a mixture of linear and branched species, there can exist higher-M branched polymers that occupy the same hydrodynamic volume as lower-M linear polymers. These two species will elute at the same retention volume in an SEC experiment, giving rise to local polydispersity in this volume. A similar effect can occur in chemically heterogeneous samples or in samples where a variety of both chemistries and architectures are present.

The detection and determination of local polydispersity in SEC/DRI/MALS/VISC analysis is discussed in detail in References 39 and 40. The general conclusion is that in order to calculate local polydispersity due to compositional or topological heterogeneity, highly accurate interdetector delay and band-broadening corrections are first needed. After this, detection of local polydispersity by multidetector SEC is possible for cases where the specific refractive index increments (see Section 9.2.1.3) of the coeluting species are extremely different from each other.

REFERENCES

1. T. H. Mourey, *Int. J. Polym. Anal. Charact.*, **9**, 97 (2004).
2. A. M. Striegel, *Anal. Chem.*, **77**, 104A (2005).
3. A. M. Striegel, ed., *Multiple Detection in Size-Exclusion Chromatography*, ACS Symp. Ser. 893, American Chemical Society, Washington, DC, 2005.
4. W. Burchard, *Adv. Polym. Sci.*, **143**, 113 (1999).
5. A. M. Striegel, in *Encyclopedia of Chromatography*, 2nd ed., J. Cazes, ed., Marcel Dekker, New York, 2005, p. 1008.
6. W. H. Stockmayer and M. Fixman, *Ann. N.Y. Acad. Sci.*, **57**, 334 (1953).
7. B. H. Zimm and W. H. Stockmayer, *J. Chem. Phys.*, **17**, 1301 (1949).
8. S. Grcev, P. Schoenmakers, and P. Iedema, *Polymer*, **45**, 39 (2004).
9. J. Roovers, in *Encyclopedia of Polymer Science and Engineering*, Vol. 2, Wiley, New York, 1985, p. 478.
10. C. Jackson, Y.-J. Chen, and J. W. Mays, *J. Appl. Polym. Sci.*, **59**, 179 (1996).
11. F. Beer, G. Capaccio, and L. J. Rose, *J. Appl. Polym. Sci.*, **80**, 2815 (2001).
12. L.-P. Yu and J. E. Rollings, *J. Appl. Polym. Sci.*, **33**, 1909 (1987).
13. (a) A. M. Striegel and M. R. Krejsa, *J. Polym. Sci. B*, **38**, 3120 (2000). (b) A. M. Striegel and M. R. Krejsa, *2000 Int. GPC Symp. Proc.*, Waters, Milford, 2000.

14. C. P. Lusignan, T. H. Mourey, J. C. Wilson, and R. H. Colby, *Phys. Rev. E*, **60**, 5657 (1999).
15. A. M. Striegel, *Polym. Int.*, **53**, 1806 (2004).
16. T. Sun, P. Brant, R. R. Chance, and W. W. Graessley, *Macromolecules*, **34**, 6812 (2001).
17. Th. G. Scholte, N. L. J. Meijerink, H. M. Schoffeleers, and A. M. G. Brands, *J. Appl. Polym. Sci.*, **29**, 3763 (1984).
18. Y. Yu, P. J. DesLauriers, and D. C. Rohlfing, *Polymer*, **46**, 5165 (2005).
19. T. A. Witten, *Rev. Mod. Phys.*, **70**, 1531 (1998).
20. B. B. Mandelbrot, *The Fractal Geometry of Nature*, updated and augmented, W.H. Freeman, New York, 1983.
21. A. M. Striegel, in Ref. 3, Chap. 4.
22. Y. Liu, S. Bo, Y. Zhu, and W. Zhang, *Polymer*, **44**, 7209 (2003).
23. J. Roovers, in *Star and Hyperbranched Polymers*, M. K. Mishra and S. Kobayashi, eds., Marcel Dekker, New York, 1999, Chap. 11.
24. S. G. Ostlund and A. M. Striegel, *Polym. Degrad. Stab.*, **93**, 1510 (2008).
25. I. A. Haidar Ahmad and A. M. Striegel, Pittcon 2008, paper 1860-8.
26. A. M. Striegel, *J. Biochem. Biophys. Methods*, **56**, 117 (2003).
27. A. M. Striegel, *Biomacromolecules*, **8**, 3944 (2007).
28. W. W. Graessley, *Polymeric Liquids and Networks: Structure and Properties*, Garland Science, New York, 2004.
29. C. E. Ioan, T. Aberle, and W. Burchard, *Macromolecules*, **33**, 5730 (2000).
30. E. de Luca and R. W. Richards, *J. Polym. Sci. B*, **41**, 1339 (2003).
31. M. J. Smith, I. A. Haidar, and A. M. Striegel, *Analyst*, **132**, 455 (2007).
32. I. A. Haidar Ahmad and A. M. Striegel, *Instrum. Sci. Technol.*, in press.
33. S. T. Balke, T. H. Mourey, D. R. Robello, T. A. Davis, A. Kraus, and K. Skonieczny, *J. Appl. Polym. Sci.*, **85**, 552 (2002).
34. C. R. Cantor and P. R. Schimmel, *Biophysical Chemistry*, Pt. III, W.H. Freeman, San Francisco, 1980.
35. P. J. Flory, *Statistical Mechanics of Chain Molecules*, Interscience, New York, 1969.
36. W. F. Reed, in Ref. 3, Chap. 2.
37. T. Mourey, K. Le, T. Bryan, S. Zheng, and G. Bennett, *Polymer*, **46**, 9033 (2005).
38. M. P. Tarazona and E. Saiz, *J. Biochem. Biophys. Methods*, **56**, 95 (2003).
39. S. T. Balke and T. H. Mourey, *J. Appl. Polym. Sci.*, **81**, 370 (2001).
40. D. Gillespie and W. W. Yau, *1998 Int. GPC Symp. Proc.*, Waters, Milford, 1999, p. 32.

12

AQUEOUS SEC

12.1 INTRODUCTION

Interest in aqueous SEC stems from a number of sources. Many analytes, principally natural polymers, dissolve preferentially in aqueous solvents, making aqueous SEC a preferred choice for analysis. Some biopolymers preserve their architectural properties in aqueous solvents, allowing higher-order structure to be studied by SEC. Solvent use and disposal costs are usually less when using aqueous solvents. Also, the frequent presence of non-size-exclusion effects in aqueous SEC often calls for additional studies.

In this chapter we describe column packings, mobile phases, mobile-phase additives, and separation techniques for aqueous SEC. Additionally, non-size-exclusion effects are discussed, all in the context of aqueous SEC [1]. As in other chapters, select examples help showcase the application range of the technique. As with its organic counterpart, aqueous SEC is used extensively for preparative applications. Preparative SEC is discussed in Section 15.2.

As described in Chapter 1, SEC was originally developed using low-pressure and aqueous mobile phases. Also known as gel filtration chromatography (GFC), the technique was used primarily for the separation of proteins. However, it was ill-suited for separating organic-soluble polymers, due to low column efficiency and long analysis times. Therefore, the development of organic SEC [or gel permeation chromatography (GPC)] then took place to allow the characterization of organic polymers. Left unaddressed, however, was the separation of water-soluble synthetic

Modern Size-Exclusion Liquid Chromatography: Practice of Gel Permeation and Gel Filtration Chromatography, Second Edition
By André M. Striegel, Wallace W. Yau, Joseph J. Kirkland, and Donald D. Bly

polymers. The first attempt at addressing the analysis of this type of macromolecule was the development and use of controlled-pore glass (CPG). This column packing is water compatible and mechanically rugged, thus permitting the use of high flow rates [2]. Unfortunately, CPG also strongly absorbs proteins, cationic polymers, and even certain neutral macromolecules. Additionally, the large particle size of CPG limited chromatographic resolution due to inefficient column packing.

The introduction of high-efficiency aqueous columns in about 1980 made it possible to operate at flow rates and pressures that allowed the separation of water-soluble synthetic polymers with acceptable resolution and in a reasonable time. These columns originally consisted of either surface-derivatized porous silica or cross-linked synthetic hydrophilic polymer gel packings. Later, high-efficiency agarose-based packings were introduced. Current aqueous SEC column packings also include cross-linked dextran, cross-linked agarose with covalently attached dextran, and hydroxylated polymethacrylates, among others.

12.2 AQUEOUS SEC COLUMNS

For aqueous SEC, column packing materials must have highly hydrophilic surfaces with minimal ionic or hydrophobic sites. Additionally, the packing material must display good chemical and mechanical stability while providing high efficiency.

Cross-linked dextran gels are among the earliest size-exclusion media and still used today, although in different forms. Dextran cross-linked with epichlorohydrin is a packing medium popular for desalting purposes. Increasing the cross-link density results in increased mechanical stability and reduced porosity. Solutions of 2, 4, and 6% agarose, in bead-formed gels, are also commonly used in aqueous SEC. Bead rigidity is directly proportional, and porosity inversely proportional, to agarose concentration. Highly cross-linked agarose gels are made from 6 and 12% agarose solutions. These provide high efficiency and reduced run times in addition to increased chemical and physical stability. Allyl dextran cross-linked with *N,N′*-methylenebisacrylamide provides a narrower particle-size distribution and more rigid particles relative to the agarose-based gels. To increase chromatographic resolution over a narrow molar mass range, cross-linked agarose with covalently attached dextran is used [3–5]. Properties of dextran gels as SEC column packing include suitability of use with most common buffers, salt additives, and organic modifiers (up to ca. 30% modifier); extreme pH resistance; and long-term pH stability [5].

Acrylate-based columns are also used in aqueous SEC. Poly(ethylene glycol) dimethylacrylate, particles afford extreme pH and high-temperature stability and can withstand up to 50% organic solvent. The particles generally have a slight negative charge (about 5 to 18 μEq/mL), so addition to the eluent of a small amount of a neutral salt (most commonly NaCl) is recommended [5,6].

Although native silicas are not normally employed in SEC for either biopolymers or polar synthetic polymers, they do form the supports for a number of bonded silica packings. Many of the different properties of these packings are due to the variety of precursor materials. Reference 7 gives 19 different silanes employed in the synthesis of hydrophilic bonded silicas. Most silica-based packings introduced around 1980 did not experience much improvement during the course of that decade. Newer silica-based columns have particles of less than 5 μm, are mechanically rugged and able to withstand a variety of additives, cover a wide molar mass range, and are available with different types of bonding. Grafting polymers onto the silica serves to mask surface silanol groups and to alter the porosity and stability of the material. A strong advantage for silica-based SEC packings is that they can be used at very high mobile-phase velocities (and pressures), and they can be used with a high concentration of (or neat) organic solvents. A comprehensive review of polymer-modified silica packings for aqueous SEC may be found in Reference 8.

Other column packing materials include poly(vinyl alcohol) gels, used mainly for saccharide analysis, and derivatized polystyrene-, poly(methyl methacrylate)-, and polyacrylamide-based packings, used successfully with carbohydrates, poly(ethylene oxide), and various proteins [9].

For poly(ethylene oxide) (PEO) and poly(vinylpyrrolidone) (PVP) analysis, a recent comparison was made of various commercial, acrylate-based, aqueous SEC columns and PEO calibration standards [10]. Columns were ranked for chromatographic efficiency in water (for high-M PEO standards) and in water–methanol (for low-M PEO standards). Significant results of this study include:

1. Using methanol as a mobile-phase modifier significantly improved separation of PVP from system peaks.
2. Columns giving the most efficient separation of PEO standards did not correspond to columns giving most efficient separation of PVP.
3. Newer column technologies did not perform significantly better than older ones for PVP analysis.
4. The column with the lowest exclusion limit also had the worst regression coefficient, r^2, for a linear calibration based on PEO standards in water–methanol. All four columns showed $r^2 > 0.99$ for the same type of PEO calibration in water.

Precautions to be taken in caring for aqueous SEC columns are similar to those for organic columns. An additional concern in aqueous SEC is the possibility of bacterial growth in the columns or other parts of the chromatographic system. This can be prevented by addition of a small amount (about 0.02%) of sodium azide to the mobile phase. If this is not possible for separations, flushing the columns and instrumentation with a 0.02% sodium azide solution, either periodically or prior to long-term storage, is recommended.

12.3 NON-SIZE-EXCLUSION EFFECTS AND MOBILE-PHASE ADDITIVES

Non-size-exclusion effects that can potentially plague aqueous SEC analysis include ion exchange, ion inclusion, ion exclusion, intramolecular electrostatic interactions, and adsorption [11–13]. Each effect is described individually below.

Ion-exchange effects arise due to the pH-induced dissociation of silanol groups into anionic groups (e.g., in glycerylpropyl-bonded silica, or residual silanol groups on polymer-based packings). The latter act as cation-exchange sites onto which cationic polyelectrolytes can adsorb, resulting in elution after the total column volume (or in no elution). Conversely, the anionic groups on the packing material may electrostatically repulse anionic polyelectrolytes, a type of *ion-exclusion effect*. This prevents these polyelectrolytes from entering the pores of the packing material and results in elution at the exclusion limit.

The preferential diffusion of counterions into the pores is responsible for the establishment of a Donnan membrane equilibrium. To maintain electroneutrality between the species inside and outside the pores, additional polymer is forced into the pores (beyond ideal SEC solute distribution). The analyte then experiences *ion inclusion* as a result of the need to balance the chemical potential difference between the stagnant and flowing mobile phases.

Adsorption effects can be due to either hydrophobic interactions, hydrogen bonding, or ion exchange. For example, sodium polystyrene sulfonate (NaPSS) interacts hydrophobically with glycerylpropyl-bonded silica gel. This hydrophobicity was ascribed both to unreacted phenyl groups (in the case on $<100\%$ sulfonation) and to backbone C–C linkages [11].

Intramolecular electrostatic interactions occur because of the fixed charges on polyelectrolytes and internal electrostatic repulsive forces that act to expand the polymer chain in solution. This expansion increases the hydrodynamic volume occupied by the polymer. Adding electrolyte to increase the ionic strength of the solution helps shield these electrostatic interactions by creating a more compact shape in solution. This compaction results in earlier elution volumes with respect to those observed in the absence of added electrolyte. Polyelectrolytes are treated in Section 12.4.4.

Adding surfactant helps reduce the ionic strength of the solution. This either eliminates or reduces hydrophobic interactions. Adding guanidine or urea to the mobile phase disrupts hydrogen-bonding effects. To deal with ion inclusion, which usually manifests itself by the appearance of a late-eluting electrolyte peak in the DRI trace, the mobile phase may be modified by adding an electrolyte such as NaCl. Also, an oligomeric-type (small pore size) column may be added to the column set for "desalting" purposes (i.e., to increase resolution between the polymer and electrolyte peaks). Alternatively, a selective UV detector may be used instead of a DRI.

Ion-exchange and ion-exclusion effects can be eliminated by adding electrolyte to the mobile phase (e.g., 10^{-2} to 10^{-1} M NaCl), or by reducing the pH of the mobile phase below 4. The latter suppresses the dissociation of surface silanol and carboxylic groups.

12.4 SELECT APPLICATIONS OF AQUEOUS SEC

The use of aqueous SEC as both an analytical and preparative-scale technique in the biological sciences is extensive, and the number of applications for water-soluble synthetic polymers is also quite high. Applications exemplifying some of this work are given in the following subsections.

12.4.1 Polysaccharides

Carbohydrates and polysaccharides such as sugars, cotton, and starch account for approximately 75% of the global annual photosynthetic biomass production (based on tons of dry matter). Lignin is responsible for about another 20%. Approximately half of the mass percentage of the polysaccharides is due to cellulose, another 20 to 25% is due to hemicelluloses, 2 to 5% is due to the starch components amylose and amylopectin, with the remaining balance made up of a large variety of species (mannans, xylans, galactans, fructans, etc.) [5]. Therefore, characterization of these products is an important commercial goal.

Although aqueous SEC might appear to be the obvious choice for determining the molar mass and structural characteristics of polysaccharides, many polysaccharides are not water soluble. An example of one such biopolymer is cellulose. This carbohydrate possesses an extensive web of inter- and intramolecular hydrogen bonding that limits solubility, without degradation or derivatization, to a small set of solvents [14]. This set is further restricted to solvents that are adequate chromatographic mobile phases and provide acceptable signal-to-noise ratios in detectors such as the static light-scattering photometer and the differential refractometer (Chapter 9).

Aqueous mobile phases containing a small amount (0.01 to 0.1 M) of potassium nitrate or sodium chloride are adequate for characterizing dextrans and pullulans. In addition, arabinogalactan, carrageenan, agarose, maltodextrins (see Figure 12.1), and derivatives of schyzophyllan and lentinan are handled by this approach [15]. Relatively narrow ($M_w/M_n \approx 1.1$ to 1.2) pullulan calibration standards are available, ranging from 1.5×10^3 to 1.7×10^6 g/mol. Pectin, guar, and certain cellulose derivatives usually dissolve in acidic (pH 3.7) media, with ionic strength adjusted to 0.3 to 0.7 with sodium sulfate [15]. Dissolution of amylose and amylopectin is usually accomplished with either 9 : 1 DMSO/H_2O or with a 0.1 to 0.5 N NaOH or 0.001 N KOH solution [3,16]. Use of KOH and NaOH with silica-based column packings can degrade these materials by slow dissolution of the silica support. NaOH can also degrade certain detector seals.

12.4.2 Proteins and Peptides

Many of the difficulties associated with SEC analysis of proteins and peptides are related to electrostatic and hydrophobic interactions between the analytes and the column packing material. Other problems have been the dependence on calibration curves for molar mass determination and the accompanying difficulties encountered in attempting to preserve the conformational integrity of the analytes in calibration-curve-dependent analyses [18,19].

Electrostatic interactions arise from negative charges on the weakly acidic silanol groups of silica-based packings. This even occurs with "capped silica" packings, due to residual, unmodified silanols or from the negative charges on carboxyl groups of polymeric packing materials. Positively charged proteins will adsorb onto the column packing and elute later than expected, whereas negatively charged proteins will be repelled from the column packing and elute earlier than expected. When a calibration curve is being used for molar mass determination, adsorption will result in

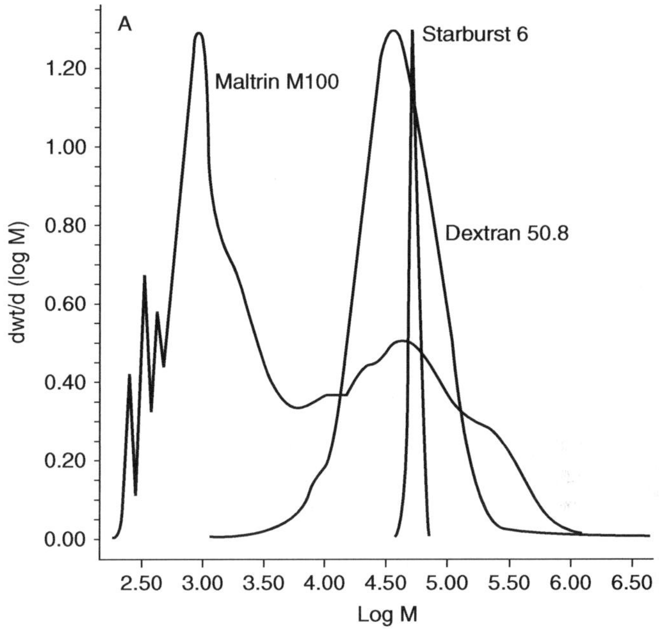

Figure 12.1 MMD overlays of dextran, maltodextrin, and dendrimer. Dextran 50.8: $M_w = 50{,}800$ g/mol, $M_w/M_n = 1.75$; Maltrin M100: maltodextrin, $M_w = 42{,}000$, $M_w/M_n = 22$. Starburst 6: generation 6 poly(amido amide) dendrimers, $M_w = 58{,}000$ g/mol. Columns, set of three analytical columns, KS-802, KS-803, KS-804, preceded by a KB-G guard column; solvent, H_2O/0.02% NaN_3; temperature, 50°C; flow rate, 1 mL/min; detectors, VISC + DRI (universal calibration; see Section 8.2.2). (Reprinted with permission from Ref. 17.)

underestimation of the molar mass of the analyte, while repulsion will cause the molar mass to be overestimated.

Electrostatic interactions are most commonly suppressed by varying the ionic strength or, if possible, the pH of the eluent. The latter may not always be an option, due to the pH-dependent stability of many proteins. Near-ideal SEC retention is characterized by ionic-strength-independent analyte elution, rather than by a minimum in elution volume as a function of ionic strength; said minimum is only indicative of the ionic strength at which the sum of the electrostatic and hydrophobic interactions is minimized [18]. At fairly high ($\geq$0.5) ionic strength, hydrophobic interactions are often manifested as an increase in elution volume with increasing ionic strength. These interactions can be reduced either by reducing the ionic strength of the mobile phase or by adding a small amount of organic solvent modifier (e.g., 1% methanol).

Problems relating to calibration-curve-dependent molar mass determination have generally been overcome with the use of online light-scattering detectors and/or multidetection schemes. Recently, SEC/MALS/UV/DRI was employed to determine molar mass, degree of glycosylation or polymer conjugation, and solution association state of glycoproteins and protein–polymer conjugates [20]. Mass spectrometry has

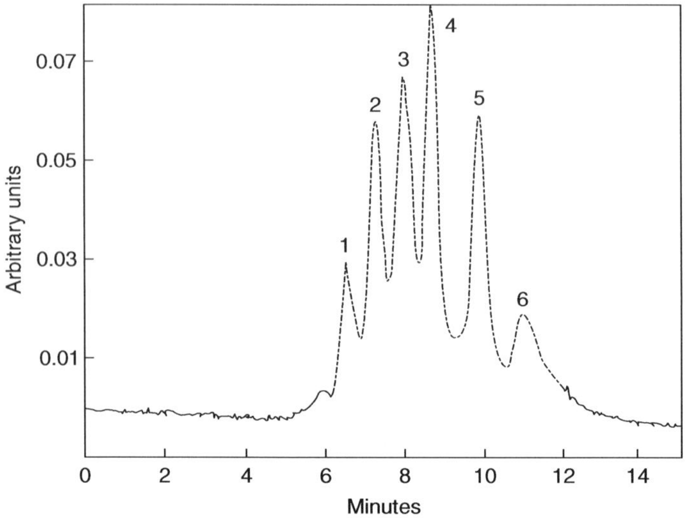

Figure 12.2 Aqueous SEC analysis of protein mixture. Analytes: 1, thyroglobulin; 2, ferritin; 3, catalase; 4, ovalbumin; 5, cytochrome *c*; 6, D,L-alanine; columns, 50 × 0.8 cm column packed with silica grafted with aminoethyl-modified dextran, Eluent, 0.07 M phosphate buffer (pH 7.0) with 0.25 M NaCl; flow rate, 2.0 mL/min; detection, UV at 230 nm. (Reprinted with permission from Ref. 23.)

also been used to overcome the limitations of calibration curves. For example, electrospray ionization and chemical reaction interface mass spectrometry (ESI-MS and CRIMS, respectively) have been coupled to SEC for the study of peptides [19,21].

The SEC separation of nucleic acids is, in many cases, experimentally similar to the analysis of proteins and peptides. Often, $MgCl_2$, which is used in DNA ligation reactions, has been found to promote interactions between the analyte and certain column packings. These interactions can be reduced or eliminated by chelation with EDTA prior to SEC analysis [3]. Reference 22 provides a review of the aqueous SEC analysis of DNA fragments, RNA, plasmids, and oligonucleotides. Figure 12.2 shows the separation of a protein mixture, using a hydrophilic stationary phase consisting of dextran-grafted silica gel. Matrix effects from the silica were minimized by silanization of the silanols, by steric shielding of the residual silanols using grafted dextrans, and by electrostatic compensation of the SiO^- groups by positively charged aminoethyl moieties introduced onto the dextran prior to grafting [23]. Additional examples of aqueous SEC analysis of proteins are shown in Figures 1.1 and 1.2.

12.4.3 Synthetic Polymers

While the majority of nonionic synthetic polymers are water insoluble, a few exist that can be analyzed under aqueous conditions. Examples of the latter are poly(vinyl alcohol), poly(ethenyl formamide), poly(vinylpyrrolidone), and select dendrimers and acrylamide copolymers.

12.4.3.1 Poly(vinyl alcohol). Both fully (98%) and partially (88%) hydrolyzed poly(vinyl alcohol) (PVOH) can be analyzed by aqueous SEC for determination of M-averages, MMD, and evidence of branching. In the case of partial hydrolyzation, the remainder of the polymer is comprised of remaining precursor vinyl acetate functional groups randomly distributed along the macromolecular backbone. These groups can be responsible for branching in PVOH. Dissolution of PVOH requires heating to at least 90°C for approximately 30 minutes; subsequent to dissolution, the polymer remains in solution at room temperature. Mark–Houwink and conformation plots show the polymer to be linear, regardless of M, polydispersity, or percent hydrolyzation. Mark–Houwink plot overlays are shown in Figure 12.3, where super-low, low, medium, and so on, refer to relative rankings of M. All slopes fall within the range 0.60 to 0.63, the only exception being super-low 88% PVOH, where the slope is slightly higher.

Aqueous SEC has also been used to study the changes that occur in PVOH while cross-linking using γ-ray irradiation. Universal calibration SEC/VISC results were

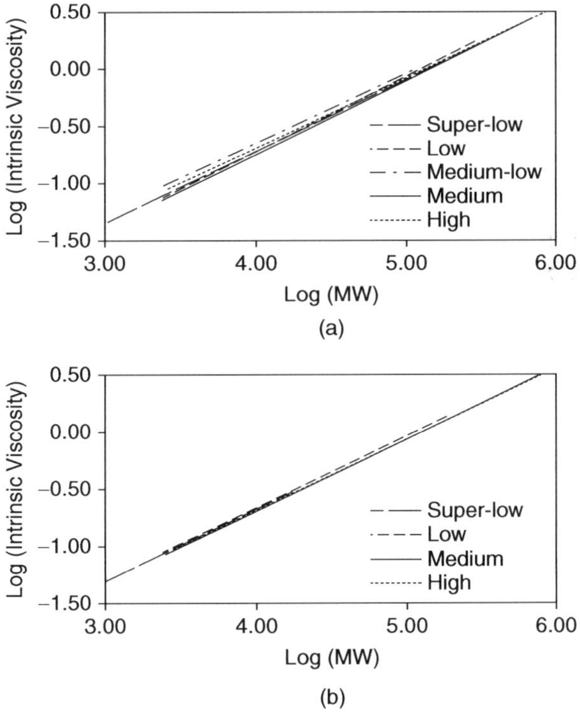

Figure 12.3 Overlay of Mark–Houwink plots of (a) partially hydrolyzed and (b) fully hydrolyzed PVOH. Line labels refer to relative ranking of M. Columns, set of four analytical columns, TSK-PW 2000, 3000, 4000, and 5000 Å; mobile phase, H_2O/0.05 M $NaNO_3$; temperature, 35°C; flow rate, 1.0 mL/min; detectors, MALS (three angle) + VISC + DRI in SEC^3 mode (see Section 9.6). (Reprinted with permission from Ref. 24.)

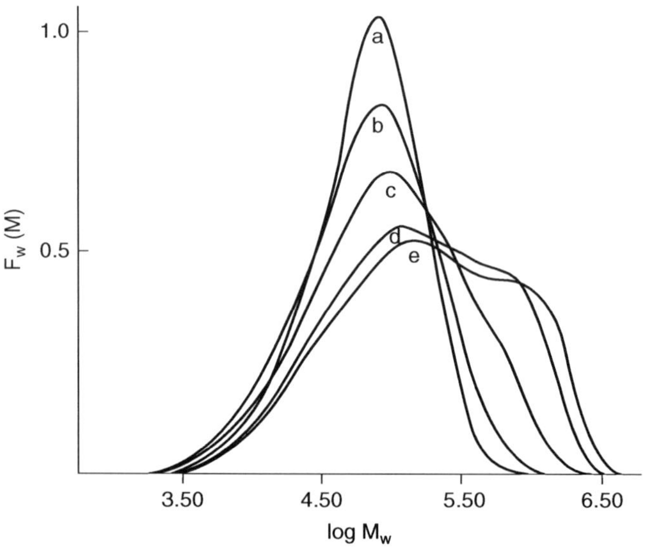

Figure 12.4 Evolution of PVOH MMD as a function of γ-ray irradiation. Irradiation dose, in Mrad: a, 0; b, 0.041; c, 0.082; d, 0.123; e, 0.191. Columns, set of two GF-7M HQ columns; solvent, H_2O/50 mM LiCl; temperature, 40°C; flow rate, 0.7 mL/min; detectors, VISC + DRI (universal calibration; see Section 8.2.2). (Reprinted with permission from Ref. 25.)

augmented with results from off-line MALS and QELS (Chapter 9) to show the change in molar mass averages, polydispersity, and distribution in PVOH as a discrete function of irradiation, leading up to gelation [25]. Results are shown in Figure 12.4. SEC results were combined with second virial coefficient and R_G and R_H data, as well as with calculation of the dimensionless ratios ρ and $V_{A_2\eta}$ and of the coil interpenetration function, Ψ^* (Section 11.4). Through this, it was demonstrated that both an increase in branching and decreased solvation occur as a function of increased exposure to irradiation.

12.4.3.2 Poly(ethenyl formamide). For the water-soluble polymer poly-(ethenyl formamide) (PEF), Figure 12.5a overlays the molar mass distributions of various samples. Analysis of the slopes of the conformation plots showed that the lower-*M* samples are linear random coils in aqueous solution. The decrease in slope with increasing molar mass was ascribed to long-chain branching in the higher-*M* samples. The conformation plot of an intermediate molar mass sample is shown in Figure 12.5b.

12.4.3.3 Dendrimers. Dendrimers, which possess a branch point at every repeat unit, are theoretically the most highly branched structures that can exist. Examples of these structures are shown in Figure 12.6. The theoretical average molar mass, *M*,

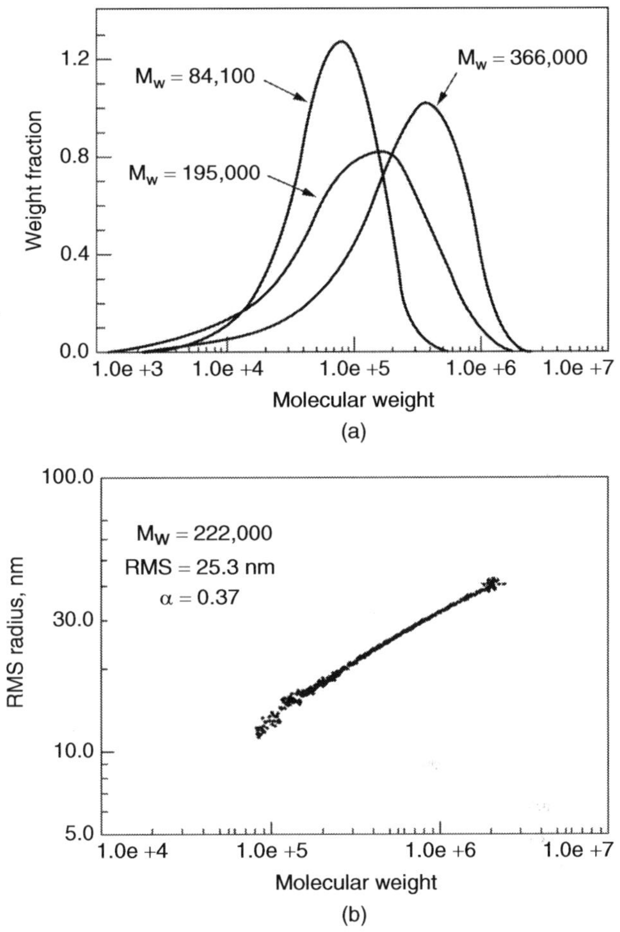

Figure 12.5 SEC/MALS of poly(ethenyl formamide) (PEF): (a) MMD overlay of three PEF samples; (b) conformation plot of intermediate molar mass PEF sample; slope (α) of 0.37 is indicative of highly compact structure in solution (see Section 11.4.1), ascribed in original reference to long-chain branching. Columns, TSK-PW; solvent, H_2O/0.05 M $NaNO_3$; temperature, 35°C; flow rate, 1.0 mL/min. (Reprinted with permission from Ref. 26.)

of dendrimers can be calculated as [17]

$$M = M_c + N_c\left(M_{RU}\frac{N_b^{G+1} - 1}{N_b - 1} + M_t N_b^{G+1}\right) \qquad (12.1)$$

where M_c, M_{RU}, and M_t are the molar masses of the initiator core, repeat units, and terminal units, respectively, N_c and N_b are the initiator core and branch juncture multiplicity, respectively, and G is the dendrimer generation. Care must be taken to

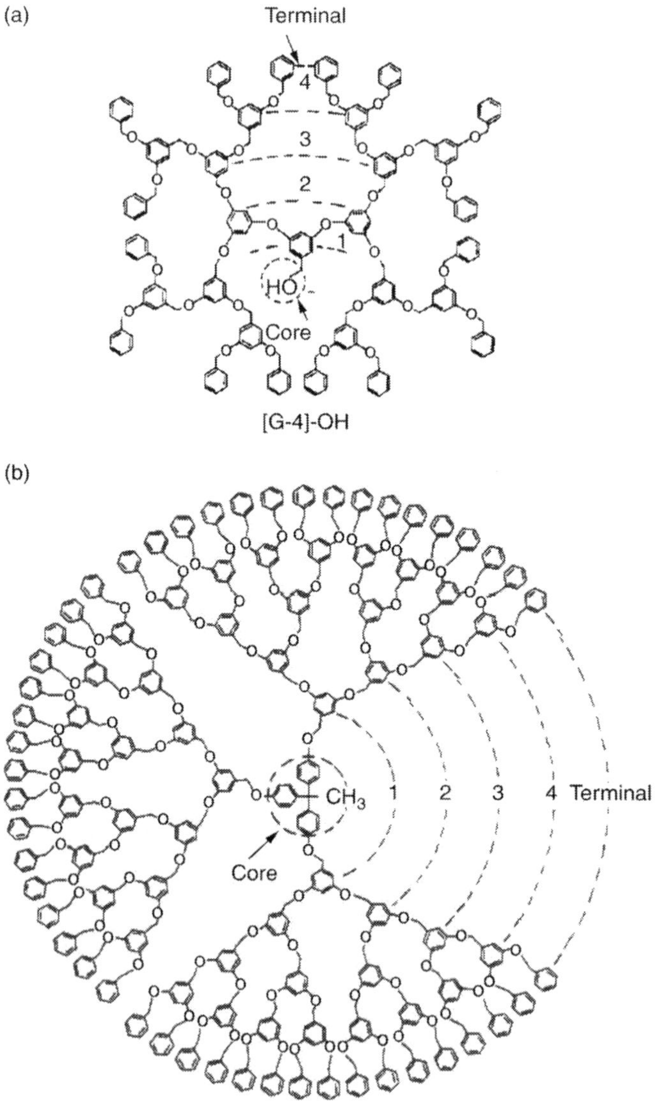

Figure 12.6 Structures of convergent-growth dendrimers: (a) fourth-generation monodendron; (b) fourth-generation tridendron. (Reprinted with permission from Ref. 27.)

notice whether the core is assigned a generation number of 0 or 1, as two different nomenclatures are often used. The degree of polymerization, or number of repeat units N_{RU}, of the dendrimers can be calculated as [17]

$$N_{RU} = N_c \frac{N_b^{G+1} - 1}{N_b - 1} \tag{12.2}$$

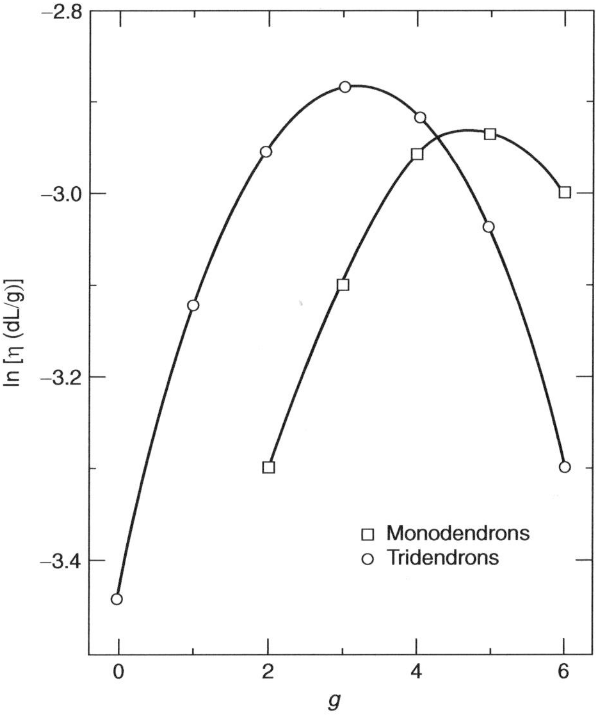

Figure 12.7 Mark–Houwink inversion in dendrimers: SEC/LALS/VISC/DRI analysis of convergent-growth dendrimers. The abscissa represents *g*, the generation number for a set of convergent-growth dendrimers. The core is given a generation number of zero (*g*0). "Monodendron" and "tridendron" refer to structures in Figure 12.6. Experimental conditions: Set of four 7.5 × 300 mm, 10-μm particle diameter, mixed-bed PLgel columns; solvent, THF; temperature, 30°C; flow rate, 1.0 mL/min. (Reprinted with permission from Ref. 27.)

Dendrimers display a curious property in solution: a decrease in intrinsic viscosity with increasing molar mass beyond a certain generation. This Mark–Houwink inversion" appears to be due to dendrimers growing faster in density than in radial growth [17,27]. In these polymers, volume increases cubically, whereas mass increases exponentially. As such, around generation 3 to 5 (G3 to G5) a morphological transformation appears to take place, from a planar, disk-like structure to a near-spherical architecture. Intrinsic viscosity will initially increase with molar mass, reaching a maximum around G3 or G5, and then $[\eta]$ steadily decreases with increasing dendrimer generation. Examples of this behavior can be seen in Figure 12.7. The molar mass distribution of a G6 starburst poly(amido amide) dendrimer is shown in Figure 12.1.

12.4.3.4 Poly(ethylene oxide) and Poly(ethylene glycol). Narrow polydispersity standards of these polymers are commercially available, ranging from 1×10^2 to 1.5×10^6 g/mol. A variety of mobile-phase additives have been used in

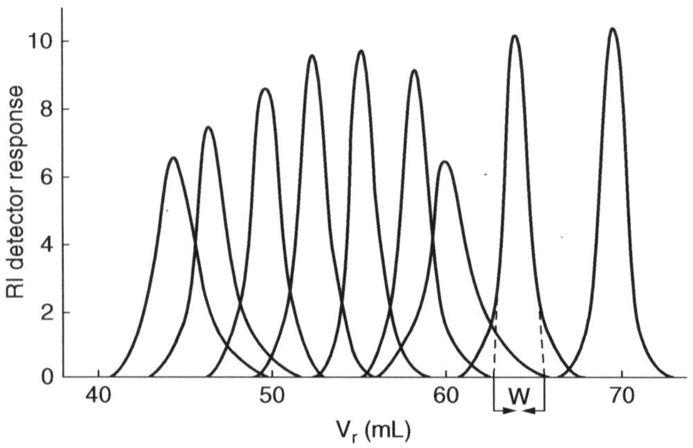

Figure 12.8 SEC/DRI elution profiles of a series of poly(ethylene oxides), ranging in M_w from 2×10^3 to 1.3×10^6 g/mol. For all samples, $M_w/M_n \leq 1.15$. Columns, set of four 7.5 × 600 mm TSK-GEL columns, one G2000PW, one G3000PW, and two G5500PW; solvent, H_2O/0.1 M NaCl; temperature, 25°C; flow rate, 1.0 mL/min; concentration; 0.2 mg/mL; detector; DRI. (Reprinted with permission from Ref. 28.)

the aqueous SEC analysis of poly(ethylene glycol) (PEG) and its higher molar-mass analog poly(ethylene oxide) (PEO). At 25°C, equivalent results have been obtained using 0.1 M NaCl, 0.02% NaN_3, or 0.08 M Tris-HCl buffer (pH 7.9). Chromatograms for a series of PEOs in aqueous NaCl solution are shown in Figure 12.8. Combining the SEC/DRI data with off-line viscosity measurements, it was concluded that PEO obeys the universal calibration principle (Section 8.2.2).

***12.4.3.5 Other Synthetics* [*3*].** Neutral synthetic polymers such as poly-(vinylpyrrolidone) (PVP) can be analyzed with glycerylpropyl-bonded silica columns using a mobile-phase consisting of 1 : 1 H_2O/methanol plus 0.1 M $LiNO_3$. The methanol is included to suppress non-size-exclusion effects due to adsorption. SEC analyses of acrylamide copolymers have been conducted on hydrophilic polymer gel columns using H2O/0.05 N $NaNO_3$ as eluent. With the same type of columns, partially hydrolyzed polyacrylamides were analyzed using a H_2O/0.1 M NaCl mobile-phase.

12.4.4 Polyelectrolytes

As with proteins and peptides, much of the work in the last decade has focused on determining experimental conditions under which polyelectrolytes elute by a strict size-exclusion mechanism. Once these conditions are found, relative or universal calibration curves (Section 8.2) based on either neutral or polyelectrolytic standards can be used to determine M averages and distributions of the analytes [29,30]. Difficulties in defining near-ideal SEC conditions for peptides/proteins and polyelectrolytes have now been eliminated by the use of online static light-scattering detectors [31].

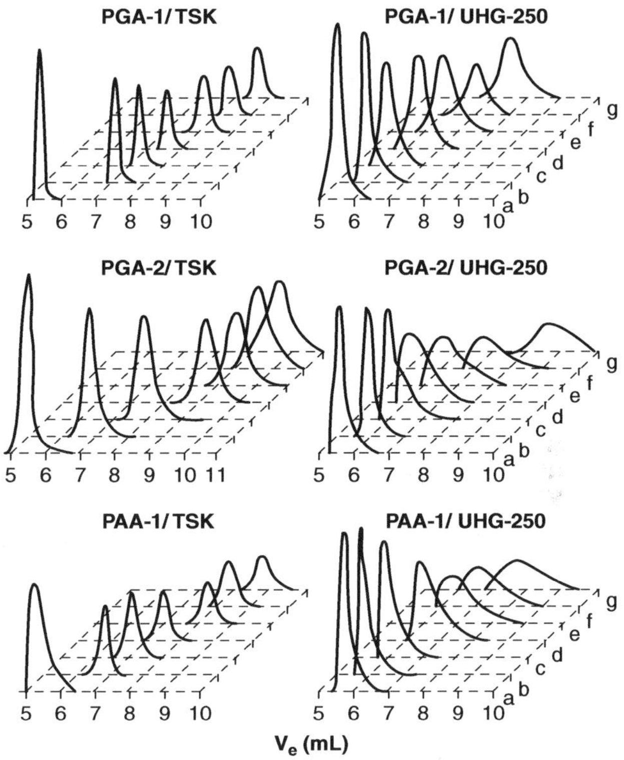

Figure 12.9 Effect of ionic strength of eluent on SEC elution of polyelectrolytes: elution profiles of poly(L-glutamic acid), PGA-1, sodium poly(styrene sulfonate), PSS-2, and poly(acrylic acid), PAA, on two different SEC columns, TSK and UNG-250, at pH 7.0 and various ionic strengths: (a) pure water; (b) 0.005 M; (c) 0.01 M; (d) 0.02 M; (e) 0.05 M; (f) 0.10 M. (Reprinted with permission from Ref. 29.)

Consequently, the extensive literature on the size-exclusion and non-size-exclusion behavior of polyelectrolytes is not reviewed here. As an example of how SEC elution of polyelectrolytes is affected by the ionic strength of the eluent, Figure 12.9 presents the elution profiles of poly(L-glutamic acid), sodium poly(styrene sulfonate), and poly(acrylic acid) on two different SEC columns as a function of ionic strength.

A dependence of K_{SEC} on $\sqrt{I}$ has been noted, with I being the ionic strength of the eluent [32,33]. This dependence was described in terms of increased thickness of the electrical double layer. Following the treatment of Reference 33, X_c is defined as the repulsion length, the difference between the geometric pore radius and the apparent (diminished) effective pore radius for polyelectrolytes. This repulsion length is a manifestation of the overlapping double layers of the charged pore and charged polymer and can be thought of as the region in which the electrostatic potential is more repulsive than in the bulk. The theories proposed show good agreement between calculated and experimental values of K_{SEC} as a function of X_c. The theories also provide for variation in the shielding constant versus X_c relationship with variation in both pore size and ionic strength. Experimental observations appeared to be well reproduced by a linearized Poisson–Boltzmann model [32]. This model predicts that the electrostatic repulsive potential within the charged pore falls off in a roughly exponential fashion with distance from the pore surface.

The dimensions of a polyelectrolyte in solution are a direct function of the ionic strength of the solvent. A decrease in ionic strength leads to an increase in intramolecular charge repulsion and to a consequent increase in the size of the macromolecule. The balance sought between the expanding electrostatic forces and the randomizing entropic forces makes the adoption of a rigid-rod structure highly unlikely. The all-*trans* form is the lowest entropy state of all possible coil conformations. For solutions of moderate ionic strengths (about 1 mM to 1 M), the most successful theories for describing the dimensions and second virial coefficients of polyelectrolytes combine local, charge-induced stiffening of the chain (electrostatic persistence length) with long-range electrostatic repulsion forces. These repulsive forces prevent distant repeat units along the macromolecular backbone from reaching proximity (electrostatic excluded volume) [31].

12.4.5 Inorganic Compounds

Although not generally regarded as a typical use for the technique, a number of niche applications of SEC to the study of inorganic compounds have developed over the years. The early aqueous SEC work in this area, leading up to 1980, is reviewed in Reference 34.

SEC has been used to determine the charge of various ions with known charge ranging from −4 to +2 (35). The method was based on values of K_{SEC} determined with two different eluents over a range of ionic strengths. The eluents employed were aqueous $NaClO_4$ and Na_2SO_4 solutions of ionic strength 0.30, 0.10, and 0.03 mol/L. Deviations between the ionic charges as determined by SEC and known ionic charges are given in Table 12.1.

Table 12.1 Deviation of SEC-determined charges from known ionic charges

Ion	SEC Deviation
Mg^{2+}	−0.12
Na^+	−0.01
TcO_4^-	+0.09
Cl^-	+0.24
SO_4^{2-}	−0.10
$Fe(CN)_6^{3-}$	+0.11
$Fe(CN)_6^{4-}$	−0.02
$Mo(CN)_8^{4-}$	−0.18

Source: Ref. 35.

An SEC protocol has been developed for determining the relative MMD of alumina, silica, and hydroxyaluminosilicate colloids [36]. These colloids are in clays such as montmorillonite, allophone, and imogolite in soils or sediments, and also in water samples. Interest stems from the fact that colloid MMD correlates to reactivity, such as the ability of colloids to adsorb contaminants. SEC/DRI analysis was performed using a 30 × 0.78 cm, 6-μm particle size, Progel-TSK G3000PXWL hydrophilic polymer gel column with residual sulfonic acid groups. These acid groups provided a slight negative charge to the stationary phase, rendering it stable over a pH range of 2 to 12. Optimal chromatographic conditions were found to be 0.1 M $NaNO_3$, pH 5, 0.8 mL/min flow rate, and colloid concentration of 400 mg/L.

REFERENCES

1. P. L. Dubin, ed., *Aqueous Size-Exclusion Chromatography*, J. Chromatogr. Library, Vol. 40, Elsevier, Amsterdam, 1988.
2. P. L. Dubin, in Ref. 1, Preface.
3. A. Williams and L. Hagel, in *Column Handbook for Size Exclusion Chromatography*, C.-S. Wu, ed., Academic Press, San Diego, CA, 1999, Chap. 2.
4. L. Hagel, in *Strategies in Size Exclusion Chromatography*, ACS Symp. Ser. 635, M. Potschka and P. L. Dubin, eds., American Chemical Society, Washington, DC, 1996, Chap. 13.
5. A. Huber and W. Praznik, in *Column Handbook for Size Exclusion Chromatography*, C.-S. Wu, ed., Academic Press, San Diego, CA, 1999, Chap. 16.
6. U. D. Neue, in *Column Handbook for Size Exclusion Chromatography*, C.-S. Wu, ed., Academic Press, San Diego, CA, 1999, Chap. 11.
7. K. K. Unger and J. N. Kinkel, in Ref. 1. Chap. 8.
8. Y. Cohen, R. S. Faibish, and M. Rovira-Bru, in *Interfacial Phenomena in Chromatography*, Surfactant Sci. Ser., Vol. 80, E. Pefferkorn, ed., Marcel Dekker, New York, 1999, Chap. 7.
9. K. Makino and H. Hatano, in Ref. 1, Chap. 9.
10. C.-S. Wu, L. Senak, D. Osborne, and T. M. H. Cheng, in *Column Handbook for Size Exclusion Chromatography*, C.-S. Wu, ed., Academic Press, San Diego, CA, 1999, Chap. 17.
11. S. Mori and H. G. Barth, *Size Exclusion Chromatography*, Springer-Verlag, Berlin, 1999, Chap. 11.
12. M. Janado, in Ref. 1, Chap. 2.
13. P. L. Dubin, in Ref. 1, Chap. 3.
14. A. F. Turbak, R. B. Hammer, R. E. Davies, and H. L. Hergert, *Chemtech*, **10**, 51 (1980).
15. S. C. Churms, in *High Performance Liquid Chromatography and Capillary Electrophoresis*, J. Chromatogr. Library, Vol. 58, Z. El Rassi, ed., Elsevier, Amsterdam, 1995, Chap. 7.
16. D. S. Jackson, *Starch/Stärke*, **43**, 422 (1991).
17. A. M. Striegel, R. D. Plattner, and J. L. Willett, *Anal. Chem.*, **71**, 978 (1999).

18. J. O. Baker, W. S. Adney, M. Chen, and M. E. Himmel, in *Handbook of Size Exclusion Chromatography and Related Techniques*, 2nd ed., revised and expanded, Chromatogr. Sci. Ser., Vol. 91, C.-S. Wu, ed., Marcel Dekker, New York, 2004, Chap. 15.
19. G. B. Irvine, *J. Biochem. Biophys. Methods*, **56**, 233 (2003).
20. B. S. Kendrick, in *Multiple Detection in Size-Exclusion Chromatography*, ACS Symp. Ser. 893, A. M. Striegel, ed., American Chemical Society, Washington, DC, 2005, Chap. 7.
21. P. Lecchi and F. P. Abramson, in *Multiple Detection in Size-Exclusion Chromatography*, ACS Symp. Ser. 893, A. M. Striegel, ed., American Chemical Society, Washington, DC, 2005, Chap. 11.
22. Y. Kato and S. Nakatani, in *Handbook of Size Exclusion Chromatography and Related Techniques*, 2nd ed., revised and expanded, Chromatogr. Sci. Ser., Vol. 91, C.-S. Wu, ed., Marcel Dekker, New York, 2004, Chap. 16.
23. M. Petro, P. Gemeiner, and D. Berek, *J. Chromatogr. A*, **665**, 37 (1994).
24. D. J. Nagy, *Am. Lab.*, **35**, 38 (2003).
25. B. Wang, S. Mukataka, E. Kokufuta, M. Ogiso, and M. Kodama, *J. Polym. Sci. B*, **38**, 214 (2000).
26. D. J. Nagy, *J. Appl. Polym. Sci.*, **59**, 1479 (1996).
27. T. H. Mourey, S. R. Turner, M. Rubinstein, J. M. J. Fréchet, and K. L. Wooley, *Macromolecules*, **25**, 2401 (1992).
28. T. Kato, T. Tokuya, and A. Takahashi, *J. Chromatogr.*, **256**, 61 (1983).
29. R. García, I. Porcar, A. Campos, V. Soria, and J. E. Figueruelo, *J. Chromatogr. A*, **655**, 191 (1993).
30. A. Dondos, *J. Polym. Sci. B*, **44**, 1106 (2006).
31. W. F. Reed, in *Strategies in Size Exclusion Chromatography*, ACS Symp. Ser. 635, M. Potschka and P. L. Dubin, eds., American Chemical Society, Washington, DC, 1996, Chap. 2.
32. M. Potschka, *J. Chromatogr.*, **441**, 239 (1988).
33. P. L. Dubin, R. M. Larter, C. J. Wu, and J. I. Kaplan, *J. Phys. Chem.*, **94**, 7243 (1990).
34. M. Shibukawa and N. Ohta, in Ref. 1, Chap. 4.
35. T. G. Tji, H. J. Krips, W. J. Gelsema, and C. L. de Ligny, *J. Chromatogr.*, **504**, 403 (1990).
36. B. A. Logue, C. Burns, and J. A. Rice, *Microchem. J.*, **55**, 151 (1997).

13

OLIGOMERIC SEC

13.1 INTRODUCTION

The study of natural and synthetic oligomeric species, by SEC and related techniques, has become extremely important in recent years. Part of this added importance can be attributed directly to the greatly enhanced possibilities for oligomer analysis afforded by the newest generations of SEC columns. These columns have an extended molar mass range, from the high polymeric to low oligomeric regions (see Figure 1.6). Of special interest are columns with concentrated resolution in the oligomeric region. An extended separation range is invaluable for quantitating the amount of oligomer in polymers. This is information required by regulatory agencies for import and export of products in the United States and Europe. Columns with concentrated resolution in the oligomeric region are generally capable of separating individual members of homologous series, and even different types of isomers.

Synthetic oligomers are used as lubricants, plasticizers, and in low-solvent-content coatings. The uses of oligosaccharides in foods, feeds, biomolecular recognition processes, and other applications are also numerous. A section on the analysis of these carbohydrates is included in this chapter. Since the first edition of this book, the number of applications of oligomeric SEC (or GPC or GFC or steric exclusion chromatography) has grown by approximately two orders of magnitude. A SciFinder search of the literature revealed less than 50 publications on this topic between 1958 and 1978. A search from 1979 to the present revealed approximately *one thousand* papers on oligomeric SEC! [1]. As with previous chapters, the aim of this chapter

Modern Size-Exclusion Liquid Chromatography: Practice of Gel Permeation and Gel Filtration Chromatography, Second Edition
By André M. Striegel, Wallace W. Yau, Joseph J. Kirkland, and Donald D. Bly

is not to provide a comprehensive or critical literature review of the applications of SEC to the analysis of oligomeric substances. Rather, we illustrate how SEC analysis of oligomers differs from the analysis of species of higher degrees of polymerization with respect to experimental parameters, to the type of information obtained, and to information needed for accurate quantitation.

For high-M polymers, SEC is used mainly to obtain molar mass averages and distributions and the type of structure-property relations described in Chapter 11. For oligomers, the technique is used quite differently. The intent in oligomeric SEC is to resolve individual components from one another and/or from the polymeric portion of the sample. Results will be used either for quantitation of the total oligomeric species, to assign molar mass values to the individual oligomers, for fingerprint identification of additives or prepolymeric species, or to obtain entropic data related to the solution conformation of analytes.

13.2 WHAT IS AN OLIGOMER?

This question could also be restated as any of the following: What is a polymer?, What distinguishes an oligomer from a polymer?, or When does a molecule become a polymer? There are probably as many definitions of what a polymer (and hence an oligomer) is as there are polymer textbooks. Most of these definitions, however, are fairly similar. A few examples, beginning with the more general (we've replaced the antiquated and inaccurate term *molecular weight* with the more accurate term *molar mass*, as done throughout the book):

1. "*Polymers* are macromolecules built up by the linking together of large numbers of smaller molecules" [2].
2. "If only a few monomer units are joined together, the resulting low-[molar mass] polymer is called an *oligomer*" [3].

Two more specific definitions are:

3. "A *polymer* may be defined as a large molecule comprised of repeating structural units joined by covalent bonds. . . . In this context, a large molecule is commonly arbitrarily regarded either as having a [molar mass] of at least 1000 or as one containing 100 structural units or more" [4].
4. "Polymers with only a few structural units (from 5 to 10) are usually called 'oligomers' " [5].

For a more utilitarian approach:

5. "This word [polymer] means 'many parts' and designates a large molecule made up of smaller units. . . . Polymers generally have [molar masses] greater

than about 5000 but no firm lower limit need be defined since the meaning of the word is nearly always clear from its use.... An oligomer is a low-[molar-mass] polymer. It will contain at least two monomer units.... The distinction between the sizes of oligomers and the corresponding polymers is left vague, however, because there is no sharp transition in most properties of interest" [6].

Finally, a legal definition:

6. "**Polymer molecule**—a molecule that contains a sequence of at least 3 monomer units, which are covalently bound to at least one other monomer unit or other reactant" and "**Oligomer**—a low [molar mass] species derived from the polymerization reaction" [7].

Certainly, definition 5 above seems to make the most sense for day-to-day use and discussion. Defining a distinction between oligomer and polymer on molar mass alone is ineffective, as a molar mass of 224 g/mol corresponds to a dimer of polystyrene but to a 16-mer of polyethylene. Definitions based on degree of polymerization (DP) or on the percentage of the repeat units of the polymer which are end groups create difficulties when examining solution properties. For example, the specific refractive index increment $\partial n/\partial c$ (see Section 9.2.1c) of polymer A may have achieved a constant value at a certain DP, while the $\partial n/\partial c$ of polymer B may not yet be constant at that same DP at the same solvent, temperature, and wavelength conditions. Moreover, the $\partial n/\partial c$ of polymer A may achieve a constant value at a certain DP at one set of experimental conditions, but the $\partial n/\partial c$ of the same polymer A may not yet be constant, for this same DP, under a different set of conditions. Similar situations exist for other dilute solution properties. These considerations will be revisited when we discuss the legal implications of accurately determining the oligomeric content of a polymer sample in Section 13.5.4.

There is a recent and promising attempt at defining a polymer and at distinguishing between polymeric and oligomeric behavior. It has been found that for both poly(dimethyl siloxane) (PDMS) and polystyrene (PS), the transition from small-molecule behavior to Rouse dynamics is a smooth function of the weight-average molar mass, M_w, when the latter is scaled to the molar mass of a random step, m_R. The molar mass dependence of the glass transition temperature, T_g, for both PDMS and PS appears to be the same when plotted as a function of M_w/m_R, as shown in Figure 13.1 [T_g(inf), also referred to in the literature as T_g^∞ and as $T_g(\infty)$, is the glass transition temperature of a polymer of infinite molar mass]. From this analysis it was concluded that a molecule approaches a Gaussian coil, signifying a transition from oligomeric to polymeric behavior, at $M_w \approx 10$ to $20m_R$. Details of the experiments, theory, and rationale behind this approach may be found in Reference 8.

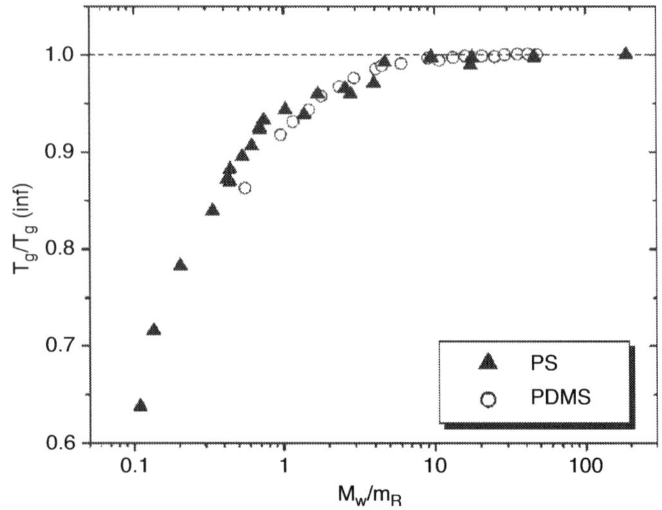

Figure 13.1 Transition from oligomeric to polymeric behavior: $T_g/T_g(\infty)$ as a function of weight-average molar mass, M_w, scaled by the mass of a random step, m_R. For PDMS, $m_R \approx 560$; for PS, $m_R \approx 5100$. For both PS and PDMS, oligomer-to-polymer transition occurs at $M_w \approx 10$ to $20m_R$. (Reprinted with permission from Ref. 8.)

13.3 PRELIMINARY CONSIDERATIONS

Although the practice of oligomeric SEC involves difficulties not encountered in its polymeric counterpart, there are also a few properties of oligomers that make them easier to study than polymers. In the next section we deal with some of the problems that might be encountered.

13.3.1 Advantages over Polymeric SEC

Depending on the situation, there are some real advantages of using SEC for the separation and analysis of oligomers.

1. *Ease of solubility*. Not only do oligomers dissolve more easily than polymers of the same chemistry, but they also become fully solvated very quickly. For ultrahigh-M polymers (i.e., $M > 10^6$ g/mol), waiting for complete solvation may take several hours beyond the dissolution stage. Additionally, a wider range of solvents is available for oligomeric SEC: If a solvent will dissolve a polymer, it will also dissolve an oligomer of the same species. The opposite of this is not true, however. For example, cellooligosaccharides are water soluble, whereas cellulose is not [9–11].

2. *Perfect molar mass monodispersity*. At the polymeric level, this can only be achieved with naturally made proteins and peptides. Otherwise, any synthetic scheme will produce macromolecules with a certain degree of polydispersity. A perfectly

monodisperse polymer is essential for studying how structural parameters affect the conformational entropy of analytes and it is also important in fundamental studies of chromatographic band broadening.

3. *Diminished concern for c* and viscous fingering*. Oligomer solutions can be made quite concentrated, if required, without having to worry about analyte entanglement or viscous fingering effects during elution (see Sections 3.4.3 and 7.4.2). For samples without appreciable polymeric content, high concentrations permit determination of molar mass, size, and intrinsic viscosity using light-scattering and viscometric detectors (Section 13.5.7).

13.3.2 Difficulties as Compared to Polymeric SEC

1. *Nonconstant $\partial n/\partial c$ and ε*. As exemplified in Figure 9.4, the specific refractive index increment of most analytes is not constant in the oligomeric region. Consequently, the response of detectors such as the differential refractometer and static light-scattering photometer, which in both cases is dependent on $\partial n/\partial c$, will also not be constant in this region. This is also seen in Figure 13.2, which shows that the molar absorptivity (extinction coefficient, ε) is likewise not constant in the oligomeric region. The molar absorptivity is necessary when using an ultraviolet (UV) detector as a concentration-sensitive detector.

2. *Negative viscosity and failure of universal calibration*. The viscosity of oligomer solutions may be negative, which at the relatively low shear rates generally encountered in SEC experiments is usually attributed to a simple mixing rule effect

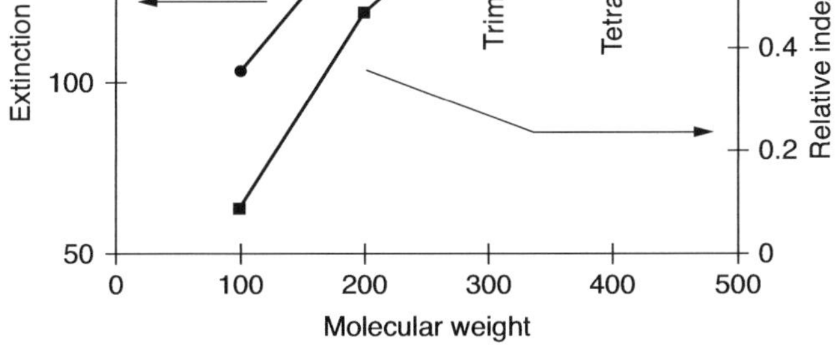

Figure 13.2 Physical properties of methyl methacrylate oligomers: molar absorptivity (extinction coefficient, ε) and relative index of refraction of DP 1-4 MMA oligomers in THF at 24°C. (Reprinted with permission from Ref. 12.)

[13]. This is seen in Figure 13.3 for neat and end-functionalized styrene monomer and for a polyethylene oligomer.

As the ordinate in a universal calibration graph is the logarithm of the product of the intrinsic viscosity and the molar mass (Section 8.2.2 and Figure 8.4), and it is impossible to take the logarithm of a negative number, the negative viscosity effect contributes to the failure of universal calibration at the oligomeric level [13, 14]. Other contributions to this failure are non-SEC behavior, discussed next, and the shape of various oligomeric configurations.

3. *Non-SEC behavior*. In Figure 13.3a, there is another problem that may arise in oligomeric SEC: analytes displaying non-size-exclusion behavior. In this figure, n-butyl-terminated styrene oligomers of DP 2-6, collectively referred to in the figure as "PS 370," and n-butyl-terminated styrene monomer (PS 162) all elute according to size. However, pure styrene monomer ("styrene") displays not only a negative viscosity in the given solvent at the given temperature but also elutes at approximately the same retention volume as an n-butyl-terminated styrene trimer. This behavior was attributed to interactions between the double bond of the vinyl group of styrene and the cross-linked polystyrene–divinylbenzene column packing material under the given solvent–temperature conditions. All other members of the oligostyrene series examined lacked the alkene functionality [13].

4. *Effect of chain ends on* A_2. Figure 13.4 shows that for a series of oligo- and poly(α-methylstyrene)s (a-PαMS), at near-theta conditions, the second virial coefficient increases rapidly with decreasing degree of polymerization. This increase occurs at a faster rate in the oligomeric than in the polymeric region. This behavior is attributed to a chain-end effect. As the degree of polymerization of a linear macromolecule becomes smaller, the percentage of the polymer chain ends become larger. The chain-end effect is also responsible for the nonconstant $\partial n/\partial c$ and ε in the oligomeric region, noted earlier in this section, as the chemical and electronic environments of the chain ends can be different from those of the repeat units of the polymer.

5. *Low response of static light-scattering (SLS) and viscometry (VISC) detectors*. As mentioned in Chapter 9, static light-scattering and viscometric detectors are molar-mass sensitive. Consequently, all other factors being equal, an oligomer will generate a weaker SLS or VISC signal than a polymer. For samples that are only oligomeric and that do not contain an appreciable quantity of polymer, this is not a problem. As described in Section 13.3.1, for strictly oligomeric samples a higher sample concentration can be used. However, if the sample contains appreciable amounts of polymeric constituent, the concentration of the sample in solution cannot be very large, as it will exceed the critical overlap concentration, c^*. Exceeding c^* means that the sample solution can no longer be considered near-infinitely dilute (Sections 3.4.3 and 7.4.2). In such cases, many of the assumptions described in Chapter 9 as underlying viscometric and static light-scattering calculations must be discarded. Also, high concentrations and/or injection volumes contribute to viscous fingering (Section 7.4.2), which can result in chromatographic peak tailing.

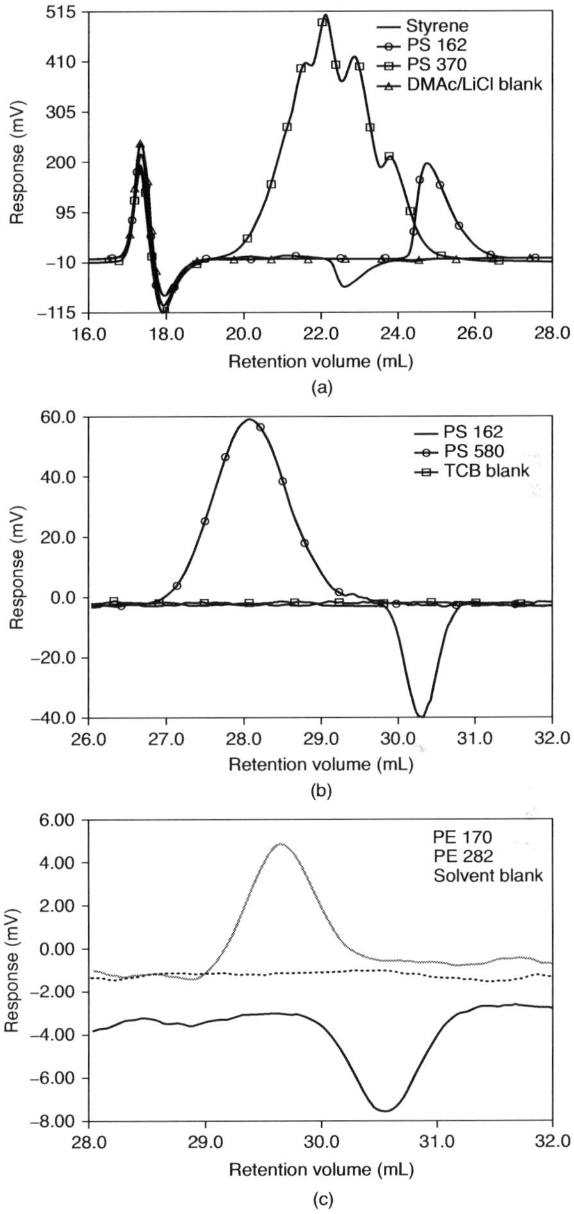

Figure 13.3 Negative viscosity of oligomer solutions: (a) *n*-butyl-terminated PS with DP 2-6 (PS 370) and *n*-butyl-terminated styrene monomer (PS 162) display positive viscosities in DMAc/0.5% LiCl at 35°C, while neat styrene monomer displays both negative viscosity and non-SEC behavior; (b) negative viscosity of PS 162 in TCB at 135°C; (c) positive viscosity of octadecane (PE 282) and negative viscosity of dodecane (PE 170) in TCB at 135°C; trace with no peak corresponds to solvent blank. Ordinates in all graphs correspond to the differential pressure trace of a differential viscometer, proportional to the specific viscosity of the solutions. (Adapted from Ref. 13.)

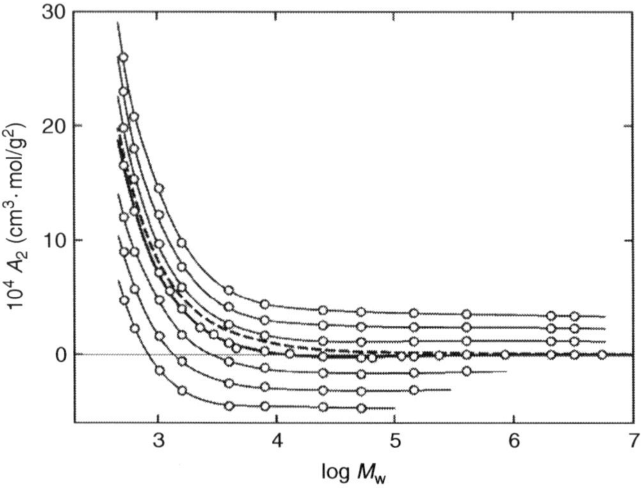

Figure 13.4 Rapid change of the second virial coefficient in the oligomeric region. Plots of A_2 versus log M_w for a-PαMS in cyclohexane at various temperatures. From top to bottom (in °C): 45.0, 40.0, 35.0, 30.5 (θ), 25.0, 20.0, and 15.0. The data points at 45.0, 40.0, and 35.0°C are shifted upward by 3×10^{-4}, 2×10^{-4}, and 1×10^{-4} cm$^3 \cdot$ mol/g^2, respectively, and those at 25.0, 20.0, and 15.0°C downward by 1×10^{-4}, 2×10^{-4}, and 3×10^{-4} cm$^3 \cdot$ mol/g^2, respectively. (Reprinted with permission from Ref. 15.)

The lack of appreciable angular light-scattering dissymmetry of oligomers makes it very difficult to measure R_G for these analytes. Again, this is especially difficult if the sample contains both polymeric and oligomeric components, for the same reasons as described in the preceding paragraph. As mentioned in Chapter 9, however, measurement of the viscometric and hydrodynamic radii may still be possible in the oligomeric region, even if R_G cannot be measured accurately.

6. *Counterintuitive solubility behavior*. If an oligomer with DP of x dissolves easily in a given solvent at a given temperature, and if a member of the homologous series with DP of $x + 2$ also dissolves easily at these condition, it is natural to expect the intermediate oligomer with DP of $x + 1$ also to readily dissolve. This is not always the case. A notable example is that of the cyclic maltooligosaccharides known as cyclodextrins (CDs), where α-, β-, and γ-CD are the common names given to the structures of DP 6, 7, and 8, respectively. In organic solvents such as DMAc and DMAc/LiCl, all three CDs dissolve readily [16]. However, in H_2O at 25°C and pH 7.00 and at 34°C and pH 7.4, β-CD is at least nine times less soluble than either α- or γ-CD [9,17]. Another seemingly counterintuitive solubility characteristic of CDs is that their solubility in H_2O is substantially higher than in D_2O. Proposed explanations for both these effects, for which quantitative data are given in Table 13.1, may be found in Reference 17.

7. *Other electromagnetic and hydrodynamic effects*. How the translational diffusion coefficient D_T scales with the DP of oligomers is different from how D_T scales

Table 13.1 Solubility (g/g solvent) of cyclodextrins in H_2O and D_2O at 25°C

Solute	H_2O	D_2O	Relative Difference (%)
α-CD	0.1295 ± 0.0007	0.0758 ± 0.0005	41
β-CD	0.0184 ± 0.0002	0.0108 ± 0.0001	41
γ-CD	0.2492 ± 0.0002	0.1988 ± 0.0006	20

Source: Ref. 17.

with DP for polymers. The same is true of the mean-square optical anisotropy of oligomers. Also, as seen in Figure 9.14, the ability of oligomers to depolarize incident radiation can be much greater than that of polymers [18].

8. *Chromatographic resolution.* The large majority of SEC applications involve measuring the MMD of polymers or measuring how a particular property, such as polymeric radius or intrinsic viscosity, changes as a function of M. In these cases, separating the individual components of a multicomponent mixture from each other is not the goal of the separation. In contrast, the objective in oligomeric SEC resembles that of traditional, enthalpically controlled separations of small molecules, where optimizing chromatographic resolution is paramount. If separation via a strict size-exclusion mechanism is desired (as in Section 13.5.6), however, SEC is handicapped by its generally low resolution and peak capacity (see Chapter 4). Optimizing resolution in SEC is discussed in Section 13.6.

13.4 OLIGOMERIC SEC COLUMNS

As seen in Figure 1.6, even column sets with ultrahigh-M exclusion limits are able to separate analytes well into the oligomeric region, sometimes down to a single repeat unit constituent or monomer. While benefits for the analysis of very broad polydispersity macromolecules are obvious, the type of separation exemplified by the calibration curve in Figure 1.6 sacrifices resolution in any particular region of the MMD. When the focus is exclusively on the oligomeric region of the MMD, SEC columns with the following features are preferred and are commercially available:

1. The lower exclusion limit is virtually always monomeric.
2. Upper exclusion limits vary from 2×10^3 to 6×10^4 g/mol, with discrete intervals between these values. The effect of pore size on oligomeric separations can be seen in Figures 13.7 and 13.16 and are discussed in Section 13.6.
3. 10-, 8-, 5-, and even 3-μm packing particle sizes are now available, allowing for increased resolution. Some of the newest 3-μm columns are said by the manufacturer to have guaranteed efficiencies of greater than 100,000 plates/meter.

4. Organic columns are generally rugged and can withstand a variety of solvents and relatively high pressures and temperatures.
5. Columns designed for aqueous mobile phases can withstand a wide pH and temperature range, a modest amount (about 20%) of organic co-solvent, and a variety of salt additives. (Silica-based SEC columns can stand much higher pressures than polymeric columns.)
6. Column packings for aqueous mobile phases are often functionalized, to maximize the resolution of particular families of analytes.

We do not review the column offerings from individual manufacturers or their specifications. These are continuously changing and current information can be found at the manufacturers' own websites or in the recent literature [19]. As a contrast to Figure 1.6, Figure 13.5 shows a calibration curve obtained by coupling three 5-μm SEC columns with nominal exclusion limits of 2×10^3, 4×10^3, and 3×10^4 g/mol. Each is guaranteed by the manufacturer to possess greater than 60,000 plates/m under the conditions of analysis. The upper exclusion limit of 3×10^4 g/mol may still be too high for some oligomeric applications, wasting resolution on a separation region where no sample elutes. However, such a column set is appropriate for characterizing of a number of tackifiers and resin prepolymers and for addressing polymer exemption legalities. These applications are described in Section 13.5.

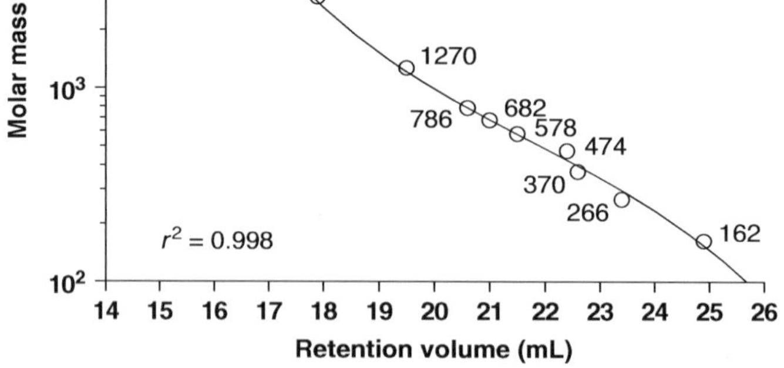

Figure 13.5 Polystyrene calibration curve on oligomeric SEC columns. Numbers of graph are peak-average molar masses (M_p) of each narrow polydispersity PS standard. Circles denote averages of triplicate determinations, with standard deviations substantially smaller than data points and, therefore, not shown. Line denotes third-order fit calibration curve. Columns, three 30 × 0.75 cm, 5-μm particle size PLgel columns; pore size 50, 100, and 500 Å; solvent, THF; temperature, 35°C; flow rate, 1 mL/min; detection, DRI. (Reprinted with permission from Ref. 20.)

Examples of oligomer separations will be given in the following sections, as we discuss the application of oligomeric SEC to individual scenarios.

13.5 SELECT APPLICATIONS OF OLIGOMERIC SEC

13.5.1 Characterization of Tackifiers, Resins, and Resin Prepolymers

Tackifiers are substances added to a base material either to increase the resistance of the material to slide against itself or another surface or to increase the resistance of the material to separate from itself. Collectively, these properties are generally referred to as *tack*. Tackifiers can be copolymers of substituted styrene, glycerol, or other esters of rosins, terpene phenolic copolymers, and so on. Characterization of tackifiers is important for both formulation and fingerprinting in the adhesives business. Figure 13.6 shows the differential refractive index (DRI) detector trace of a commercial tackifier composed of a low-M copolymer of styrene and α-methylstyrene.

One approach to quantitating tackifier components begins with the assumption that for a given tackifier, all components will always be present in the same

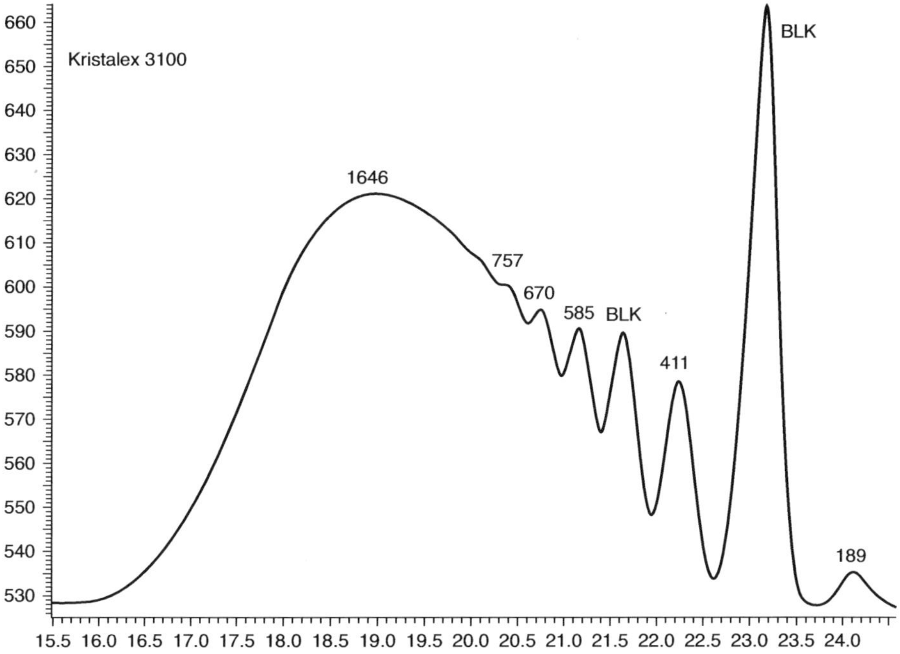

Figure 13.6 SEC analysis of Kristalex 3100 tackifier. Abscissa corresponds to retention volume (mL), ordinate to DRI response (arbitrary units). Numbers on graph represent PS-relative M_p of each mode; BLK denotes peaks from solvent blank or peaks common to tackifier and solvent blank. Experimental conditions the same as those in Figure 13.5.

proportions. If true, for a given set of experimental conditions (column set, solvent, temperature, type and wavelength of detector) the total area under all the component curves should remain constant for a given concentration of tackifier sample in solution. By accurately determining the concentrations of both the tackifier of interest and of a reference compound (which itself may be multicomponent), the relative detector response RDR of the tackifier with respect to the reference compound can be determined using a concentration-sensitive detector:

$$\mathrm{RDR} = \frac{A_{\mathrm{tack}} C_{\mathrm{ref}}}{A_{\mathrm{ref}} C_{\mathrm{tack}}} \tag{13.1}$$

A_{tack} corresponds to the total area for the tackifier of interest, C_{tack} to the solution concentration of this tackifier, and A_{ref} and C_{ref} correspond to the same values for the reference compound. Once the RDR has been determined, the concentration of the unknown (with respect to the reference) may be calculated in future determinations by analyzing a given concentration of the reference compound and comparing the total peak area of this reference sample to that of the tackifier. A list of RDRs can be compiled for various tackifiers and identification of individual tackifiers may be done by combining nuclear magnetic resonance (NMR) or infrared (IR) data with fingerprinting data from oligomeric SEC experiments, such as that shown in Figure 13.6. A similar approach to this can also be used with many of the compounds discussed below, such as resin prepolymers, antioxidant lubricant additives, and plasticizers.

Due to the highly cross-linked, network structure of epoxy resins, these compounds are by definition insoluble and will, at best, swell in select solvents. However, the so-called "prepolymers" of these resins, containing 1,2-epoxy groups which react with curing agents to form the network are soluble in most common SEC solvents, including tetrahydrofuran (THF). Figure 13.7 gives examples of the separation of two different epoxy resin prepolymers, Epikote 1007 and Epikote 1004, on four column sets differing in exclusion limit (and, hence, pore size) [21]. In each case, two 50-cm columns of each type were used. Exclusion limits were as follows. Figure 13.7a (these SEC columns will be referred to herein as column A): 1500 g/mol; Figure 13.7b (column B): 5000 g/mol; Figure 13.7c (column C): 70,000 g/mol; Figure 13.7d (column D): 400,000 g/mol. Columns A and B did not separate the higher-*M* portions of the samples. Analysis using column D showed decreased resolution compared to the separation using column C. For these prepolymers, column C appears to be the best of the four choices.

Melamine resin prepolymers present a similar case to that of the epoxy resin prepolymers. However, the melamine compounds are not soluble in as many solvents as are the epoxy compounds. Generally, dimethylformamide (DMF) or dimethyl sulfoxide (DMSO) has been used for the separation of melamine prepolymers.

Phenol–formaldehyde resins that are prepared via acid-catalyzed condensation polymerization are referred to as *novolak resins*. If the polymerization is base catalyzed, the resins are referred to as *resols*. Figure 13.8 gives examples of the SEC analysis of resins of either type.

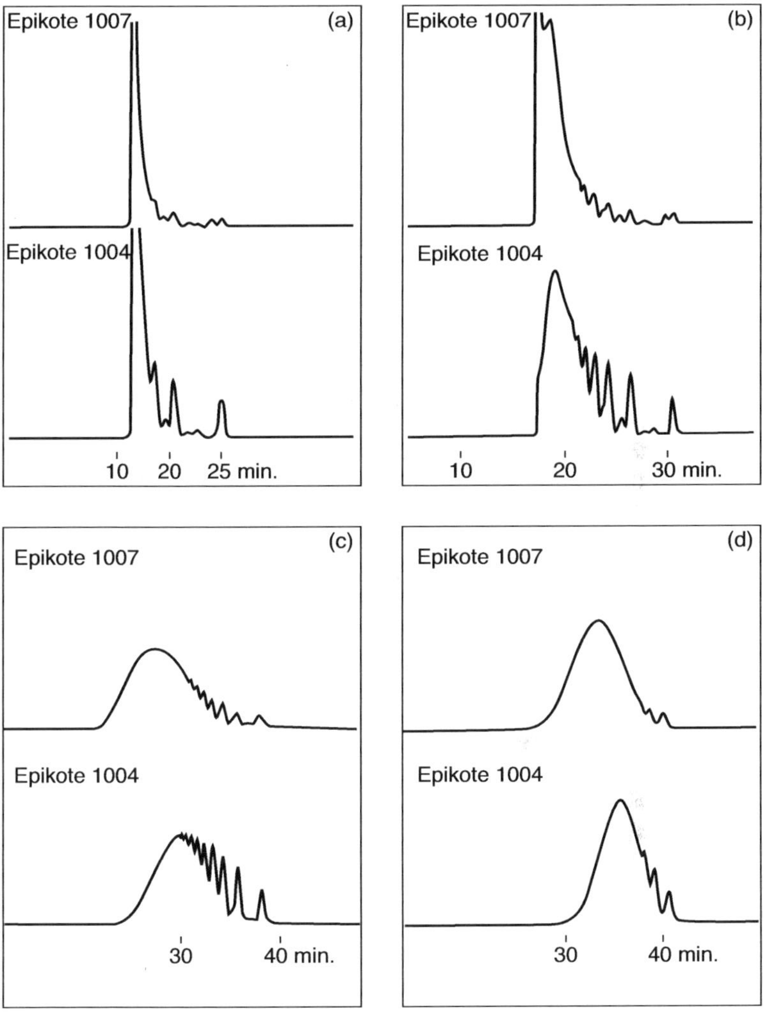

Figure 13.7 Effect of pore size on separation of epoxy resin prepolymers: Separation of two epoxy resin prepolymers, Epikote 1007 and Epikote 1004, on SEC columns with varying exclusion limits. Exclusion limits are: (a) A801, 1500 g/mol; (b) A802, 5000 g/mol; (c) A803, 70,000 g/mol; (d) A804, 400,000 g/mol. In each case, a set of two columns of each type was used. Solvent, THF; temperature, room temperature; flow rate, 1 mL/min; detector, UV (254 nm). (Courtesy of Showa Denko KK.)

13.5.2 Characterization of Antioxidant Lubricant Additives

Antioxidants are used in the lubricant industry to inhibit oxidation, bearing corrosion, and wear by inhibiting oxidation of oil and, sometimes, by forming protective films on metal surfaces. The alkylation reactions used in the preparation of methylene- and sulfur-bridged hindered phenolic antioxidants result in complex mixtures of

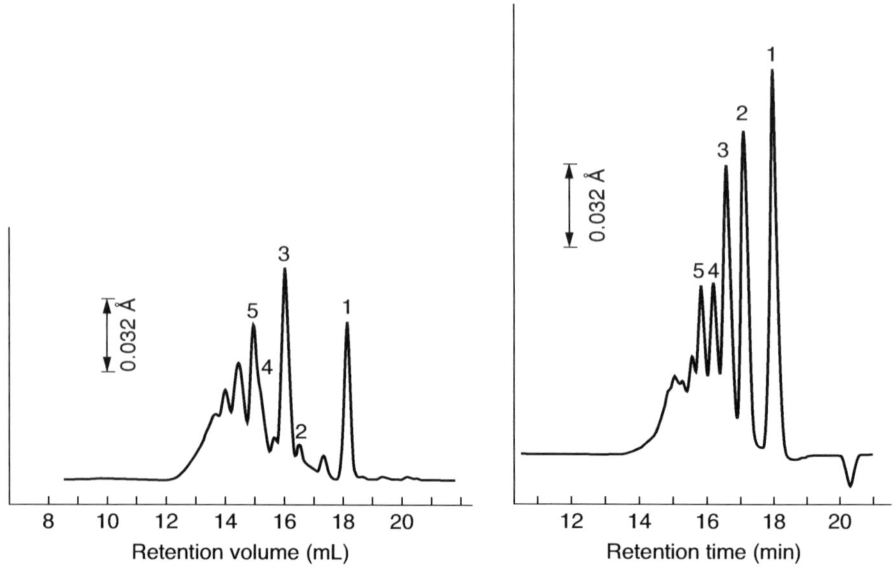

Figure 13.8 SEC analysis of phenol–formaldehyde resin prepolymers: (a) acid-catalyzed; (b) base-catalyzed. Columns, two KF802 50-cm, 5000-g/mol exclusion limit columns; mobile phase, THF; flow rate, 1 mL/min; detector, UV (254 nm). (Reprinted with permission from Ref. 22.)

oligomers, including a variety of hindered phenolic isomers. Examples are shown in Figure 13.9, along with the structures of alkylated diphenylamine antioxidants. Figure 13.10 shows the SEC separation of these various types of oligomers, with peak assignments made by comparison to chemical standards.

13.5.3 Characterization and Quantitation of Plasticizers

Plasticizers are used to lower the processing temperature of polymers and to increase the softness and flexibility of the final product, without increasing the tack. In general, plasticizers must conform to the following requirements:

1. Nonvolatility. This generally correlates with an $M \geq 300$ g/mol.
2. Similar solubility parameter (see Section 7.2.1) to that of the polymer being plasticized. The solubility parameters of several common plasticizers are given in Table 13.2.
3. Be capable of some specific interaction with the polymer, should the latter have a tendency to crystallize.
4. If point 3 above is not met, the plasticizer should not be a crystalline solid at room temperature.

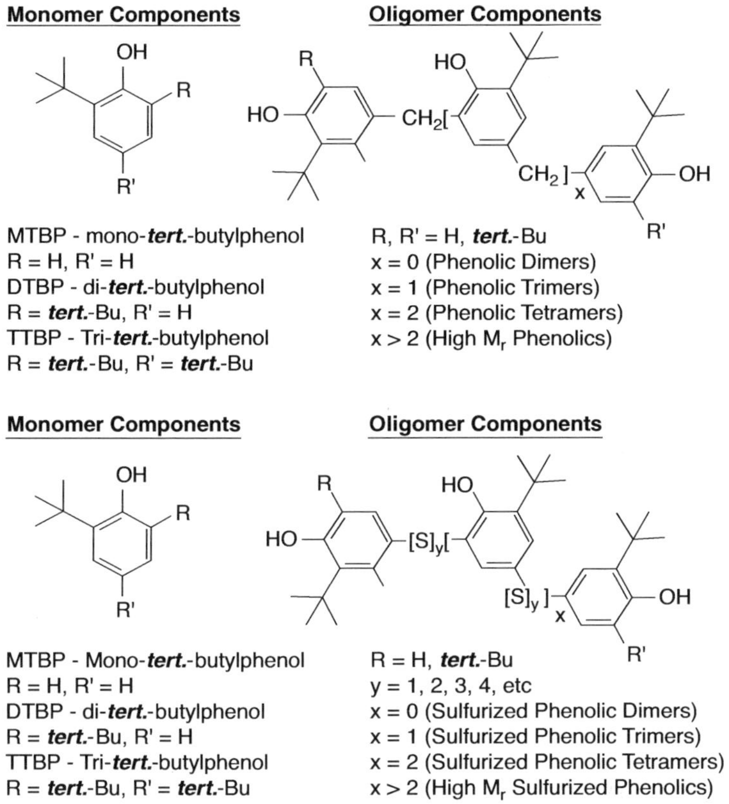

Figure 13.9 Structures of antioxidant lubricant additives: (a) methylene-bridged *tert*-butyl phenolics; (b) sulfur-bridged *tert*-butyl phenolics. (Reprinted with permission from Ref. 23.)

For poly(vinyl chloride) (PVC), the most common plasticizers are esters of phthalic acids, epoxidized soy oils, and esters of dibasic alkyl acids. To characterize a mixture of these plasticizers, dual DRI and UV detection is recommended. All plasticizers will generate a DRI response, but only esters with phthalate aromatic moieties give a 254-nm UV response.

Quantitating the plasticizer content in a sample can usually be achieved by either knowing the specific refractive index increments of both the polymer and the plasticizer or through dual dose–response calibration curves for both of these species. In many cases, plasticization is a physical (i.e., not chemical) phenomenon. Plasticizer molecules insert themselves between polymer molecules, reducing polymer–polymer contacts and increasing the free volume. Because SEC separates analytes based on physical characteristics (by size differences between analytes in solution versus column pore size, not by selective sorption of analytes onto the column packing), the small-molecule plasticizer can often be separated from the high polymer it is plasticizing using SEC. Then the relative areas of the

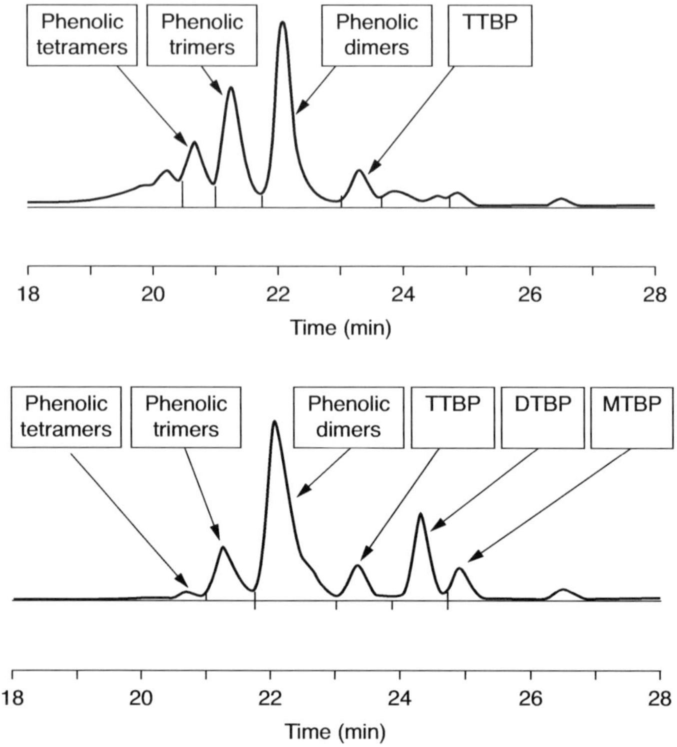

Figure 13.10 SEC analysis of antioxidant lubricant additives: oligomeric SEC analysis of compounds shown in Figure 13.9. Columns, three PLgel 30-cm, 5-μm particle size, 100-Å pore size columns; solvent, THF; temperature, room temperature; flow rate, 1 mL/min; detector, DRI. (Reprinted with permission from Ref. 23.)

chromatographic peaks of these two species are measured using a concentration-sensitive detector. After normalizing peak areas for differences in $\partial n/\partial c$, the relative area of the plasticizer peak with respect to the high polymer peak can provide the percent of plasticizer in the sample. Prior to arriving at this conclusion, the accuracy of the method should be tested using samples with a known amount of plasticizer. This will not work if there are specific interactions between plasticizer and polymer molecules. An example of this is the hydrogen-bonding interactions between PVC and tritolyl phosphate. PVC is a polar polymer and tritolyl phosphate is highly polar and a strong proton acceptor.

13.5.4 Polymer Exemption Data

Determining the oligomeric content of a sample is important for the purposes of premanufacture notification (PMN) regulations for new chemical substances as well as for import and export purposes. Many polymers are exempted from these regulations and, as such, determining whether a substance can be classified as a "polymer," from

Table 13.2 Solubility parameters of some common plasticizers

	δ	
Plasticizer	$(cal/cm^3)^{1/2}$	$MPa^{1/2}$
Paraffinic oils	7.5[a]	15.3[a]
Aromatic oils	8.0[a]	16.4[a]
Camphor	7.5	15.3
Diisooctyl adipate	8.7	17.8
Dioctyl sebacate	8.7	17.8
Diisodecyl phthalate	8.8	18.0
Dibutyl sebacate	8.9	18.2
Di(2-ethylhexyl) phthalate	8.9	18.2
Diisooctyl phthalate	8.9	18.2
Di-2-butoxyethyl phthalate	9.3	18.9
Dibutyl phthalate	9.4	19.2
Triphenyl phosphate	9.8	20.0
Tritolyl phosphate	9.8	20.0
Trixylyl phosphate	9.9	20.2
Dibenzyl ether	10.0	20.4
Triacetin	10.0	20.4
Dimethyl phthalate	10.5	21.4
Santicizer 8	11.0[a]	22.4[a]

Source: Ref. 24. Data obtained by Small's method [P. A. Small, *J. Appl. Chem.*, **3**, 71 (1953)], except for that of Santicizer 8, which was estimated from boiling-point measurements.

[a] Denotes approximate value.

a legal standpoint, is of great economic interest. *For information purposes only,* we list here the three exemptions given to polymers by the U.S. Environmental Protection Agency [7]:

1. *The (e)(1) exemption.* This concerns polymers with 1000 g/mol $\leq M_n < 10{,}000$ g/mol. Oligomer content must be less than 10% by weight below 500 g/mol and less than 25 wt% below 1000 g/mol. Polymers must also meet certain functional group criteria not discussed here.
2. *The (e)(2) exemption.* This concerns polymers with $M_n \geq 10,000$ g/mol and oligomer content less than 2% below 500 g/mol and less than 5% below 1000 g/mol.
3. *The (e)(3) exemption.* This concerns certain polyesters composed solely of listed monomers and reactants.

Other rules may apply in conjunction with the criteria above: for example, in the case of cationic or degradable polymers.

The need to determine M_n and to quantitate the percentage of the MMD below a certain molar mass value makes SEC an invaluable tool, almost ideally suited to this task. However, it must be remembered that the response of concentration-sensitive detectors is generally nonconstant in the oligomeric region (see Section 13.3.2). If a generic polymer has an M_n of 5000 g/mol, 1% of the MMD below 500 g/mol, and 3% of the MMD below 1000 g/mol, it is highly likely to fall under the (e)(1) exemption, regardless of differences in $\partial n/\partial c$ between the polymeric and oligomeric regions. However, all other factors being equal, if now 22% of the MMD appears to be below 1000 g/mol, differences in $\partial n/\partial c$ become a concern and this parameter may need to be measured for the species of interest. Alternatively, the use of dose–response calibration curves may become necessary.

13.5.5 SEC of Oligosaccharides

Size-dependent separations of oligosaccharides have usually been performed in aqueous mobile phases that include additives. In these cases, the separations are probably accompanied by non-size-exclusion effects. The latter do not normally alter the elution order, and larger DP species continue to elute earlier than lower DP oligomers of a given homologous series. Nonetheless, avoiding non-SEC effects is crucial when deriving conformational entropies from SEC data, as discussed in the next section. In this section we review the SEC analysis of oligosaccharides briefly. For other small molecules such as polypeptides and amino acids, recent reviews can be found in References 25 and 26.

If the separation is being conducted for preparative purposes only, a variety of column–eluent combinations exist, depending on the oligosaccharide series of interest. Many of these combinations are listed in Reference 27. An application of SEC to the separation of pullulan enzymatic degradation products is shown in Figure 13.11. Analytical separations of a series of maltooligosaccharides, in the

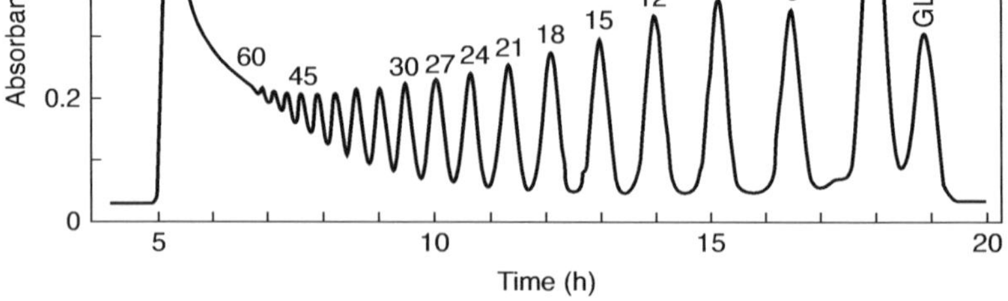

Figure 13.11 SEC of polymaltotrioses: SEC analysis of pullulan degradation products produced by the action of the enzyme pullulanase. Numbers above peaks indicate number of glucose units in each analyte. Glucose added as marker. Solvent, H_2O; temperature, 60°C; flow rate, 25 mL/h. One Bio-Gel P-6 197.6 × 1.8 cm column, 47 μm particle size. (Reprinted with permission from Ref. 28.)

polar aprotic solvent *N,N*-dimethyl acetamide (DMAc) and under aqueous conditions, are shown in Figure 13.12. In each of the latter two cases, the sugars eluted with near-ideal SEC behavior (i.e., non-SEC effects were virtually absent), as described in the next section. Also, as most manufacturers now offer high–resolution oligomeric columns for both aqueous and organic separations, up-to-date information on appropriate columns can be found on company websites.

Evaporative detectors (see Section 9.2.3) have also been used in SEC analysis of oligosaccharides. Figure 13.13 shows results of the SEC/ELSD analysis of a series of galacturonic acid (GA) oligomers. The chromatographic method was employed to quantitate the concentration of oligomers in solutions, with results checked against those from a colorimetric method.

13.5.6 Determining the Solution Conformational Entropy of Oligomers

In Chapter 2, SEC was defined as an entropically controlled technique. It stands to reason that it should thus be able to provide entropic information about analytes. SEC is a solution-based method, and as mentioned in Section 2.3, there are fewer conformational degrees of freedom for an analyte inside the pores of a column packing than outside the pores. For these reasons, the type of entropy measured by SEC is a solution conformational entropy.

Solution conformational entropy data is of great interest in a number of fields. Many molecular recognition functions of the lock-and-key type are based on flexibility differences among similar ligands or target molecules. Small molecules, such as surface carbohydrates on bacteria or parasites, with the appropriate conformation or flexibility can access target sites in a host molecule such as a protein. The interaction between the bacterial or parasitic surface sugars and the appropriate protein surface glycosilation site can initiate infection in the host organism. Traditionally, many of these problems have been approached using computer modeling at various levels of approximation. Quantitative, experimentally determined data concerning the flexibility of given analytes in solution are relatively sparse.

To date, this application of SEC has occurred only for oligosaccharides, largely due to the availability of high-resolution oligomeric columns. In oligosaccharides, the solution conformational entropy ΔS can be influenced by anomeric configuration, glycosidic linkage, degree of polymerization, or intramolecular hydrogen bonding, among others. The manner in which these properties affect ΔS has a consequent effect on structure–property interactions of the carbohydrates. These interactions include the selectivity of DNA aptamers for one disaccharide over other, seemingly similar ones; the structural targeting of recognition events in glycolipids and glycoproteins; and generation of the sweet response by maltose, by selective accession of this disaccharide to the sweetness receptor on the tongue.

Calculation of the standard conformational entropy difference between mobile and stationary phases for oligomers in solution is based on the retention times of the

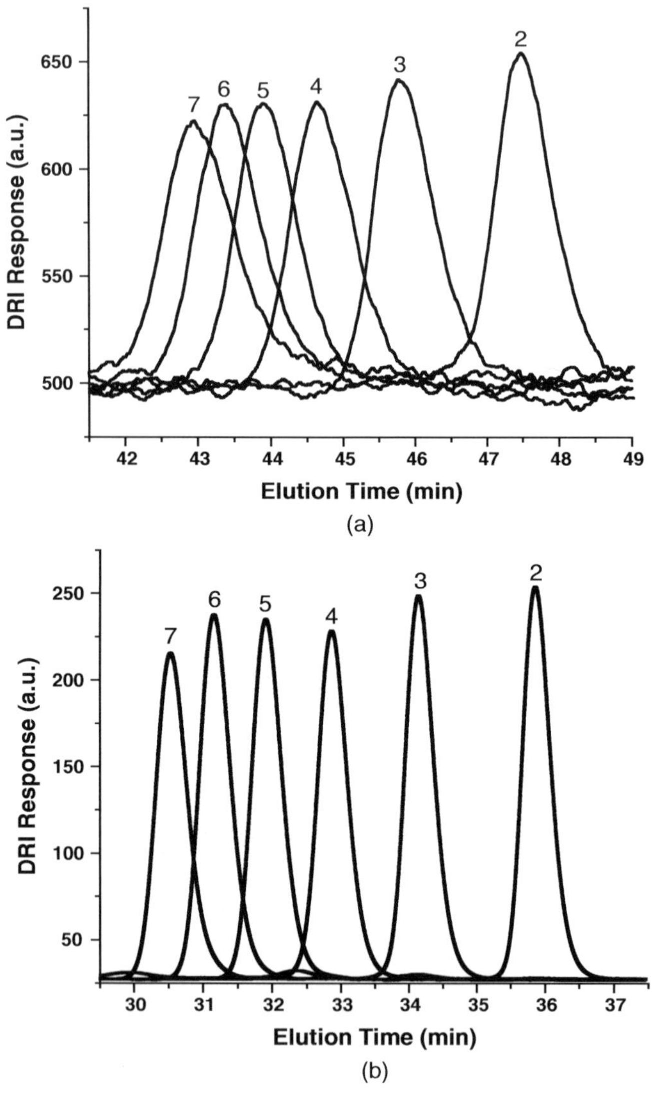

Figure 13.12 SEC of maltooligosaccharides in organic and aqueous solvents: overlay of chromatograms for the series maltose (DP 2) through maltoheptaose (DP 7), where the numbers above the elution profiles correspond to the DP of each sugar. (a) Organic conditions: columns, four PLgel 30 × 0.75 cm, 5-μm particle size, 50-Å pore size columns; solvent, *N,N*-dimethylacetamide; flow rate, 0.5 mL/min; temperature, 50°C; detector, DRI. (b) Aqueous conditions: columns, four Ultrahydrogel 30 × 0.75 cm, 6-μm particle size, 120-Å pore size columns; solvent, H_2O; temperature, 37°C, pH, 7.4; flow rate, 1.0 mL/min; detector, DRI. (Adapted from Refs. 9 and 16.)

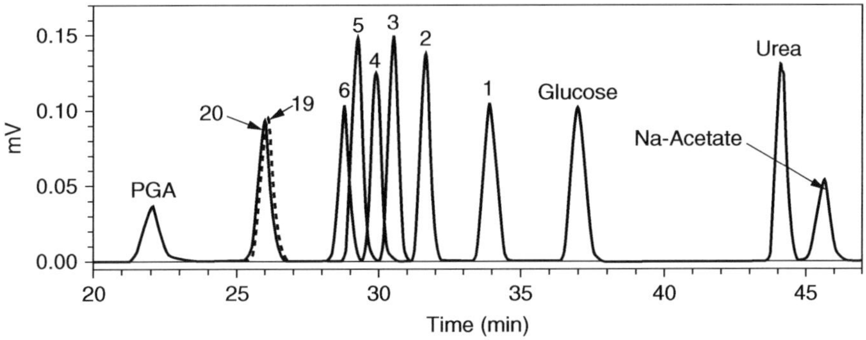

Figure 13.13 SEC/ELSD analysis of galacturonic acid (GA) oligomers: overlay of SEC/ELSD traces of 1% urea, 0.1% Na-acetate, glucose, and GA oligomers. Numbers above peak indicate degree of polymerization of GA oligomers. PGA, polygalacturonic acid. Columns, two TSK-Gel SE G3000PW 60 × 0.75 cm, 10-μm particle size columns, plus one 7.5 × 0.75 cm, 12.5-μm particle size guard column; flow rate, 1 mL/min; ELSD nebulizer and evaporation chamber temperatures, 40 and 55°C, respectively; solvent, 40 mM ammonium acetate; pH, 3.7. (Reprinted with permission from Ref. 29.)

peak maxima, V_R, as well as on the solute distribution coefficient, K_{SEC}. As seen in Section 2.3, these parameters are related via

$$K_{SEC} = \frac{V_R - V_0}{V_i - V_0} \tag{13.2}$$

where V_0 is the void volume of the columns (measured with one or more totally excluded analytes) and V_i is the total column volume (measured with one or more analytes which exhibit total permeation). The internal pore volume of the system, V_p, is defined as the difference between V_i and V_0.

To obtain valid ΔS data, analytes must elute at near-ideal SEC conditions (i.e., in the virtual absence of enthalpic effects such as adsorption, partition, etc.). To check for this, separations should be run at different temperatures. If enthalpic effects are absent, K_{SEC} should be invariant with changing temperature over a wide range of the latter. If K_{SEC} varies appreciably with changing temperature, enthalpic effects are present and separation has not proceeded by a strict size-exclusion mechanism. Note that it is important to use a flow-rate marker (e.g., toluene, acetone) in these types of experiments to correct for minor flow-rate variations due to pump pulsations and other factors.

The relation between ΔS and the solute distribution coefficient K_{SEC} is given by

$$\Delta S = R \ln K_{SEC} \tag{13.3}$$

where R is the molar gas constant in J/mol · K. The standard entropy difference, which will be negative (i.e., $-\Delta S$), denotes the difference between the conformational entropy of the analytes in the flowing mobile phase outside the pores of the SEC column packing versus the entropy of the analytes in the stagnant mobile phase

Table 13.3 Solution conformational entropy of mono- and disaccharides in DMAc/0.5% LiCl, as determined by SEC[a]

Disaccharide	Anomeric Configuration and Glycosidic Linkage	$-\Delta S$ (J/mol · K)
Maltose	α-(1 → 4)	15.51 ± 0.01
Isomaltose	α-(1 → 6)	15.73 ± 0.01
Melibiose	α-(1 → 6)[b]	15.67 ± 0.02
Trehalose	β,β-(1 → 1)	16.07 ± 0.01
Cellobiose	β-(1 → 4)	15.57 ± 0.01
Gentiobiose	β-(1 → 6)[c]	16.10 ± 0.02
Glucose	α	12.92 ± 0.01
Galactose	α	12.77 ± 0.01

Source: Ref. 30.

[a]Using a series of four PLgel 30 × 0.75 cm, 5-μm particle size, 50-Å pore size column; flow rate, 0.500 mL/min; temperature, 80°C; detector, DRI.

[b]Melibiose is a galactopyranosyl glucopyranose; all other disaccharides are glucopyranosyl glycopyranoses.

[c]Mixture of 91% β-anomer, 8% α-anomer, as reported by the manufacturer.

inside the pores. As explained in Section 2.4, the use of the negative sign stems from the fact that solute permeation in SEC is associated with a decrease in conformational entropy, due to the more limited mobility of analytes inside the pores.

The above method can be applied only to perfectly monodisperse species, where the width of the eluted peak is due only to chromatographic band broadening and not to analyte polydispersity. Calculating ΔS from SEC data should only be done for discrete oligomers or for perfectly monodisperse proteins or peptides, assuming that all of these can be separated under near-ideal SEC conditions. Converting the distribution of retention times into a "ΔS distribution" cannot be done accurately. This is because it is impossible to unambiguously decouple analyte polydispersity from chromatographic band broadening.

Table 13.3 illustrates the influence of anomeric and epimeric configuration and of glycosidic linkage on ΔS of select mono- and disaccharides [30]. Figure 13.14 shows how $-\Delta S$ varies with degree of polymerization for a series of linear malto- and cellooligosaccharides [9,16]. In accordance with the temperature-independent separation mechanism described above, the K_{SEC} of glucose, maltose, and cellobiose in DMAc/0.5% LiCl changed by 1 part per thousand or less when the temperature of this experiment was varied from 50 to 80°C in DMAc/LiCl, or by less than 4 parts per hundred when the temperature was varied from 37 to 50°C in aqueous solvent (see Table 2.1).

13.5.7 Determining Molar Masses of Oligomers by SEC/MALS

SEC/MALS (Section 9.3.1) has been applied to the separation of very low M oligomers of styrene, and has even been able to detect a single, monomeric styrene unit (M = 104 g/mol), as shown in Figure 13.15. (Note that under the particular

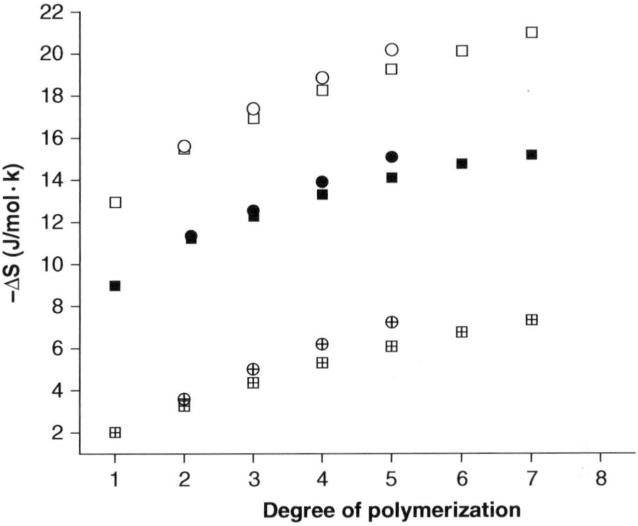

Figure 13.14 Using SEC to measure solution conformational entropy $-\Delta S$ as a function of degree of polymerization. Analytes: linear maltooligosaccharides (circles) and cellooligosaccharides (squares). Open symbols denote results in DMAc/0.5% LiCl, filled symbols denote results in neat DMAc, hatched symbols denote results under aqueous conditions described in Figure 13.12. Organic solvent data obtained at 50°C; otherwise, conditions for DMAc and DMAc/LiCl experiments are identical to those for data in Table 13.3. Results are averages of at least six replicates, with standard deviations substantially smaller than data points and therefore not shown. (Adapted from Refs. 9 and 16.)

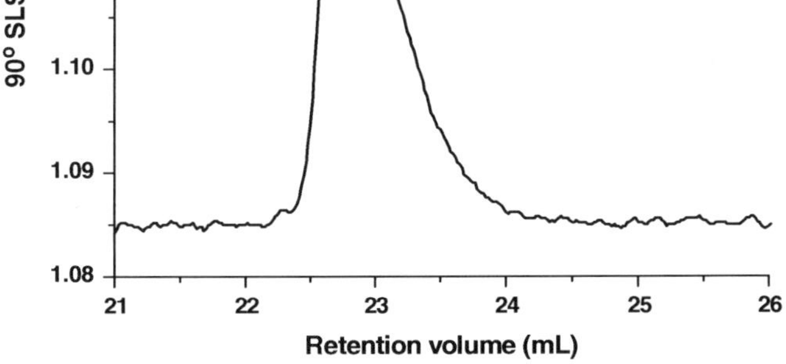

Figure 13.15 SEC/MALS of styrene monomer: 90° static light-scattering signal from a MALS photometer of the SEC analysis of nonfunctionalized styrene monomer, $M = 104$ g/mol. Columns, series of two OligoPore 30 × 0.75 cm, 6-μm particle size columns; solvent, DMAc/0.5% LiCl; temperature, 35°C; flow rate, 1.0 mL/min; injected mass, 10 mg. (Adapted from Ref. 13.)

Table 13.4 Comparison of SEC/MALS, MALDI-TOF-MS, VPO, and ^{1}H-NMR for molar mass determination of poly(diisopropyl trimethylene-1,1-dicarboxylate) oligomers[a]

Sample Number	M_n (× 10^3 g/mol)				M_w/M_n	
	NMR	SEC/MALS	VPO	MALDI	SEC/MALS	MALDI
1	3.04	2.49	2.42	—	1.13	—
2	6.04	5.96	5.99	5.6092	1.08	1.029
3	7.91	6.56	6.56	6.6249	1.08	1.027
4	9.2	8.31	8.31	8.2898	1.05	1.021
5	>10.0	9.10	9.10	9.4172	1.06	1.018

Source: Ref. 31.

[a]See Reference 31 for details of oligomer synthesis. SEC analysis conducted in chloroform at 40°C and 1 mL/min flow rate, using a K80M and K802.5 column set.

experimental conditions employed, styrene monomer exhibited non-size-exclusion behavior, as explained in Section 13.3.2 and shown in Figure 13.3.)

While the above example serves as proof of principle that SEC/MALS may be used in the oligomeric region, it does not address the accuracy of SEC/MALS in determining the molar mass of oligomers. In this regard, several recent studies have compared the molar-mass-determination capability of SEC/MALS to that of other techniques. In one study, M_n determined by SEC/MALS were compared to values from ^{1}H-NMR, vapor-pressure osmometry (VPO), and MALDI-TOF-MS (Section 10.2.2) for a number of poly(diisopropyl trimethylene-1,1-dicarboxylate) oligomers. While experimental error of up to 20% was found in the NMR data, SEC/MALS results were in excellent agreement with those from VPO and MALDI-TOF-MS measurements. Results of these experiments are given in Table 13.4.

A more recent investigation with SEC/MALS used certified polystyrene reference materials in the range 500 to 2400 g/mol. The benchmark method for obtaining molar masses of oligomers was supercritical fluid chromatography (SFC). This method had a relative standard uncertainty of less than 1%. Against this were compared SEC/MALS, SEC with PS-relative calibration (Section 8.2.1), off-line MALS (Section 9.3.3), ^{1}H-NMR, MALDI (Section 10.2.2), and values calculated for an ideal Poisson distribution. Results for these polystyrene standards are given in Table 13.5. SEC with PS-relative calibration proved to be quite accurate, even without band broadening and $\partial n/\partial c$ corrections. It should be noted, however, that this result is conditional upon the calibrant having the same chemistry and architecture as the analyte. SEC/MALS also proved highly accurate, although disagreements between this technique and SFC increased with decreasing molar mass, presumably due to $\partial n/\partial c$, density scattering, and optical anisotropy effects. Although off-line MALS is considered a yardstick technique for nonaggregating high-M polymers, it did not fare very well in this comparison at the oligomeric level. MALDI-TOF-MS appeared to provide accurate molar masses of the larger DP oligomers, but overestimated molar masses at lower DP. Also, polydispersities from MALDI were found to be narrower than the accepted values. Finally, M_n results by ^{1}H-NMR were found to

Table 13.5 Average molar masses and polydispersity of certified PS reference materials by SFC, SEC, SEC/MALS, MALS, MALDI-TOF-MS, and ^{1}H-NMR

Analytical	PS 500			PS 1000			PS 2400		
Method	M_n	M_w	M_w/M_n	M_n	M_w	M_w/M_n	M_n	M_w	M_w/M_n
SFC[a]	433.2 ± 3.9	497.7 ± 4.9	1.149 ± 0.022	881.0 ± 5.5	1007.2 ± 6.8	1.144 ± 0.014	2307 ± 9	2423 ± 10	1.050 ± 0.008
SEC[a,b]	426.5 ± 3.8	494.9 ± 4.1	1.160 ± 0.002	868.4 ± 4.8	996.7 ± 5.4	1.148 ± 0.002	2292 ± 15	2415 ± 16	1.054 ± 0.001
SEC/MALS[c]	538 ± 39	553 ± 26	1.053 ± 0.034	832 ± 16	917 ± 9	1.102 ± 0.027	2265 ± 33	2522 ± 17	1.113 ± 0.013
MALS[d]	—	—	—	—	935	—	—	2412	—
MALDI[c]	584 ± 15	602 ± 19	1.032 ± 0.007	1020 ± 11	1086 ± 15	1.065 ± 0.007	2281 ± 74	2345 ± 77	1.0280 ± 0.0003
^{1}H-NMR[c]	447.4 ± 0.2	—	—	885.0 ± 0.8	—	—	2357 ± 2	—	—
Poisson[e]	—	—	1.174	—	—	1.103	—	—	1.043

Source: Ref. 32.

[a]Uncertainty is expressed as standard deviation combined with uncertainty evaluated under ISO GUM.

[b]Applying first-order PS-relative calibration curve. Uncertainty from band broadening not included. Columns, set of two TSKgel-G2500H_{XL} columns; solvent, THF; temperature, 40°C; flow rate, 1.0 mL/min; detector, DRI; injection volume, 100 μL; injection concentration, 1 mg/mL.

[c]Uncertainties reflect experimental standard deviations from repeat measurements.

[d]M_w values corrected for relative $\partial n/\partial c$ coefficient β, where $\beta(M) = 1 - (88.53/M)$.

[e]M_w/M_n calculated for an ideal Poisson distribution.

be quite accurate. Results from this technique, however, are contingent upon accurate chemical characterization of the analytes.

13.6 OPTIMIZING RESOLUTION IN OLIGOMERIC SEC

As mentioned in Section 13.3.2, oligomeric SEC in many ways resembles traditional small-molecule, enthalpically controlled liquid chromatography. Whereas the latter is a high-resolution, high-peak-capacity technique, SEC generally is not. However, many of the same approaches toward optimizing resolution in traditional HPLC can be used in oligomeric SEC when separating individual components or families of components in a multicomponent mixture.

Factors affecting resolution in oligomeric SEC, many of which were discussed in Chapter 4, are:

1. *Choice of solvent.* Often limited because of the analyte, the column packing material, the detection method, or a combination of these.
2. *Temperature.* Often limited by the choice of solvent, the equipment, or both.
3. *Flow rate.* Can be limited by the backpressure generated. Can also be limited by the possibility of generating turbulence in the columns, although this is unlikely to happen at the flow rates commonly employed in analytical SEC.
4. *Particle size and particle-size distribution.* The smallest possible particle size should be used to get the highest possible column efficiency and resolution. (As with many of the factors listed, this is not necessarily true of polymeric SEC, especially for ultrahigh-*M* polymers, where flow-induced degradation may occur during elution [33]; see Section 7.2.3.) The particle-size distribution is not usually a factor over which the user has much control. Some manufacturers will provide this information upon request.
5. *Packing of column.* Again, for commercially available columns this is not a factor that the user can control. However, as in polymeric SEC, care should be taken to utilize the columns in the direction of flow specified by the manufacturer. This is usually the direction in which the columns were packed. Failure to do so may cause irreparable damage to the columns.
6. *Pore size and pore-size distribution.* A pore size or pore sizes covering the range of interest as closely as possible to should be chosen. This prevents wasting separation power on a region of the separation space that is not occupied by the analytes. Even "individual" (i.e., not "mixed") pore-size columns still possess a distribution of pore sizes around the average pore size advertised. The user cannot normally control the pore-size distribution, and manufacturer information regarding the breadth and/or modality of the distribution is relatively scarce.
7. *Column length and number of columns.* Analytical columns come in discrete lengths, most commonly 30 and 50 cm. An increased number of columns of

the same type (length, pore size, particle size) will result in increased resolution, due to the increase in number of theoretical plates. The choice of column length can be limited by the equipment available (e.g., a 50-cm column may not fit into a particular column oven) and the number of columns used will be limited by the backpressure generated as well as by practicality issues such as analysis time.

The effect of pore-size distribution on resolution can be seen in Figure 13.16a and b. Figure 13.16a shows the analysis of an epoxy resin prepolymer on a column with a broad mix of pore sizes. This column separates over a broad molar mass range of approximately 200 to 2,000,000 g/mol. Figure 13.16b shows the analysis of the same compound on a column with a narrower mix of pore sizes. This second column covers a narrower molar mass range of approximately 200 to 30,000 g/mol, more appropriate to the particular analyte. Higher resolution is achieved by using a column with a larger concentration of pore sizes in the region of interest. Another example of the effect of pore-size distribution on resolution is seen in Figure 13.7.

The effect of number of columns used on resolution is seen in Figure 11.16b to d. This figure shows the analysis of an epoxy resin prepolymer using one, two, and three identical columns. The increase in resolution when going from one to two columns is relatively large, less so when going from two to three columns.

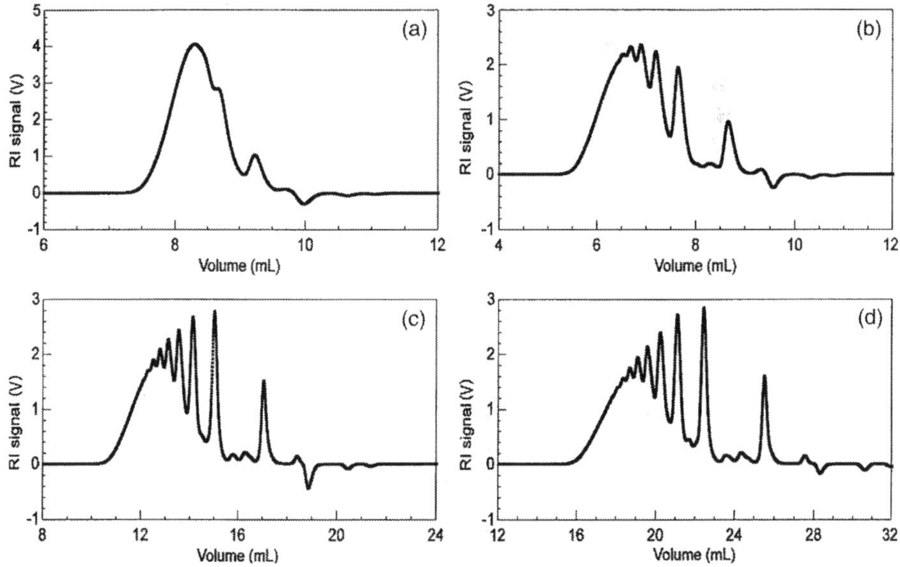

Figure 13.16 Effect of pore-size distribution and number of columns on resolution: separation of an epoxy resin prepolymer on mixed-pore-size columns. (a) Using one PLgel Mixed-C column with broad separation range, about 200 to 2,000,000 g/mol; (b–d) using one, two, and three PLgel Mixed-E columns, respectively, each with a separation range of about 200 to 30,000 g/mol. All columns 30 × 0.75 cm; detector, DRI. (Reprinted with permission from Ref. 34.)

Optimizing resolution in oligomeric SEC is discussed in References 25 and 35. Recently, a combined virtual-modeling and multivariate-optimization approach has been applied to HPLC separations [36]. This approach takes into account many of the factors mentioned above. Applying this to the oligomeric SEC separation of multicomponent mixtures appears feasible.

REFERENCES

1. A. M. Striegel, *Anal. Bioanal. Chem.*, **390**, 303 (2008).
2. G. Odian, *Principles of Polymerization*, 3rd ed., Wiley-Interscience, New York, 1991.
3. M. P. Stevens, *Polymer Chemistry: An Introduction*, 2nd ed., Oxford University Press, Oxford, UK, 1990.
4. K. J. Saunders, *Organic Polymer Chemistry*, Chapman & Hall, London, 1973.
5. B. Vollmert, *Polymer Chemistry*, Springer-Verlag, New York, 1973.
6. A. Rudin, *The Elements of Polymer Science and Engineering*, Academic Press, New York, 1982.
7. *US EPA Polymer Exemption Guidance Manual*, EPA 744-B-97-001, June 1997.
8. Y. Ding, A. Kisliuk, and A. P. Sokolov, *Macromolecules*, **37**, 161 (2004).
9. M. A. Boone, H. Nymeyer, and A. M. Striegel, *Carbohydr. Res.*, **343**, 132 (2008).
10. A. M. Striegel, *Carbohydr. Polym.*, **34**, 267 (1997).
11. A. M. Striegel, *J. Chil. Chem. Soc.*, **48**, 73 (2003).
12. A. A. Gridnev, S. D. Ittel, and M. Fryd, *J. Polym. Sci. A*, **33**, 1185 (1995).
13. A. M. Striegel and D. B. Alward, *J. Liq. Chromatogr. Rel. Technol.*, **25**, 2003 (2002). See the Erratum in *J. Liq. Chromatogr. Rel. Technol.*, **26**, 157 (2003), in which there is a typo: The value of $[\eta]_w$ for PE 282 in TCB at 135°C should be +0.0036 dL/g).
14. R. R. Chance, S. P. Baniukiewicz, D. Mintz, G. ver Strate, and N. Hadjichristidis, *Int. J. Polym. Anal. Charact.*, **1**, 3 (1995).
15. T. Kawaguchi, M. Osa, T. Yoshizaki, and H. Yamakawa, *Macromolecules*, **37**, 2240 (2004).
16. M. A. Boone and A. M. Striegel, *Macromolecules*, **39**, 4128 (2006).
17. E. Sabadini, T. Cosgrove, and F. do Carmo Egídio, *Carbohydr. Res.*, **341**, 270 (2006).
18. A. M. Striegel, in *Multiple Detection in Size-Exclusion Chromatography*, ACS Symp. Ser. 893, A. M. Striegel, ed., American Chemical Society, Washington, DC, 2005, Chap. 4.
19. C.-S. Wu, ed., *Column Handbook for Size Exclusion Chromatography*, Academic Press, San Diego, CA, 1999.
20. S. Dong, R. Rodgers, A. G. Marshall, and A. M. Striegel, in preparation.
21. H. Suzuki and S. Mori, in Ref. 19, Chap. 6.
22. S. Mori, *J. Liq. Chromatogr.*, **9**, 1329 (1986).
23. S. V. Greene and V. J. Gatto, *J. Chromatogr. A*, **841**, 45 (1999).
24. J. A. Brydson, *Plastics Materials*, 5th ed., Butterworth, London, 1989, Chap. 5.
25. S. S. Huang, in *Handbook of Size Exclusion Chromatography and Related Techniques*, 2nd ed., C.-S. Wu, ed., Marcel Dekker, New York, 2004, Chap. 17.

26. A. J. Alpert, in Ref. 19, Chap. 8.
27. S. C. Churms and A. M. Stephen, in *Methods in Carbohydrate Chemistry,* Vol. IX, J. N. BeMiller, R. L. Whistler, and D. H. Shaw, eds., Wiley, New York, 1993, Chap. 9.
28. M. John, J. Schmidt, C. Wandrey, and H. Sahm, *J. Chromatogr.*, **247**, 281 (1982).
29. R. G. Cameron, A. T. Hotchkiss, S. W. Kauffman, and K. Grohmann, *J. Chromatogr. A*, **1011**, 227 (2003).
30. A. M. Striegel, *J. Am. Chem. Soc.*, **125**, 4146 (2003). See the Erratum in *J. Am. Chem. Soc.*, **126**, 4740 (2004).
31. T. Xie, J. Penelle, and M. Verraver, *Polymer*, **43**, 3973 (2002).
32. T. Saito, M. A. Lusenkova, S. Matsuyama, K. Shimada, M. Itakura, K. Kishine, K. Sato, and S. Kinugasa, *Polymer*, **45**, 8355 (2004).
33. A. M. Striegel, *J. Liq. Chromatogr. Rel. Technol.*, **31**, 3105 (2008).
34. S. Podzimek, in *Multiple Detection in Size-Exclusion Chromatography*, ACS Symp. Ser. 893, A. M. Striegel, ed., American Chemical Society, Washington, DC, 2005, Chap. 5.
35. F. V. Warren, Jr., B. A. Bidlingmeyer, H. Richardson, and J. Ekmanis, in *Size Exclusion Chromatography*, ACS Symp. Ser. 245, T. Provder, ed., 1984, Chap. 11.
36. T. L. Chester and S. O. Teremi, *J. Chromatogr. A*, **1096**, 16 (2005).

14

SEC IN 2D-LC SEPARATIONS

14.1 INTRODUCTION

As seen in Figures 1.3 and 1.4 and Table 1.1, macromolecules can possess distributions in a number of properties. Moreover, several of these distributions may coexist in a polymer simultaneously. While the individual distributions may contribute independently to the end-use properties of the material, these distributions may also combine synergistically. The latter is, for example, the case with the distribution of chemical composition as a function of molar mass (CCD × MMD), which can affect blending, plasticization, and mechanical properties of polymers. Indeed, understanding of these synergistic combinations is still in its infancy.

Two-dimensional (2D) chromatographic techniques are used to build understanding of the synergies involved. The premier status of SEC for determining molar mass distributions and, more generally, for separating sample components according to size in solution, has made SEC a mainstay in 2D-LC separations of polymers. In this chapter we review the fundamentals of 2D-LC polymer analysis, in which size exclusion is one of the separation dimensions, covering data interpretation and design of experiments. Select applications help exemplify many of the techniques that couple to SEC and the wealth of information these couplings provide.

Modern Size-Exclusion Liquid Chromatography: Practice of Gel Permeation and Gel Filtration Chromatography, Second Edition
By André M. Striegel, Wallace W. Yau, Joseph J. Kirkland, and Donald D. Bly

14.2 PRINCIPLES OF 2D POLYMER SEPARATIONS

This chapter deals solely with the information obtained when SEC is used in "comprehensive" 2D-LC [1,2]. Here, every fraction from the first separation dimension is transferred to the second dimension using, most commonly, an automated switching valve. Advantages of the comprehensive methods include maximal sample use; quantitative interpretation of results; increased resolution, peak capacity, and precision; and no need for fraction reconcentration and/or manual handling. When polymers are prone to oxidation or degradation, the latter advantage weighs most heavily.

A recent comprehensive approach, which we also discuss, is the use of a stop-flow valve. We do not delve here into off-line approaches or into the linear approaches in which one or a few "heart cuts" from the first dimension are collected and subsequently introduced into the second-dimension column. The off-line approach suffers from potential problems with sample contamination, loss, and/or degradation, from concerns regarding repeatability, and from the fact that the methods are quite labor intensive and time consuming. Linear methods only serve to characterize discrete portions of a polymeric sample, but do not provide the more complete characterization provided by the comprehensive methods. It should be noted that the linear ("heart-cutting") approach is usually abbreviated as LC-LC, whereas the comprehensive approach is usually denoted by the term LC $\times$ LC. The nomenclature and conventions of comprehensive multidimensional chromatography are treated in Reference 3. A main drawback of the LC $\times$ LC methods is that they are often quite lengthy.

In addition to the molar mass distribution, a polymer may possess distributions in one or more other properties, such as chemical composition, block length, and functionality type [4]. To "deconvolute" the various distributions from each other, multidimensional separations are usually needed. Ideally, but not necessarily, each dimension of the separation is selective toward one particular property. To maximize deconvolution, each separation dimension should not only be selective but also specific. For example, SEC separates analytes based on their size in solution (selectivity with respect to size). Although for homopolymers this size can be related to M, this is not necessarily the case for copolymers. For complex polymers, SEC separations are not M-specific. In a sample of random copolymer A_xB_y, two chains of different lengths and also with different stoichiometries may coelute in a SEC experiment. A technique that is selective toward the A/B ratio will separate polymeric components based on chemical composition, provided that the separation is M-independent. This is because all components with the same chemical composition need not have the same molar mass. Further separation of each group of components individually using SEC will provide a "true" MMD, without a chemical composition bias. Combining the information from both techniques will not only provide both the chemical composition distribution (CCD) and the MMD but will also show how these distributions depend on each other within the polymer, a combined CCD $\times$ MMD of the analyte. Obviously, each different chemistry (or functionality, or block length, etc.) present in the polymer will result in an additional component (peak) that needs to be separated from the other components (peaks) in the first dimension and then transferred to the second dimension for further separation based on a different property.

Multidimensional separations provide a benefit not only in terms of being able to distinguish individual components of a multicomponent sample based on analyte dimensionality, but also in terms of multiplicatively increased peak capacity [5]. For N-dimensional sequential separations (i.e., for an ND system), we propose here that this extended peak capacity be expressed quantitatively by

$$n_{\text{total}} = n_1 \prod_{i=2}^{N} n_i \sin \vartheta_{i-1,i} \qquad \text{for } N \geq 2 \text{ and } i-1 \neq i+1 \tag{14.1a}$$

where n_{total} is the total peak capacity of the system (theoretical maximum or theoretical upper limit of the system peak capacity), n_i the peak capacity in dimension i, and $\vartheta_{i\text{-}1,i}$ the separation angle between two consecutive dimensions. Separation dimensions (of which there must be at least two, i.e., $N \geq 2$) must be coupled to each other sequentially, and separation dimension $i - 1$ must be different from separation dimension $i + 1$. For a 2D system, Equation 14.1a reduces to

$$n_{\text{total}} = n_1 n_2 \sin \vartheta \tag{14.1b}$$

How to assign value quantitatively to the separation angle ϑ is discussed in Section 14.2.1.

Figure 14.1 shows a generic example of the advantages of 2D-LC separations [2]. As can be seen, 2D analyses have the ability to differentiate between samples that appear identical when examined by either of the individual methods. Because different chromatographic techniques separate components based on criteria specific to each technique, their coupling allows the determination of a number of macromolecular distributions, many as a function of one another. Types of couplings, and the information obtained therefrom, are given in Table 14.1. In Section 14.6 we provide specific examples of various applications of 2D-LC of polymers where SEC is one of the dimensions of separation.

With the notable exception of the recent stop-flow valve experiments discussed in Section 14.5, the use of LC in the first dimension and SEC in the second dimension (i.e., LC × SEC) is preferred over SEC × LC. This is due to reasons related to the experimental setup (e.g., difficulties in running fast LC gradients in the second dimension) as well as to selectivity. For the latter, because SEC separates analytes based on hydrodynamic volume, eluting species are not monodisperse with respect to chemical composition, functionality type, and so on. As such, a first dimension non-SEC separation based on one of these properties (property X), followed by an SEC separation based on hydrodynamic volume, affords the best opportunity to determine the (property X distribution) × (molar mass distribution), with minimized bias. The advantages and disadvantages of placing SEC as the first dimension versus the second dimension are summarized in Table 14.2. References 1, 2, and 6 to 8 provide recent reviews of 2D-LC analysis of synthetic polymers.

14.2.1 Separation Angle and Percent Synentropy

Assigning a value to the separation angle $\vartheta_{i-1,i}$ in Equation 14.1 is not straightforward. In what follows we omit the subscript $i - 1,i$, with the understanding that ϑ

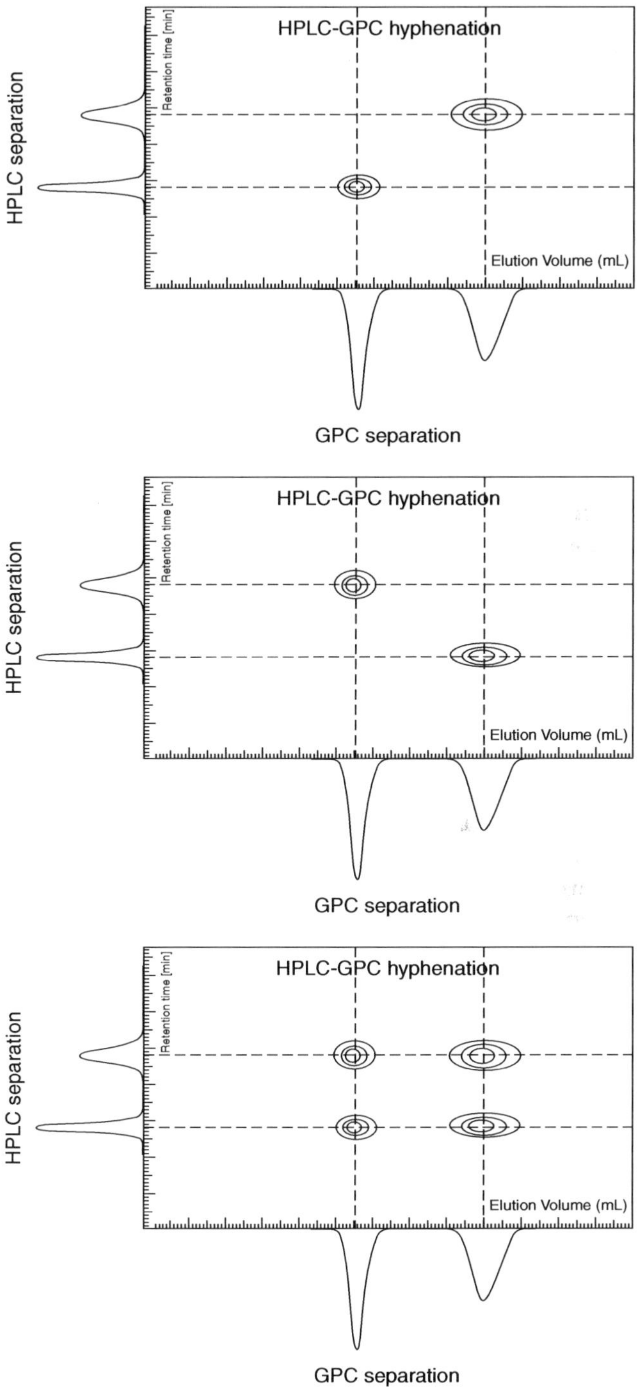

Figure 14.1 Advantage of 2D-LC separations. The 2D-LC separations (shown as a 2D contour map) show sample differences not evidenced by the individual HPLC or SEC (GPC) separations (shown as chromatograms on the ordinates and abscissas, respectively). (Reprinted with permission from Ref. 2.)

Table 14.1 2D-LC of polymers (with SEC as one dimension) for characterizing macromolecular distributions

First Dimension[a]	Second Dimension[a]	Macromolecular Distribution[b]
HPLC	SEC	MMD × CCD
(NP- or RP-)		MMD × Topology
		MMD × FTD
LCCC	SEC	MMD × FTD
		MMD × LCBD
		MMD × CCD
		MMD × BLD[c]
PFC	SEC	BLD[c]
		BND[c]
HOPC	SEC	TGD[d]
		MMD
TREF (or SEC)	SEC (or TREF)	MMD × LCBD × SCBD[e]
GPEC	SEC	MMD × CCD
		FTD[d]

[a]HPLC, high-performance liquid chromatography; NP, normal phase; RP, reversed phase; LCCC, liquid chromatography at the critical condition; PFC, phase fluctuation chromatography; HOPC, high-osmotic pressure chromatography; TREF, temperature-rising elution fractionation; GPEC, gradient polymer elution chromatography.

[b]MMD, molar mass distribution; CCD, chemical composition distribution; FTD, functionality type distribution; LCBD, long-chain branching distribution; BLD, block length distribution; BND, block number distribution; TGD, terminal group distribution.

[c]For block copolymers.

[d]For telechelic polymers.

[e]For polyolefins, using MALS/VISC/DRI/IR detection.

denotes the separation angle between continuous dimensions, in other words, between dimension $i-1$ and dimension i. It is understood that a ϑ value of 90° corresponds to two methods that are completely independent of one another (i.e., the property by which one method separates the sample does not influence the separation by the other method). Conversely, a separation angle of 0° corresponds to two separations that occur via identical retention mechanisms. The problem, however, lies in quantitatively relating ϑ values between 0° and 90° to the degree of orthogonality of two- or multidimensional separations (for simplicity, we refer in this chapter to 2D separations only, with the exception of Section 14.7).

We believe that the degree of orthogonality of a 2D separation is related to the separations angle ϑ through a quantity known in information theory as *percent synentropy* (% synentropy), which has been applied to yielding informational orthogonality for 2D chromatographic analyses [9]. A % synentropy of 0% describes a 2D chromatographic system with very different dimensional retention mechanisms. Conversely, a % synentropy of 100% describes a 2D chromatographic system with identical dimensional mechanisms. It would appear that the % synentropy therefore provides a means of quantitatively assigning a value to the separation angle ϑ. We

Table 14.2 Advantages and disadvantages of LC × SEC and of SEC × LC

SEC × LC	LC × SEC
Advantages	
High-resolution SEC possible	High-resolution (gradient) LC possible
Possible focusing on top of second-dimension column	Choice of detectors (second dimension isocratic)
Possible to exclude ("heavy") part of sample	Finite time of analysis in second-dimension
	Change first-dimension LC conditions without need to reoptimize second-dimension conditions
	LC system not easily overloaded
Disadvantages	
Second-dimension analysis time not limited	Limited resolution in (fast) second-dimension SEC
"Breakthrough" peaks in second dimension difficult to avoid	"Breakthrough" peaks in first dimension must be avoided
Gradients in second-dimension highly impractical	
Overloading and adsorption must be avoided in first dimension	
Limited choice of detectors	

Source: Ref. 1.

propose the following relationship:

$$\vartheta = \left(\frac{100 - \%\ \text{synentropy}}{100}\right) 90^\circ \tag{14.2}$$

whereby, for example, % synentropy values of 0%, 33.3%, 50%, 66.7%, and 100% would correspond, respectively, to ϑ values of 90°, 60°, 45°, 30°, and 0°.

To obtain the % synentropy, scaled (normalized) retention factors, X_a, for each separation dimension must first be calculated according to

$$X_a = \frac{R_{t,i} - R_{t,0}}{R_{t,f} - R_{t,0}} \tag{14.3}$$

where $R_{t,i}$ denotes the retention time of the solute of interest, $R_{t,f}$ is the retention time of the latest-eluting component in common with each dimension of the 2D chromatogram, and $R_{t,0}$ is the retention time of an unretained component. When the scaled retention factors of each dimension are plotted against each other, the % synentropy is a measure of the 2D informational entropy clustered along the diagonal on the normalized retention plots. (As will be seen below and in Figure 14.2, this is not the same as the correlation coefficient.) This provides a means of comparing the retention mechanisms of the separation systems being investigated. The % synentropy is calculated by dividing the informational entropy from the diagonally aligned data on the normalized retention plot by the total 2D informational entropy. As many of the quantities discussed may be unfamiliar, some detail follows.

The informational entropy of a measurement, I, is a probabilistic quantity described by

$$I = \sum_k (-\rho_k \log_2 \rho_k) \tag{14.4}$$

where ρ_k is the probability of the incidence of a single possible result, k, out of n possible results. In a chromatogram that may exhibit n possible, mutually exclusive peaks, each peak has probability ρ_k, where $k = 1, 2, 3, \ldots, n$. Consequently, ρ_k is the probability of the appearance of a peak k at a particular retention time. The sum of all ρ_k values should equal 1.

To compare two different types of chromatographic separations or columns, the informational entropy ("similarity" of data between dimensions) is initially calculated from the normalized retention time data for the first separation (first dimension), $I(k)$, and then for the second separation (second dimension), $I(k, l)$, where k and l denote the two columns (dimensions) being compared. This is done by adding the informational entropy for each normalized retention time X_a. For example, for a 1D chromatographic separation of nine solutes comprised of three X_a factors of 0.5, four X_a factors of 0.6, and two X_a factors of 0.7, the total informational entropy for dimension 1 would be

$$I(1) = -\left[\frac{3}{9}\log_2\frac{3}{9} + \frac{4}{9}\log_2\frac{4}{9} + \frac{2}{9}\log_2\frac{2}{9}\right] \tag{14.5}$$

For the case where no correlation exists between n independent variables, the informational entropy is given by

$$I(1, 2, 3, \ldots, n) = \sum_{j=1}^{n} I(j) \tag{14.6}$$

whereas if correlation exists, the total informational entropy will be less than the sum of the individual informational entropies; that is,

$$I(1, 2, 3, \ldots, n) < \sum_{j=1}^{n} I(j) \tag{14.7}$$

The informational entropy of a correlated information state is then given by

$$I(1, 2, 3, \ldots, n) = \sum_{j=1}^{n} I(j) - I(1; 2; 3; \ldots; n) \tag{14.8}$$

where $I(1;2;3;\ldots;n)$ is the mutual information that represents correlation.

To calculate the informational entropy of two chromatographic columns or dimensions, k and l, the fractional informational content, $h(k, l)$ is calculated according to

$$h(k, l) = 1 - \frac{I(k; l)}{I(k, l)} = 1 - \frac{\text{mutual information}}{\text{total 2D informational entropy}} \tag{14.9}$$

where $I(k;l)$ is the mutual information between chromatographic columns (dimensions) k and l, and $I(k,l)$ is the total informational entropy of the 2D system.

Finally, the informational similarity, $H(k,l)$ of the two columns or dimensions can be calculated via

$$H(k,l) = \sqrt{1 - h^2(k,l)} \tag{14.10}$$

$H(k,l)$ is a measure of the degree of solute crowding (dimensional saturation) of sample components being separated on a normalized 2D retention plot. An $H(k,l)$ value of 1.00 indicates complete solute crowding or dimensional saturation, whereas an $H(k,l)$ value of 0.00 indicates no solute crowding, and therefore utilization of all separation space.

Again, the % synentropy measures retention mechanism equivalency by measuring the informational entropy equally contributed from both separation dimensions. It should not be confused with the correlation coefficient, however. As seen in Figure 14.2, which is a normalized retention plot of two C18 HPLC columns, the retention correlation coefficient (r^2) is 0.80 but the % synentropy is only 34% [10]. This large difference between r^2 and % synentropy is due to the fact that the former measures how well retention data of each dimension match exactly, whereas the latter measures how well the retention data of each dimension cluster along the diagonal. A primer on informational entropy of measurements is provided by the seminal work of Shannon and Weaver [11].

Recently, Watson et al. introduced the orthogonality-independent concept of percent coverage as a metric to quantify how well the 2D separation space is filled.

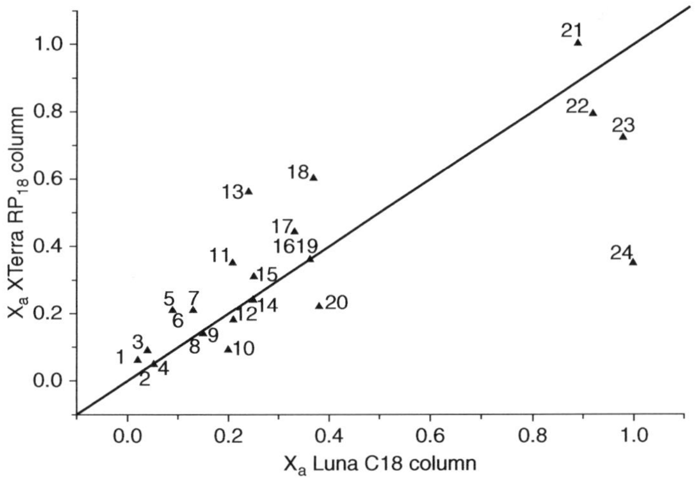

Figure 14.2 Normalized retention plot of a Luna C18 HPLC column versus an XTerra RP_{18} column. Numbering is done according to elution order on the Luna column. For this plot, the correlation coefficient is 0.80, but the % synentropy is only 34%. (Reprinted with permission from Ref. 10.)

A discussion of this approach, and of the accompanying caveats, may be found in Reference 12.

14.3 DESIGNING AN EXPERIMENTAL 2D-LC PROTOCOL

Figure 14.3 shows the protocol suggested in Reference 13 for designing comprehensive 2D-LC separations. Application of this protocol to the design of several LC × SEC systems, beginning with different diameters of the first dimension column, yielded results that were in close agreement with those obtained using current best practices in the field. Due to the promising nature of this protocol, we proceed to discuss the determination of each of the variables and parameters included therein.

The first parameters chosen are the maximum allowable total time of analysis as well as, for the first dimension column, the maximum allowable pressure and smallest viable diameter. An attractive approach to establishing the performance limits of separation systems in general and of SEC systems in particular are *Poppe plots* [14], which plot, in log-log fashion, the number of theoretical plates (N or N_{req}) on the abscissa and the plate time (t_p) on the ordinate (i.e., the required time per plate is plotted against the total plate count). *Plate time* is the time required to realize one theoretical plate and is related to the plate height (H) and the interstitial flow rate (u_0) via

$$t_p = \frac{H}{u_0} = \frac{L/N}{L/t_0} = \frac{t_0}{N} \tag{14.11}$$

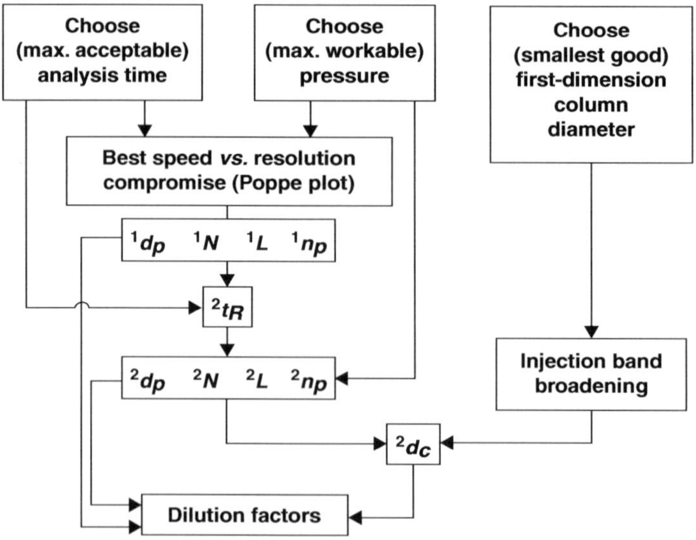

Figure 14.3 Protocol for designing comprehensive 2D-LC separations. (Reprinted with permission from Ref. 13.)

Because of the relationship between plate height, column length (L), plate number (N), flow rate, and the elution time of an unretained solute (t_0), t_p can also be related to these parameters as shown in Equation 14.11. For the ordinate of a Poppe plot, the following relationship can be used for fast separations by conventional HPLC, due to the limiting nature of the van Deemter mass transfer (C) term in fast separations:

$$t_p = \frac{H}{u_0} \approx C \tag{14.12}$$

However, because the diffusion coefficient (D_m) is substantially smaller for the types of high-molar-mass polymers analyzed by SEC than for the low-molar-mass analytes typically characterized by conventional HPLC, the Poppe plot ordinate for SEC cannot be written using Equation 14.11 but instead needs to be written in terms of dimensionless parameters: namely, the reduced plate height, h, and the dimensionless velocity, υ_0, as given by

$$h = \frac{H}{d_p} \tag{14.13}$$

and

$$\upsilon_0 = \frac{u_0 d_p}{D_m} \tag{14.14}$$

Figure 14.4 shows overlays of Poppe plots for conventional HPLC and for SEC [14]. As can be seen, the difference between the plots is striking. For HPLC, fast

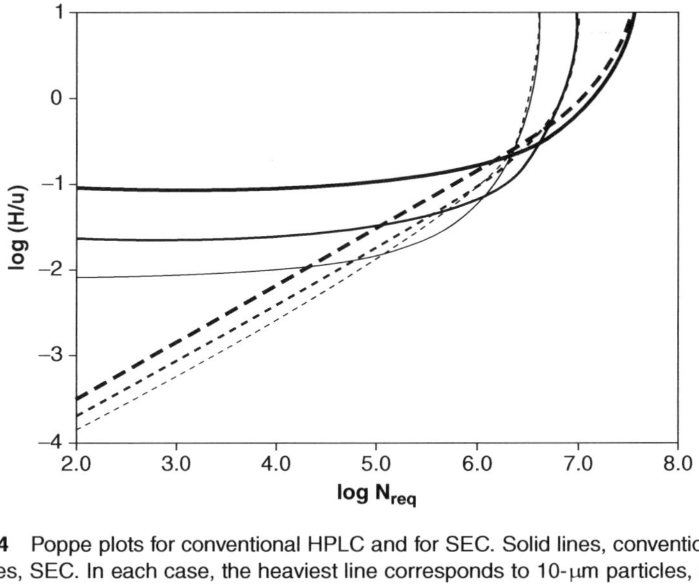

Figure 14.4 Poppe plots for conventional HPLC and for SEC. Solid lines, conventional HPLC; dashed lines, SEC. In each case, the heaviest line corresponds to 10-μm particles, middle line to 5-μm particles, and thinnest line to 3-μm particles. See Ref. 14 for experimental details. (Reprinted with permission from Ref. 14.)

separations (large values of u_0) lead to essentially constant plate times, whereas for SEC the possibility of fast separations with low plate counts becomes attractive due to the continuous decrease in plate time with increasing flow rate. For high-molar-mass analytes, "quick and dirty" SEC separations can be performed at relatively high flow rates [14].

Poppe plots yield the best attainable values for first-dimension plate number, ${}^1N$, and first-dimension column length, ${}^1L$, for every possible value of first-dimension particle size, ${}^1d_p$. (In this chapter, a preceding superscript denotes the corresponding dimension, in accordance with standard multidimensional separation conventions.) For isocratic LC, peak capacity for the first dimension can be estimated as

$$n_p = \frac{\sqrt{{}^1N}}{4R_s} \ln \frac{1+k_\omega}{1+k_\alpha} + 1 \tag{14.15}$$

where n_p is the number of peaks that can be separated, with resolution R_s, between a starting retention factor k_α and a final retention factor k_ω. For gradient-elution LC, peak capacity, $(n_p)_G$, is given by

$$(n_p)_G = \frac{t_G}{4(\sigma_t)_G} = \frac{t_G\sqrt{{}^1N}}{4t_m(1+k_e)} \tag{14.16}$$

where t_G is the gradient time, $(\sigma_t)_G$ the band dispersion in gradient elution, t_m the mobile-phase hold-up time, and k_e the retention factor at the moment of elution. For a reasonable gradient, $t_G = 10t_m$ and $k_e \approx 3$, so that

$$(n_p)_G \approx \frac{\sqrt{{}^1N}}{1.6} \tag{14.17}$$

More general interrelations between the various factors in Equation 14.16 may be found in Reference 13.

Once best estimates of ${}^1d_p$, ${}^1N$, ${}^1L$, and ${}^1n_p$ have been arrived at, the maximum analysis time in the second dimension, ${}^2t_R$, is determined. Assuming that at least four second-dimension chromatograms are needed in order for each first-dimension peak to achieve truly comprehensive separation, ${}^2t_R$ can be calculated from

$${}^2t_R = {}^1\sigma_t = \frac{{}^1t_R}{\sqrt{{}^1N}} = \frac{{}^1t_m(1+{}^1k)}{\sqrt{{}^1N}} \tag{14.18}$$

where ${}^1\sigma_t$ is the band dispersion in the first dimension, ${}^1t_R$ the retention time in the first dimension, ${}^1t_m$ the mobile-phase hold-up time in the first dimension, and ${}^1k$ the retention factor in the first dimension.

After determining ${}^2t_R$, suitable data for the eluent viscosity (η) and analyte diffusion coefficient (D_m) are entered and a maximum allowable second dimension pressure drop (${}^2\Delta P_{\max}$) is selected. If necessary, the latter value may be calculated from the modified form of Darcy's law which incorporates the effects of particle size:

$$\Delta P = \frac{u_0 \phi \eta L}{d_p^2} = \frac{u_0 \phi \eta N H}{d_p^2} \tag{14.19}$$

When the porosity of the column packing varies significantly, the Kozeny–Carman equation may prove more useful than Darcy's law [5]. Also, because different instrumentation is used for each dimension, ${}^1\Delta P_{\text{max}}$ need not be equal to ${}^2\Delta P_{\text{max}}$.

At this point, the process used earlier for the first dimension can be reapplied to the second dimension to obtain ${}^2L$, ${}^2d_p$, and ${}^2n_p$. What is left is to decide on suitable internal column diameters for the first- and second-dimension columns and on suitable flow rates and injection volumes. ${}^1L$ and ${}^2L$ are known, as are ${}^1t_R$ and ${}^2t_R$, such that ${}^1u_0$ and ${}^2u_0$ will therefore depend only on the first- and second-dimension column diameters, ${}^1d_c$ and ${}^2d_c$, respectively, and on the porosity of each column. Choosing a value for ${}^1d_c$ automatically decides the injection volume for the second dimension, ${}^2V_{\text{inj}}$, which can itself be used to select a value of ${}^2d_c$ that produces an acceptable amount of extra-column band broadening. The contribution from the injection volume, V_{inj}, to band broadening is denoted as σ_{inj} and can be estimated from

$$\sigma_{\text{inj}}^2 = \frac{V_{\text{inj}}^2}{\delta^2} \tag{14.20}$$

where for an ideal chromatographic band, $\delta = \sqrt{12}$ but, in reality, $\delta \approx 4$ to 5. The fractional contribution of the injection band broadening, θ_{inj}, vis-à-vis the true chromatographic band broadening in the column, σ_{col}, is given by

$$\theta_{\text{inj}} = \frac{\sqrt{\sigma_{\text{inj}}^2 + \sigma_{\text{col}}^2}}{\sigma_{\text{col}}} - 1 \tag{14.21}$$

and it is θ_{inj} that should be kept within acceptable limits.

At the bottom of the protocol in Figure 14.3 are dilution factors (DFs). Analyte dilution during a chromatographic separation directly affects detectability. For isocratic LC, DF may be calculated using

$$\text{DF} = \sqrt{\frac{2\pi}{N}} \frac{V_m}{V_{\text{inj}}}(1 + k) \tag{14.22}$$

where V_m is the volume of mobile phase in the column.

14.4 ELUENT TRANSFER IN 2D-LC

Figure 14.5 shows a generic experimental setup for a 2D-LC experiment [7]. The setup shown is fairly typical, coupling a first-dimension HPLC column to a second-dimension SEC column. Two pump systems are used, a gradient system for the first

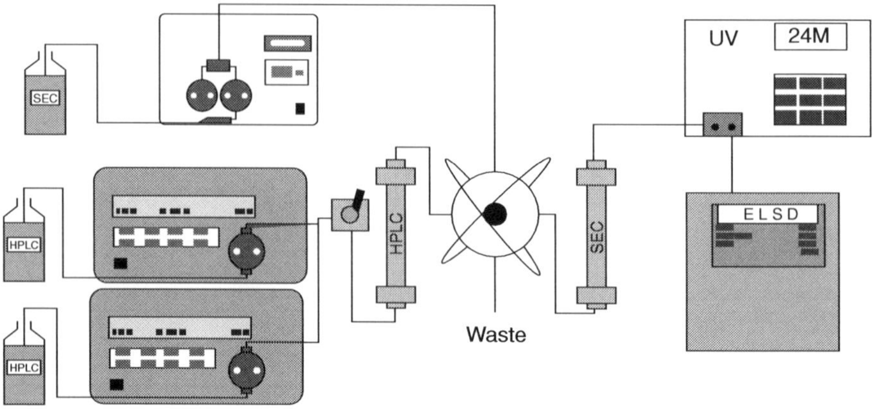

Figure 14.5 Generic setup for a 2D-LC experiment. (Reprinted with permission from Ref. 7.)

dimension and an isocratic pump for the second, as are two detectors, in this case UV and ELSD (discussed, respectively, in Sections 9.2.2. and 9.2.3).

Fractions are transferred between dimensions (columns) using electrically or pneumatically actuated 8- or 10-way dual-loop switching valves where each loop is used, alternately, to store the effluent from the first-dimension column and to inject it into the second-dimension column. A 10-way valve is shown schematically in Figure 14.6, in both asymmetrical (a) and symmetrical (b) configurations [1]. For comprehensive 2D-LC separations, the symmetrical configuration was found superior to the asymmetrical arrangement. The former allowed for partial filling of the injection loop and for accurate quantitation. Virtually identical retention times and peak shapes were obtained for even and odd injections. This was not the case for the asymmetrical configuration, where each of the loops is used differently when injecting into the second column, one loop emptying in forward-flush mode and the other in backward-flush mode. The asymmetrical configuration resulted in retention time shifts and variations in peak width and asymmetry.

14.5 STOP-FLOW SEC × LC

Distinct from the setup outlined earlier, SEC can be used as the first dimension and reversed-phase LC (RPLC) as the second dimension (i.e., SEC × LC) in comprehensive 2D experiments. This approach, described in Reference 15, also incorporates a stop-flow valve, as shown in Figure 14.7.

While the approach described in Section 14.3 did not set any a priori boundary conditions, the stop-flow method is different. Here, it is assumed that there is already an existing one-dimensional separation and the desire is to increase peak capacity without reoptimizing the original separation. The increase in peak capacity is realized by adding an additional and different type of column to the system.

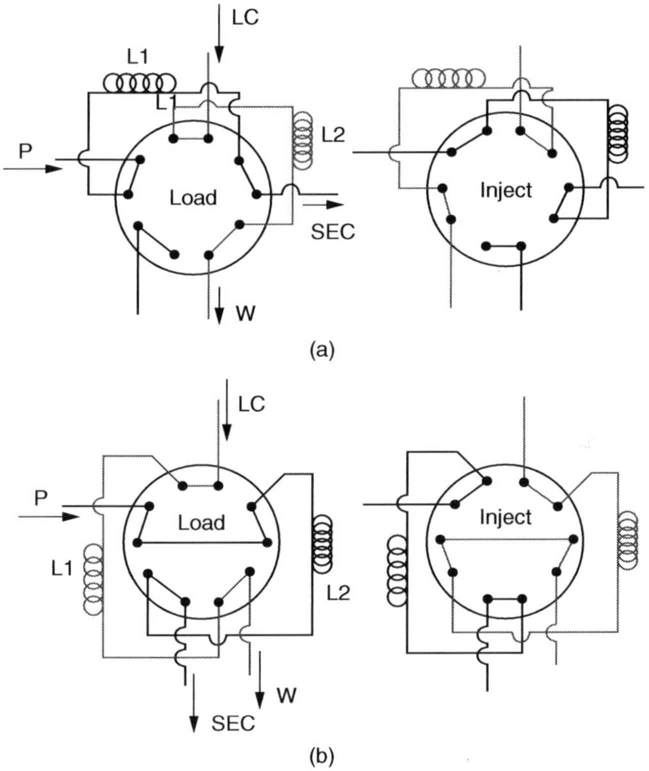

Figure 14.6 Pneumatically actuated 10-port, two-way switching valve for 2D-LC: (a) asymmetrical and (b) Symmetrical configurations. L1, loop 1; L2, loop 2; P, pump; W, waste; LC, from LC; SEC, to SEC. (Reprinted with permission from Ref. 1.)

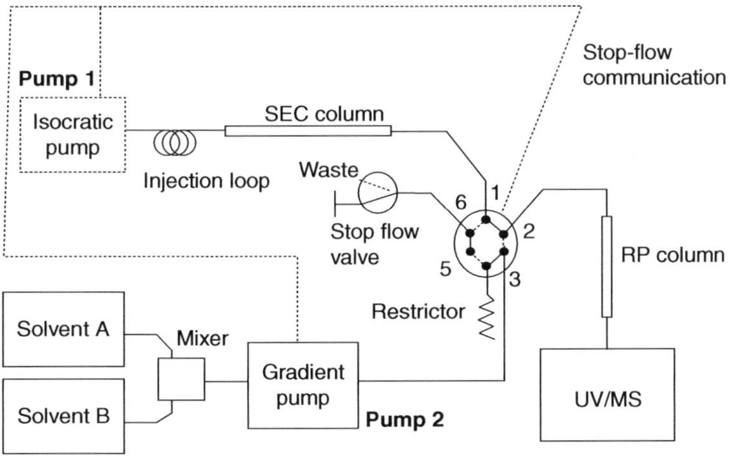

Figure 14.7 Stop-flow SEC × RPLC system. (Reprinted with permission from Ref. 15.)

The protocol begins with the familiar criterion that the optimum number of fractions that need to be transferred from the first to the second dimension is on the order of four per peak. Therefore, the standard deviation of the narrowest first-dimension peak (in units of time), ${}^1\sigma_t$, is related to the total analysis time in the second dimension, ${}^2t_a$, as follows:

$$ {}^1\sigma_t = {}^2t_a \tag{14.23} $$

Equation 14.23 links the operation of the first-dimension separation to that of the second-dimension separation, as ${}^2t_a$ is the total analysis time necessary for each fraction transferred from the first-dimension column to be run on the second-dimension column.

Next, the required number of plates for the first-dimension separation, ${}^1N$, is determined. From the desired total peak capacity and the known peak capacity for the second-dimension separation, using Equation 14.1 and assuming a separation angle $\vartheta = 90°$ (% synentropy = 0%), the peak capacity for the first-dimension separation, ${}^1n_c$, can be calculated. For isocratic LC conditions and resolution $R_s = 1$, this ${}^1n_c$ value can then be used to calculate ${}^1N$:

$$ {}^1N = \frac{16({}^1n_c - 1)^2}{[\ln({}^1t_{\max}/{}^1t_{\min})]^2} \tag{14.24} $$

where ${}^1t_{\max}$ and ${}^1t_{\min}$ are the time limits within which first-dimension peaks can elute. For SEC, $t_{\max}/t_{\min} \approx 2$. The column length is then calculated from the plate number and plate height, where the latter is a function of the chromatographic velocity, u, of the peak of interest:

$$ {}^1L = {}^1N\,{}^1H(u) \tag{14.25} $$

Because

$$ {}^1N = \frac{{}^1t_R^2}{{}^1\sigma_t^2} \tag{14.26} $$

and

$$ {}^1t_R = \frac{{}^1L}{{}^1u} \tag{14.27} $$

then

$$ {}^1L = \frac{{}^2t_a^2\,{}^1u^2}{{}^1H(u)} \tag{14.28} $$

where

$$ {}^{1}u = \frac{A + \sqrt{A^2 - 4\left[\left({}^{2}t_a \big/ \sqrt{{}^{1}N}\right) - C\right](-B)}}{2\left[\left({}^{2}t_a \big/ \sqrt{{}^{1}N}\right) - C\right]} \tag{14.29} $$

where A, B, and C correspond to the appropriate coefficients of the van Deemter equation. Substituting ${}^{1}u$ from Equation 14.29 into the van Deemter equation gives ${}^{1}H$, from which the desired value for the first-dimension column length, ${}^{1}L$, can be calculated.

A surprising result of the calculations above is that for a given $H(u)$ curve for the first-dimension column, only one column length can be used at one specific velocity. Neither a different column length nor a different velocity can simultaneously yield the desired total peak capacity and the desired four second-dimension separation runs over a peak from the first-dimension separation. Only one set of conditions can be adopted for the first-dimension separation, with all others resulting in nonoptimal comprehensive separations. This should also hold for other modes of LC $\times$ LC not involving SEC as one of the dimensions.

Rather than perform the first-dimension separation at the very low continuous flow rates, ${}^{1}u$, needed to achieve ${}^{2}t_a$ and ${}^{1}N$, as is usually done, intermittent high flow–zero flow conditions (i.e., stop-flow methodology) are used. The stop-flow approach was used, as shown in Figure 14.7, in conjunction with the theory outlined above, in a study of a complex mixture of peptides [15]. Operating the system in stop-flow mode rather than continuous-flow mode did not result in additional band broadening but instead appears to facilitate instrument design, as no storage loops are needed. Rather, a simple transfer valve can be used to transfer fractions from the first to the second dimension.

14.6 SELECT APPLICATIONS OF 2D-LC

As seen in Table 14.1, a number of chromatographic techniques can couple to SEC in order to yield information about various macromolecular distributions, often as a function of analyte molar mass. Couplings can be classified either by technique(s) or by the type of information produced. Here we opt for the former, with each subsection giving representative examples of the coupling of a different technique with SEC.

14.6.1 HPLC

HPLC $\times$ SEC has been used to characterize the combined CCD $\times$ MMD of polymers, such as in monitoring the grafting of PMMA onto a polybutadiene (PB) backbone [16]. Figure 14.8 shows the SEC/DRI trace of the polybutadiene backbone before the grafting process and the SEC/DRI/UV traces of the graft product after a

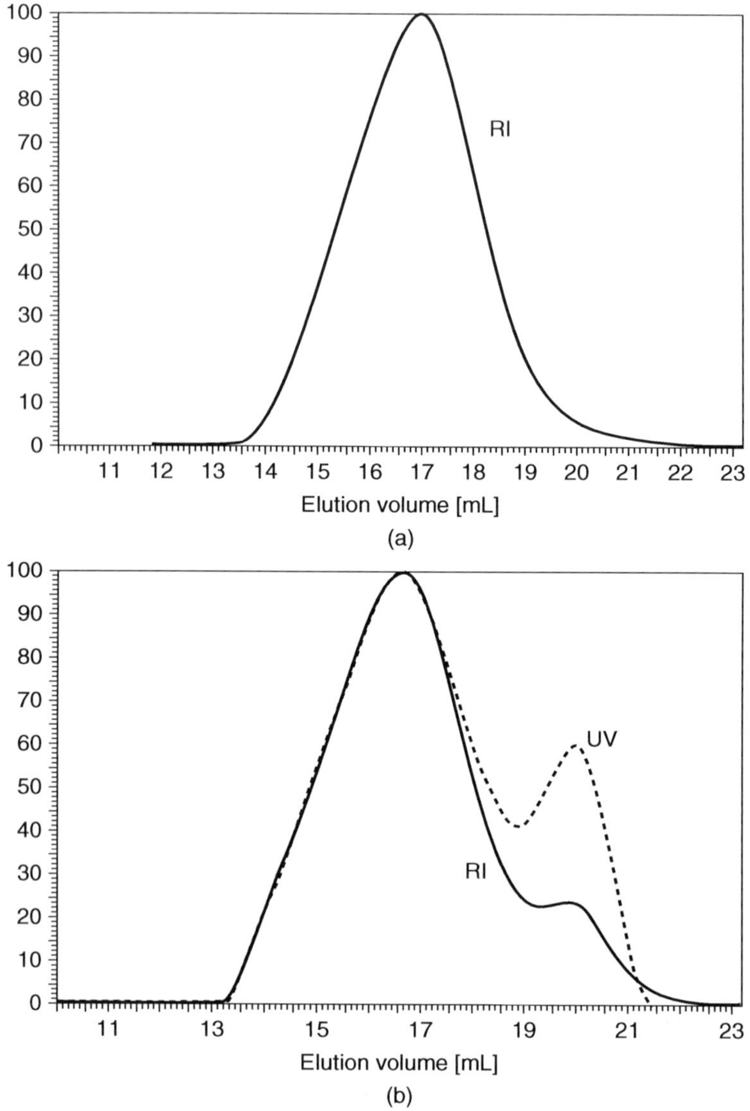

Figure 14.8 Monitoring grafting of PMMA onto PB backbone using SEC: (a) SEC/DRI of PB backbone; (b) SEC/DRI/UV ($\lambda_0 = 239$ nm) of graft product after 480-min reaction time. Solvent, THF. (Reprinted with permission from Ref. 16.)

reaction time of 480 minutes. In Figure 14.8b the bimodality was taken as an indication of the presence of a higher-molar-mass reaction product and a lower-molar-mass by-product. In this case the chemical heterogeneity (% MMA as a function of sample molar mass) of the sample could have been determined, at least in theory, by comparing the signals from the DRI and UV detectors, as MMA absorbs more strongly in

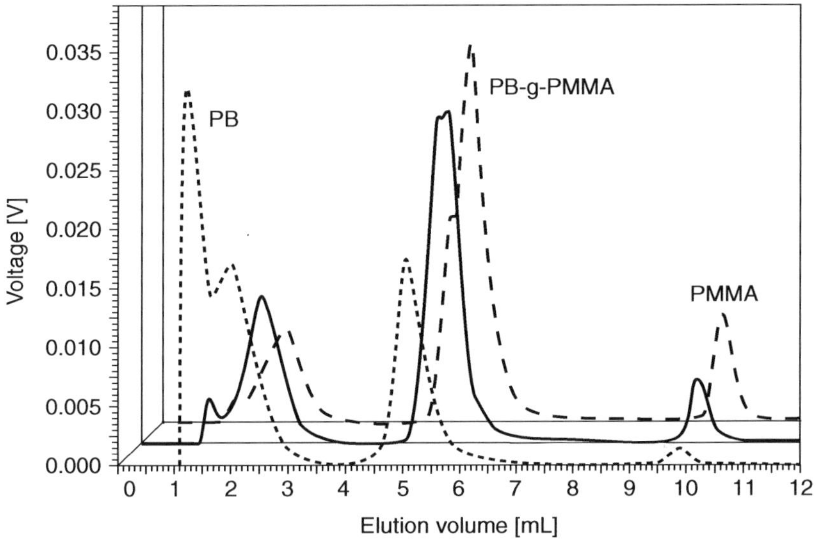

Figure 14.9 Monitoring grafting of PMMA onto PB backbone using gradient HPLC: gradient HPLC/ELSD chromatograms of graft products after 60 min (dashed line), 240 min (solid line), and 480 min (dashed-dotted line). See Ref. 16 for experimental details. (Reprinted with permission from Ref. 16.)

the UV ($\lambda_0 = 239$ nm) than does PB. What was done, rather, was actually more complete, as the sample was then analyzed by gradient HPLC/ELSD, which, as shown in the back trace of Figure 14.9 for a 480-minute reaction time (and, indeed, for all three reaction times plotted), presented evidence of three different species coexisting in the sample: PB, PB-*g*-PMMA, and PMMA. A 2D experiment using HPLC × SEC with ELSD yielded the combined CCD × MMD of the 480-minute sample, shown as a contour plot in Figure 14.10. In this plot, the ordinate represents the HPLC separation and the abscissa is the molar mass distribution relative to narrow PMMA standards. The intensities of the peaks in the contour plots represent the relative concentrations of the components, as indicated by the shaded bar on the right. Addition of an FTIR detector to this type of setup permits construction of contour plots for non- or weakly-UV-absorbing groups that may be present in the sample.

A variant on the SEC × HPLC setup consisted of using a set of six analytical SEC columns (7.8 × 300 mm each) coupled to an LC/LC interface which used two RPLC columns (nonporous C18 modified silica, 4.6 × 33 mm) in parallel, rather than using storage loops [17]. A four-port valve connected to the SEC outlet alternated connection between the two RPLC columns, thereby serving as the injector. The detectors in this setup, which is shown in Figure 14.11, were UV and MS connected in series. The system was used to analyze tryptic digests of ovalbumin and serum albumin. The interface design allows for the use of conventional HPLC and SEC columns, facilitating system assembly and maintenance.

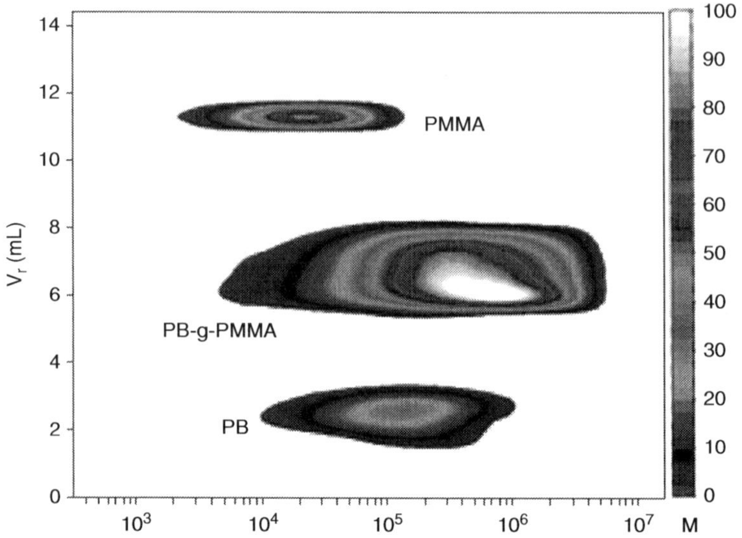

Figure 14.10 CCD × MMD of graft product of PMMA and PB: Gradient HPLC × SEC separation of product of grafting PMMA onto a PB backbone, after 480-min reaction time. Contour plots represent the combined CCD × MMD of the product. Relative concentrations of individual components are obtained from peak intensities, as determined from the right ordinate color/gray scale bar. Detector, ELSD. (Reprinted with permission from Ref. 16.)

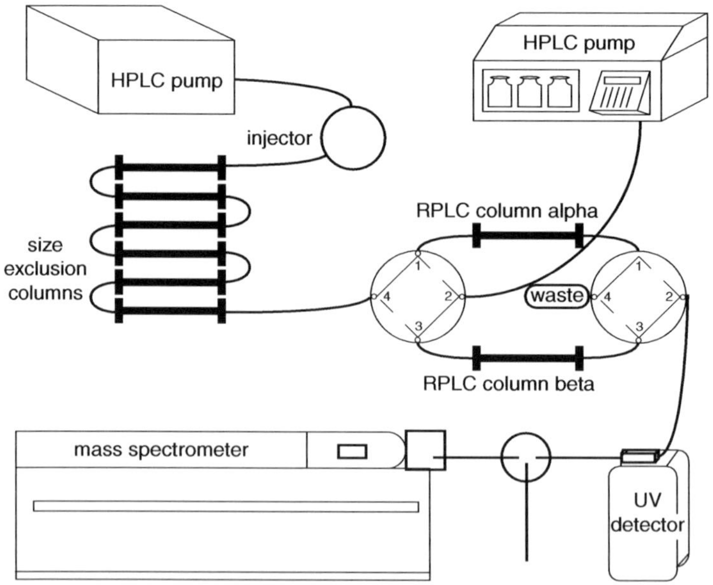

Figure 14.11 SEC × LC/LC system: 2D chromatography using six analytical SEC columns connected to a parallel column LC/LC interface with dual UV and MS detection. (Reprinted with permission from Ref. 17.)

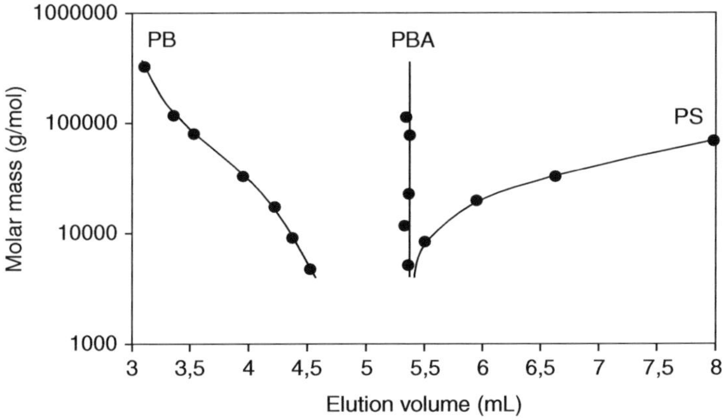

Figure 14.12 LCCC of poly(butyl acrylate): elution behavior of polystyrene (PS), polybutadiene (PB), and poly(butyl acrylate) (PBA) at the critical condition for PBA. Eluent, 15.5:84.5 (v/v); THF, cyclohexane. (Reprinted with permission from Ref. 19.)

14.6.2 Liquid Chromatography at the Critical Condition

Liquid chromatography at the critical condition (LCCC), also referred to as *liquid chromatography at the critical adsorption point* or as *liquid chromatography at the critical partition point*, is used to determine polymer functionality or chemical composition distributions. In LCCC, the functionality or chemical composition distribution of a polymer is characterized by initially identifying the mobile-phase condition (if it exists) at which homologs of the nonfunctionalized polymer coelute. This corresponds to $\Delta G_{\text{transfer}} = 0$ between the mobile phase and the stationary phase. At that critical, thermodynamically pseudoideal condition, the functionalized polymer is separated based solely on the interaction of the chemical functionality of interest with the stationary phase, with the chromatogram at the critical condition being a reflection of the functionality type distribution (FTD) of the polymer. Finding the exact solvent condition that corresponds to the critical condition is a nontrivial task which usually involves the use of mixed solvents or even of enhanced-fluidity mobile phases [6,18].

A representative example of the use of LCCC × SEC is provided by the grafting of butyl acrylate (BA) onto a poly(styrene-*b*-butadiene) backbone [19]. Figure 14.12 shows the behavior of the individual homopolymers at the critical condition of poly(butyl acrylate) (PBA), while Figure 14.13 shows the LCCC chromatogram of the graft product, as monitored by both UV and ELSD detection. The second peak, eluting around 5.0 to 5.5 minutes, corresponds mainly to PBA, while the earlier eluting peak was assigned to coeluting graft copolymer and ungrafted block copolymer. The later eluting peak and the tail were assigned to block copolymer fractions with an increased styrene/butadiene ratio, where the PB block was partially degraded.

The 2D LCCC × SEC experiment yields the combined CCD × MMD, shown in Figure 14.14. Here, peaks 5 and 4 correspond to PBA and to a stabilizer added

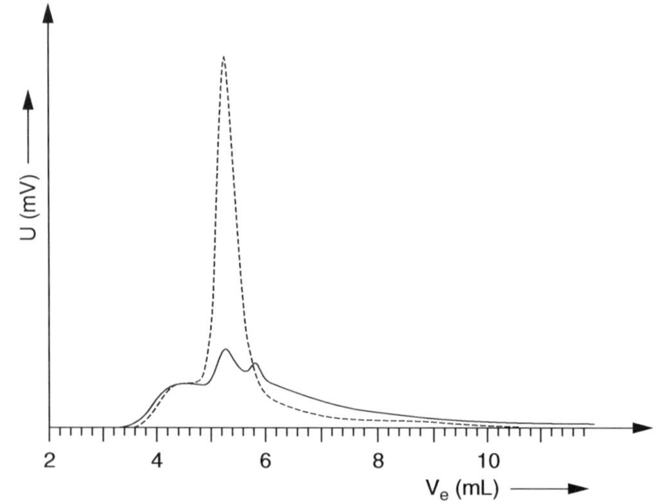

Figure 14.13 LCCC of butyl acrylate grafted onto PS-*b*-PB backbone. Solid curve, UV detection ($\lambda_0 = 254$ nm); dashed curve, ELSD. (Reprinted with permission from Ref. 19.)

to the block copolymer for long-term storage stabilization. Peak 2 has the highest intensity of all the peaks and also the highest molar mass. It corresponds to the graft copolymer, whereas peak 1 can be assigned to the ungrafted PS-*b*-PB copolymer. The tail in position 3 is assigned to graft and block copolymer fractions with partially degraded PB blocks. Peak 6 is a low-concentration product fraction detected only with the ELSD. The molar mass of this component is on the same order of magnitude as the molar mass of PBA, but the higher elution volume of peak 6 indicates a higher polarity relative to PBA. This peak corresponds to fractions of copolymer of BA and maleic anhydride, a small amount of the latter having been added during the grafting reaction of BA onto the PS-*b*-PB backbone.

Addition of an off-line continuous FTIR detector (see Section 10.3) to the system permitted determination of the chemical heterogeneity of the system: that is, of the average amount of styrene (S), butadiene (B), or butyl acrylate (BA) present as a function of molar mass or elution volume (see, e.g., Figure 10.12). Online flow cell FTIR (Section 10.3) has also been used as a second-dimension detection method in 2D separations [20].

14.6.3 Other Methods

A variety of other chromatographic methods have been coupled to SEC. Most of these couplings show great potential, although their application has thus far been limited. Gradient polymer elution chromatography [21–23] has been coupled to SEC to compare the CCD × MMD of α,ω-dihydroxypolystyrene prepared by atom transfer radical polymerization versus that prepared by atom transfer radical coupling

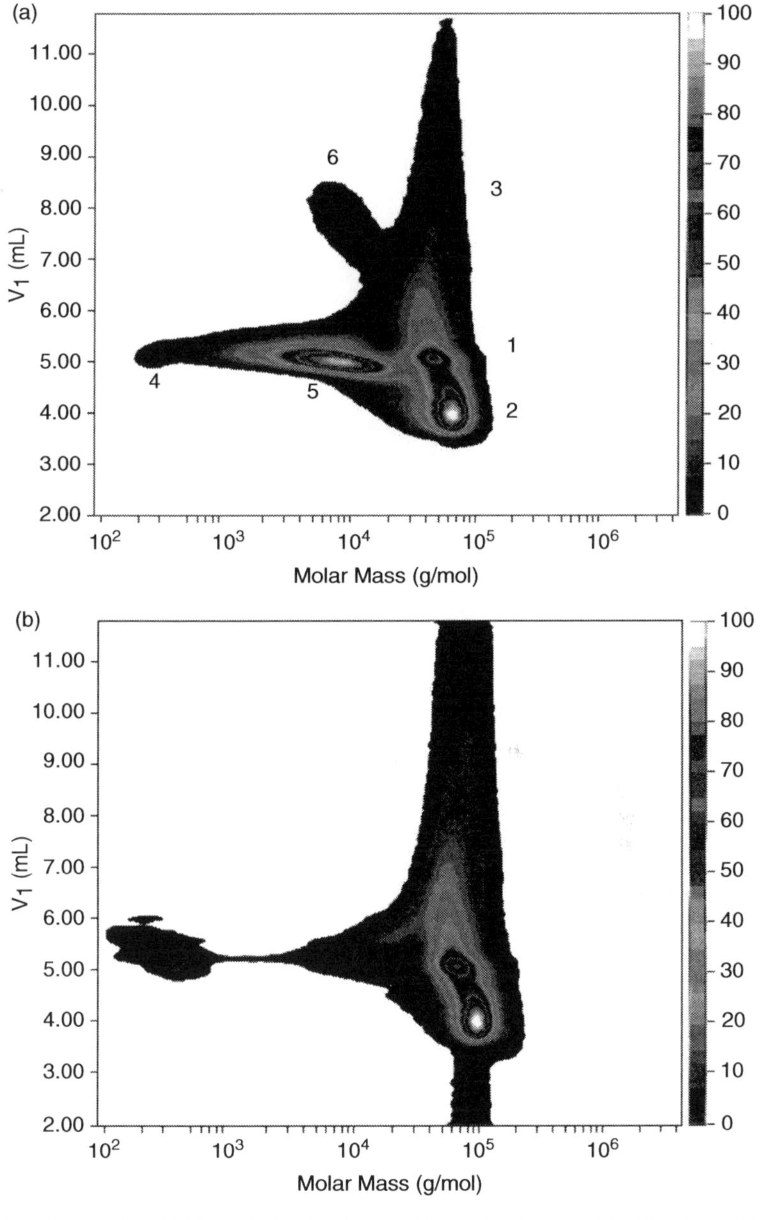

Figure 14.14 LCCC × SEC analysis of butyl acrylate grafted onto PS-*b*-PB: (a) ELSD; (b) UV detection. CCD × MMD of graft product, LCCC at critical condition for PBA. Relative concentrations of individual components are obtained from peak intensities, as determined from the right ordinate color/gray scale bar. (Reprinted with permission from Ref. 19.)

[24]. The coupling of techniques such as high-osmotic-pressure chromatography and phase fluctuation chromatography to SEC can provide the distribution of block number or block length in block copolymers (via PFC × SEC) and the terminal chemistry distribution of telechelic polymers (via HOPC × SEC) [8].

Other chromatographic couplings include SEC × GC/MS for the analysis of polymer additives [25], SEC/TREF for the study of polyolefins [26], and SEC × SEC to study chromatographic band broadening [27]. The latter technique is of limited use in polymer analysis, but if narrow fractions can be collected from the first-dimension SEC column, band broadening in the second-dimension SEC will be due only to chromatographic dispersion and not to the selectivity of the SEC mechanism (i.e., to band breadth due to sample polydispersity).

14.7 SEC IN 3D SEPARATIONS

The 2D approaches described in this chapter have been taken one step further with the coupling SEC × RPLC × CZE, the latter in fast mode [28]. This 3D method separates based on size, hydrophobicity, and electrophoretic mobility. The setup used is that shown in Figure 14.15. As can be seen, this is a particularly time-consuming technique, the limiting step being the 6-hour SEC analysis. Length of the SEC runs was dictated by the slow, 11-μL/min flow rate, necessary because of the geometry and flow incompatibilities between the SEC and RPLC columns. Highly concentrated samples were used to overcome dilutory factors inherent in the SEC/RPLC interface, and rigorous temperature control in all three dimensions was necessary to avoid retention/migration time drift. The gain in peak capacity by adding the SEC dimension was a factor of 5. An example of the type of results obtained, for a sample of tryptic digest of ovalbumin, is seen in Figure 14.16.

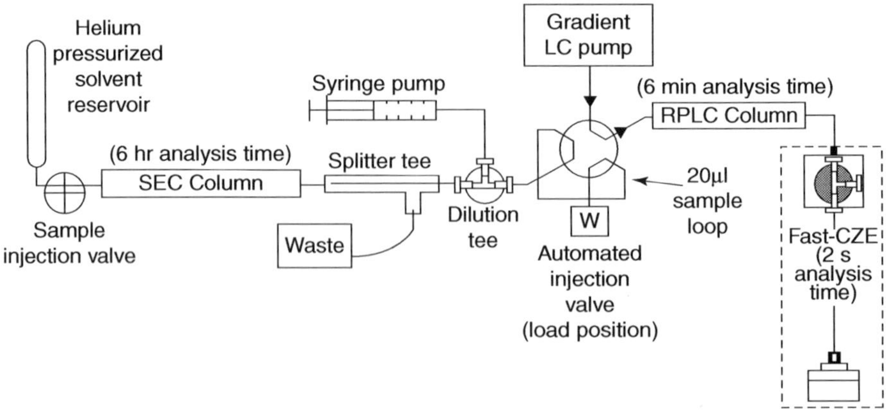

Figure 14.15 3D SEC × RPLC × CZE system. (Reprinted with permission from Ref. 28.)

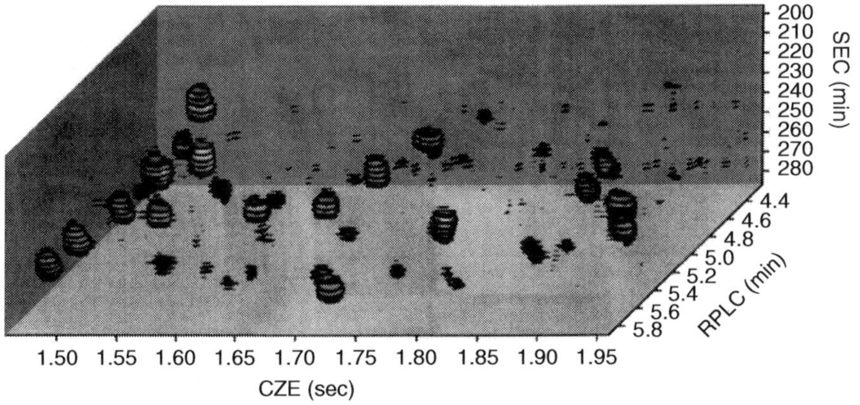

Figure 14.16 SEC × RPLC × CZE analysis of tryptic digest of ovalbumin: 3D separation using the system shown in Figure 14.16. A series of planar slices through each data volume shows the peaks, which in 3D have the appearance of "stacks" of disks or ellipsoids. (Reprinted with permission from Ref. 28.)

REFERENCES

1. A. van der Horst and P. J. Schoenmakers, *J. Chromatogr. A*, **1000**, 693 (2003).
2. P. Kilz, *Chromatographia*, **59**, 3 (2004).
3. P. Schoenmakers, P. Marriott, and J. Beens, *LC-GC Eur.*, **16**, 1 (2003).
4. A. M. Striegel, in *Multiple Detection in Size-Exclusion Chromatography*, ACS Symp. Ser. 893, A. M. Striegel, ed., American Chemical Society, Washington, DC, 2005, Chap. 1.
5. J. C. Giddings, *Unified Separation Science*, Wiley, New York, 1991.
6. D. Berek, in *Handbook of Size Exclusion Chromatography and Related Techniques*, 2nd ed., revised and expanded, C.-S. Wu, ed., Marcel Dekker, New York, 2004, Chap. 18.
7. H. Pasch, in *Multiple Detection in Size-Exclusion Chromatography*, ACS Symp. Ser. 893, A. M. Striegel, ed., American Chemical Society, Washington, DC, 2005, Chap. 14.
8. I. Teraoka, in *Multiple Detection in Size-Exclusion Chromatography*, ACS Symp. Ser. 893, A. M. Striegel, ed., American Chemical Society, Washington, DC, 2005, Chap. 15.
9. P. J. Slonecker, X. Li, T. H. Ridgway, and J. G. Dorsey, *Anal. Chem.*, **68**, 682 (1996).
10. T. J. Whelan, M. J. Gray, P. J. Slonecker, R. A. Shalliker, and M. A. Wilson, *J. Chromatogr. A*, **1097**, 148 (2005).
11. C. E. Shannon and W. Weaver, *The Mathematical Theory of Communication*, University of Illinois Press, Champaign-Urbana, IL, 1963.
12. N. E. Watson, J. M. Davis, and R. E. Synovec, *Anal. Chem.*, **79**, 7924 (2007).
13. P. J. Schoenmakers, G. Vivó-Truyols, and W. M. C. Decrop, *J. Chromatogr. A*, **1120**, 282 (2006).
14. S.-T. Popovici and P. J. Schoenmakers, *J. Chromatogr. A*, **1073**, 87 (2005).
15. F. Bedani, W. Th. Kok, and H.-G. Janssen, *J. Chromatogr. A*, **1133**, 126 (2006).
16. A. Siewing, B. Lahn, D. Braun, and H. Pasch, *J. Polym. Sci. A*, **41**, 3143 (2003).
17. G. J. Opiteck, J. W. Jorgenson, and R. J. Anderegg, *Anal. Chem.*, **69**, 2283 (1997).

18. S. Phillips and S. V. Olesik, *Anal. Chem.*, **74**, 799 (2002).
19. J. Adrian, E. Esser, G. Hellmann, and H. Pasch, *Polymer*, **41**, 2439 (2000).
20. S. J. Kok, Th. Hankemeier, and P. J. Schoenmakers, *J. Chromatogr. A*, **1098**, 104 (2005).
21. B. Klumperman and H. J. A. Philipsen, *LC-GC*, **17**, 118 (1999).
22. A. M. Striegel, *J. Chromatogr. A*, **996**, 45 (2003).
23. A. M. Striegel, *J. Chromatogr. A*, **971**, 151 (2002).
24. H. Gao, D. J. Siegwart, N. Jahed, T. Sarbu, and K. Matyjaszewski, *Designed Monom. Polym.*, **8**, 533 (2005).
25. H. J. Cortes, G. E. Bormett, and J. D. Graham, *J. Microcol. Sep.*, **4**, 51 (1992).
26. W. W. Yau and D. Gillespie, *Polymer*, **42**, 8947 (2001).
27. S. T. Popovici, A. van der Horst, and P. J. Schoenmakers, *J. Sep. Sci.*, **28**, 1457 (2005).
28. A. W. Moore, Jr., and J. W. Jorgenson, *Anal. Chem.*, **67**, 3456 (1995).

15

SPECIAL TECHNIQUES

15.1 INTRODUCTION

In this chapter we discuss techniques that are variations on the theme of SEC. These include preparative, inverse, recycle, vacancy, and differential SEC as well as size-exclusion electrochromatography (SEEC). Each technique has found specialized applications, some more widespread than others. Inverse SEC, for example, is used almost exclusively to measure pore-size distributions of column packing materials, but its application has extended to measuring pore sizes of other materials such as wood pulps and hemodialysis membranes. Methods such as vacancy and differential SEC have found only limited application, while preparative and recycle SEC have recently expanded into the area of nanoparticle analysis. The combination of pressure- and electro-driven flows that occurs in SEEC gives efficiencies comparable or higher than in SEC while consuming orders of magnitude less sample. If mass selectivity issues can be resolved and commercially available SEEC columns manufactured, this technique might find more widespread use.

15.2 PREPARATIVE SEC

In other sections of the book we emphasize the data-acquisition aspects of SEC, in which either qualitative or quantitative information about a sample is obtained. In this section we discuss another aspect of SEC, the preparative technique. Preparative

Modern Size-Exclusion Liquid Chromatography: Practice of Gel Permeation and Gel Filtration Chromatography, Second Edition
By André M. Striegel, Wallace W. Yau, Joseph J. Kirkland, and Donald D. Bly

SEC is effective and convenient for isolating relatively large amounts of purified components for molar mass standards, testing, materials characterization, and so on. While sufficient quantities of materials for identification may often be obtained with analytical systems, the larger quantities of purified samples needed for other studies must normally be prepared with large-diameter columns and with conditions that are different from those used for analytical SEC. Sample capacity is the goal of preparative SEC, and variations of technique and equipment from those employed for SEC analysis must be used.

15.2.1 Experimentation

Commonly, large-diameter, low-pressure columns and lower-cost column packings are employed for preparative SEC studies. In this section we describe some of the specialized aspects of preparative SEC; more extensive general treatments of preparative LC are given in References 1 and 2.

15.2.1.1 Columns. In preparative SEC, sample capacity is increased by using columns of larger internal diameter (i.d.). Increasing the diameter of SEC columns does not necessarily reduce chromatographic resolution (Chapter 6). In fact, separation efficiencies with large-diameter columns are frequently superior to those obtained with narrow-bore columns of the same column length, provided that the same ratio of sample weight to cross-sectional area is maintained (for nonoverloaded systems). The practical upper limit of column i.d. for SEC has not yet been established. Laboratory columns of 5.7 cm i.d. have been reported [3], and production apparatus of up to 2500-L capacity has long been commercially available for carrying out preparative separations of biologically important substances by aqueous SEC [4]. The design of a preparative column is shown in Figure 6.5. An illustration of the superior performance of large-diameter columns is shown in Figure 15.1, where resolution units per minute based on a separation of ethylene glycol (*M* 62 g/mol, total permeation) versus pentaethylene glycol (*M* 238 g/mol) is shown for columns of 20- to 44-μm organic gels. In this study, resolution steadily increased with increasing column diameter, as has been noted with columns of other LC packings [6]. As with analytical columns (Chapter 6), preparative SEC columns prepared in straight sections are connected when higher efficiency units are needed.

Both organic-gel and porous silica column packings for preparative SEC are listed in Tables 6.1 and 6.2. The choice of packing material is based on the same considerations as for analytical SEC separations. In aqueous SEC, preparative columns of semirigid gels are also used, with the specific pressure limitations noted in Chapter 6.

Sample capacity increases proportionally to cross-sectional area, regardless of particle size. In cases that require the highest resolution, preparative SEC separations are carried out with columns of fine (e.g., $\leq$ 10 μm) column packings. However, because of high cost, large-diameter columns of small-particle column packings have not been widely used, and columns of small particles have been used only to prepare relatively small amounts (e.g., 100 to 300 mg per run) of purified materials. Most preparative SEC separations are made with longer columns of larger particles

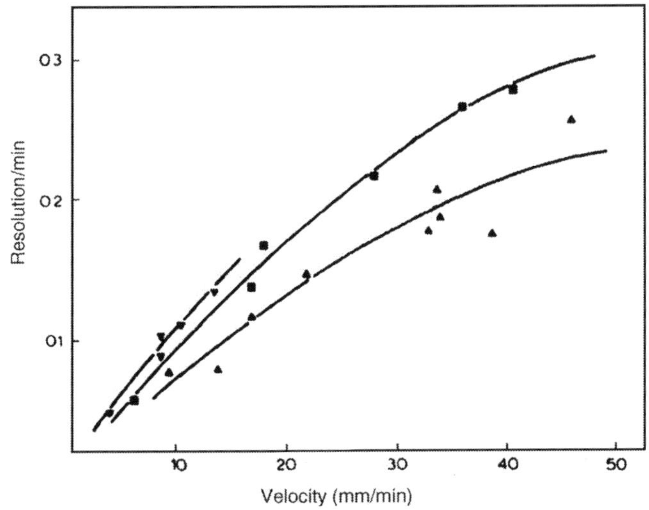

Figure 15.1 Effect of column internal diameter and velocity on SEC resolution. Columns, 16.5 cm long, 20- to 44-μm Sephadex G-25; mobile phase, water; temperature, ambient; sample, 10 μL ethylene glycol/pentaethylene glycol (1 : 1 : 5) by volume in water; columns, 6 (▲), 10 (■), and 21 (▼) mm. i.d. (Reprinted with permission from Ref. 5.)

(e.g., 30 to 60 μm). Studies on other LC methods [7] suggest that at high sample loading, long, large-diameter columns filled with coarse, uniform particles should be employed. Higher sample capacity is primarily a result of the larger amount of packing available in larger-diameter or longer columns. It should be noted that columns of larger particles require relatively low mobile-phase velocities for efficient operation; therefore, for the same column volume, significantly longer separation times are required for long columns of large particles relative to shorter columns of smaller particles.

To summarize, SEC columns packed with small particles should be used for rapid isolation of relatively small, highly purified samples at high resolution; columns packed with large particles should be used to prepare large amounts of the purified material. The resolution afforded by each approach can be made equivalent (at comparable sample size) by adjusting column length, but separation time is longer for larger particles, for which longer columns are needed for equivalent resolution.

15.2.1.2 Equipment. Equipment requirements for satisfactory preparative SEC are not as critical as for analytical SEC; lower-cost, less sophisticated systems are generally used. However, to optimize preparative SEC separations, it is necessary to use different pumps, sampling systems, and detectors than those that are normally required for analytical work.

Pumping systems should deliver solvent up to 100 mL/min for large-bore (e.g., ≥ 2 cm i.d.) columns. Very high pressure capability is not required in preparative studies, and pressure limits of about 2000 psi (140 bar) are usually adequate.

Since analytical information from the preparative chromatogram is of less interest, the precision and accuracy specifications of the pump are not as critical, as discussed in Section 5.4. Pneumatic-amplifier and reciprocating pumps provide satisfactory pumping rates and a continuous solvent output. Since the pumping systems of commercial analytical LC instruments often deliver no more than about 10 mL/min, separations with this equipment use only the narrower-bore preparative columns (e.g., ≤ 0.8 cm i.d.). Larger-i.d. columns used with analytical pumping systems require very long separation times because of the low volume flow rates. Pulsations from certain pumps (e.g., reciprocating) which affect detector baselines are usually not a serious disadvantage in preparative applications. To supply the large volume of mobile phase used in preparative SEC, relatively large solvent reservoirs are required.

In preparative SEC, solute concentrations are generally high, and highly sensitive detectors are not required. High sample concentrations can cause problems, since it may be difficult to determine whether overlapping peaks are due to column overload or to a nonlinear detector response. The DRI detector is generally suitable for preparative SEC; UV detectors with a short-path-length cell are also useful. Using both the UV and DRI detectors in series helps to ensure that all the components of interest are monitored. UV spectrophotometric detectors are often "detuned" from the wavelength of solute absorption maximum to decrease detection sensitivity and reduce the potential for a nonlinear detector response.

Preparative SEC differs from other preparative LC techniques. In SEC, the chromatogram represents the broad distribution of molar masses (or sizes) normally encountered in polymeric samples. This distribution can be thought of as an overlay of numerous individual chromatograms of monodisperse species. In traditional LC, a discrete number (two or more) of different chemical species are separated. Preparative LC columns can be overloaded without much concern for resolution. In preparative SEC, however, an increase in the sample amount injected leads to decreased resolution between individual distributions and to higher polydispersity in the fractions collected. This situation is illustrated in Figure 15.2. Because performance in preparative SEC is a compromise between polydispersity and yield, one must consider not only how much of a substance is needed but also how polydisperse it can be.

Use of low dead-volume tubings and fittings is not as critical in preparative SEC as in analytical applications, because of the relatively large internal volumes of large-diameter preparative columns. High-volume flow rates are needed for wide-diameter columns, and detectors for handling this flow should not be constructed from narrow-bore tubing, which can severely limit the flow of mobile phase and cause excessive backpressure. (Both DRI and UV detectors with larger-bore tubing are commercially available.) Alternatively, a stream splitter on the exit of the large-diameter column can be used with an analytical detector.

Sample volumes up to about 10 mL are conveniently introduced into preparative SEC columns with a sampling valve and without interrupting the mobile-phase flow. For very large sample volumes (e.g., >100 mL) the sample can be loaded into the column by means of a low-volume sample-metering pump, using a sampling valve in the stop-flow mode. The pump is attached to the sample loop with the sampling metering pump turned off and the valve in the "inject" position (Figure 5.7). After the

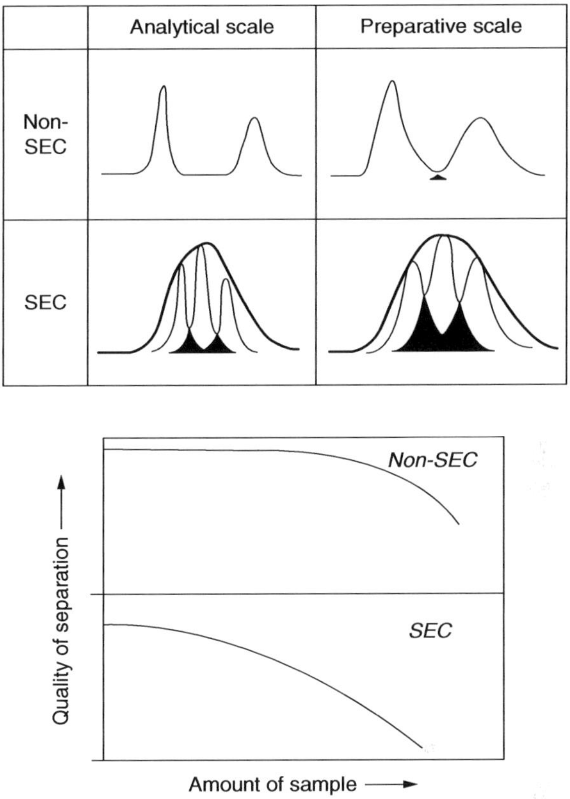

Figure 15.2 Effect of overloading in analytical and preparative LC and SEC. (Reprinted with permission from Ref. 8.)

required sample volume has been pumped through the loop into the column inlet, the valve is then rotated to the bypass position and the mobile-phase pump is restarted.

When only a few components are to be isolated, manual collection of the fraction is adequate, particularly when fast, small-particle columns are employed. However, when long, repetitive runs are needed, it is more convenient to use automatic fraction collectors (Section 5.6).

15.2.1.3 Operating Variables. Volumetric flow rates for preparative columns must be increased linearly with cross-sectional area, to maintain the same linear flow velocity as in analytical SEC columns. Table 15.1 shows typical column diameters in both analytical and preparative SEC and the corresponding volumetric flow rates at equivalent linear flow velocities.

The dispersive effects of sample injection in preparative SEC are not well understood. Loading the sample across the entire column cross section is preferred, since this permits more effective use of the total column packing with reduced column overload (Section 7.4). When possible, samples should be injected as

Table 15.1 Typical column diameters in SEC and corresponding flow rates (mL/min)[a]

Analytical Column, $\frac{3}{8}$ in. o.d.	Preparative Column, 1 in. o.d.	Preparative Column, 2.5 in. o.d.
0.1	0.8	5.5
1.0	7.7	55
5.0	38.2	275
10.0	76.4	549

Source: Ref. 9.
[a] At equivalent mobile-phase velocities.

relatively large volumes of a lower concentration rather than as smaller volumes of more concentrated solutions. Improved sample loadability and column performance result from this approach, since the effect of overloading the packing at the column inlet is minimized.

The volume of sample that can be introduced into a preparative column will depend on column internal diameter and length, the solute and solute solubility, the mobile phase/stationary phase combination, and the resolution required. As in analytical SEC (Section 7.4), for highest resolution, sample volume should not exceed about one-third the volume of a totally permeating monomer. However, much larger sample volumes are often used in preparative SEC if resolution permits (e.g., >20 mL for a 3.7-cm-i.d. column).

The sample loading limit in SEC is dependent on solute M. The loading capacity of columns can be increased by increasing column length or diameter. However, increasing column length also increases the solute resolution and retention volume; this requires additional separation time and mobile phase.

Figure 15.3 and the discussion in Section 7.4 indicate that plate heights increase (and resolution decreases) as sample volumes are increased, to the point where band dispersion occurs largely because of the sample volume alone. Column performance can also be affected by sample weight if the column is overloaded. Experimental results suggest that 0.1 to 1 mg of a solute per gram of packing can be injected into a column without significant change in either retention volume or plate height [10]. However, for preparative studies, much larger sample weights are often employed to obtain the desired weight of isolated component at the required purity. As noted in Section 7.4, the column overload phenomenon in SEC is not well understood. However, in preparative SEC, as well as for the other LC methods [1,2], sample loads should be increased to the point where there is just adequate resolution, even though such sample loads affect both retention volume and column efficiency. Because under overload conditions solute retention is a function of sample size, it is important to maintain a constant concentration when performing preparative SEC separation of polymer fractions. Only in this manner can the expected M (as determined by analytical SEC) be obtained for the isolated polymer fractions.

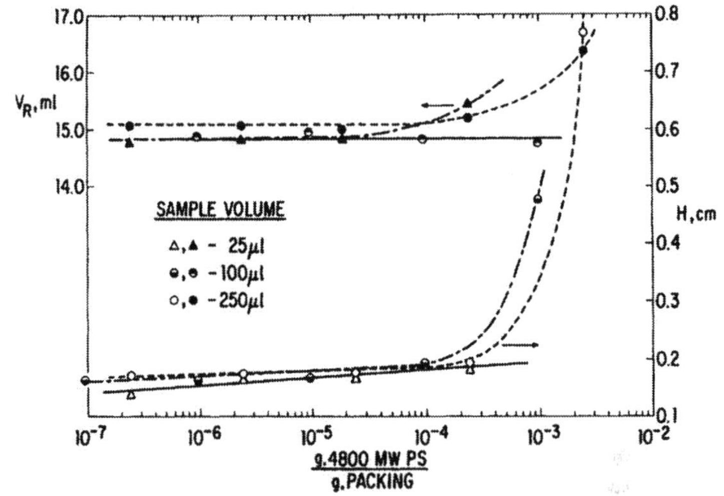

Figure 15.3 Effect of polymer sample size on column performance. Column, 100 × 0.62 cm porous silica microspheres, 47 Å (silanized); mobile phase, THF; temperature, 22°C; flow rate, 1.5 mL/min; detector, UV, 254 nm; solute, 4.8 × 10^3 g/mol polystyrene in tetrahydrofuran. Upper curve, effect of sample loading on retention volume; lower curve, effect of sample loading on plate height. (Reprinted with permission from Ref. 10.)

The effect of flow rate on preparative LC column efficiency is the same as for analytical SEC (Section 7.2.3) when columns are operated at small sample loads. However, at high sample loadings (e.g., with the column in overload) the effect of flow rate is less important [6]. In sample-overloaded preparative columns, mobile-phase velocity can be increased substantially to reduce separation time without significant sacrifice in resolution. Thus, preparative SEC separations should be carried out at the highest practical mobile-phase velocity that still allows adequate resolution.

As in analytical SEC, mobile phases of relatively low viscosity are favored in preparative SEC to maintain high column efficiency. The solvent must be compatible with the detector and should be volatile for convenient removal from isolated fractions. Of special importance is that the mobile phase be highly purified, since nonvolatile impurities are concentrated when the solvent is removed from a fraction and significant contamination results. This problem is minimized by using freshly distilled or "distilled-in-glass" solvent. Higher column temperatures usually enhance solubility if needed, but high-temperature operation is less convenient.

Isolated fractions can be concentrated by evaporation of solvent under a stream of pure, dry nitrogen while warming (e.g., with an infrared lamp). Operations that tend to condense water in the isolate (e.g., heating on a steam bath) should be avoided. Large volumes of mobile phase can often be conveniently removed with a rotary vacuum evaporator. Freeze drying is effective for some solvents, such as water, dioxane, and benzene.

The cut points that are used to collect the fraction largely determine purity and yield. For two overlapping components of equal amounts, Figure 15.4a shows that

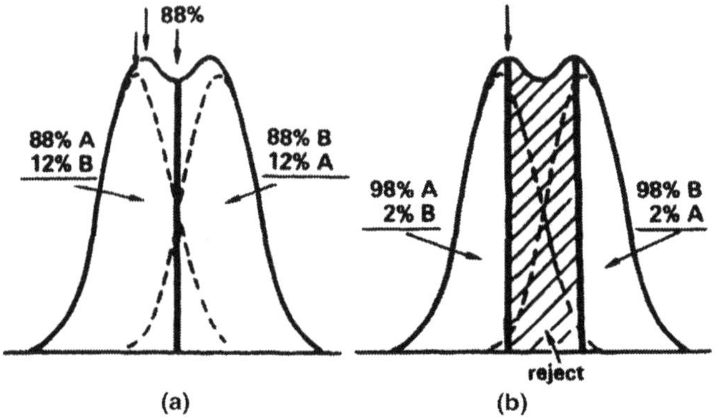

Figure 15.4 Effect on fraction purity by rejecting overlapped peaks. (Reprinted with permission from Ref. 12.)

at the equal purity cut point (i.e., valley between peaks, $R_s = 0.6$ in this example), fractions obtained are 88% A and 12% B, and 88% B and 12% A, respectively. If it is desired to improve the purity of both components, the overlapping center portion of the bands can be rejected (crosshatched in Figure 15.4b). For the fraction collection in Figure 15.4b, composition of the peak on the left would be 98% A, 2% B, while the composition of the peak on the right would be 98% B, 2% A (assuming equivalent detector response). Total yield of purified material obtained by this technique is about 61% of that injected, compared to 100% yield in the previous example. If required, the rejected overlapping fraction may be rechromatographed to obtain components of the same high purity.

For major components it is often desired to use a *heart-cut technique*, which produces a highly purified component with a modest yield loss. This approach is illustrated by the hypothetical separation in Figure 15.5, where a heart cut of a major component overlapped by unwanted contaminants is selected. Here, by rejecting the impure "wings" of the main peak, overall product yield is decreased, but with a significant improvement in product purity. If insufficient amounts of purified material are obtained from any collection approach, it may be necessary to make replicate runs and accumulate the desired fractions. Following the final collection, purity should be analyzed by SEC or other appropriate analytical technique. If the isolated component is not of the desired quality, it can be fractionated again for higher purity.

15.2.2 Applications

Occasionally, narrow-MMD or size distribution polymer standards are prepared by fractionation of a broadly distributed polymer. Preliminary runs on the broad MMD polymer determine the cut points that give fractions with the desired *M* and MMD. Figure 15.6 shows MMD curves of fractions from NIST 706 polystyrene. The

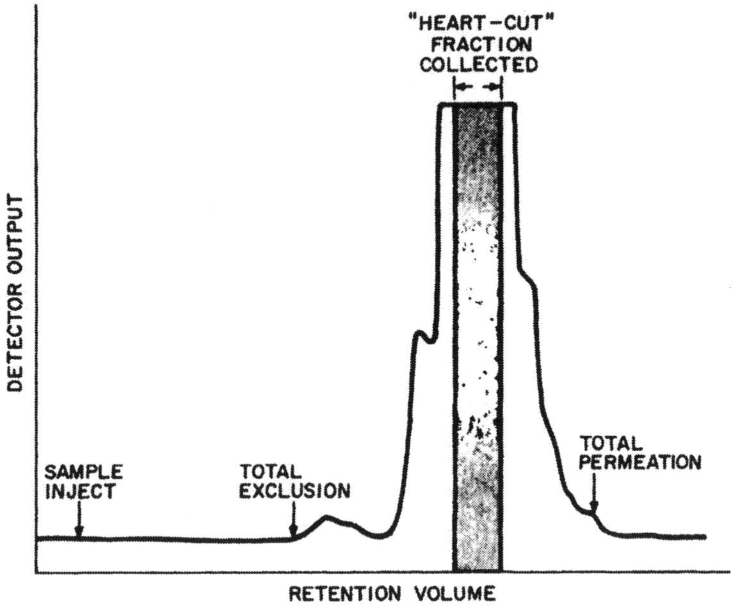

Figure 15.5 Heart-cut technique in preparative SEC.

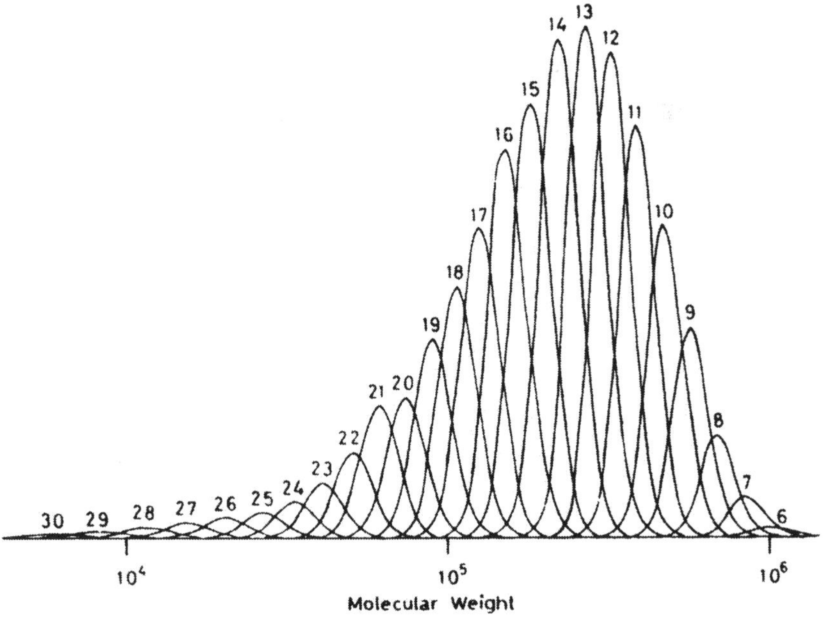

Figure 15.6 MMD curves of fractions from standard polystyrene NIST 706. Fraction numbers are indicated. Column, 240 × 0.8 cm total TSK-GEL Type G, 10 μm particle size, 10^6, 10^4, and 10^3 Å porosity; mobile phase, methyl ethyl ketone/methyl alcohol (88.7 : 11.3-theta solvent at 25°C); flow rate, 8.5 mL/min; temperature, 25°C; detector, DRI; sample, 20 mL, 6.5 mg/mL of polymer in mobile phase. (Reprinted with permission from Ref. 11.)

polymer fractions obtained in this preparation showed polydispersities of 1.017 to 1.035, which represents material suitable for molar mass calibration by the peak position calibration method (Section 8.2; see also Section 15.3 for a method used to determine the polydispersity of very narrow MMD fractions).

Figure 15.7 shows the analytical and preparative chromatograms of a sample of cadmium sulfide (CdS), a semiconductor colloid. The size (diameter) polydispersity of the original sample was in the range 1.5 to 1.7. Size polydispersity of the fractions obtained by preparative SEC ranged from 1.04 to 1.1, with standard deviations of 11 to 18%. Using columns of 32 mm i.d. allowed for injections of up to 200 mg of sample.

In addition to preparing narrow MMD standards of polymers, high-resolution preparative SEC is useful to prepare purified low-*M* compounds (e.g., monomers). As illustrated in Figure 15.8, 150 mg of three components (molar mass indicated on peaks) was injected into a small-particle analytical SEC column. Such a sample load is easily fractionated at high resolution with this column. Larger amounts of purified samples can be obtained by increasing sample size and using the heart-cut technique.

Preparative SEC is often valuable for isolating and identifying trace concentrations of high-*M* additives in a low-*M* matrix, or for measuring a low-*M* additive (e.g., plasticizer) in a polymer. For example, Figure 15.9 shows a chromatogram in which a polymeric additive in lubricating oil is well separated and readily available for collection and subsequent characterization by a suitable auxiliary technique. In favorable cases, parts per million of such additives can be isolated and identified in a single run.

Preparative SEC has been carried out on a wide variety of water-soluble macromolecules. Figure 15.10a shows the preparative chromatograms of three industrial pectin polysaccharides from different agricultural sources. Analysis of each individual pectin fraction was then conducted using analytical columns; the latter chromatograms are shown in Figure 15.10b.

It has been shown that complexes of plasmid DNA and the biodegradable polymer poly(2-dimethylaminoethylamino)phosphazene [p(DMAEA)-ppz] mediate tumor-selective gene expression after intravenous administration in mice. Preparative SEC allowed for isolation of narrow-polydispersity (1.1 to 1.3) fractions of p(DMAEA)-ppz, ranging in molar mass from 130 to 950 kg/mol. For various polymer-to-DNA (polyplex) ratios, in vitro toxicity positively correlated with polymer molar mass. For example, administration of polyplexes using low-*M* (130 kg/mol) p(DMAEA)-ppz showed no signs of toxicity and resulted in tumor-selective gene expression. Preparative and analytical SEC chromatograms of the starting polymer and of an intermediate-*M* (290 kg/mol) fraction are shown in Figure 15.11a and b, respectively.

SEC can also be carried out on a process scale to prepare commercial quantities of purified materials. For example, certain water-soluble biological compounds (e.g., enzymes) have long been purified by process-scale aqueous SEC using relatively soft hydrophilic gels [4]. The theory and apparatus for continuous preparative

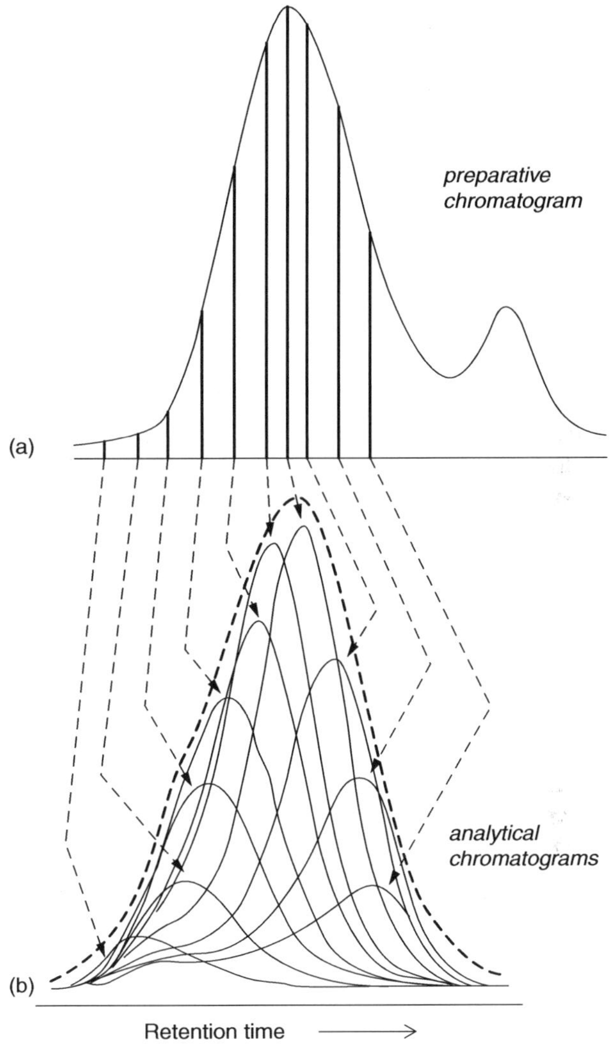

Figure 15.7 Preparative SEC of a CdS colloidal solution. (a) Preparative chromatogram (the small, late retention time peak is due to polyphosphate). Ten fractions (thick vertical lines) were collected. Column, two 32 × 125 mm Knauer columns, first column packed with 5-μm Nucleosil 500 C4, second column packed with 5-μm Nucleosil 1000 C4; solvent, solution of 1×10^{-3} M cadmium perchlorate and 6×10^{-3} M sodium polyphosphate; flow rate, 4.5 mL/min. (b) Analytical chromatograms. Dashed line, diluted original solution (peak height normalized); solid lines, analytical chromatograms from the 10 fractions collected in preparative mode. Time scales for preparative and analytical experiments are different. Columns are the same as in (a), but of dimensions 4 × 125 mm. (Reprinted with permission from Ref. 8.)

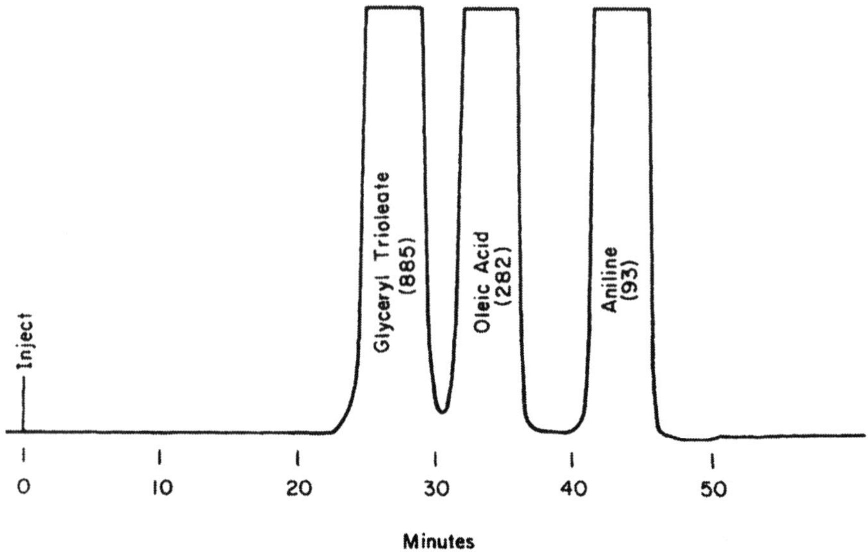

Figure 15.8 Preparative SEC of small molecules. Column, 150 × 0.8 cm μ-Styragel 100 Å; mobile phase, THF; flow rate, 1.0 mL/min; detector, DRI; sample, 150 mg of each compound. (Reprinted with permission from Ref. 13.)

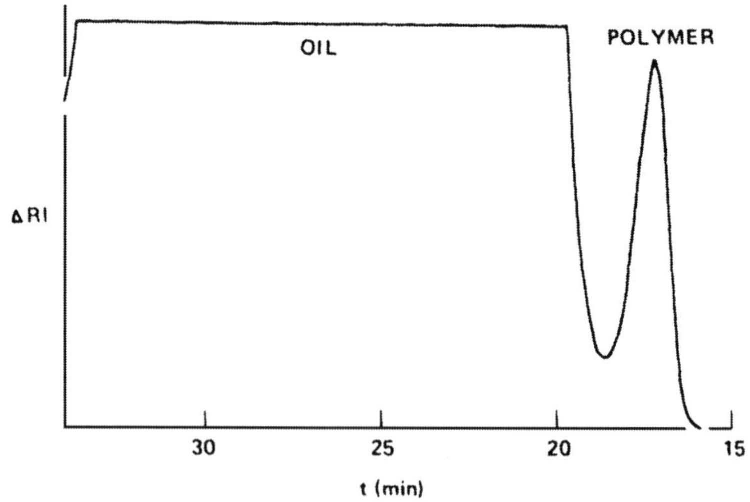

Figure 15.9 Polymer additive in lubricating oil. Column, 120 × 0.8 cm Styragel 2–10^2 Å, 1–10^3 Å, 1–10^4 Å; mobile phase, THF; temperature, room; detector, DRI. (Reprinted with permission from Ref. 14.)

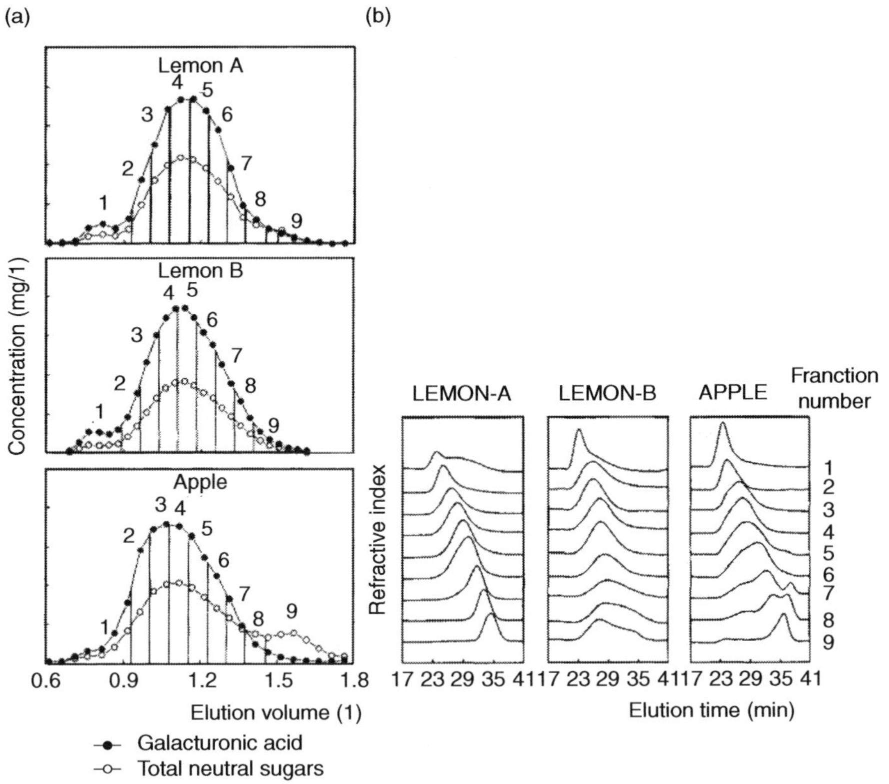

Figure 15.10 Analytical and preparative SEC of industrial pectins. (a) Preparative analysis on a 5 × 90 cm Fractogel TSK HW 55(S)/75(S) mixed-bed column. Solvent, 0.1 M sodium succinate buffer, pH 4.8; flow rate, 120 mL/h; injected mass, 150 mg of each pectin. (b) Analytical SEC using a series of 7.5 × 300 mm Bio-Gel TSK columns (60XL, 40XL, 30 XL) plus a 7.5 × 75 mm Bio-Gel TSK guard column. Solvent, 0.4 M sodium acetate buffer, pH 3.0; flow rate, 0.8 mL/min; temperature, 30°C; detector, DRI (at 40°C). (Reprinted with permission from Ref. 15.)

chromatography of a binary mixture of polymers by SEC have also been developed [17,18].

15.3 RECYCLE SEC

Very high resolution is needed occasionally for certain SEC separations, and the increase in resolution required can be obtained by adding many extra columns to the system. Resolution increases linearly with the square root of column length, but for well-packed columns the backpressure increases linearly with length. Therefore, in practice there is a finite restriction on maximum column length as a result of pressure limitations. One solution to these problems is to recycle the sample through the same

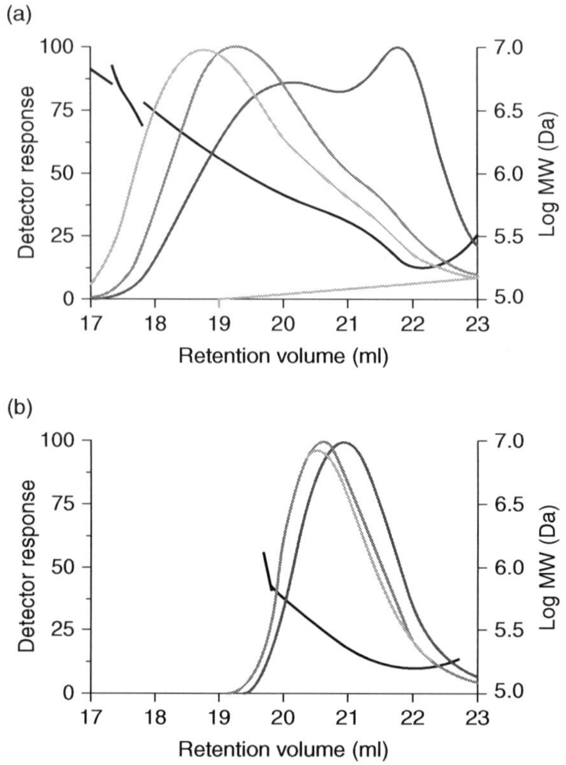

Figure 15.11 Analytical and preparative SEC of biodegradable polymer used for gene delivery to tumor cells. (a) Preparative analysis of starting p(DMAEA)-ppz polymer. Columns, preparative OHpak SB-LG and SB-2006M; solvent, 0.3 M sodium acetate, pH 5.0; flow rate, 1.5 mL/min. (b) Analytical SEC of polymer fraction with $M_w = 290$ kg/mol, $M_w/M_n = 1.2$. Left ordinate denotes detector responses for a triple-detector system: low-angle light scattering (bright gray), differential viscometry (medium gray), DRI (dark gray). Right ordinate is molar mass (plotted on a logarithmic scale), as determined by SEC^3 (Section 9.6): darkest line in figures corresponds to M as a function of retention volume. (Reprinted with permission from Ref. 16.)

column set one or more times to increase the effective column length. Increased resolution by recycle is obtained just as if extra column lengths were added, but the attendant increase in pressure is not experienced. There are several advantages and pitfalls of recycle:

Advantages:

1. Additional resolution is obtained without the need for additional columns.
2. With some arrangements the sample profile is recorded by the detector after each pass through the column.
3. The method can be made semiautomatic.

Disadvantages or pitfalls:

1. Some commercial equipment will permit recycle only after substantial modifications.
2. Extracolumn effects must be minimized more carefully (particularly if recycling is carried out through the pump).
3. Complex or broad-MMD samples permit very few cycles before the front edge of the retention curve in one cycle overtakes the trailing edge of the retention curve from the preceding cycle.

The recycle method can be used for many applications to (1) increase the accuracy of molar masses calculated from the chromatogram, (2) determine true M_w/M_n values of very narrow MMD standards, (3) purify fractions of materials for other studies, and (4) increase column resolution to bring out the fine features of a sample (e.g., to distinguish individual oligomers).

15.3.1 Theory

Chromatographic peak separation (i.e., the distance between peak retention volumes, $V_{R2} - V_{R1}$ or ΔV_R) in SEC is linearly proportional to the column length L. In addition, $D_2 = \Delta \ln M / \Delta V_R$, which is proportional to the slope of the molar mass calibration curve. For a change in column length, $D_2 \propto 1/L$. In recycle, for n passes through the column,

$$D_{2,n} = \frac{D_2}{n} \tag{15.1}$$

However, peak or band spreading as measured by the peak width σ varies with $\sqrt{L}$ so that for n passes through the column, σ varies as $\sqrt{n}$.

SEC column resolution, R_s, makes use of the parameters for peak separation (ΔV_R) and broadening (σ) and is expressed as (Section 4.2)

$$R_s = \frac{\Delta V_R}{4\sigma} = \frac{\Delta \ln M}{4 D_2 \sigma} \tag{15.2}$$

For n passes through the column (recycle), the column resolution becomes

$$R_{s,n} = \frac{n\ \Delta V_R}{\sqrt{n}\, 4\sigma} = \sqrt{n}\, R_s \tag{15.3}$$

If a more rigorous expression of resolution is needed, the σ terms must be broken down into the components. The values of σ due to injection are not repeated on additional passes through the column, and values of σ due to pump mixing do not exist in the first pass but are introduced in subsequent cycles.

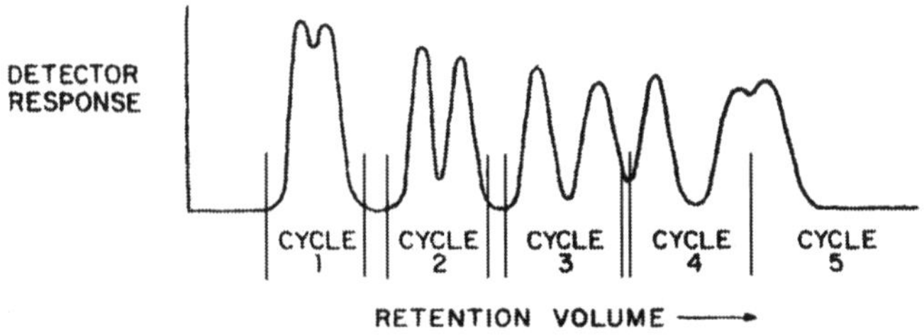

Figure 15.12 Separation of tyranine and dopamine using recycle SEC. Biogel P-2 columns, UV detector, 254 nm. (Reprinted with permission from Ref. 19.)

One main limitation to the recycle method is that the fastest-moving peak eventually overtakes the slowest, and remixing occurs. An example of this is shown in Figure 15.12, where optimum separation occurs in the third cycle and remixing is evident in the fourth cycle. For separating two components, the optimum number of cycles, n_{opt}, is given by [19]

$$n_{\text{opt}} = \frac{V_{R1}}{2(V_{R2} - V_{R1})} \tag{15.4}$$

where V_{R2} is the slower-moving peak and n_{opt} is a round-off integer. For example, if the relative peak distance is 0.25, $n_{\text{opt}} = 2$, whereas for a relative peak distance of 0.05 (5% separation), $n_{\text{opt}} = 10$. Optimum resolution for n passes through the column is calculated by [19]

$$R_{s,\text{opt}} = \frac{1}{5.7} N^{1/2} \left(\frac{V_{R2} - V_{R1}}{V_{R1}} \right)^{1/2} \tag{15.5}$$

where in this case the values of V_R and N are those obtained at the n_{opt} pass through the column. As described below, for multipeak samples a "draw-off" procedure can be used to eliminate unwanted materials, permitting increased separation of other components by additional cycles.

15.3.2 Equipment

There are two approaches to carrying out recycle SEC. The first and simplest is the closed-loop method shown in Figure 15.13, for which the solvent-flow options are indicated by the arrows. The sample is passed through the column and detector and back through the pump in a closed loop for the required number of times. Each pass is monitored by the detector, and a switching valve (V_2) then permits the operator to collect or discard peaks as they emerge. Although very simple in concept, this approach is sometimes difficult in practice. The position of several valves must be carefully coordinated, and the detector must be capable of withstanding the high operating pressure of the system without leaking (many commercial detectors have not been so constructed). Additionally, peak dispersion due to the connector tubing

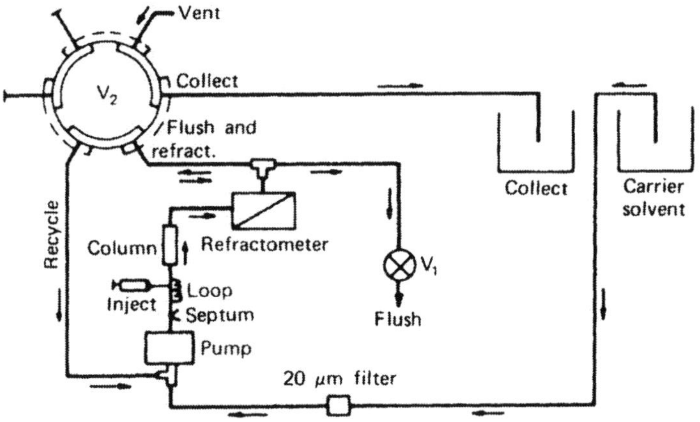

Figure 15.13 Closed-loop method of recycle operation. (Reprinted with permission from Ref. 20.)

and pump chamber must be carefully minimized or the advantages of recycle will not be obtained.

The other recycle method employs alternate pumping of dual columns, which has the advantage that the sample does not pass repeatedly through the pump chamber, thus minimizing peak broadening. An understanding of the details of this method is aided by reference to Figure 15.14. The sample containing the peaks to be resolved is introduced to the system via the injection valve (six-port). As the sample passes through column 1, it is monitored by cell 1 of a dual-cell UV photometer before it passes into column 2. In the first valve position, cell 1 is at high pressure while cell 2 is at ambient pressure. While the sample is in column 2, the valve is switched to divert the flow back into column 1. In this valve position, cell 2 becomes the high-pressure cell. As the sample emerges from column 2, the output from cell 2 provides a record to indicate whether adequate resolution has been obtained. A sample can be cycled through such a column system until the peaks broaden and completely occupy one column volume, or until adequate resolution has been obtained. Actually, it is not essential to monitor the peaks as they emerge from each column if matched-performance columns of essentially the same elution volume are used. In this case, the switching cycle of one column can be calculated based on data obtained using the other matched column.

15.3.3 Uses of the Recycle Method

15.3.3.1 Improvement of Molar Mass Accuracy. Recycle can be used to decrease the errors in calculated molar mass averages. The errors in M_n and M_w are given by Equations 4.18 and 4.19. Because for n passes through the column the resolution increases by $\sqrt{n}$ (Equation 15.3), the molar mass *error* becomes

$$M^*_{w,n} \text{ or } M^*_{n,n} = e^{\pm(1/2n)(\sigma D_2)^2} - 1 \qquad (15.6)$$

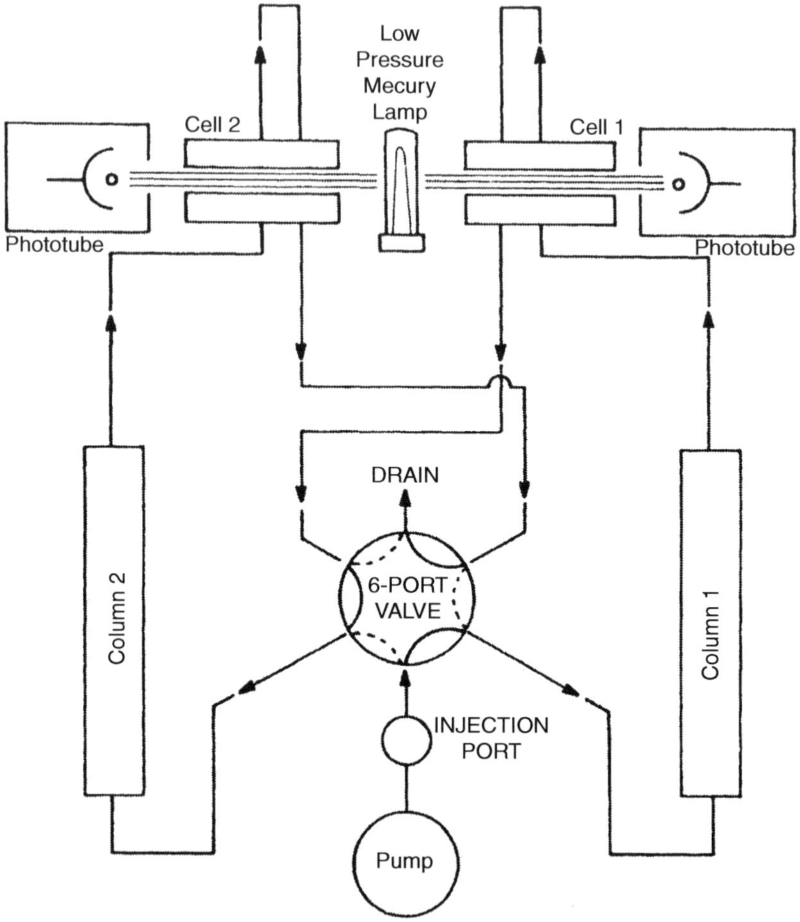

Figure 15.14 Dual column, alternate pumping recycle method. (Reprinted with permission from Ref. 21.)

This relationship shows that for low values of σD_2 (typical values for SEC range from 0.2 to 0.5), the molar mass error decreases nearly linearly with n. For example, M_w^* and M_n^* for $\sigma D_2 = 0.4$ are 8%, 4%, 3%, and 2% for $n = 1, 2, 3$, and 4, respectively. Thus, a molar mass error based on a single pass is halved by a second pass through the column. However, the use of recycle to improve molar mass accuracy assumes that no additional extracolumn peak broadening occurs.

When using the recycle method to improve accuracy, either the peak position method or GPCV2 or GPCV3 described in Section 8.3 must be used for molar mass calibration. Also, the values of σ and D_2 determined for the standards must be for the same number of passes n as for the unknown samples. Increasing column length can also improve molar mass accuracy, but this approach requires higher pressures and additional column inventory. On balance, it is usually more convenient to add column length than to use recycle if well-packed columns are available and the pressure is not excessive.

15.3.3.2 Determination of the MMD of a Very Narrow MMD Material. Recycle chromatography for accurately determining the molar mass distribution of a very narrow MMD material takes advantage of the increased resolution of the sample peak in each pass through the column and permits extrapolation of the band broadening to zero so that the final peak width is due to MMD alone. As discussed in Section 3.1, the variance of the total chromatographic curve width is given by the sum of variances of each of the contributors:

$$\sigma_T^2 = \sigma_{\text{inj}}^2 + \sigma_{\text{disp}}^2 + \sigma_{\text{MMD}}^2 + \sigma_{\text{ex.col.}}^2 \tag{15.7}$$

where σ_T^2 is the total peak dispersion, σ_{inj}^2 is that dispersion due to sample injection, σ_{disp}^2 is the chromatographic band dispersion, σ_{MMD}^2 is the spreading due to the natural MMD of the sample, and $\sigma_{\text{ex. col.}}^2$ consists of the spreading caused by all extracolumn sources.

In optimum recycle SEC experiments, σ_{inj}^2 becomes insignificant and $\sigma_{\text{ex. col.}}^2$ is also minimized, so that only the chromatographic and molar mass dispersions are important. Because σ_{disp} is proportional to $\sqrt{n}$ and σ_{MMD} is proportional to n,

$$\sigma_{T,n}^2 \simeq n\sigma_{\text{disp}}^2 + n^2\sigma_{\text{MMD}}^2 \tag{15.8}$$

or

$$\frac{\sigma_{T,n}^2}{n^2} \simeq \frac{\sigma_{\text{disp}}^2}{n} + \sigma_{\text{MMD}}^2 \tag{15.9}$$

A plot of $\sigma_{T,n}^2/n^2$ versus $1/n$ yields a straight line with the intercept σ_{MMD}^2. (This corresponds to an extrapolation to an infinite number of cycles, i.e., infinite resolution.) To obtain values of M and MMD from the σ_{MMD}^2 data, the molar mass calibration curve for the columns is required and the shape of the MMD must be assumed. Figure 15.15 presents the data for a very narrow MMD in-house-fractionated polystyrene. Using a peak-position calibration curve and assuming a Gaussian distribution of molar masses for the sample, a polydispersity value of 1.00248 was obtained [22]. Further verification of the method is provided by Figure 15.16, which shows the data obtained for hexane. Here the sample is monodisperse and the extrapolated $\sigma_{T,n}^2/n^2$ value approaches the expected value of zero.

15.3.3.3 Preparative SEC Separations by Recycle. Recycle SEC can be used to obtain the additional resolution needed to separate materials of nearly the same size in preparative SEC (Section 15.2). Using the equipment shown in Figure 15.13, Figure 15.17 illustrates the recycle and draw-off method for isolating various components in Triton X-45, a complex surfactant based on alkylaryl polyether

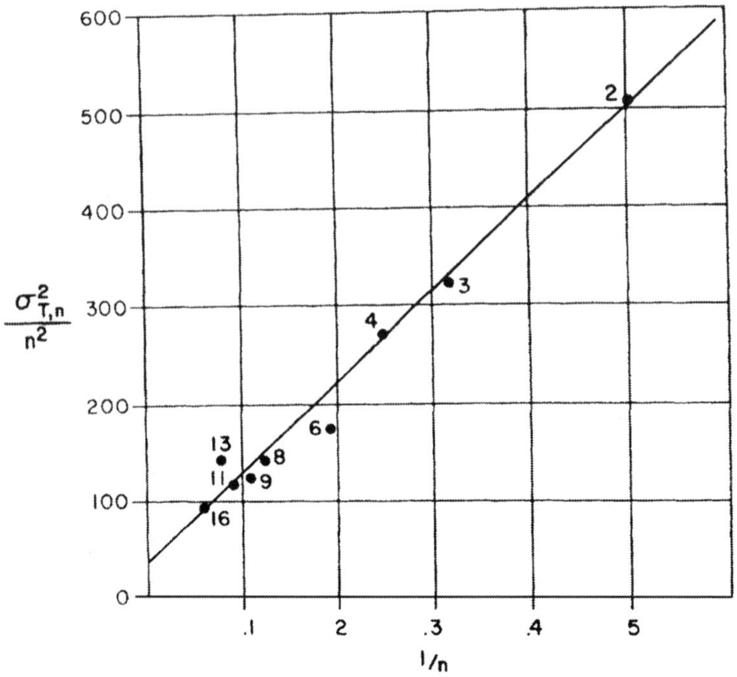

Figure 15.15 Relationship between number of cycles and total curve broadening for a narrow-MMD polystyrene sample in the recycle SEC mode. (Reprinted with permission from Ref. 22.)

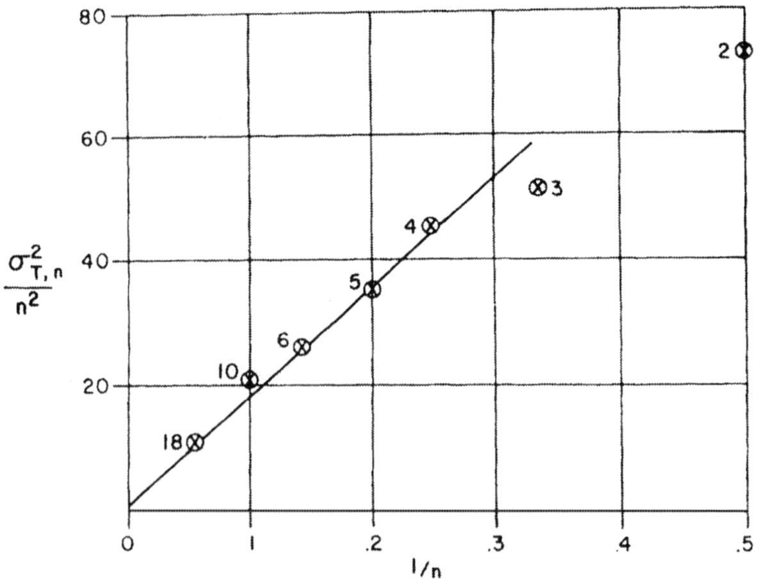

Figure 15.16 Relationship between number of cycles and total curve broadening for hexane in the recycle SEC mode. (Reprinted with permission from Ref. 22.)

alcohols, sulfonates, and sulfates. In this case there were six cycles, with the low-M end resolved first. The increase in resolution with cycle number is apparent, and the characterization of the components could probably be accomplished after cycle 3 or 4 because adequate resolution was obtained. To collect purified fractions, peak 1 could be drawn off in cycle 3; peaks 2 and 3 must be drawn off at cycle 5 since they overtake the highest-M peak in the next cycle. Collection and further recycling of peak 3 five times (Figure 15.17) indicates that 92% of peak 3 and 8% of peak 4 were obtained in the original peak 3 cut.

Similarly, preparative SEC followed by recycle (alternate pumping method) has been used to separate vinyl chloride (VC) oligomers from poly(vinyl chloride) (PVC) polymer. Figure 15.18a shows the preparative chromatogram of the diethyl ether Soxhlet extract of PVC. The next four figures in the series (b to e) show the recycle chromatograms after the first pass and after the first, second, and third recycles. The recycle method was shown to provide high-resolution separation of VC oligomers and was able to isolate sufficient quantity of material for subsequent analysis by GC-MS.

15.3.3.4 Closed-Loop Versus Alternate Pumping Methods. The two recycle SEC approaches, closed-loop (Figure 15.13) and alternate pumping (Figure 15.14), were recently compared for the analysis and separation of nanocrystalline gold (i.e., of gold nanocrystals stabilized by thiols). A broad distribution sample of 1-decanethiol-Au nanoparticles was analyzed by both recycle methods. Figure 15.19 shows the closed-loop recycle chromatogram. The initial, broad peak is observed to resolve into two peaks; however, resolution is quickly lost due to the extensive broadening introduced by the closed-loop method. Results from the alternate pumping method are shown in Figure 15.20. Figure 15.20a shows the chromatogram of cycle 2, with the small peak at 11.6 minutes corresponding to larger gold particles. The reproducibility and stability of the system are seen in Figure 15.20b, the overlay of four chromatograms after cycle 8. In Figure 15.20c, the evolution of the recycle SEC data is shown as a function of cycle number. The efficiency (plate number, N) and resolution (ratio) of the alternate pumping method are shown in Figure 15.21, which plots N and resolution ratio as a function of the square root of the cycle number. Peaks 2 and 3 in Figure 15.20b were used for the comparison. For peaks 2 and 3, the resolution ratio is given by

$$\frac{R(n)}{R(2)} = \frac{\sqrt{n}}{\sqrt{1+p^2}} \tag{15.10}$$

where $R(2)$ and $R(n)$ are the resolutions at cycles 2 and n (for $n \geq 2$), respectively; n is the number of cycles; and p is the fractional increase in peak width after each pass through the system. The parameter $p = \sigma_v/\sigma$, where σ_v is the bandwidth due to the tubing and switching valve, and σ is the bandwidth due to the stationary phase. Consequently, p can be considered a measure of the degree of inefficiency of the recycling process. The nearly 100% efficiency of the alternate pumping method were indicated by the high correlation coefficient, $r^2 = 0.97$, of the resolution line.

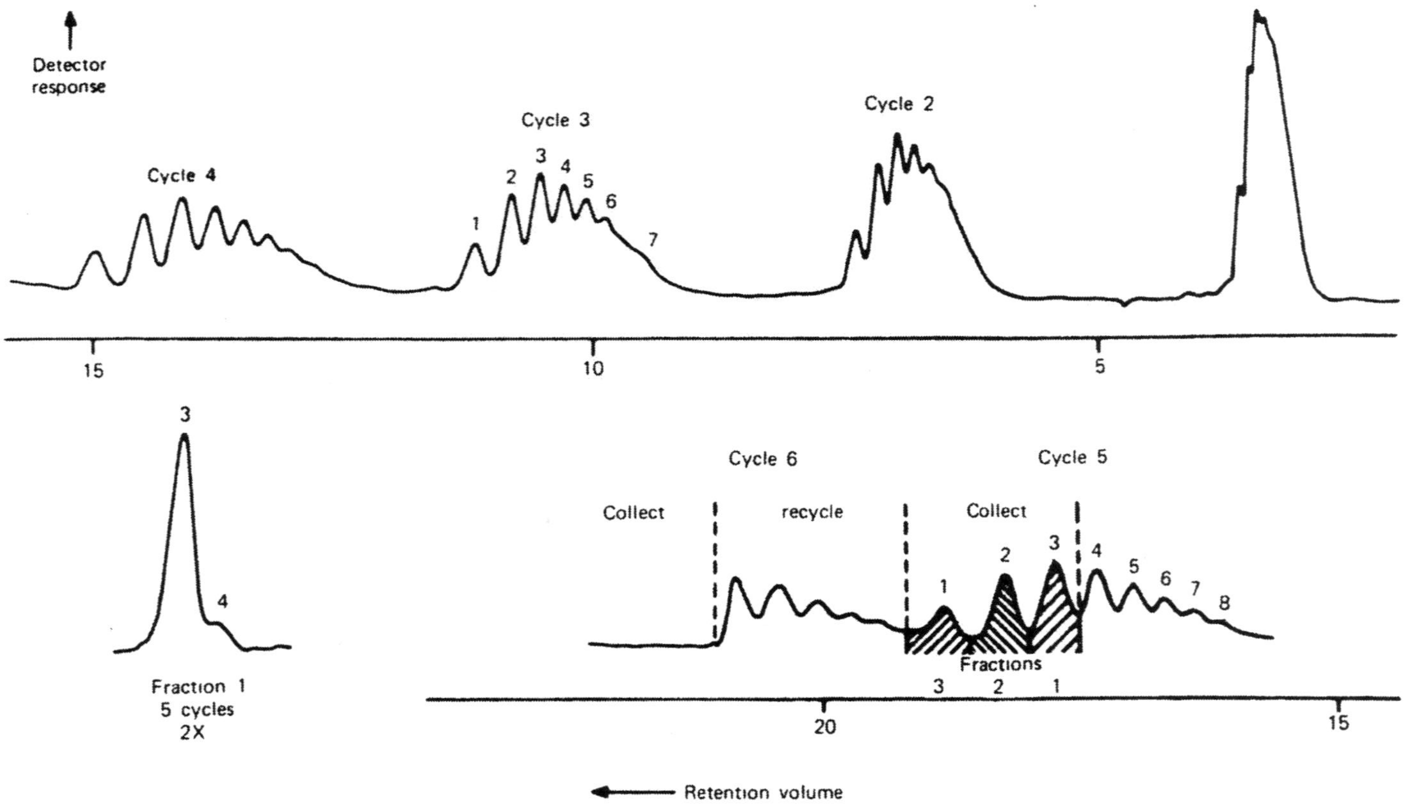

Figure 15.17 Separation of a complex mixture by recycle SEC. Columns, 450 × 0.8 cm Styragel 60 Å; mobile phase, tetrahydrofuran; temperature, 25°C; flow rate, 0.48 mL/min; sample, 50% Triton X-45, 30 μL; detector, DRI. (Reprinted with permission from Ref. 20.)

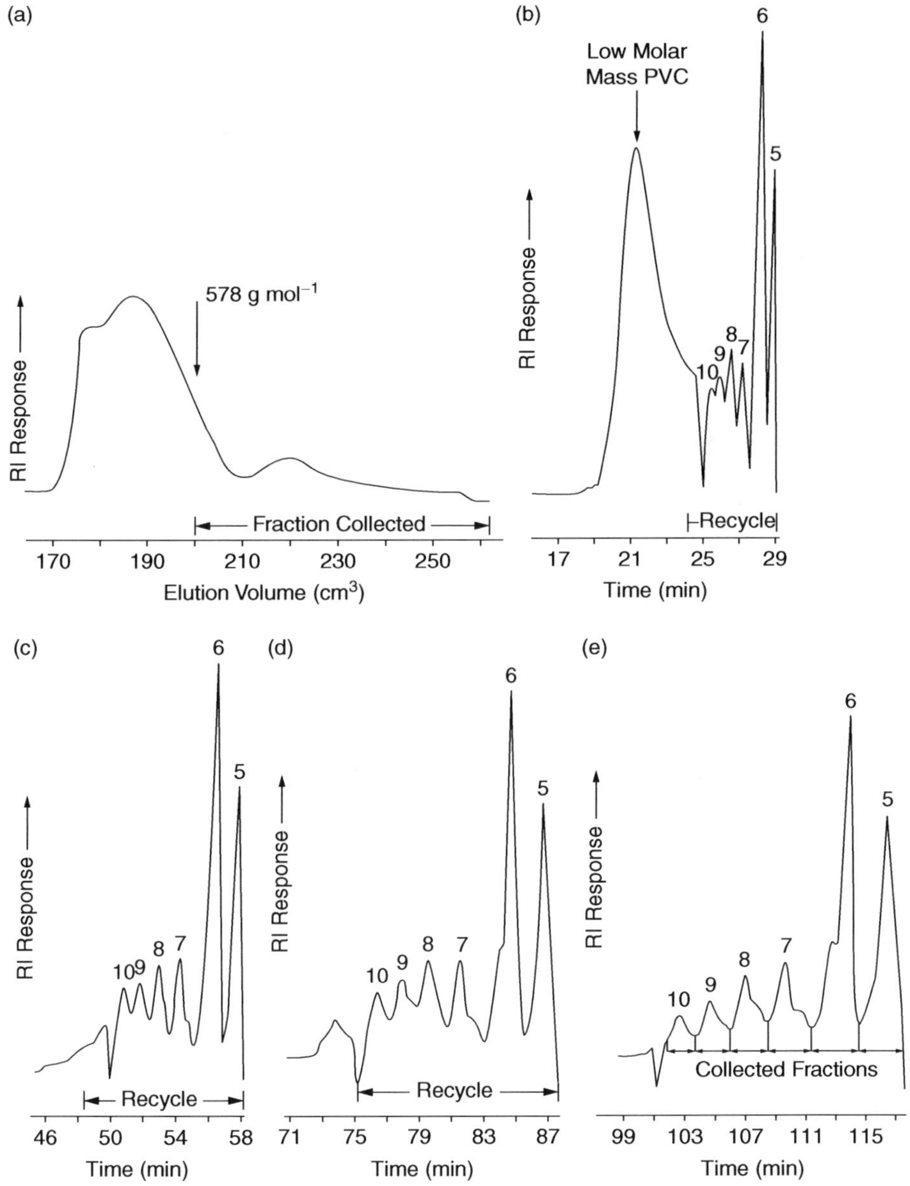

Figure 15.18 Preparative SEC of PVC and recycle SEC of VC oligomers. (a) Preparative SEC: column, 2.5 × 88 cm Bio-Beads S-X8; solvent, toluene; flow rate, 3.5 mL/min. (b–e) Alternative pumping method recycle SEC: columns, two 60 cm, 5-μm particle size, 50-Å pore size PLgel; solvent, toluene; flow rate, 1 mL/min; detector, DRI. (b) First pass; (c–e) first through third recycle passes. (Reprinted with permission from Ref. 23.)

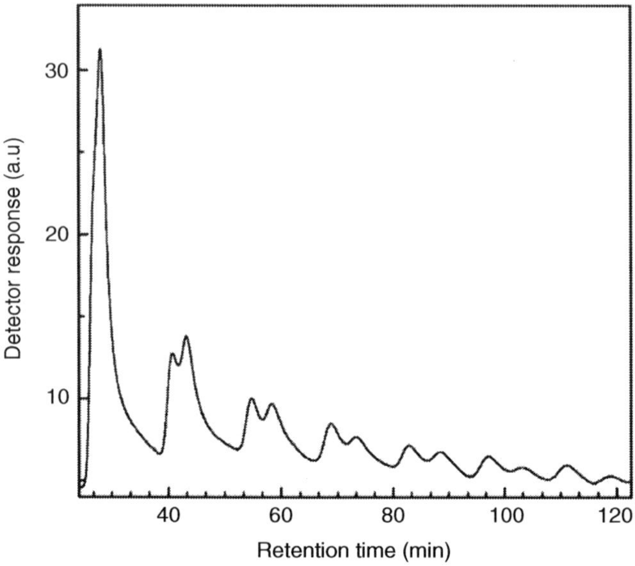

Figure 15.19 Closed-loop recycle SEC of nanocrystalline gold. Column, 6.5 × 300 mm, 5-μm particle size, 1000-Å pore size PLgel; solvent, toluene; flow rate, 0.5 mL/min; detector, UV/VIS. (Reprinted with permission from Ref. 24.)

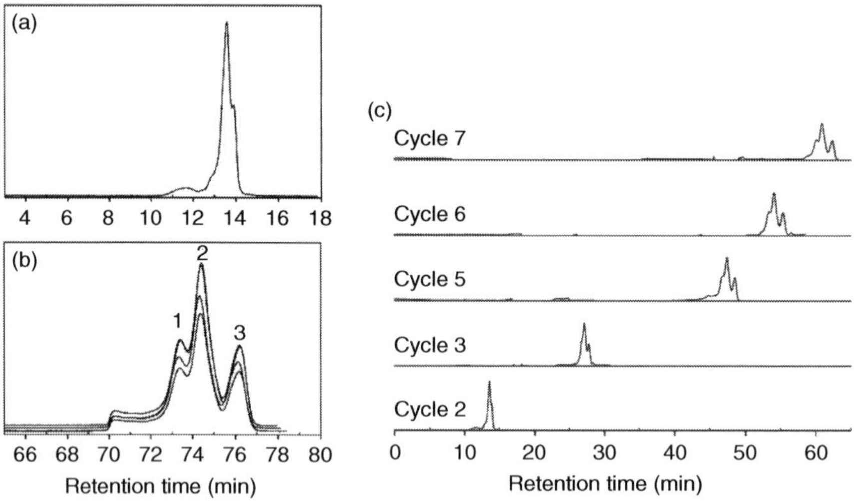

Figure 15.20 Alternate pumping recycle SEC of nanocrystalline gold: (a) chromatogram of cycle 2; (b) overlay of four chromatograms after cycle 8; (c) evolution of recycle SEC data as a function of cycle number. Experimental conditions are the same as for Figure 15.19, but two columns were used and the flow rate was 1.0 mL/min. (Reprinted with permission from Ref. 24.)

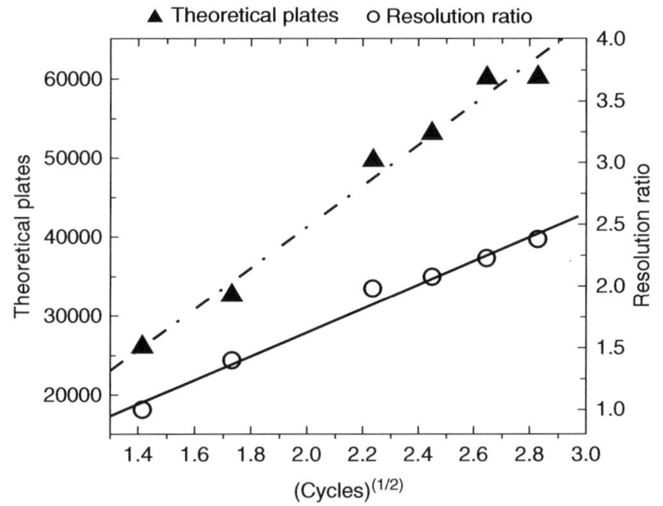

Figure 15.21 Efficiency and resolution ratio of alternate pumping recycle SEC. For analysis of nanocrystalline gold particles, comparing peaks 2 and 3 in Figure 15.20b. Efficiency is given as plate number N, resolution as resolution ratio (Eq. 15.10). (Reprinted with permission from Ref. 24.)

15.4 HIGH-SPEED SEC

The need for high-speed separations is driven by increases in high-throughput screening and analysis, especially in the area of combinatorial research, and by the increased demands in quality control and quality assurance (QC/QA) in industrial production. High-speed SEC has been a subject of interest for nearly as long as SEC itself has existed (see Figure 2.5) [25,26]. It is only during the last decade or so that the technique has become more widespread, due to the commercial availability of high-speed columns and/or systems. In general, commercial approaches to high-speed SEC have proceeded along four fronts [27]: (1) faster flow rates using conventional columns, (2) shorter columns, (3) narrower columns, and (4) columns with different aspect ratio (short, wide-bore columns). Comparisons are with respect to conventional columns operated at "normal" analytical flow rates.

Figure 15.22 is a chromatogram of a series of narrow-polydispersity PS standards analyzed under standard conditions (i.e., using a 300 × 8 mm, 5-μm particle size, styrene/divinylbenzene column run at 1 mL/min in tetrahydrofuran at room temperature). This chromatogram will serve as a benchmark to which results from the various high-speed approaches will be compared [28]. The plate number for this analysis [for the butylated hydoxytoluene (BHT) peak] was calculated as 92,500 plates/m and the specific resolution as 5.2.

1. *Using conventional columns at elevated flow rates.* As discussed in Chapter 3, band broadening in SEC is controlled predominantly by the mass transfer term

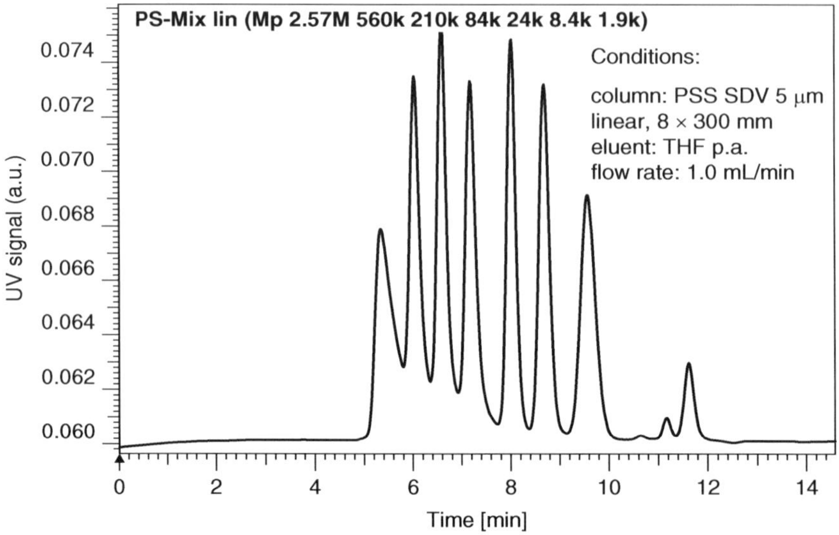

Figure 15.22 Conventional SEC analysis of narrow PS standards. *M* (in g/mol) of PS standards: 1.9×10^3, 8.4×10^3, 2.4×10^4, 8.4×10^4, 2.10×10^5, 5.60×10^5, 2.57×10^6; column, 300 × 8 mm, 5-μm particle size PSS SDV linear; solvent, THF; temperature, ambient; flow rate, 1.0 mL/min; detector, UV. (Courtesy of PSS Polymer Standards Service.)

(*C*-term) of the van Deemter equation (Equation 3.14). All other conditions being equal, chromatographic efficiency in SEC (outside the oligomeric region) is expected to decrease continuously with increasing flow velocity. This decrease in efficiency can be seen in Figure 15.23, which shows the results of replicating the experiment from Figure 15.22 at 4 mL/min. With this fourfold increase in flow rate, the plate number has been reduced by a factor of 2, to 43,000 plates/m. Additionally, tailing of the 2.57×10^6 g/mol PS standard can be observed in the high-speed SEC chromatogram. This tailing is probably due to the on-column, flow-induced degradation of the high-*M* analyte (Section 7.2.3 and Figure 7.6) [29].

2. *Using shorter columns*. All other factors being equal, retention time t_R is directly proportional to the first power of the column length L, while resolution is proportional to the square root of L (i.e., while $t_R \propto L$, $R_s \propto \sqrt{L}$; Section 2.2 and Equations 4.3 and 3.11). Reducing L by a factor of 2 cuts analysis time in half while decreasing resolution by a factor of 1.4. Because of the decrease in pore volume available for separation that accompanies substantial decreases in column length, however, large gains in analysis time usually involve unacceptable losses in resolution. Some ground can be recovered through hardware optimization, the application of band-broadening correction algorithms, and the use of broad standard calibration techniques in lieu of peak-position calibrations using narrow standards. An example of the loss in resolution caused by reducing L is given in Figure 15.24, which replicates the analysis in Figure 15.22 using a column one-sixth the length. While run time has been reduced to less than 3 minutes, resolution is substantially lower and the three system peaks (retention times of about 10.6, 11.2,

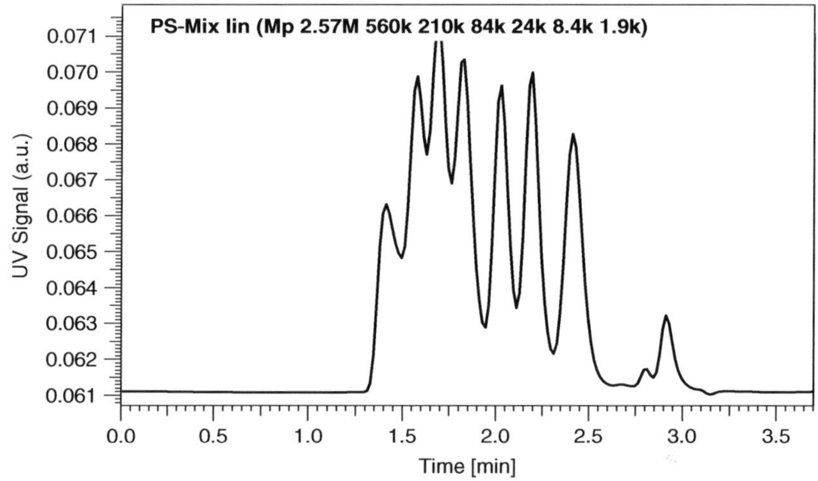

Figure 15.23 High-speed SEC analysis using conventional column and increased flow rate. Flow rate, 4 mL/min. All other conditions the same as in Figure 15.22. (Courtesy of PSS Polymer Standards Service.)

and 11.6 minutes in Figure 15.22) have merged into a single peak (retention time of about 1.9 minutes in Figure 15.24).

Short columns for use at elevated flow rates, in both aqueous and organic solvents, are available from several manufacturers. Quoted molar mass ranges are from oligomeric to about 10 million g/mol.

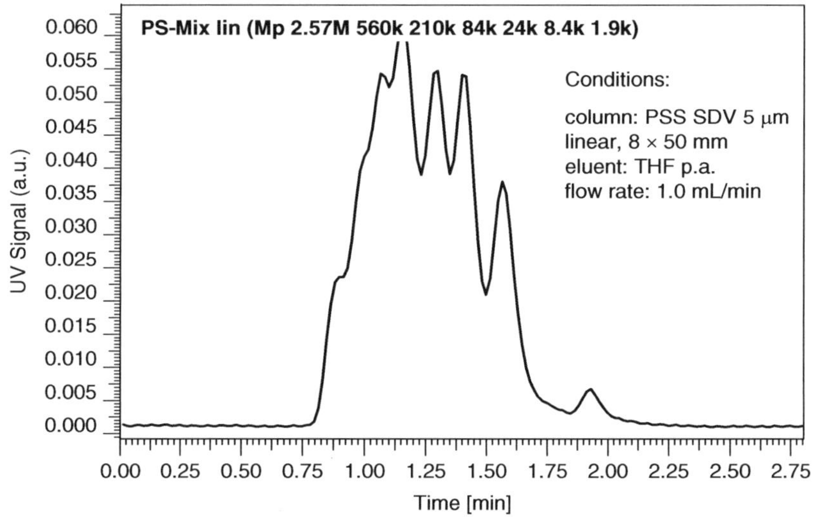

Figure 15.24 High-speed SEC using short columns. Column dimensions, 50 × 8 mm. All other conditions the same as in Figure 15.22. (Courtesy of PSS Polymer Standards Service.)

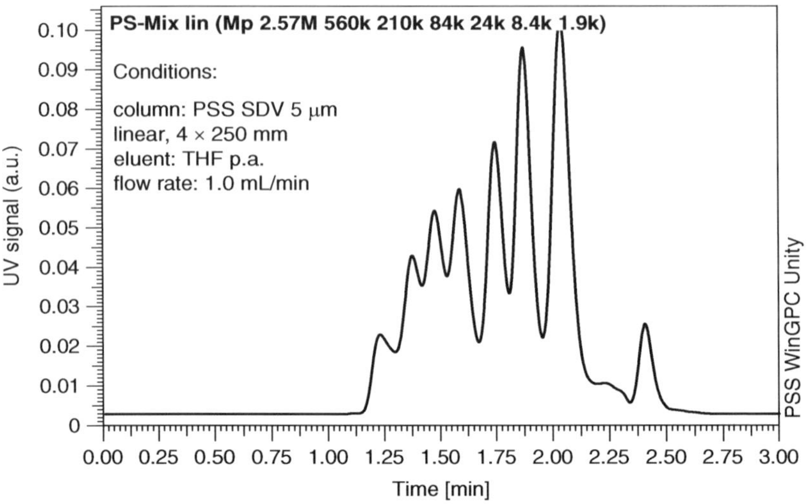

Figure 15.25 High-speed SEC using reduced-diameter columns. Column dimensions, 250 × 4 mm. All other conditions the same as in Figure 15.22. (Courtesy of PSS Polymer Standards Service.)

3. *Using narrower columns*. As can be seen in Figure 15.25, when all other factors are maintained constant, reducing column diameter leads to reduced analysis time. As with shorter columns, the reduction in pore volume available for separation also leads to a reduction in resolution. This is seen when comparing Figures 15.22 and 15.25. The peak heights of the standards relative to one another are also seen to change with a decrease in column diameter. For the high-*M* standards, the lowering and broadening of the peaks is probably due to on-column, flow-induced degradation (Section 7.2.3 and Figure 7.6) [29]: While the flow rate was maintained at 1 mL/min, the linear flow velocity inside the column increases with decreasing column diameter, leading to increased stress on the polymer during its passage through the chromatographic medium. As was the case with short columns, the three system peaks have coalesced into a single peak which is not well separated from the peak of the lowest-*M* standard.

A distinct advantage of the three high-speed approaches discussed thus far (using fast flow rates, short columns, or narrow columns) is that all methods substantially reduce solvent consumption and, thus, waste generation. This is not so for the next method, which employs short, wide-bore columns to effect high-speed separations.

4. *Using short, wide-bore columns (modified aspect ratio method)*. Two limitations of the previous three methods have been (1) the loss of resolution that accompanies the reduction in pore volume resulting from using either shorter or narrower columns, and (2) the degradation of high-*M* analytes that occurs when using conventional-diameter columns at elevated flow rates or when using narrow-diameter columns. An alternative approach to high-speed SEC uses elevated flow rates with columns of modified aspect ratio (i.e., short, wide-bore columns). The columns are

packed with open-pore particles and equipped with special inlet frits for quick and even sample distribution across the radial dimension of the column. Pore volume is maintained constant relative to analyses using conventional flow rates and columns of standard dimensions. The wide column diameter compensates for the elevated flow rates in relation to the shear stresses imposed on the polymer (in particular, on high-*M* polymers) during analysis. With high flow rates and wide columns, the column length can now be shortened, resulting in a reduction in time of analysis. Results from this method are shown in Figure 15.26. When compared to Figure 15.22, peak heights are maintained relative to one another, and high-*M* analytes do not appear to have degraded during analysis. The three system peaks have again merged into a single peak, although this peak does appear well resolved from that of the lowest-*M* standard. Relative to the benchmark method (Figure 15.22), the modified aspect ratio approach suffered from about a 20% loss in resolution (from 5.2 to 4.3) and about a 40% loss in efficiency (from 92,500 plates/m to 58,500 plates/m).

Modified-aspect-ratio columns for high-flow-rate analysis are commercially available for use with either aqueous or a variety of organic solvents, including columns for high-temperature analysis and for use with hexafluoroisopropanol (HFIP). Several particle sizes are available, ranging from 5 to 20 μm. The manufacturer claims a molar mass range of 2000 to over 100 million g/mol. While the modified-aspect-ratio approach appears quite promising and has shown good results thus far (including in 2D-LC separations; see Chapter 14), unlike the other three high-speed methods, there is no reduction in solvent consumption or waste generation.

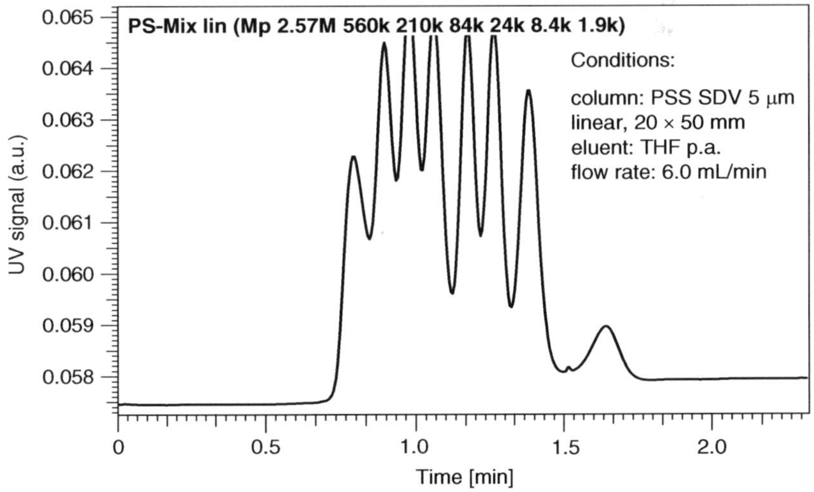

Figure 15.26 High-speed SEC using high flow rates and modified-aspect-ratio columns. Column dimensions, 50 × 20 mm; flow rate, 6.0 mL/min. All other conditions the same as in Figure 15.22. (Courtesy of PSS Polymer Standards Service.)

If reduction in solvent consumption or waste generation is of added importance in high-speed analysis, the user should first make sure the sample is not expected to contain high-M components. Then, the resolution of methods 1 to 3 should be compared to that of method 4. If the loss in resolution of the former with respect to the latter is acceptable, methods 1 to 3 should be compared to each other with respect to resolution, plate count, peak symmetry, and molar mass range. If applicable, methods 1 to 3 should also be compared with respect to separation of the system peak(s) from the peak of the lowest-M (or smallest) analyte. It will usually be the case that a compromise must be found between gain in time of analysis and loss in resolution [30].

5. *Using high temperatures*. As mentioned above, SEC columns with modified aspect ratio are commercially available for high-temperature, high-speed analysis. In this case, however, it is the modified aspect ratio of the columns, not the elevated temperature, which primarily enables high-speed analysis.

It has been shown that the reduction in solvent viscosity that accompanies operating at high temperature can result in a fourfold reduction in analysis time and a four- to sixfold increase in separation efficiency at 150°C versus room temperature. An example of this is shown in Figure 15.27.

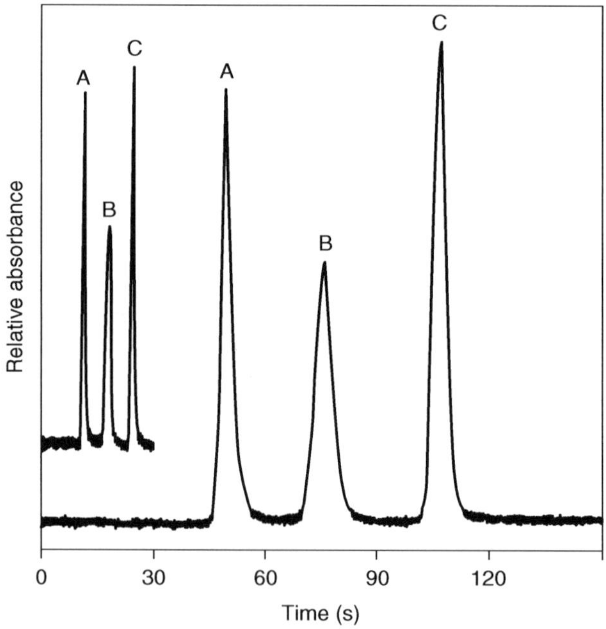

Figure 15.27 High-speed SEC through use of high temperature: (A) PS with $M_p = 1.13 \times 10^6$ g/mol, $M_w/M_n = 1.06$; (B) PS with $M_p = 9 \times 10^3$ g/mol, $M_w/M_n = 1.04$; (C) toluene, $M =$ 92 g/mol. Lower trace: column, 250 × 1 mm, 5-μm particle size, 50-Å pore size silica-based microbore column; solvent, CH_2Cl_2; temperature, 25°C; flow rate, 100 μL/min; pressure, 4100 psig. Upper trace: temperature, 150°C; flow rate, 400 μL/min; pressure, 5100 psig; all other conditions the same as for the lower trace. (Reprinted with permission from Ref. 31.)

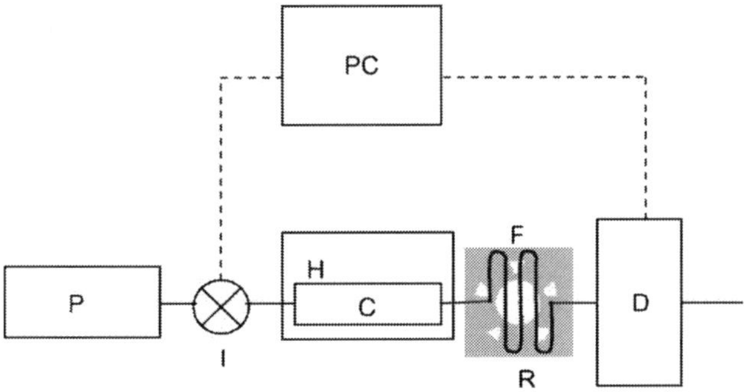

Figure 15.28 High-speed SEC using high temperature and an in-line restrictor. P, pump; I, injector; H, column oven; C, Column; F, cooling fan; R, restrictor tubing; D, detector; PC, personal computer. (Reprinted with permission from Ref. 32.)

A novel approach for high-temperature operation consists of inserting a section of narrow-bore tubing (a restrictor) between the column and the detector [32]. With a column heated to high temperatures, well above the boiling point of the solvent, the restrictor increases column backpressure and prevents solvent boiling. The restrictor also serves to cool the effluent to room temperature before it reaches the detector. This allows use of detectors without the need for modifications for high-temperature use. A schematic of this type of system is shown in Figure 15.28. The effluent temperatures measured at the restrictor outlet, over an 80°C range, are given in Table 15.2. For the system operating in tetrahydrofuran (THF) at 2 mL/min, a 370 × 0.18 mm i.d. restrictor creates about 0.7 MPa of backpressure, increasing the boiling point of THF from 66 to 135°C. Even for short dwell times in the restrictor (0.28 s at 2 mL/min), as seen in Table 15.2, fast cooling to room temperature was achieved in all cases. An example of the application of this approach to the analysis of a series of narrow PS standards is given in Figure 15.29.

Table 15.2 High-speed SEC using a restrictor to elevate temperature

Effluent Column Temperature (°C)	Temperature at Restrictor Outlet[a] (°C)
50	22
60	22
70	22
80	22
90	23
100	23
110	23
120	24
130	24

Source: Ref. 32.

[a] Measured at a flow rate of 2.0 mL/min.

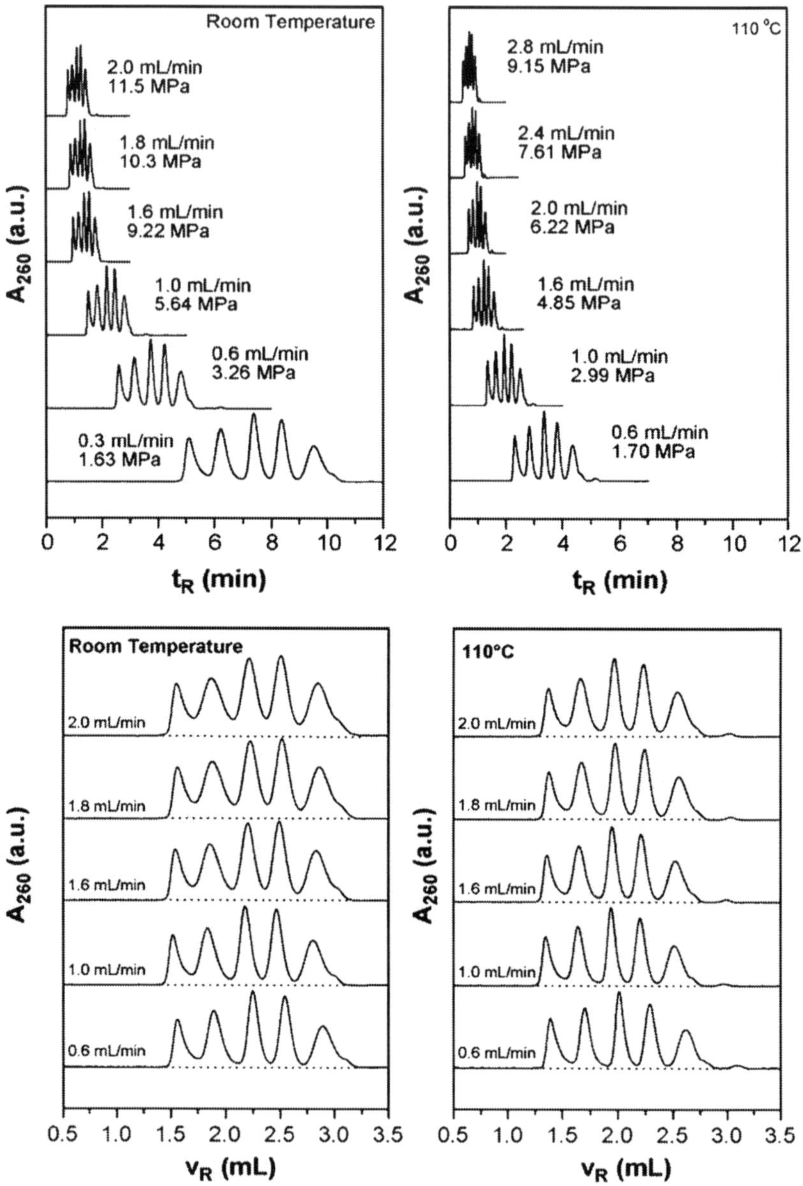

Figure 15.29 High-temperature, high-speed SEC using restrictor tubing, for PS analysis: chromatograms of five narrow polydispersity ($M_w/M_n \leq 1.07$) PS standards at room temperature (figures at left) and at 110°C (figures at right), at various flow rates. Top figures, elution plotted as retention time; bottom figures, elution plotted as retention volume. Column, 250 × 4.6 mm, 5-μm particle size PolyPore; restrictor, 370 × 0.18-mm i.d. narrow-bore stainless steel tubing; solvent, THF; detector, UV/visible. M (in g/mol) of PS standards: 1.8×10^6, 2.0×10^5, 30.9×10^4, 5.05×10^3, 690. (Reprinted with permission from Ref. 32.)

15.5 INVERSE SEC

Pores of column packings are classified into three categories, depending on their size range: (1) Micropores are pores of diameter less than 2 nm; (2) mesopores are pores with diameter between 2 and 50 nm; and (3) macropores or flow-through pores are larger than 50 nm. Inverse size-exclusion chromatography (ISEC) is often used to determine the pore-size distribution of column packing materials. The ISEC method works best for characterizing micro- and mesopores. In this size range, ISEC compares favorably with other methods, such as mercury intrusion porosimetry or nitrogen adsorption. A particular advantage of ISEC over other pore sizing methods is its ability to characterize packing materials under the conditions of analysis (i.e., in a "wet, swollen" state). As described below, ISEC depends on the availability of well-characterized narrow polydispersity standards. Because of this limitation, it is not well suited to the characterization of macropores [33,34].

The ISEC method is based on monitoring the column residence times of various probe molecules of varying size. Ideally, a monodisperse solute is injected onto a column packed with porous material. The solute then probes the pores of the packing material that are larger than the solute. This procedure is repeated for a series of solutes of increasing size, with each solute measuring the average pore size within the pore range up to the next larger probe solute. As the smallest solute probes the smallest pores and so on, an infinite number of monodisperse probe molecules will, in principle, provide the continuous pore-size distribution of the packing materials. Potential interferences can arise from solute adsorption, longitudinal or intraparticle diffusion, mass transfer processes, differences in the structure or density of liquid near particle surfaces as compared to free solution, and from the fact that truly monodisperse probe species are not usually available [35].

The ISEC technique relies on the relationship (see Section 2.2)

$$V_R = V_o + K_{\text{SEC}} V_i \tag{15.11}$$

where V_R is the retention volume of the probe solute species, V_o the void (interstitial) volume of the column, V_i the internal pore volume of the column packing material, and K_{SEC} the solute distribution coefficient, the ratio of the average solute concentration inside the pores to the concentration outside the pores. Because K_{SEC} represents the fraction of pore volume accessible to a probe solute of a given size, a plot of solute size versus K_{SEC} (Figure 15.30) will show the accessibility of a given porous material to a given set of probes. This information is then used to derive the pore-size distribution of the material. The solute size plotted is usually the viscometric radius R_η (see Section 9.5.3 and Table 9.2). The consequences of the probe materials not being truly monodisperse but merely narrowly polydisperse have generally been ignored.

The pore-size distributions obtained are usually the differential, cumulative, and integral distributions. If $f(r)$ is defined as the differential pore-size distribution as a function of radius r, and $f(r)\,dr$ corresponds to the fraction of pore volume in

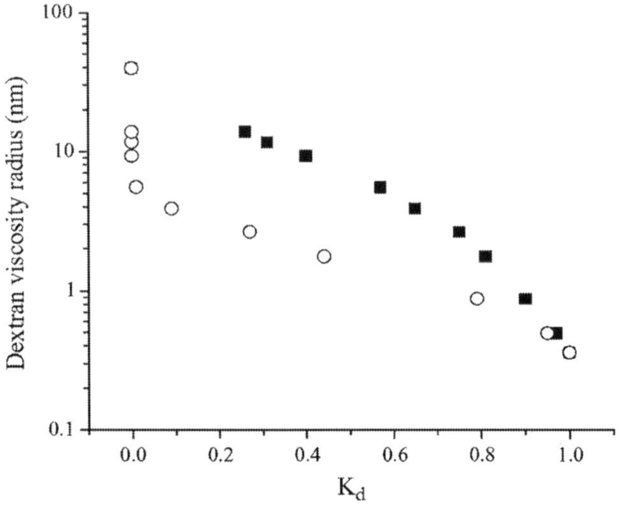

Figure 15.30 Dextran calibration curve for inverse SEC analysis. Columns: 30Q (■) and Q Sepharose XL (○). Viscosity radius according to Equation 9.48 (Table 9.2). K_d is the solute distribution coefficient K_{SEC}. (Reprinted with permission from Ref. 36.)

the range r to $r + dr$, the cumulative and integral pore-size distributions, $g(r_0)$ and $F(r_0)$, respectively, can be calculated as

$$g(r_0) = 1 - \int_0^{r_0} f(r)\,dr \tag{15.12}$$

$$F(r_0) = \int_0^{r_0} f(r)\,dr \tag{15.13}$$

Graphical representations of the various distributions are shown in Figure 15.31 for the same packing material as in Figure 15.30.

Some debate exists as to how best to characterize the size of the smallest pores from data such as those in Figure 15.30. One school of thought advocates extrapolating the slope of the bottom part of the curve in Figure 15.30, with the y-axis intercept of the extrapolated line corresponding to the radius of the smallest pores. An alternative approach uses the data for the probe molecule with the largest value of K_{SEC} (the largest K_d in Figure 15.30) as the value for the smaller pores. When the pore-size distribution is monodisperse, both methods give the same result. This is not the case, however, for a polydisperse pore-size distribution such as that in Figure 15.31. Because at the total permeation volume $K_{SEC} = 1$, regardless of solute size (and in the absence of enthalpic effects), we recommend that the radius of the largest probe species with $K_{SEC} = 1$ be considered the radius of the smallest pores in the pore-size distribution.

ISEC has been used to determine the pore-size distribution not only of SEC column packings but also of HPLC and cation-exchange packing materials, wood pulps,

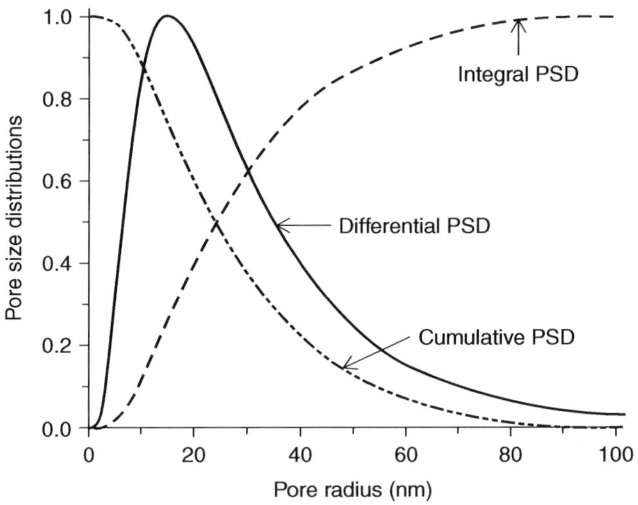

Figure 15.31 Differential, integral, and cumulative pore-size distributions from fitting the dextran data from Figure 15.30 for column 30Q. (Reprinted with permission from Ref. 36.)

alginate/(L-lysine)/alginate microcapsules, and hemodialysis membranes, among others. Reference 37 provides a review of the applications of ISEC to the measurement of pore sizes. Applications of ISEC to the measurement of pore structure are not reviewed here, as results are highly model dependent (see, e.g., Reference 38).

15.6 VACANCY AND DIFFERENTIAL SEC

Advantages of using the vacancy or differential SEC methods occur (1) in process control, where the control or reference material sample is used in the mobile phase and the test sample is then injected, and (2) in problem systems, where the column packing surface can be deactivated by the solute-containing mobile phase. Few applications of the vacancy method have been reported [39–41]. The differential method has experienced some resurgence in recent years, in the brand identification of lubricant oils [42] and in the analysis of gelatin–polyelectrolyte complexes [43].

In vacancy size-exclusion chromatography, a sample of pure mobile phase is injected into columns that have been equilibrated with a dilute mobile-phase solution of the solute to be analyzed [39]. Under ideal conditions, a chromatogram is obtained that is nearly the exact mirror image of a conventional SEC chromatogram. Figure 15.32 illustrates that the conventional and vacancy chromatograms for a polystyrene polymer are very similar but not identical mirror images. The same is true of the calibration curves obtained by the two methods, as shown in Figure 15.33. At lower concentrations (e.g., 3×10^{-4} g/mL of PS in THF), vacancy chromatograms that are exactly the mirror image of the regular SEC chromatograms are obtained and the vacancy and regular SEC calibration curves are identical to each

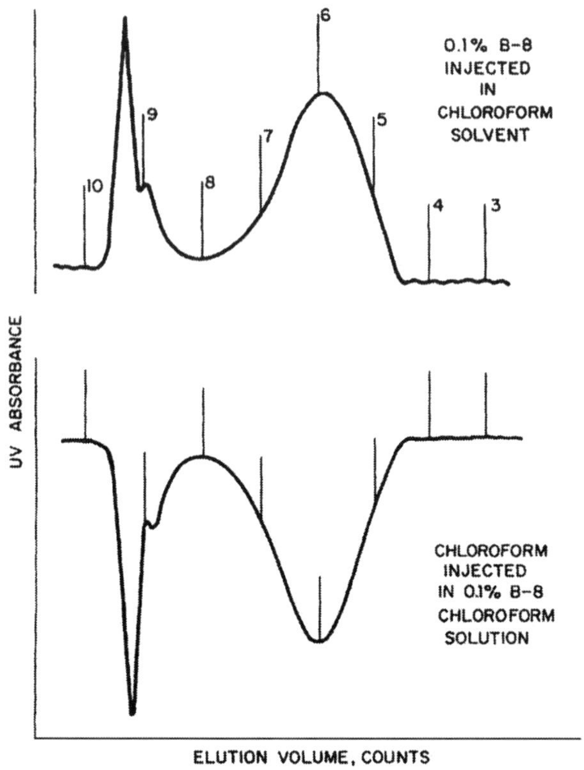

Figure 15.32 Comparison of conventional and vacancy size-exclusion chromatograms for polystyrene. Upper curve: conventional chromatogram; column, 120 × 0.8 cm Styragel 10^4 Å; mobile phase, chloroform; temperature, 25°C; flow rate, 1 mL/min; sample, polystyrene Dow B-8, 0.1% (amount injected unspecified); detector, UV. Lower curve: vacancy chromatogram; same as upper curve, except mobile phase is 0.1% Dow B-8 polystyrene in chloroform; sample, pure chloroform. (Reprinted with permission from Ref. 39.)

other [40]. Thus, discrepancies between elution volumes in vacancy and regular SEC can be attributed to a concentration dependence rather than to column performance. In the vacancy mode, solute polymer is already present inside the pores of the packing material. In regular SEC, there is initially no polymer inside the pores. As the concentration of polymer in solution increases, macromolecular crowding occurs and chains begin to contract relative to their size in near-infinitely dilute solution. A greater concentration dependence is expected in the vacancy mode, where macromolecular crowding is more pronounced due to the preexistence of polymer inside the pores.

In the differential SEC technique, a sample solution is injected into columns equilibrated with a dilute mobile-phase reference solution. Any small difference between the sample solution and the reference mobile-phase solution is detected in the differential chromatogram. Figure 15.34 shows the conventional and differential

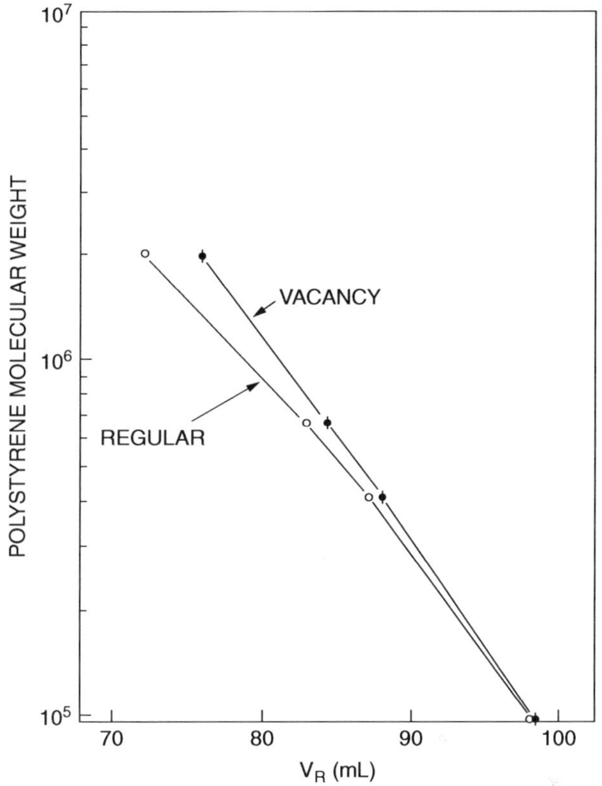

Figure 15.33 Comparison of calibration data for conventional and vacancy SEC on swellable column packing. Conventional: columns, set of four 30 × 0.8 cm each of 10^6-, 10^5-, 10^4-, and 10^3-Å Styragel, respectively; mobile phase, chloroform, 25°C; flow rate, 1.5 mL/min; samples, narrow-MMD polystyrenes, total solids 0.1%, 2 mL; detector, DRI. Vacancy: same as conventional except mobile phase, total solids 0.01% in chloroform; sample, pure chloroform. (Reprinted with permission from Ref. 41.)

chromatograms of gelatin. In conventional mode (curve a), the mobile phase is the solvent. In differential mode (curve b), the mobile phase is the solvent plus the gelatin (at a concentration of 0.1 mg/mL). In curve a of Figure 15.34, even though gelatin was injected, no gelatin peak is observed, indicating that gelatin has adsorbed onto the column packing material. In curve b, the differential SEC chromatogram, an injection of the same concentration of gelatin produces a signal, indicating that adsorption has been reduced. Performing the same type of experiment with various concentrations of gelatin, of the polyelectrolyte sodium poly(styrene sulfonate) or NaPSS, and with various gelatin/NaPSS ratios allowed for determination of the stoichiometry of gelatin–polyelectrolyte complex formation. Results from the differential SEC method were in excellent agreement with those obtained from static light-scattering (see Section 9.3) experiments.

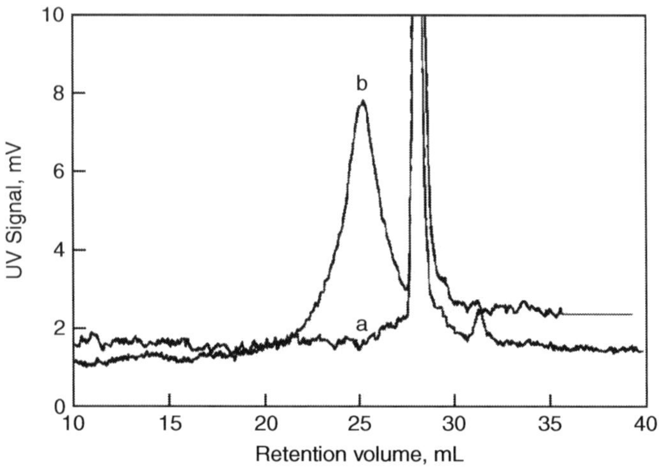

Figure 15.34 Conventional and vacancy SEC of gelatin. (a) Conventional chromatogram of gelatin; columns, three 300 × 7.8 mm Synchropak columns (pore size 4000, 1000, and 100 Å); mobile phase, 10 mM sodium acetate, pH 5.6; temperature, 40°C; detector, UV (230 nm); gelatin concentration, 0.4 mg/mL. (b) Differential chromatogram of gelatin: conditions same as for (a) except mobile phase is 0.1 mg/mL gelatin in 10 mM sodium acetate, pH 5.6 (40°C). (Reprinted with permission from Ref. 43.)

15.7 SIZE-EXCLUSION ELECTROCHROMATOGRAPHY

Size-exclusion electrochromatography (SEEC), a largely research-based technique, combines pressure- and electro-driven flow for transport and separation of analytes across a capillary column packed with porous, noninteracting material. Separation voltages of about 20 to 30 kV are used and injections can be performed electrokinetically (e.g., by applying 5 kV for 10 s). Solvents with a high dielectric constant (e.g., water, dimethylformamide) are needed to generate the electroosmotic flow. The addition of electro-driven flow to pressure-driven flow provides increased separation efficiency. This appears to be the result of the improved mass transfer that results from intraparticle electroosmotic flow: With electroosmotic flow, the flow velocity distribution across the column cross section is virtually homogeneous. As a result, it is also possible to generate and control intraparticle flow, resulting in an increase in mass transfer. Combined, these provide efficiencies comparable to, and sometimes grater than, those obtained by pressure-driven SEC using similar stationary phases [44]. Because of the dimensions of SEEC columns (typically, about 0.25 to 0.5 m in length by 50 to 100 μm i.d.), sample and solvent consumption can be reduced by a factor of about 10^4 over conventional SEC. The mass selectivity of SEEC has been found to be lower than in conventional, pressure-driven SEC. This presumably results from the reduced retention window in SEEC, where, unlike SEC, the mobile phase inside the pores is not stagnant [45]. The predictive error of SEEC calibration

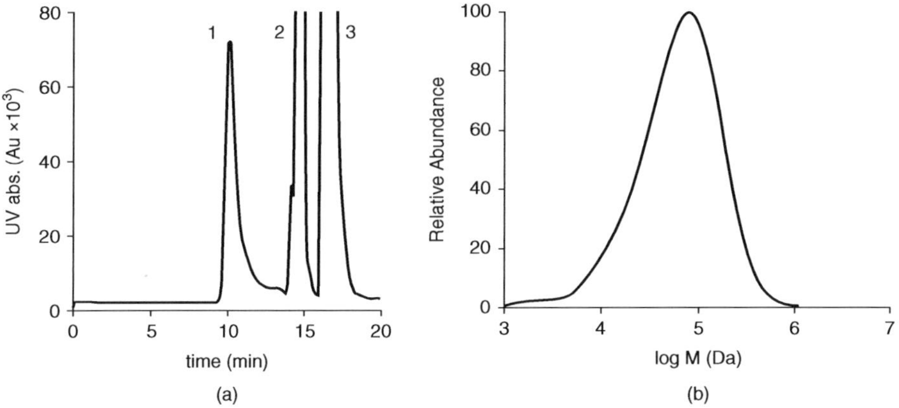

Figure 15.35 Size-exclusion electrochromatography analysis of cellulose tricarbanilate (CTC). (a) Column, 500 × 0.1 mm Nucleosil 300-5; mobile phase, acetone + 0.1 mM tetrabutylammonium tetrafluoroborate; temperature, 20°C; detector, UV; applied field strength, 30 kV (600 V/cm); peak 1, CTC; peaks 2 and 3, residual methanol/phenyl isocyanate and pyridine, respectively, from cellulose carbanilation reaction. (b) MMD of CTC, based on Mark–Houwink calibration curve. (Reprinted with permission from Ref. 47.)

curves for M determination has been found to be two to four times higher than the same error for SEC curves [46].

Figure 15.35a shows the SEEC chromatogram of a cellulose tricarbanilate (CTC) sample. Peak 1 in Figure 15.35a corresponds to the CTC, while peaks 2 and 3 correspond to residual methanol/phenyl isocyanate (peak 2) and pyridine (peak 3) from the carbanilation reaction of cellulose. Figure 15.35b shows the MMD of cellulose tricarbanilate obtained from SEEC analysis, applying a Mark–Houwink calibration (see Section 8.2.3) based on a series of PS, PMMA, and derivatized pullulan standards.

To date, there are no commercially available SEEC columns.

REFERENCES

1. L. R. Snyder and J. J. Kirkland, *Introduction to Modern Liquid Chromatography*, 2nd ed., Wiley-Interscience, New York, 1979, Chap. 15.
2. J. J. DeStefano and J. J. Kirkland, *Anal. Chem.*, **47**, 1103A (1975); **47**, 1193A (1975).
3. A. R. Cooper, A. J. Hughes, and J. F. Johnson, *J. Appl. Polym. Sci.*, **19**, 435 (1975).
4. J. Curling, in *Chromatography of Synthetic and Biological Polymers*, Vol. 2, R. Epton, ed., Ellis Horwood, Chichester, UK, 1978, Chap. 6.
5. T. A. Maldacker and L. B. Rogers, *Sep. Sci.*, **6**, 747 (1971).
6. J. J. DeStefano and H. C. Beachell, *J. Chromatogr. Sci.*, **10**, 654 (1972).
7. A. Wehrli, *Z. Anal. Chim.*, **277**, 289 (1975).
8. C.-H. Fischer, *J. Liq. Chromatogr.*, **17**, 3593 (1994).

9. A. R. Cooper, A. J. Hughes, and J. F. Johnson, *Chromatographia*, **8**, 136 (1975).
10. J. J. Kirkland and P. E. Antle, *J. Chromatogr. Sci.*, **15**, 137 (1977).
11. Y. Kato, T. Kametani, K. Furukawa, and T. Hashimoto, *J. Polym. Sci. A-2*, **13**, 1695 (1975).
12. L. R. Snyder and J. J. Kirkland, *Introduction to Modern Liquid Chromatography*, Wiley-Interscience, New York, 1974, Chap. 3.
13. A. P. Graffeo, Association of Official Analytical Chemists' Meeting, Washington, DC, Oct. 19, 1977.
14. L. R. Snyder and J. J. Kirkland, *Introduction to Modern Liquid Chromatography*, Wiley-Interscience, New York, 1974, Chap. 10.
15. T. P. Kravtechko, G. Berth, A. G. J. Voragen, and W. Pilnik, *Carbohydr. Polym.*, **18**, 253 (1992).
16. H. K. de Wolf, M. de Raad, C. Snel, M. J. van Steenbergen, M. H. A. M. Fens, G. Storm, and W. E. Hennik, *Pharm. Res.*, **24**, 1572 (2007).
17. P. E. Barker, F. J. Ellison, and B. W. Hatt, in *Chromatography of Synthetic and Biological Polymers*, Vol. 1, R. Epton, ed., Ellis Horwood, Chichester, UK, 1978, Chap. 13.
18. P. E. Barker, B. W. Hatt, and A. N. Williams, *Chromatographia*, **10**, 377 (1977).
19. H. Kalasz, J. Nagy, and J. Knoll, *J. Chromatogr.*, **107**, 35 (1975).
20. K. J. Bombaugh and R. F. Levangie, *Sep. Sci.*, **5**, 751, (1970).
21. R. A. Henry, S. H. Byrne, and D. R. Hudson, *J. Chromatogr. Sci.*, **12**, 197 (1974).
22. J. L. Waters, *J. Polymer Sci. A-2*, **8**, 411 (1970).
23. J. V. Dawkins, M. J. Forrest, and M. J. Shepherd, *J. Liq. Chromatogr.*, **13**, 3001 (1990).
24. A. M. Al-Somali, K. M. Krueger, J. C. Falkner, and V. L. Colvin, *Anal. Chem.*, **76**, 5903 (2004).
25. J. N. Little, J. L. Waters, K. J. Bombaugh, and W. J. Pauplis, *J. Polym. Sci. A-2*, **7**, 1775 (1969).
26. W. W. Yau, J. J. Kirkland, D. D. Bly, and H. J. Stoklosa, *J. Chromatogr.*, **125**, 219 (1976).
27. A. M. Striegel, *Anal. Bioanal. Chem.*, **390**, 303 (2008).
28. P. Kilz, in *Handbook of Size Exclusion Chromatography and Related Techniques*, 2nd ed., C.-S. Wu, ed., Marcel Dekker, New York, 2004, Chap. 19.
29. A. M. Striegel, *J. Liq. Chromatogr. Rel. Technol.*, **31**, 3105 (2008).
30. S. T. Popovici and P. J. Schoenmakers, *J. Chromatogr. A*, **1099**, 92 (2005).
31. C. N. Renn and R. E. Synovec, *Anal. Chem.*, **64**, 479 (1992).
32. S. Park, H. Cho, Y. Kim, S. Ahn, and T. Chang, *J. Chromatogr. A*, **1157**, 96 (2007).
33. J. Urban, S. Eeltink, P. Jandera, and P. J. Schoenmakers, *J. Chromatogr. A*, **1182**, 161 (2008).
34. M. Thommes, R. Skudas, K. K. Unger, and D. Lubda, *J. Chromatogr A*, **1191**, 57 (2008).
35. M. Goto and B. J. McCoy, *Chem. Eng. Sci.*, **55**, 723 (2000).
36. Y. Yao and A. M. Lenhoff, *J. Chromatogr. A*, **1037**, 273 (2004). See the Erratum in *J. Chromatogr. A*, **1113**, 259 (2006).
37. A. Revillon, *Stud. Surf. Sci. Catal.*, **87**, 363 (1994).
38. B. A. Grimes, R. Skudas, K. K. Unger, and D. Lubda, *J. Chromatogr. A*, **1144**, 14 (2007).
39. C. P. Malone, H. L. Suchan, and W. W. Yau, *J. Polym. Sci. B*, **7**, 781 (1969).

40. M. Ye, Y. Ding, J. Mao, and L. Shi, *J. Chromatogr*, **518**, 238 (1990).
41. E. P. Otocka and M. Y. Hellman, *J. Polym. Sci. B*, **12**, 439 (1974).
42. W. Qingguo, C. Lixing, D. Thompson, Y. Xinkui, C. Zhiliang, C. Junbo, and Q. Qiaomei, *J. Liq. Chromatogr. Rel. Technol.*, **24**, 317 (2001).
43. C. A. Harrison and J. S. Tan, *J. Polym. Sci. B*, **37**, 275 (1999).
44. R. Stol, W. Th. Kok, and H. Poppe, *J. Chromatogr. A*, **914**, 201 (2001).
45. W. Th. Kok, *J. Chromatogr. A*, **1044**, 145 (2004).
46. Y. Vander Heyden, S. T. Popovici, and P. J. Schoenmakers, *J. Chromatogr. A*, **957**, 127 (2002).
47. R. Stol, J. L. Pedersoli, Jr., H. Poppe, and W. Th. Kok, *Anal. Chem.*, **74**, 2314 (2002).

16

HIGH-TEMPERATURE SEC AND RHEOLOGICAL CONNECTIONS

16.1 INTRODUCTION

In this chapter we discuss high-temperature SEC (HT-SEC) and also several links between rheology and SEC. Because our intent is to examine some of the ways in which SEC and rheology complement each other, we do not provide an exhaustive review of rheological methods. Moreover, this chapter is unique in that some knowledge by the reader of rheological terminology, techniques, and theory is assumed. Many excellent texts on rheology exist; notable among these for its thoroughness and readability is Reference 1.

Most of the applications of HT-SEC are in the analysis of polyolefins. This coincides with rheological approaches intended to reconstruct the MMD of linear and branched polymers in general, but which have focused principally on polyolefins. Other areas of intersection between SEC and rheology are the determination and quantitation of long-chain branching (LCB) and the use of SEC-derived data to predict rheological properties of branched polymers.

16.2 HIGH-TEMPERATURE SEC

As mentioned above, the principal use of HT-SEC has been in the analysis of polyolefins. High-molar-mass polyolefins are relatively difficult to handle in SEC, primarily because of solubility problems. With high-M solid polyethylenes (PE),

Modern Size-Exclusion Liquid Chromatography: Practice of Gel Permeation and Gel Filtration Chromatography, Second Edition
By André M. Striegel, Wallace W. Yau, Joseph J. Kirkland, and Donald D. Bly

polypropylenes (PP), poly(ethylene/vinyl acetates) (EVAs), and similar polymers, large percentages of the polymer molecules are in the crystalline state. The degree of crystallinity decreases with increasing branch and/or copolymer contents, but there are no known solvents for many of these polymers at room temperature. It usually becomes necessary to heat the polymer almost to its melting point (in the solvent) to break up the crystalline bond forces before dissolution occurs. Typically PE, PP, and EVA polymers are analyzed by SEC in 1,2,4-trichlorobenzene (TCB) or *o*-dichlorobenzene (ODCB) at 130 to 150°C.

There are problems with handling hot solutions in SEC analyses, because the solutions can cause severe thermal burns. Thus, polymer dissolution should be accomplished in stirred bottles in a metal (not glass) oil bath, or in round-bottomed flasks heated with a mantle and stirred with a magnetic bar. The operator should wear rubber gloves, protective clothing, and a face shield or safety glasses when handling such hot solutions. While an antioxidant (e.g., Santonox-R, 3-*tert*-butyl-4-hydroxy-5-methylphenyl sulfide) is usually added to prevent oxidative degradation of the polymer during the several-hour dissolution process, the temperature should be kept as low as possible to minimize polymer degradation. The sample solution must be injected hot, or the polymer will precipitate and plug the injector or the column-inlet frits. Oven-preheated syringes should be used to fill the heated valve injectors. Since high-M polyolefin solutions can be very viscous even at high temperatures, sample solutions should be dilute [typically, 0.1% (w/v)] to facilitate injection and passage through SEC columns.

While analyses by HT-SEC at 130 to 150°C are usually feasible, certain technical limitations are imposed. Organic-gel columns can be used since the gels swell readily in these hot solvents. However, if the column is cooled, the gel may shrink, with attendant consolidation of the packing. This shrinking may lead to the creation of channels in the packed bed and to reduced resolution. Loss of resolution is especially noticeable with high-performance columns, even when small changes in the packing structure occur. Because the original packed structure is not recovered in the next heat-up, organic gel columns are best kept hot and with solvent flowing through them even when not in use. Solvent flow during extended periods of inactivity can be at a reduced (e.g., 0.1 mL/min) flow rate; alternatively, the outlet tubing from the instrument can be directed into the inlet reservoir, so that solvent flow (and hence column pressure) is maintained at the same rate as during experiments, without additional solvent consumption. Recycling should *never* be performed during analysis, as this will direct analyte into the inlet reservoir and lead to cross-contamination, an exception being when performing recycle SEC experiments (see Section 15.3).

Rigid column packings (e.g., silica) do not suffer from a swelling problem, but adsorption of the polyolefins sometimes occurs, which can bias the SEC separation mechanism. Such difficulties can be overcome by using silanized packings (Section 6.2.2) or by modifying the mobile phase to neutralize the active sites on the column packing (e.g., by adding about 1% Carbowax-200 to ODCB as a modifier). It is more convenient to operate without the modifier, if possible, since its presence may contribute to baseline noise and interference in subsequent characterization steps (e.g., IR); however, at times it is essential for successful analyses.

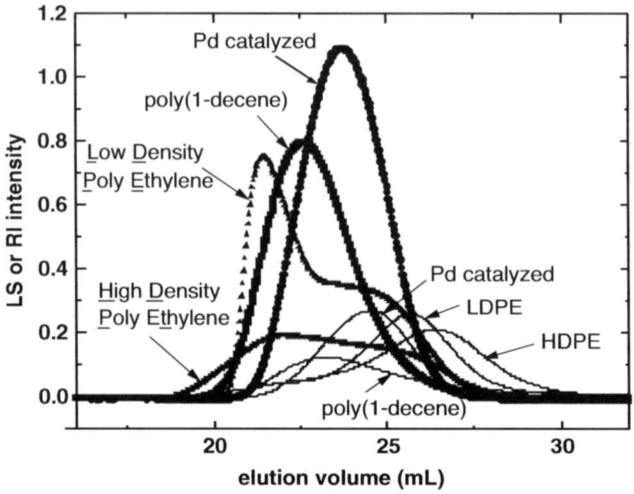

Figure 16.1 HT-SEC analysis of linear and branched polyethylenes: DRI traces (solid lines) and 90° SLS traces (symbols) for a linear HDPE and three branched polyethylenes. Columns, set of four 30-cm length, 10-μm particle size, PLgel Mixed-B columns; solvent, TCB (with 0.05% BHT); temperature, 150°C. (Reprinted with permission from Ref. 2.)

Figure 16.1 shows the DRI and SLS (90° scattering angle) HT-SEC chromatograms of a series of polyolefins, both linear and branched. Similar amounts of each polymer were injected onto the columns, as evidenced by the DRI traces (solid lines) in Figure 16.1. From the SLS traces (symbols), it can be seen that the branched polyolefins scatter substantially more light than does the linear high-density polyethylene (HDPE). This corresponds to a higher M, at equal elution volumes, for the branched species as compared to the HDPE (see Section 9.3).

16.2.1 HT-SEC Instrumentation

Only a few commercial high-temperature instruments are currently manufactured. Three of these are integrated systems (PL-GPC 120 and 220 from Varian/Polymer Labroatories, GPC and GPCV 2000 from Waters, GPCIR from Polymer Char) and one is a modular system (HT-GPC from Malvern/Viscotek). Product specifications for these are given in Table 16.1.

Except for the GPCIR, all commercial high-temperature instruments include a DRI detector and can include a differential viscometer, both from the instrument manufacturer. The HT-GPC instrument also includes a LALS/RALS static light-scattering (SLS) detector from the manufacturer that can be used in either LALS or RALS mode, but not both. In RALS mode, the SLS detector combines with the viscometer and refractometer in SEC^3 mode (see Section 9.6). The HT-GPC can also interface with an IR detector. The PL-GPC 120 and 220 and the GP-CIR units have room for up to four detectors within the instrument, and these can

Table 16.1 Product specifications of commercial high-temperature SEC instruments

Feature	GPC 2000 GPCV 2000	PL-GPC 120 PL-GPC 220	HT-GPC	GPCIR
Manufacturer	Waters	Varian/Polymer Laboratories	Malvern/Viscotek	Polymer Char
Temperature range (°C)	30–180	30–120 (PL-GPC 120), 30–220 (PL-GPC 220)	30–175	30–220
Instrument heating	Zone heating	Zone heating	Modular	Zone heating
Online/ in-line options	Degasser, filter	Degasser, filter	Degasser, filter[a]	Degasser, filter[a]
Maximum number of columns in oven	6 (30 or 50 cm)	6 (30 cm)	4 (30 cm)	4 (30 cm)
Detectors	DRI (GPC 2000); DRI, VISC[b] (GPCV 2000)	DRI, VISC[c]	DRI, LALS/RALS, VISC[c]	VISC, IR (dual wave-length)
Detector interfacing options	MALS,[d] QELS	LALS, MALS,[d] QELS, FTIR, ELSD, (TREF)[e]	IR	MALS, QELS, (TREF)[e]
Pump	Gradient	Gradient	Isocratic	Gradient or isocratic
Sample preparation	Heated vial spin, agitation, and filtering within instrument	Off-line heated agitation, shaking, filtering	Off-line heated stirring	Heated shaking

[a] Self-cleaning filter.
[b] Triple-capillary Waters design (see Section 9.5).
[c] Dual-capillary Viscotek design (see Section 9.5).
[d] Two- and three-angle MALS detectors fit in heated column/detector compartment, eighteen-angle MALS detector necessitates heated transfer line (see Section 9.3.1.1).
[e] TREF; temperature-rising elution fractionation (see Refs. 6b,c).

include LALS, two- or three-angle MALS (both of which can include a QELS unit as well), FTIR, and ELSD. The PL-GPC and the GPCIR instruments can also interface with a temperature-rising elution fractionation (TREF) unit. The PL-GPC, GPC or GPCV 2000, and GPCIR units can interface with a high-temperature version of the eighteen-angle MALS/QELS detector (see Sections 9.3.1.1 and 9.4.1). This detector will not fit within the instruments, so eluent transfer from the instrument to the detector must be carried out using a heated transfer line.

Instead of a DRI detector, the GPCIR instrument has a dual-wavelength IR detector. Measurements at one wavelength are used to determine the concentration of analyte in each slice eluting from the columns. Measurements at the second wavelength are used either to determine the chemical heterogeneity of samples or the short-chain branching distribution of polyolefins (see Section 10.3).

All high-temperature instruments include an online degasser and in-line filter, the filters in the HT-GPC and GPCIR being self-cleaning. The GPC, GPCV, PL-GPC, and GPCIR instruments all have heating zones, so that the temperatures in the pump, autosampler, injector, and column compartments can be programmed individually. Heating zones help keep samples in solution, while reducing the risk of thermal degradation from extended exposure to high temperatures (of increasing concern the farther back a sample is placed in the run queue). In the GPC, GPCV, and GPCIR units, the samples can be spun and/or agitated to assist in dissolution while in the autosampler, prior to injection. Solutions can also be filtered by the instrument. In the PL-GPC and HT-GPC instruments, off-line instrument options exist for sample solution heating, stirring, agitation, and filtering.

16.3 COMPLEMENTARITY OF SEC AND RHEOLOGY

In the remainder of this chapter we examine some of the ways in which rheological and SEC methods are complementary. Topics covered are the determination of polydispersity and long-chain branching in polymers, the use of multidetector SEC data to predict rheological properties, rheological determination of the MMD, and the study of oligomer solutions. Because determining the MMD of polymers through rheology also remains the principal use of SEC, we begin by briefly reviewing the theory associated with extracting MMD information from data obtained rheologically.

16.3.1 Obtaining the MMD from Rheological Measurements

An empirical method for determining the MMD rheologically relies on using viscoelastic data to infer the relaxation spectrum of a polymer [3,4]. The relaxation spectrum is decomposed into two components, one due to entanglement coupling (reptation) modes and one due to Rouse modes. The entanglement part of the spectrum is then used to determine the MMD. The procedure works as follows.

The complex modulus $G^*(\omega)$ is calculated using a standard rheological method such as low-amplitude oscillatory measurement, where the response of a polymer to a small, sinusoidally oscillating mechanical stimulus is measured. To convert $G^*(\omega)$ into an MMD, a linear relaxation spectrum $H(\tau)$ must be computed. [In the nomenclature of rheology, the MMD is often referred to as $w(M)$].

When a small strain is applied to a sample over a very brief time interval, a stress results (stress = force per unit area). As this stress relaxes, so does the linear relaxation modulus $G(t)$, which is the stress divided by the strain. Sample relaxation results from the different processes or modes of a sample, each mode contributing a particular strength and time scale to the overall sample relaxation. As shown in Figure 16.2, the linear relaxation spectrum $H(\tau)$ represents the strength of the

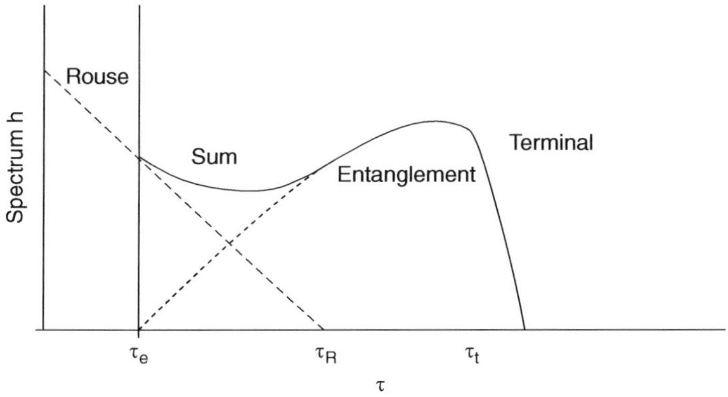

Figure 16.2 Relaxation time spectrum of a generic polymer. Rouse portion of the spectrum is shown as a dashed line, entanglement portion of the spectrum as a dotted line, sum of both spectra as a solid line. Axes are plotted on logarithmic scales. (Reprinted with permission from Ref. 3.)

relaxation at each time scale. Two main types of modes contribute to $H(\tau)$. At short time scales (i.e., at high frequencies), Rouse modes dominate sample relaxation. These modes are due to the segmental motions of individual polymer chains. At longer time scales (i.e., at low frequencies), whole-molecule motions give rise to reptation modes. Only the portion of $H(\tau)$ due to the reptation modes is used to obtain $w(M)$. Because of this the Rouse modes, which are only weakly dependent on $w(M)$, must be subtracted from the spectrum.

The transformation of $H(\tau)$ into $w(M)$ begins with an approximation based on the generalized mixing rule:

$$G_{\text{rept}}(t) = G_N \left[\int_{M_e}^{\infty} F(M, t)^{1/\beta} w(M) \frac{dM}{M} \right]^{\beta} \tag{16.1}$$

where G_N is the plateau modulus, M_e the entanglement molar mass ($M_e \approx M_c/2$, where M_c is the critical molar mass; values of M_e for many common polymers are available in the literature), $F(M, t)$ the relaxation kernel function describing the relaxation behavior of a fraction with molar mass M, and β a parameter that characterizes the mixing behavior. For a linear mixing rule, $\beta = 1$; for a quadratic mixing rule, $\beta = 2$. The subscript "rept" in $G_{\text{rept}}(t)$ is meant to indicate that only contributions from reptation dynamics of the whole polymer are being considered (i.e., the dynamics from the Rouse modes are not considered).

The relationship between the linear relaxation spectrum $H(\tau)$ and the linear relaxation modulus $G(t)$ is given by

$$G(t) = \int_{-\infty}^{\infty} H(\tau) \exp(-t/\tau)\, d \ln \tau \tag{16.2}$$

where τ is the relaxation time and t is the experimental time. For $G_{\text{rept}}(t)$ this becomes

$$G_{\text{rept}}(t) = \int_{-\infty}^{\infty} H_{\text{rept}}(\tau) \exp(-t/\tau)\, d \ln \tau \tag{16.3}$$

where $H_{\text{rept}}(\tau)$ is the part of the relaxation spectrum due only to reptation dynamics. Before conversion to $w(M)$ is effected, the Rouse mode contributions must be subtracted from $H(\tau)$, which will be done shortly.

The molar mass M in Equation 16.1 is related to the relaxation time τ in Equation 16.2 by

$$\tau = k_\tau M^\alpha \tag{16.4}$$

where, in most cases, $\alpha \approx 3.4$. Because values of k_τ are not usually available, the parameter k_η is used instead, where

$$\eta_0 = k_\eta M^\alpha \tag{16.5}$$

η_0 is the zero-shear viscosity and k_τ and k_η are related via

$$k_\tau = \frac{k_\eta}{G_N} \tag{16.6}$$

Values of k_η are temperature dependent. If the $G^*(\omega)$ and $H(\tau)$ data are obtained at a temperature different from that at which k_η was determined, k_η must be adjusted accordingly. This can be done using the Arrhenius or Williams–Landel–Ferry (WLF) models. Traditionally, the weight-average molar mass M_w has been used in Equation 16.4 or 16.5. This, along with the effects of molar mass polydispersity, is addressed in Section 16.3.2.

The plateau modulus G_N can be calculated from material properties using the expression

$$G_N = \frac{\rho RT}{M_e} \tag{16.7}$$

where ρ is the sample density at the experimental temperature, R the universal gas constant (in J/mol · K), and T the experimental temperature (in K). Alternatively, G_N can also be calculated from $H(\tau)$:

$$G_N = \int_{\tau_e}^{\infty} \frac{H(\tau)}{\tau} d\tau \tag{16.8}$$

The entanglement relaxation time τ_e can be obtained by inserting the entanglement molar mass M_e into Equation 16.4.

As mentioned above, the Rouse mode contributions must be subtracted from the relaxation spectrum to obtain $w(M)$:

$$H_{\text{rept}}(\tau) = H(\tau) - H_{\text{Rouse}}(\tau) \tag{16.9}$$

The Rouse part of the relaxation spectrum can be modeled by a simple power law as

$$H_{\text{Rouse}}(\tau) \propto \left(\frac{\tau}{\tau_e}\right)^S \exp\left[-\left(\frac{\tau}{\tau_R}\right)^{S_c}\right] \tag{16.10}$$

The power-law exponent S is given by Rouse theory as -0.5 and by Zimm theory as -0.67. The Rouse time τ_R can be derived from

$$\tau_R = \frac{k_\eta}{G_N}(2M_e)^{\alpha-2}(M_n)^2 \tag{16.11}$$

where M_n is the number-average molar mass. S_c in Equation 16.10 is a cutoff parameter that governs the smoothness of the exponential cutoff at $\tau = \tau_R$; the larger the value of S_c, the sharper the cutoff. For a monodisperse sample, $S_c = \infty$. For most real-world polymers, $1 < S_c < 2$.

To summarize, the following information is needed for calculating the MMD or $w(M)$ by rheology:

1. M_e, the entanglement molar mass (from the literature)
2. k_η, at the experimental temperature (from the literature) or adjusted accordingly (using the Arrhenius of WLF models)
3. G_N, the plateau modulus (from Equations 16.7 or 16.8)
4. α, the exponent in Equations 16.4 and 16.5 (usually, $\alpha \approx 3.4$)
5. β, the generalized mixing parameter in Equation 16.1 ($\beta = 1$ for linear mixing rule, $\beta = 2$ for quadratic mixing rule; if the value of β is unknown, it is recommended that the quadratic value be used)

An example of the application of this method to determining the MMD of a PS blend is shown in Figure 16.3. The sample consisted of a blend of 80% PS 1.77×10^5 g/mol and 20% PS 6.0×10^4 g/mol. The solid curve is the MMD as measured by SEC, the curve with error bars is the MMD derived from rheological data, with $\beta = 2.2$.

The limitations associated with this approach toward obtaining the MMD from rheological data are:

1. The theory was developed for amorphous, linear homopolymers.
2. The theory does not consider the effects of crystallinity or long-chain branching.

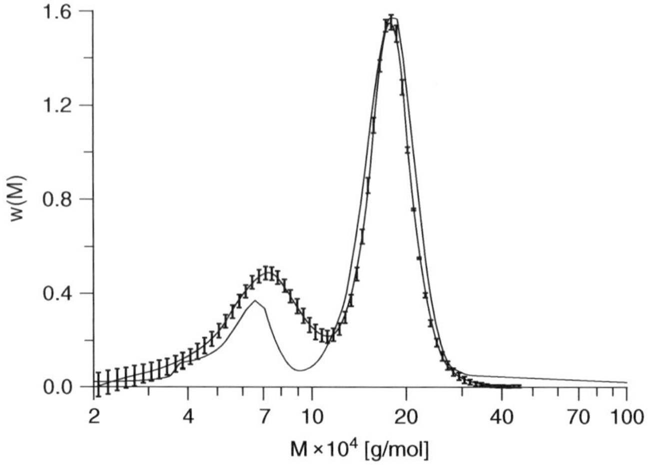

Figure 16.3 MMD of a PS blend measured by SEC and estimated by rheology. 80:20 blend of PS177K:PS60K; solid line, MMD by SEC; line with error bars, MMD from rheological data, using $\beta = 2.2$. (Reprinted with permission from Ref. 3.)

3. The polymer sample is assumed to be free of diluents, fillers, plasticizers, significant levels of residual initiator, and so on.
4. To date, it is unknown whether the theory is applicable to copolymers or to mixtures of different types (chemistries) of polymers.
5. The range of M (between M_{max} and M_{min}) over which the theory can be used is given by the maximum and minimum relaxation times of the material, τ_{max} and τ_{min}, respectively. If $(\tau_{max}/\tau_{min}) = R$, then $(M_{max}/M_{min}) = R^{1/\alpha}$, following Equation 16.4. Because $\alpha \approx 3.4$, each decade in M requires approximately 3.5 decades of τ.
6. The technique should be considered an empirical method for decomposing the relaxation spectrum into two parts, the reptation (entanglement) part and the Rouse part. A firm theoretical basis for the method has not yet been developed [5].

16.3.2 Obtaining Rheological Properties from SEC Measurements

16.3.2.1 Effect of Molar Mass Polydispersity. If the molar mass distribution [MMD or $w(M)$], is already available from SEC experiments, the calculations above can be performed in the reverse direction to obtain the linear relaxation modulus $G(t)$ of the sample [4]. Converting the MMD into a $G(t)$ is considered more mathematically straightforward than converting the linear relaxation modulus into a molar mass distribution and allows the application of alternative relaxation kernel functions, $F(M, t)$. Examples of the latter are the Tuminello kernel, the single-exponential kernel, the Doi kernel, the des Cloizeau kernel, and the Baumgaertal–Schausberger–Winter (BSW) kernel.

As mentioned in Section 16.3.1, M_w has generally been used as the molar mass average in Equation 16.5 relating the zero-shear viscosity η_0 to M. This relation does not take into account the breadth of the MMD. While this breadth is normally characterized using the polydispersity index M_w/M_n, this value is not representative of the high-M end of the MMD. A high-M polydispersity index can be defined as the ratio of the z-average to weight-average M (i.e., as M_z/M_w).

M_x-Based Theory of Zero-Shear Melt Viscosity Power Law on M [6a]. Yau generated a series of MMDs with the generalized exponential (GEX) distribution model, ranging in M_w/M_n from 2 to 20 and in M_z/M_w from 1.50 to 6.75 [6a]. These are shown in Figure 16.4.

A procedure similar to inverting the method in Section 16.3.1 was then followed to obtain η_0. The effects of polydispersity (M_w/M_n and M_z/M_w) on η_0 are shown in Figure 16.5, where PD $= M_w/M_n$; for the parameters in Equation 16.5, $\alpha = 3.6$ and $k_\eta = 3.40 \times 10^{-14}$ poise (a literature value for polyethylene at 190°C); and for molar mass, M_w was used.

To compensate for the effect of polydispersity on η_0, Yau proposed a new statistical average of M, based on the value of M_x:

$$M_x \equiv \left(\frac{\sum_i w_i \times M_i^x}{\sum_i w_i} \right)^{1/x} \tag{16.12}$$

For Equation 16.12, when $x = 1$, $M_x = M_w$, and when $x = 2$, $M_x = [M_w \times (M_z/M_w)^{0.5}]$. This results in the following relation between η_0 and M:

$$\eta_0 \propto M_x^\alpha \approx M_w^\alpha \left(\frac{M_z}{M_w} \right)^{\alpha \times y} \tag{16.13}$$

For $x = 1$, $y = 0$; for $x = 2$, $y = 0.5$.

In the melt, the radius of gyration R_G of a linear polymer is expected to scale with the 0.5 to 0.6 power of M. Because hydrodynamic volume (HV) scales with the third power of R_G, for a linear polymer in the melt the following should apply:

$$HV_i \propto \left(R_{G,i}\right)^3 \propto \left(M_i^{0.5}\right)^3 \text{ to } \left(M_i^{0.6}\right)^3 \propto M_i^{1.5} \text{ to } M_i^{1.8} \tag{16.14}$$

The average proposed to compensate for high-M polydispersity is thus M_x with $x =$ 1.5 to 1.8 and was termed the hydrodynamic volume average molar mass M_{HV}:

$$M_{\text{HV}} \equiv \left(\frac{\sum_i w_i \times M_i^x}{\sum_i w_i} \right)^{1/x} \quad \text{with } x = 1.5 \text{ to } 1.8 \tag{16.15}$$

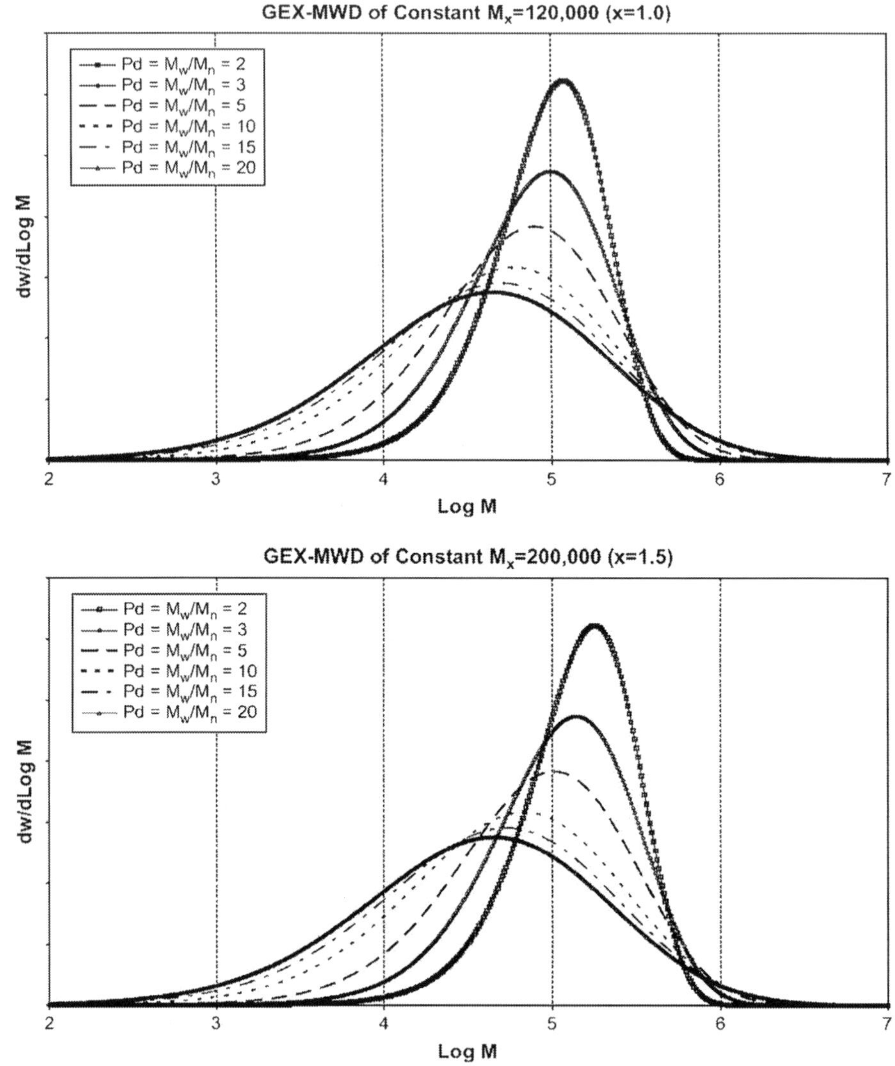

Figure 16.4 GEX-MMD of constant M_x. MMD curves generated using the generalized exponential (GEX) distribution model: (a) $x = 1$, $M_x = M_w = 120{,}000$ g/mol; (b) $x = 1.5$, $M_x = 200{,}000$ g/mol. In both cases, as M_w/M_n increases from 2 to 20, M_z/M_w increases from 1.50 to 6.75. (Reprinted with permission from Ref. 6.)

For example, in the case of $x = 1.5$, the y term in Equation 16.13 equals 0.2635. If $\alpha = 3.6$ and $x = 1.5$, then

$$\eta_0 \propto M_x^{3.6} = M_{\mathrm{HV}}^{3.6} \approx M_w^{3.6} \left(\frac{M_z}{M_w} \right)^{0.95} \tag{16.16}$$

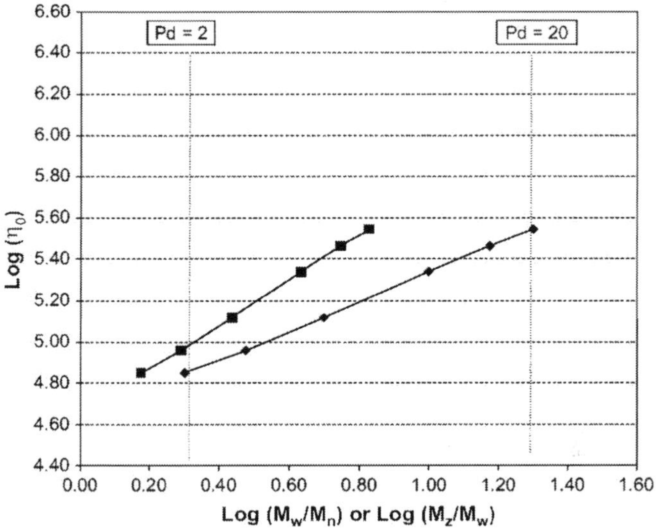

Figure 16.5 Effect of molar mass polydispersity on zero-shear viscosity. PD $= M_w/M_n$. (■) log η_0 vs. log(M_z/M_w); (♦) log η_0 vs. log(M_w/M_n). For $\eta_0 = k_\eta M^\alpha$, $k_\eta = 3.40 \times 10^{-14}$ poise, $\alpha =$ 3.6, $M = M_w$. (Reprinted with permission from Ref. 6.)

As can be seen in Figure 16.6, for $\alpha = 3.2$, 3.4, or 3.6, using M_x with $x = 1.5$ compensates for polydispersity much better than does using M_x with $x = 1$ (i.e., the traditional power law based on the weight-average molar mass M_w).

The M_{HV} term of M_x with $x = 1.5$ to 1.8, as used in the discussion above, is meant to apply only to linear homopolymers, not to long-chain branched polymers or to copolymers. Long-chain branching effects are discussed in Section 16.3.2.2.

M_x-Based Theory of the Mixing Rule of Melt Viscosity [7]. The M_x concept has been extended by Yau et al. to the development of a new mixing rule of rheological parameters.

We first review the derivation of the current (traditional) M_w-based mixing rule, as shown in Equation 16.17. The traditional power law relating η_0–M_w is

$$\eta_0 = KM_w^a, \quad \text{or} \quad M_w = \left(\frac{\eta_0}{K}\right)^{1/a} \tag{16.17a}$$

For the M_w of a binary mixture,

$$\begin{aligned} M_{w_{12}} &= (W_1 M_{w_1} + W_2 M_{w_2}) = \left[W_1\left(\frac{\eta_{0,1}}{K}\right)^{1/a} + W_2\left(\frac{\eta_{0,2}}{K}\right)^{1/a}\right] \\ &= (K)^{-1/a}\left(W_1\eta_{0,1}^{1/a} + W_2\eta_{0,2}^{1/a}\right) \end{aligned} \tag{16.17b}$$

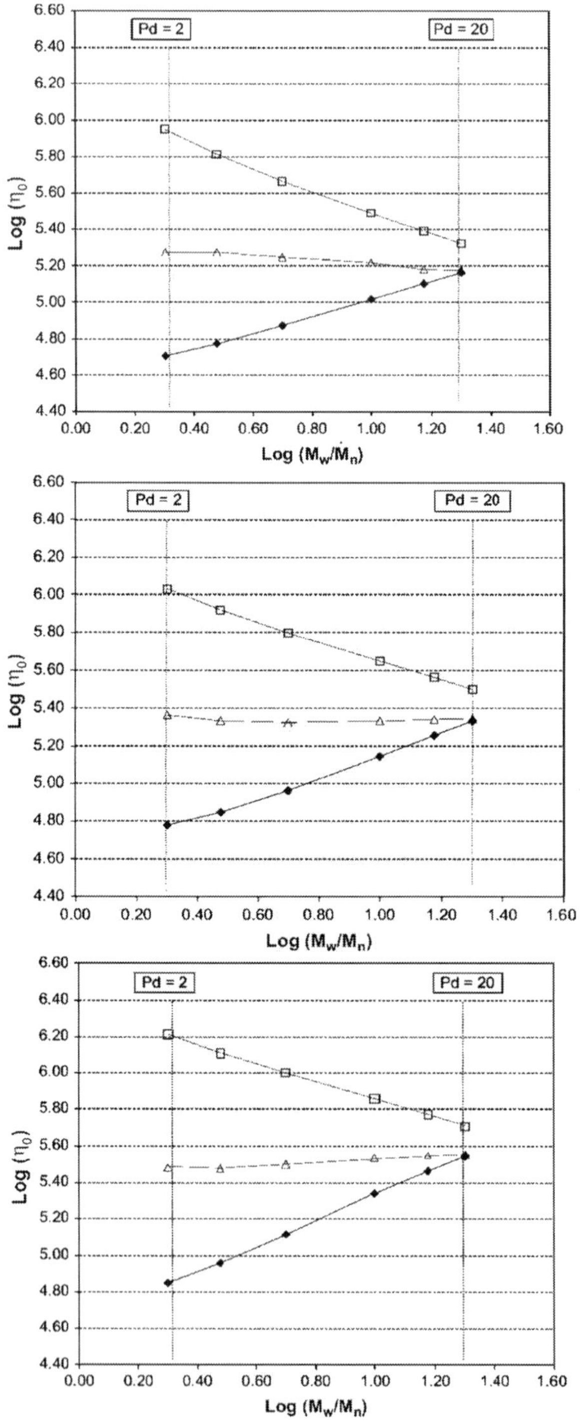

Figure 16.6 Compensating for effect of polydispersity on zero-shear viscosity by using M_{HV}. Applying Equation 16.13, with proportionality constant $k_\eta = 3.40 \times 10^{-14}$ poise and $\alpha = 3.2$ (a), 3.4 (b), and 3.6 (c). PD $= M_w/M_n$. (□) M_x, with $x = 2$; (Δ) M_x with $x = 1.5$ (i.e., $M_x = M_{HV}$); (♦) M_x with $x = 1$ (i.e., $M_x = M_w$). (Reprinted with permission from Ref. 6.)

This leads to the traditional η_0 mixing rule [8]:

$$\begin{aligned}\eta_{0,12} &= KM_{w_{12}}^a = K\langle (K)^{-a/a}\rangle \left(W_1\eta_{0,1}^{1/a} + W_2\eta_{0,2}^{1/a}\right)^a \\ \eta_{0,12} &= \left(W_1\eta_{0,1}^{1/a} + W_2\eta_{0,2}^{1/a}\right)^a \qquad \text{with} \quad a = 3.2 \text{ to } 3.6\end{aligned} \tag{16.17c}$$

For $a = 3.6$,

$$\eta_{0,12} = \left(W_1\eta_{0,1}^{1/3.6} + W_2\eta_{0,2}^{1/3.6}\right)^{3.6} \tag{16.17d}$$

The derivation of the new M_x-based mixing rule now follows, beginning with Equation 16.18. Starting with the M_x-based η_0 melt viscosity power law,

$$\eta_0 = KM_x^a \quad \text{or} \quad M_x = \left(\frac{\eta_0}{K}\right)^{1/a} \tag{16.18a}$$

leads to the M_x-averaged M of a binary mixture:

$$\begin{aligned}M_{x_{12}} &= \left(W_1M_{x_1}^x + W_2M_{x_2}^x\right)^{1/x} = \left[W_1\left(\frac{\eta_{0,1}}{K}\right)^{x/a} + W_2\left(\frac{\eta_{0,2}}{K}\right)^{x/a}\right]^{1/x} \\ &= (K)^{-1/a}\left(W_1\eta_{0,1}^{x/a} + W_2\eta_{0,2}^{x/a}\right)^{1/x}\end{aligned} \tag{16.18b}$$

This leads to the final result of the new M_x-based η_0 mixing rule:

$$\begin{aligned}\eta_{0,12} &= KM_{x_{12}}^a = K\langle (K)^{-a/a}\rangle \left(W_1\eta_{0,1}^{x/a} + W_2\eta_{0,2}^{x/a}\right)^{a/x} \\ \eta_{0,12} &= \left(W_1\eta_{0,1}^{x/a} + W_2\eta_{0,2}^{x/a}\right)^{a/x} \qquad \text{with} \quad x = 1.5 \text{ to } 1.8\end{aligned} \tag{16.18c}$$

For $a = 3.6$ and $x = 1.8$,

$$\eta_{0,12} = \left(W_1\eta_{0,1}^{1.8/3.6} + W_2\eta_{0,2}^{1.8/3.6}\right)^{3.6/1.8} = \left(W_1\eta_{0,1}^{1/2} + W_2\eta_{0,2}^{1/2}\right)^2 \tag{16.18d}$$

In addition to the mixing rule of zero-shear viscosity, the M_x-based mixing rule can be extended to the melt index (MI) measurement of polymer melts. The laboratory MI measurement records the grams of extruded polymer melt under an applied weight. Because the MI value is an inverse of melt viscosity, a generalized M_x mixing rule can be derived by a procedure similar to that outlined above. Start with the M_x-based power law for MI dependency on M,

$$\mathrm{MI} = K'M_x^{-a} \quad \text{or} \quad M_x = \left(\frac{\mathrm{MI}}{K'}\right)^{-1/a} \tag{16.19a}$$

For the M_x-averaged M of a binary mixture, this gives

$$M_{x_{12}} = \left(W_1 M_{x_1}^x + W_2 M_{x_2}^x\right)^{1/x} = \left[W_1\left(\frac{\mathrm{MI}_1}{K'}\right)^{-x/a} + W_2\left(\frac{\mathrm{MI}_2}{K'}\right)^{-x/a}\right]^{1/x}$$
$$= (K')^{1/a}\left(W_1 MI_1^{-x/a} + W_2 MI_2^{-x/a}\right)^{1/x} \qquad (16.19b)$$

This leads to the final result of the new M_x-based MI mixing rule:

$$\mathrm{MI}_{12} = KM_{x_{12}}^{-a} = K\langle (K)^{-a/a}\rangle \left(W_1\mathrm{MI}_1^{-x/a} + W_2\mathrm{MI}_2^{-x/a}\right)^{-a/x}$$
$$\mathrm{MI}_{12} = \left[W_1\mathrm{MI}_1^{-x/a} + W_2\mathrm{MI}_2^{-x/a}\right]^{-a/x} \quad \text{with} \quad x < 1.8 \qquad (16.19c)$$

For $x = 1$, the equation reduces to the traditional MI mixing formulation:

$$\mathrm{MI}_{12} = \left(W_1\mathrm{MI}_1^{-1/a} + W_2\mathrm{MI}_2^{-1/a}\right)^{-a} \quad \text{with } a = 3.2 \text{ to } 3.6 \qquad (16.19d)$$

The shear rate of an MI experiment varies with the MI value of the polymer sample. Because of this, the influence of the polymer components of different M in the sample varies. The mixing rule is expected to vary for different MI products according to Equation 16.19b and therefore requires the use of different x-values. For polymers having MI ≈ 1, the default formula (Equation 16.19d) for the traditional MI-M_w mixing rule of $x = 1$ is expected to work reasonably well. For higher MI products, however, fractional x-values below 1 are required. For samples with fractional MI, better results are obtained using x-values below 1 [7].

M_x-Based Theory of Correlating SEC and Rheology PDI Measurements [9]. Two commonly used rheological methods of measuring polymer polydispersity are (1) the PDI method of crossover modulus [10], and (2) the method of modulus separation (ModSep) [11]. There are two outstanding challenges in using these polydispersity indices (PDI). First, the crossover method does not apply to high-MI samples. Also, the attempt to convert the ModSep result to an equivalent crossover PDI is not always reliable. Second, neither crossover PDI nor ModSep results provide good correlation with the M_w/M_n polydispersity measured in an SEC experiment.

To close these gaps, Yau et al. have developed a generalized M_x-based polydispersity parameter for the SEC experiment:

$$\mathrm{PD}_{x_{12}} = \frac{M_{x_2}}{M_{x_1}} = \frac{M_{x_2}}{M_{x_2-D_x}} \quad \text{where} \quad x_2 = x_1 + D_x \qquad (16.20)$$

The use of $\mathrm{PD}_{x_{12}}$, as defined in Equation 16.20, provides the needed flexibility for finding the x-value that is appropriate for the shear rate obtained in the rheology experiment. The choices of conventional SEC M-polydispersity are limited to the ratios of M_z, M_w, and M_n. This limitation provides no flexibility when trying to account for shear rate effects in the rheology PDI measurements.

$PD_{x_{12}}$ from Equation 16.20 with $D_x = 2$ and $x_2 = 1$ gives the default conventional SEC polydispersity of M_w/M_n. With $D_x = 1$ and $x_2 = 2$, the conventional $M_z/M_w = (PD_{x_{12}})^2$ is obtained. The use of $PD_{x_{12}}$ from SEC to study rheology PDI of polypropylene samples [9] provides the following conclusions:

1. There can be no simple universal conversion formulation between ModSep and crossover PDI, because of the varying degree of shear rate effects for samples of different MI values.
2. However, the PDI from both ModSep and crossover can be correlated successfully with the $PD_{x_{12}}$ values from the SEC experiment.
3. By using $PD_{x_{12}}$, SEC can be used as a highly effective complementary test to rheology for the QC control of polymer polydispersity during production.
4. Most important, the basic understanding obtained from this study resolves the concern, and the perception, for the lack of correlation among these PDI tests. Results of this study should provide considerable help in closing the gap between SEC and rheology.

16.3.2.2 Effect of Long-Chain Branching. Three methods of predicting the dynamic viscosity of branched polyethylenes from SEC data were compared in Reference 12: (1) the multiple linear regression model, based on relating molar mass averages to parameters in the Cross viscosity equation; (2) a mixing rule (method of Pedersen and Ram); and (3) a method based on the similarity between a cumulative molecular property distribution and a plot of dimensionless viscosity versus frequency. Because the third method (the curve similarity method) provided the most promising results, it is the only one discussed here.

The cumulative molecular property in the curve similarity method is a combination of the cumulative MMD and the cumulative g' distribution. This is shown in Figure 16.7 for a series of polyethylenes with varying degrees of long-chain branching (LCB).

The contraction factor g', based on Equations 11.8 and 11.9, is meant to account for the effects of LCB on the viscosity ratio η/η_0, and is given by

$$g'_i = \left(\frac{M_{i,\text{LIN}}}{M_{i,\text{LS}}}\right)^{a+1}_{V=V_i} \tag{16.21}$$

where M_{LIN} is the molar mass obtained from a linear polyethylene calibration curve derived from a linear polystyrene universal calibration curve; M_{LS} the molar mass obtained using SEC with MALS detection; a the exponent in the Mark–Houwink equation (Equation 8.2), V the SEC elution volume; and the subscript i denotes that data are for individual elution slices. Accordingly,

$$\log g'_i = (a + 1)\,(\log M_{i,\text{LIN}} - \log M_{i,\text{LS}}) \tag{16.22}$$

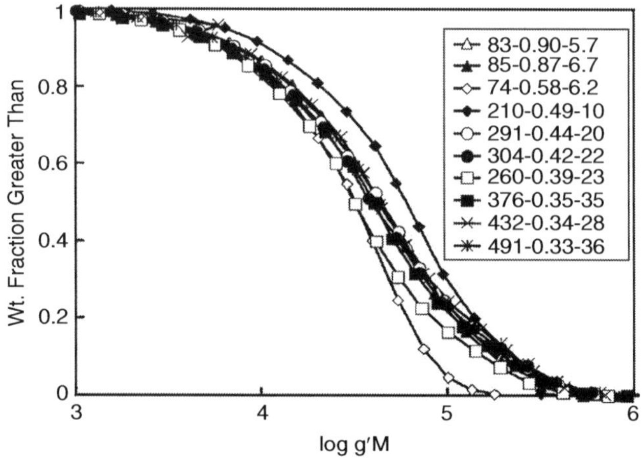

Figure 16.7 Cumulative distribution of $g'M$ for various polyethylene samples. Numbers in inset box are sample identifiers. (Reprinted with permission from Ref. 12.)

The cumulative g' distribution is given by a plot of the cumulative area under the DRI chromatogram versus log g'. Plotting the same cumulative area versus log $M_{i,\mathrm{LS}}$ provides the cumulative MMD. The goal of the curve similarity method is to superimpose these cumulative distributions onto a plot of η/η_0 versus log ω. This is achieved through the transformation of molar mass and branching data into frequency using the relation

$$\log \omega = \delta_0 + \delta_1 \log M_{n,\mathrm{LIN}} + \delta_2 \log M_{i,\mathrm{LS}} + \delta_3 \log g_i' \tag{16.23}$$

$M_{n,\mathrm{LIN}}$ is the number-average molar mass of the whole polymer (i.e., not the M_n of an individual elution slice), obtained from the linear polyethylene calibration curve. Zero shear viscosity data can be obtained either from creep experiments or by fitting the viscosity versus frequency data to a model. In Reference 12 the Cross model was used, which relates the dynamic viscosity η^* to the zero-shear viscosity and the frequency via

$$\eta^*(\omega) = \frac{\eta_0}{1 + K^*\omega^a} \tag{16.24}$$

To obtain values for the δ constants, a two-step fitting procedure was used. First, values for δ_0, δ_1, and δ_2 were searched for by minimizing a standard, unweighted least-squares objective function to match the cumulative molecular property distribution heights of the samples involved in the viscosity ratios. For each sample, the molecular property data consist of the cumulative distribution heights, where for each height the values of log M and log g' are known. In the first step, only data corresponding to samples with $g' = 1$ were used. In the second step, a value for δ_3 was obtained by using all of the data for the samples studied.

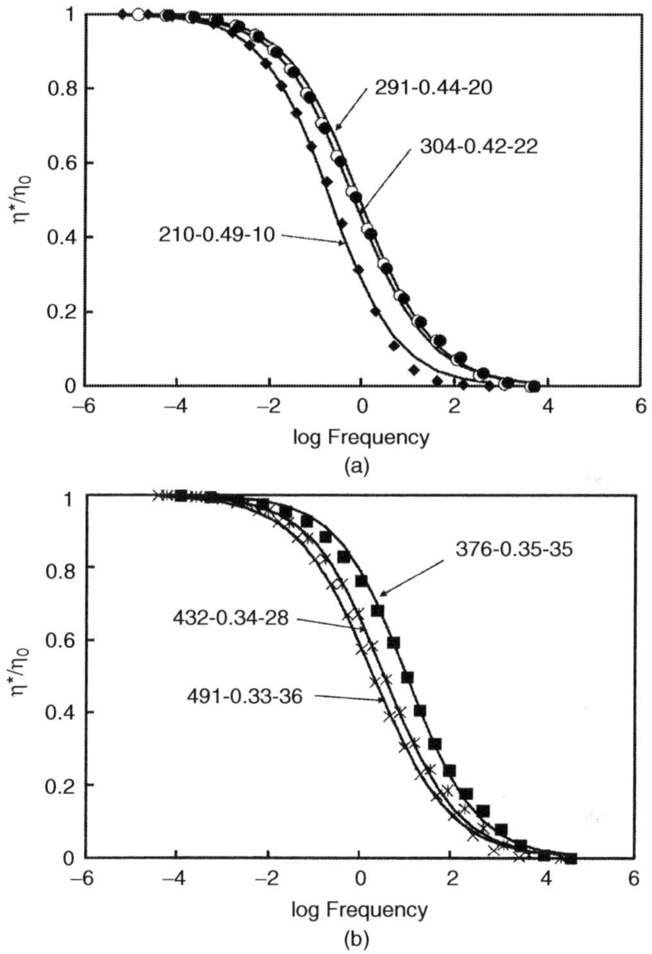

Figure 16.8 Fits of viscosity data for polyethylenes, using the curve similarity method. The symbols in the plot are heights from a cumulative distribution of log $g'M$ (see Figure 16.7), the lines are Cross viscosity equation fits to the viscosity ratio data. (a) Branched polyethylenes; (b) highly branched polyethylenes. Viscosity data for each set of polyethylenes (branched or highly branched) were fit using individual sets of δ values in Equation 16.23. Numbers in figures are sample identifiers. (Reprinted with permission from Ref. 12.)

A set of 10 polyethylenes was analyzed, two HDPEs and eight LDPEs. The latter were divided into branched and highly branched samples. Individual sets of δ values were able to fit each class of samples: linear, branched, and highly branched. This is shown in Figure 16.8 for branched and highly branched samples, where the symbols are the heights from the cumulative distribution of log $g'M$, shown in Figure 16.7, and the lines are the viscosity ratio data as fitted by the Cross model, Equation 16.24.

Several other methods for predicting the linear viscoelastic behavior or for determining the level of LCB of polydisperse, long-chain branched homopolymers have been proposed. All of these methods rely, to a greater or lesser extent, on information derived from SEC measurements. This information might simply be M_w, or the parameters needed might extend to the MMD, the average M between long-chain branches (see Section 11.2.3), the average M of a long-chain branch (see Reference 13 for a proposed method to obtain this value using SEC with M-sensitive detectors), and so on.

For example, for randomly branched polymers it has been proposed that [14]

$$\eta_0 = \begin{cases} AM_w & \text{for } M_b < M_c \\ AM_w\left[1 + \left(\dfrac{M_w}{Mc}\right)^{2.4}\right] & \text{for } M_c < M_w < M_b \\ AM_b\left[1 + \left(\dfrac{M_b}{M_c}\right)^{2.4}\right]\left(\dfrac{M_w}{M_b}\right)^{s/\gamma} & \text{for } M_c < M_b < M_w \end{cases} \tag{16.25}$$

where M_b is the average M between branch points or between the last branch point and the chain end, and M_c is the critical M for entanglement of random branches. The prefactor A has units of viscosity and is polymer- and temperature-specific. The exponent s/γ is obtained through empirical adjustments to a theory for the viscoelasticity of randomly branched polymers in the vulcanization class. This exponent is meant to account for the η_0 of branched polymers being either greater or less than the η_0 of their linear counterparts of the same M_w.

Another proposed method for combining rheology and SEC information to determine polymer LCB relies on converting complex viscosity data into a *viscosity MMD* [15]. The latter is then compared to the MMD obtained from SEC experiments, and the level of LCB is correlated to the difference in the location of the peaks of the viscosity and SEC MMDs.

Combinatorial dilution rheology methods have also been proposed for predicting the linear viscoelasticity of polydisperse, long-chain branched homopolymers [16]. In dilution rheology, first a branched polymer is sequentially diluted with increasing amounts of a linear polymer of the same chemistry (*diluent polymer*). Alternatively, a polymer of unknown branching status can be diluted sequentially with a polymer of the same chemistry and with well-defined branching (e.g., with a well-characterized star, H-, or pom-pom polymer of the same chemistry). Then the relationships between either the storage or loss modulus (G' and G'', respectively) and the frequency ω are determined for the various blends of branched and linear polymers. The rheological data are combined with the values of M_w and M_b from SEC to obtain plots of η_0 versus volume fraction of the polymer of interest. Because blended components can be chosen arbitrarily, arbitrarily large numbers of data sets can be obtained combinatorially and used to infer the type and level of branching present in a polymer. As can be seen from Figure 16.9, different curves for η_0 versus volume fraction of star or comb polymer are obtained when linear polybutadiene (PBd) is used as diluent for

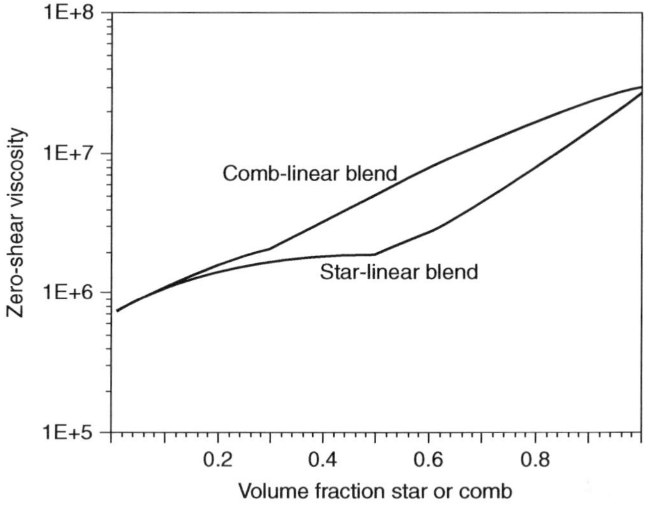

Figure 16.9 Dilution rheology of branched polymer melts: predicted zero-shear viscosities, η_0, of blends of star/linear and comb/linear polybutadienes, as a function of volume fraction of star or comb polymer in blend. Star is a four-arm star, comb is a six-arm comb. Linear, star, and comb polymers each has a total $M = 200,000$ g/mol. (Reprinted with permission from Ref. 16.)

either a four-arm star PBd or a six-arm comb PBd, where the linear, star, and comb polymers each have the same total molar mass.

16.3.3 Behavior of Dilute Oligomer Solutions

In Section 13.3.2 it was noted that in select cases, dilute solutions of oligomers may have viscosities which are lower than the viscosities of the neat solvent itself [17], as witnessed by SEC/VISC measurements (see Figure 13.3). In these cases, the specific viscosity of the solution, η_{sp}, and the intrinsic viscosity of the oligomers in solution, $[\eta]$, are both negative (i.e., $\eta_{sp} < 0$ and $[\eta] < 0$) [17]. This can lead to a failure of universal and Mark–Houwink calibrations (Section 8.2) at the oligomeric level and to negative values for the viscometric radii R_η of oligomers (Table 9.2) [18].

In Figure 13.3, the specific viscosity of a solution of styrene monomer dissolved in *N,N*-dimethyl acetamide (DMAc) with 0.5% LiCl is seen to be lower than the viscosity of the DMAc/LiCl solvent itself. Steady shear rheometry experiments of DMAc/LiCl with varying volume fractions of styrene monomer showed that upon addition of styrene to the solvent, the viscosity of the solution becomes lower than that of the solvent and becomes increasingly negative as more styrene is added. These results are shown in Figure 16.10.

The seemingly paradoxical result of having a solution with a viscosity lower than that of the neat solvent was explained successfully, in quantitative fashion, using a predictive rule for one-phase binary mixtures:

$$\eta = \eta_A\phi_A + \eta_B\phi_B + P\phi_A\phi_B \qquad (16.26)$$

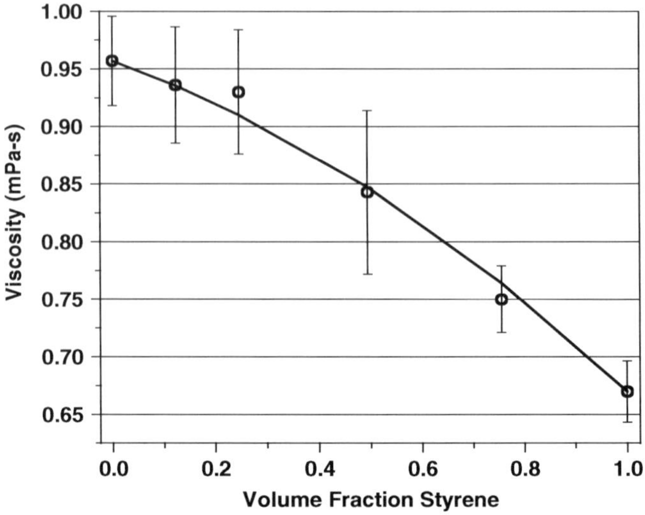

Figure 16.10 Negative viscosity of dilute styrene solutions. Solvent, DMAc/0.5% LiCl. Viscosities obtained using a steady shear rheometer with Couette geometry at 35°C. Solid line represents prediction from Equation 16.26. (Adapted from Ref. 17.)

where η is the viscosity of the solution, η_A and η_B are the viscosities of the oligomer and solvent, respectively, and ϕ_A and ϕ_B are the volume fractions of oligomer and solvent, respectively, in the mixture. The P term is an interaction term that can be either positive or negative, depending on the system. The last term on the right-hand side of Equation 16.26 represents the concentration-dependent interaction between two different chemical species. The solid line in Figure 16.10 represents the least-squares fit of the rheological data for styrene monomer dissolved in DMAc/LiCl, modeled using Equation 16.26. The seemingly abnormal behavior of the monomer and oligomer solutions examined was thus ascribed to fundamental hydrodynamic properties of solutions, without the need to invoke more esoteric effects or theories.

REFERENCES

1. C. W. Macosko, *Rheology: Principles, Measurements, and Applications*, Wiley-VCH, New York, 1994.
2. P. M. Cotts, in *Multiple Detection in Size-Exclusion Chromatography*, ACS Symp. Ser. 893, A. M. Striegel, ed., American Chemical Society, Washington, DC, 2005, Chap. 3.
3. W. Thimm, C. Friedrich, and M. Marth, *J. Rheol.*, **44**, 429 (2000).
4. *AR 2000 Rheometer Operator's Manual*, TA Instruments, New Castle, DE, 2005, Appendix on "Calculation of Polymer Molecular Weight Distribution."
5. J. M. Dealy, *J. Rheol.*, **45**, 603 (2001).
6. (a) W. W. Yau, *Polymer*, **48**, 2362 (2007); (b) B. Monrabal, J. Sancho-Tello, N. Mayo,

and L. Romero, *Macromol. Symp.* **257**, 71 (2007); (c) W. W. Yau, *Macromol. Symp.* **257**, 29 (2007).

7. W. W. Yau et al., *Proceedings of the 4th International Symposium on Engineering Plastics (EP 2009)*, August 23–26, 2009, Dalian, Liaoning, China.
8. J. M. Dealy and R. G. Larson, *Structure and Rheology of Molten Polymers*, Hanser Gardner, Cincinnati, OH, 2006, p. 136.
9. W. W. Yau, J. Wang, R. Cong, D. Gillespie, and J. Huang, *Proceedings of SPE ANTEC at NPE 2009*, June 22–24, Chicago, IL, USA.
10. G. R. Zeichner and P. D. Patel, *Proceedings of the 2nd World Congress of Chemical Engineering*, Montreal, Quebec, Canada, 1981.
11. H. J. Yoo, *SPE ANTEC Tech. Pap.*, **39**, 3037 (1993).
12. S. T. Balke, T. H. Mourey, and C. P. Lusignan, *Int. J. Polym. Anal. Charact.*, **11**, 21 (2006).
13. A. M. Striegel, *Polym. Int.*, **53**, 1806 (2004).
14. J. Janzen and R. H. Colby, *J. Mol. Struct.*, **485–486**, 569 (1999).
15. P. M. Wood-Adams and J. M. Dealy, *Macromolecules*, **33**, 7481 (2000).
16. R. G. Larson, *Macromolecules*, **34**, 4556 (2001).
17. A. M. Striegel and D. B. Alward, *J. Liq. Chromatogr. Rel. Technol.*, **25**, 2003 (2002). See the Erratum in *J. Liq. Chromatogr. Rel. Technol.*, **26**, 157 (2003), in which there is a typo: The value of $[\eta]_w$ for PE 282 in TCB at 135°C should be +0.0036 dL/g.
18. M. J. Smith, I. A. Haidar, and A. M. Striegel, *Analyst*, **132**, 455 (2007).

SYMBOLS

Minor symbols (e.g., those used only once or used as lesser constants) are not shown.

Symbol	Definition	SI or cgs Units (Common Units)
A	Eddy-diffusion contribution to plate height	cm
A	Peak area	—
A_2	Second virial coefficient of polymer solution	mol · mL/g^2
a	Exponent constant for the Mark–Houwink relation, $[\eta] = KM^a$	—
$\bar{a}$	Effective pore radius, $\bar{a} = 2 \times$ pore volume/ pore surface	nm (Å)
a_c	Inside radius of a cylindrical pore	nm (Å)
B	Coefficient for longitudinal molecular diffusion contribution to plate height	cm^2/s
B_n, B_w	Number- and weight-average number of branch points in a polymer	—
B/A	Peak asymmetry factor (use A/B for fronting peaks)	—
C	Coefficient for mobile-phase mass transfer, lateral diffusion contribution to plate height	s^{-1}
C_M	Interparticle C-term coefficient (Sec. 3.2)	s^{-1}
C_n	Characteristic ratio (Eq. 11.29)	—
C_S	LC stationary-phase C-term coefficient (Sec. 3.2)	s^{-1}
C_{SM}	Stagnant mobile phase, "SEC stationary phase," C-term coefficient	s^{-1}
C_1, C_2	Intercept and slope of linear SEC calibration, $V_R = C_1 - C_2 \log \text{MW}$	cm^3 (mL)
C_1', C_2'	Effective linear calibration constants (Sec. 9.3)	cm^3 (mL)
C_∞	Asymptotic value of the characteristic ratio C_n (Eq. 11.32)	—

Modern Size-Exclusion Liquid Chromatography: Practice of Gel Permeation and Gel Filtration Chromatography, Second Edition
By André M. Striegel, Wallace W. Yau, Joseph J. Kirkland, and Donald D. Bly

Symbol	Definition	SI or cgs Units (Common Units)
$C^v(R(\theta))$	Cabannes factor, at angle θ, for vertically polarized incident radiation.	—
c	Concentration of sample solution	g/cm^3 (g/mL)
c_m	Solute concentration in mobile phase (Sec. 3.3)	g/cm^3 (g/mL)
c_s	Solute concentration in stationary phase (Sec. 3.3)	g/cm^3 (g/mL)
c^*	Critical overlap concentration	g/cm^3 (g/mL)
$c^*_{A_2}$	Critical overlap concentration based on second virial coefficient A_2	g/cm^3 (g/mL)
c^*_η	Critical overlap concentration based on intrinsic viscosity $[\eta]$	g/cm^3 (g/mL)
D	Diffusion coefficient of analyte in solution	cm^2/s
De	Deborah number	—
D_E	Eddy-diffusion coefficient (Sec. 3.3)	cm^2/s
D_M	Solute-diffusion coefficient in interparticle mobile phase	cm^2/s
D_S	Solute-diffusion coefficient in LC stationary phase (Sec. 3.2)	cm^2/s
D_{SM}	Solute-diffusion coefficient in stagnant mobile phase (i.e., "SEC stationary phase")	cm^2/s
D_T	Translational diffusion coefficient of analyte in solution	cm^2/s
D_1	Intercept of true linear SEC calibration, $M = D_{1\,\exp}(-D_2V_R)$	g/mol
D_2	Slope of true linear SEC calibration, $M = D_{1\,\exp}(-D_2V_R)$	cm^{-3} (1/mL)
D'_1, D'_2	Effective linear calibration constants (Sec. 9.3)	g/mol, cm^{-3} (1/mL)
d_f	Film thickness of LC stationary phase (Sec. 3.2)	μm (cm)
d_f	Fractal dimension	—
d_p	Particle diameter	μm (cm)
d_T	Topological or Euclidean dimension	—
F	Eluent volume flow rate	cm^3/s (mL/min)
F, $F(V_R)$, $F(V)$	Experimental SEC elution curve height as a function of retention volume	—
f	Branching functionality; number of branches emanating from a common point	—
G	Dedrimer generation	—
$G(V_R - y)$	Instrument-column-dispersion function which describes the weight fraction of a solute that should have been at the retention volume y but is actually dispersed and detected at retention volume V_R (Sec. 4.3)	—
ΔG	Free energy of mixing (Sec. 7.2)	cal
ΔG°	Standard free energy difference (Sec. 2.4)	cal
g	Ratio of mean-square radii of branched and linear polymers (Eq. 11.1)	—

Symbol	Definition	SI or cgs Units (Common Units)
g'	Ratio of intrinsic viscosities of branched and linear polymer (Eq. 11.7)	—
H, HETP	Height equivalent to a theoretical plate	cm (mm)
H_F	Eddy-diffusion plate height (Sec. 3.2)	cm
H_L	Longitudinal-diffusion plate height (Sec. 3.2)	cm
H_M	Plate height due to interparticle mobile-phase effects (Sec. 3.2)	cm
$H_{\min}$	Minimum value of H in the plate height vs. velocity plot (Sec. 3.2)	cm
H_S	Plate height due to LC stationary-phase effect (Sec. 3.2)	cm
H_{SM}	Plate height due to stagnant-mobile-phase (SEC stationary-phase) effects	cm
ΔH	Enthalpy of mixing (Sec. 7.2)	cal
ΔH°	Standard enthalpy difference (Sec. 2.4)	cal
h, h_t	Chromatogram height	—
h	Ratio of the hydrodynamic radii of branched and linear polymers (Eq. 11.20)	—
$\mathbf{h}$	Reduced plate height, $\mathbf{h} = H/d_p$	—
h_p	Peak height at apex	—
I	Ionic strength	—
I_L	Linearity index; goodness of the linear fit to SEC calibration (Sec. 4.5)	—
I_R	Separation range index; MW separation range of the SEC calibration curve	—
K	A proportionality constant for the Mark–Houwink relation, $[\eta] = KM^a$	m^3/kg (dL/g)
K, K_e	Equilibrium solute distribution coefficient between two chromatographic phases (Sec. 2.4)	—
K_{LC}	LC solute distribution coefficient (Sec. 2.2)	—
K_{SEC}	Solute distribution coefficient in SEC	—
k	Boltzmann's constant (Sec. 2.4)	erg/deg molecule^{-1} (J/K)
k'	LC solute capacity factor, $k' = K_{\mathrm{LC}} V_S / V_M$	
L	Column length	cm (ft)
L_c	Contour length of polymer chain	nm (Å)
L_p	Persistence length	nm (Å)
M	Molar mass	g/mol
$M_{\mathrm{seg}}^{\mathrm{ave}}$	Average molar mass between branch points of a randomly branched polymer	g/mol
M_L	Molar mass per unit contour length	g/mol · nm (g/mol · Å)
M_n	Number-average molar mass	g/mol
$(M_n)_{\mathrm{true}}$, $(M_w)_{\mathrm{true}}$ or $M_n(t)$, $M_w(t)$	True values of number- and weight-average molar mass	g/mol

Symbol	Definition	SI or cgs Units (Common Units)
$(M_n)_{\mathrm{exp}}$, $(M_w)_{\mathrm{exp}}$ or $M_n(u)$, $M_w(u)$	Experimental values of number- and weight-average molar mass uncorrected for instrument spreading	g/mol
$M_n(V)$, $M_w(V)$	Actual M_n and M_w (as in a detector cell) as a function of retention volume (Sec. 8.4)	g/mol
$M_t(V)$	True calibration curve in Hamielec and GPCV2 methods	—
M_w/M_n	Molar mass polydispersity index	
M_p	Peak-average molar mass	g/mol
M_w	Weight-average molar mass	g/mol
M_z	z-average molar mass	g/mol
M_v or M_η	Viscosity-average molar mass	g/mol
M^*	Average of absolute M_n^* and M_w^* (Sec. 4.3)	—
M_n^*, M_w^*	Percent error in $(M_n)_{\mathrm{exp}}$ and $(M_w)_{\mathrm{exp}}$ due to instrument spreading (Sec. 4.3)	—
N	Column plate number, number of theoretical plates	—
N_A	Avogadro's number (Sec. 3.4)	molecules/mol
N_i, N_t	Plate count of ith column and column set, respectively (Sec. 7.10)	—
N_X	Number fraction of chains with degree polymerization X (Sec. 1.3)	—
n	General sequential indexing integer	—
n	Peak capacity (Sec. 4.1)	—
n	Refractive index of dilute polymer solution	—
n_{D}^{20}	Refractive index at 20°C at the sodium D line (doublet)	—
n_p	Refractive index of polymer	—
n_{opt}	Optimum number of recycles (Sec. 15.3)	—
n_0	Refractive index of neat solvent	—
P	Pressure	Pa (psi, bar)
PS	Pore size (Sec. 4.5)	nm (Å)
PSD	Pore-size distribution; standard deviation of the lognormal PSD curve (Sec. 4.5)	—
$P(\theta)$	Particle scattering factor	—
p	Extent of reaction (Sec. 8.3)	—
Q	Volumetric flow rate	m^3/s
q	Scattering vector (Eq. 9.32)	nm^{-1} ($Å^{-1}$)
R	Gas constant	cal/deg · mole (J/mol · K)
R_G	Radius of gyration (root-mean-square radius) of solute molecules, $R_G = kM^\alpha$ (Table 9.2)	nm (Å)
R_H	Hydrodynamic (Stokes') radius of solute molecules (Table 9.2)	nm (Å)
R_p	Pore diameter	μm (nm or Å)
R_s	Resolution of two peaks	—
$R_{s,\mathrm{opt}}$	Optimum resolution in recycle SEC (Sec. 15.3)	—

Symbol	Definition	SI or cgs Units (Common Units)
R_{sp}	Specific resolution, $R_{sp} = R_s/\Delta \log M$	—
R_{sp}^*	Packing resolution factor, $R_{sp}^* = R_{sp}\Big/\sqrt{L}$	$cm^{-1/2}$
R_T	Thermodynamic radius of solute molecules (Table 9.2)	nm (Å)
R_η	Viscometric radius of solute molecules (Table 9.2)	nm (Å)
$R(\theta)$	Excess Rayleigh ratio	
$r, \bar{r}$	Radius and average radius of hard-sphere solutes (Sec. 2.4)	nm (Å)
r_e	Equivalent radius of a polymer solute (Sec. 2.5)	nm (Å)
$\langle r^2 \rangle$	Mean-square radius	$(nm)^2$ $(Å^2)$
$\langle r^2 \rangle^{1/2}$	Root-mean-square radius, R_G	(nm) (Å)
ΔS	Entropy of mixing (Sec. 7.2)	cal/K
ΔS°	Standard entropy difference (Sec. 2.3)	cal/K
s	Surface area per unit pore volume (Sec. 2.4)	cm^{-1} (cm^2/mL)
T	Temperature	K (°C)
T_c	Consolute temperature (Sec. 7.2)	K (°C)
T_g	Glass transition temperature	K (°C)
t	Time	s (min)
t_R	Retention time	s (min)
t_0	Retention time of unretained peak (Sec. 2.2)	s (min)
V	Volume	cm^3 (mL)
Var	Variance, Var $= \alpha_x^2$ (Sec. 3.1)	cm^6 (mL^2)
$V_{A_2\eta}$	Dimensionless parameter relating the polymer draining and coil interpenetration functions	—
V_e	Total permeation volume (Sec. 13.5)	cm^3 (mL)
V_h	Hydrodynamic volume of an equivalent sphere (Sec. 9.2)	$(nm)^3$ $(Å^3)$
V_i	Total accessible liquid volume contained within the pores of the SEC packing	cm^3 (mL)
V_i	Variable retention volume used in the integral-MMD calibration method (Sec. 8.3)	cm^3 (mL)
V_{inj}	Injected sample volume (Sec. 7.4)	cm^3 (mL)
V_M	Total liquid volume, $V_M = V_0 + V_i$	cm^3 (mL)
V_R	Retention volume	cm^3 (mL)
$\langle V_R \rangle$	Average retention volume (Sec. 8.4)	cm^3 (mL)
V_s	Equivalent liquid volume of a LC stationary phase (Sec. 2.1)	cm^3 (mL)
V_0	Volume of mobile phase in the interstices between the SEC packing particles	cm^3 (mL)
v	Mobile-phase velocity	cm/s
$\mathbf{v}$	Reduced velocity, vd_p/D_M (Sec. 3.2)	—
$\bar{v}$	Superficial solvent velocity	cm/s
v_{opt}	Optimum velocity at $H = H_{min}$ (Sec. 3.2)	cm/s
W, $W(V_R)$	True SEC elution curve height at ideal infinite resolution as a function of V_R	—

Symbol	Definition	SI or cgs Units (Common Units)
$W^{1/2}$	Peak width measured parallel to baseline at one-half of the peak height	cm^3 (mL)
W_b	Peak width at the base, the distance between the baseline intercepts of lines drawn tangent to the points of inflection of the elution peak trace	cm^3 (mL)
W_X	Weight fraction of chains with degree of polymerization X (Sec. 1.3)	—
X	Degree of polymerization, number of repeating monomer units in a polymer chain (Sec. 1.3)	—
X_i	$\Delta W/\Delta \log M$, differential weight fraction (Sec. 10.3)	—
X_n, X_w	Number- and weight-average degree of polymerization	—
α	Separation factor, $\alpha = k_2'/k_1'$ (Sec. 4.1)	—
α	Exponent constant in the molar mass dependence of R_G, $R_G = kM^\alpha$ (Sec. 2.4)	—
α	Expansion factor of polymer solute R_G (Sec. 7.2)	—
γ	Peak skew (Sec. 3.5)	—
γ_{SM}	Peak skew due to stationary mass transfer effect (Sec. 3.4)	—
δ	Solubility parameter (Sec. 7.2)	cal/cm^3 (J/cm^3)
δ_d	Solubility parameter due to dispersion force (Sec. 7.2)	cal/cm^3 (J/cm^3)
δ_h	Solubility parameter due to hydrogen bonding (Sec. 7.2)	cal/cm^3 (J/cm^3)
δ_p	Solubility parameter due to polar force (Sec. 7.2)	cal/cm^3 (J/cm^3)
δ_s, δ_m	Solubility parameters of solvent and macromolecules, respectively (Sec. 7.2)	cal/cm^3 (J/cm^3)
δ_v	$\sqrt{\delta_d^2 + \delta_p^2}$ (Sec. 7.2)	cal/cm^3 (J/cm^3)
δ_0	Total solubility parameter, $\delta_0 = \sqrt{\delta_d^2 + \delta_h^2 + \delta_p^2}$	cal/cm^3 (J/cm^3)
$\partial n/\partial c$	Specific refractive index increment (Sec. 7.4)	cm^3/g (mL/g)
ε	Polymer solution parameter; $\varepsilon = (2a - 1)/3$ (Eq. 9.52), where a is the exponent in the Mark–Houwink equation	—
ε	Molar absorptivity	mol/L · cm
ε	Viscosity shielding ratio (Eq. 11.8)	—
η	Viscosity	dyne · s/cm^2 (cP)
$[\eta]$	Intrinsic viscosity $[\eta] = \lim_{c\to 0} \frac{\eta_{sp}}{c} = \lim_{c\to 0} \ln \frac{\eta_{rel}}{c}$	m^3/kg (dL/g or mL/g)
η_k	Kinematic viscosity	St (cm^2/s)
η_{rel}	Relative viscosity $= \eta/\eta_0$ (Sec. 9.5)	—
η_{sp}	Specific viscosity $= \eta_{rel} - 1$ (Sec. 9.5)	—
η_0, η	Solvent and solution viscosity	dyne · s/cm^2 (cP)

Symbol	Definition	SI or cgs Units (Common Units)
Θ	Flory theta temperature (Sec. 7.2)	K (°C)
λ	Solute diameter/pore diameter *or* (in HDC) solute radius/flow channel radius	—
λ	Branching index; number of branching points per unit M (Eq. 11.6)	—
λ	Wavelength of radiation (light) in medium (Eq. 9.3)	nm (Å)
λ_0	Wavelength of radiation (light) in vacuum	nm (Å)
μ_1	First moment (peak retention) (Sec. 3.3)	cm^3 (mL)
μ_2	Second moment (peak variance) (Sec. 3.3)	cm^6 (mL^2)
μ_3	Third moment (peak skew) (Sec. 3.3)	cm^9 (mL^3)
ρ	Density	kg/m^3 (g/cm^3)
ρ	Ratio of R_G to R_H (Eq. 11.14)	—
ρ_θ^v	Depolarization ratio, at angle θ, for vertically polarized incident radiation	—
σ	Standard deviation of a Gaussian instrument spreading function	cm^3 (mL)
σ_{disp}	σ value due to column dispersion (Sec. 15.3)	cm^3 (mL)
σ_{inj}	σ value due to sample injection (Secs. 7.4, 15.3)	cm^3 (mL)
σ_{MMD}	σ value due to sample MMD (Sec. 15.3)	cm^3 (mL)
σ_t^2	Peak variance of a column set (Sec. 8.6)	cm^6 (mL^2)
σ_x	Standard deviation of SEC elution peaks of any shape	cm^3 (mL)
τ	Peak skew parameter (decay constant of the exponential modifier for a skewed σ-τ peak model)	cm^3 (mL)
τ	Tortuosity factor (Sec. 3.4)	—
Φ	Volume fraction of extraparticle solvent volume of the total liquid volume in the column (3.3)	—
Φ	Flory's universal constant, Φ_0, after undergoing correction for non-theta solvent/temperature conditions (Eq. 9.51)	mol^{-1}
Φ	Polymer draining function (Eq. 11.15)	mol^{-1}
Φ_0	Flory's universal constant ($= 2.87 \times 10^{23}$)	mol^{-1}
$\Phi(r)$	Volume of pores with radius r (Sec. 2.5)	cm^2 (mL/cm)
Ψ^*	Coil interpenetration function (Eq. 11.16)	—

ABBREVIATIONS

ACS	American Chemical Society
ASTM	American Society for Testing and Materials
A-term	Eddy-diffusion contribution to plate height
AU; AUFS	Absorbance units; absorbance unit full scale
A_2	Second virial coefficient of polymer solution
BBC	Band-broadening correction
BET	Brunauer–Emmett–Teller gas adsorption equation
BHT	Butylated hydroxytoluene or 2,6-di-*tert*-butyl-4-methoxyphenol
B-term	Longitudinal diffusion contribution to plate height
CCD	Chemical composition distribution
CD	Cyclodextrin
CPG	Controlled-pore glass
CTC	Cellulose tricarbanilate
C-term	Mass transfer (lateral diffusion) contribution to plate height
DLS	Dynamic light scattering [or quasielastic light scattering (QELS)]
DMAc	*N,N*-Dimethylacetamide
D-MALS	Depolarized multiangle light scattering
DMF	*N,N*-Dimethyl formamide
DMSO	Dimethyl sulfoxide
DNA	Deoxyribonucleic acid
DP	Degree of polymerization
DP	Viscometer differential pressure
DRI	Differential refractive index
DSTD	Dynamic surface tension detection
EDTA	Ethylenediaminetetraacetic acid
ELSD; EMD	Evaporative light-scattering detector; evaporative mass detector

Modern Size-Exclusion Liquid Chromatography: Practice of Gel Permeation and Gel Filtration Chromatography, Second Edition
By André M. Striegel, Wallace W. Yau, Joseph J. Kirkland, and Donald D. Bly

EPDM	Ethylene–propylene–diene monomer
ESI	Electrospray ionization
EVA	Poly(ethylene-*co*-vinyl acetate)
FFF	Field-flow fractionation
FT	Fourier transform
GA	Galacturonic acid
GAG	Glucoseaminoglycan
GC	Gas chromatography
GFC	Gel filtration chromatography
GPC	Gel permeation chromatography
GPCV2	GPC linear calibration method, version 2
GPCV3	GPC linear calibration method, version 3
GPEC	Gradient polymer elution chromatography
HDC	Hydrodynamic chromatography
HDPE	High-density polyethylene
HETP	Height equivalent to theoretical plate
HFIP	1,1,1,3,3,3-Hexafluoroisopropanol
ICP	Inductively coupled plasma
i.d.	Inside diameter
IP	Viscometer inlet pressure
IR	Infrared
ISEC	Inverse size-exclusion chromatography
IV	Intrinsic viscosity, $[\eta]$
LALS	Low-angle light scattering
LC	Liquid chromatography
LCB	Long-chain branching
LCBD	Long-chain branching distribution
LCCC	Liquid chromatography at the critical condition
LDPE	Low-density polyethylene
LEC	Liquid exclusion chromatography (SEC)
LED	Light-emitting diode
LLC	Liquid–liquid chromatography
LS	Light scattering
M	Molar mass
MALDI	Matrix-assisted laser desorption/ionization
MALS	Multiangle light scattering
M_H	Hamielec linear calibration plot
MI	Melt index
MMA	Methyl methacrylate
MMD	Molar mass distribution
M_p	Peak-average molar mass
MS	Mass spectrometry
M_t	True linear calibration plot
m/z	Mass-to-charge ratio
NaPSS	Sodium poly(styrene sulfonate) (PSS)

NATFAT	Sodium trifluoroacetate
NIST	National Institute of Standards and Technology
NMR	Nuclear magnetic resonance
O.D.	Outside diameter
ODCB	*o*-Dichlorobenzene
PAA	Poly(acrylic acid)
PAN	Polyacrylonitrile
PAN-S	Polyacrylonitrile with sulfonate groups
PBLG	Poly(γ-benzyl-L-glutamate)
p(DMAEA)-ppz	Poly(2-dimethylamino ethylamino)phosphazene
PDMS	Poly(dimethyl siloxane)
PE	Polyethylene
PEF	Poly(ethenyl formamide)
PEG	Poly(ethylene glycol)
PEMA	Poly(ethyl methacrylate)
PEO	Poly(ethylene oxide)
PET	Poly(ethylene terephthalate)
PGA	Poly(L-glutamic acid)
PHIC	Poly(*n*-hexyl isocyanate)
PMMA	Poly(methyl methacrylate)
PMN	Premanufacture notification
PP	Polypropylene
PS	Polystyrene
PS	Pore size
PSBr	Brominated polystyrene
PSD	Pore-size distribution
PS-DVB	Polystyrene–divinylbenzene
psi	Pounds per square inch
PSM	Porous silica microsphere
PSS	Sodium poly(styrene sulfonate) (NaPSS)
PTFE	Polytetrafluoroethylene (Teflon)
PVA or PVAc	Poly(vinyl acetate)
PVB	Poly(vinyl butyral) or poly(vinyl butyral-*co*-vinyl alcohol-*co*-vinyl acetate)
PVC	Poly(vinyl chloride)
PVCz	Poly(*N*-vinylcarbazole)
PVOH	Poly(vinyl alcohol)
PVP	Poly(vinylpyrrolidone)
QA/QC	Quality assurance and quality control
QELS	Quasielastic light scattering [or dynamic light scattering (DLS)]
RAFT	Reversible addition fragmentation chain transfer
RALS	Right-angle light scattering
RGD	Rayleigh–Gans–Debye
RI	Refractive index

RIG	Refractive index gradient
RIU	Refractive index unit
RNA	Ribonucleic acid
SBF	Separation by flow
SC	Slalom chromatography
SCB	Short-chain branching
SCBD	Short-chain branching distribution
S/DVB	Styrene/divinylbenzene
SEC	Size-exclusion chromatography
SEC^3	Combination of RALS, VISC, and DRI (Section 9.6)
SEEC	Size-exclusion electrochromatography
SFC	Supercritical fluid chromatography
SLS	Static light scattering
S/N	Signal-to-noise ratio
TCB	1,2,4-Trichlorobenzene
TFFF	Thermal field-flow fractionation
THF	Tetrahydrofuran
TLC	Thin-layer chromatography
TOF	Time-of-flight mass analyzer
UC	Universal calibration
UV	Ultraviolet
VC	Vinyl chloride
VISC	Viscometry
VPO	Vapor-pressure osmometry
2D	Two-dimensional
3D	Three-dimensional
μ-MMS	Microscale molecular mass sensor

INDEX

Modern Size-Exclusion Liquid Chromatography: Practice of Gel Permeation and Gel Filtration Chromatography, Second Edition
By André M. Striegel, Wallace W. Yau, Joseph J. Kirkland, and Donald D. Bly

Printed in Poland
by Amazon Fulfillment
Poland Sp. z o.o., Wrocław
05 December 2020

0ec28162-b577-4983-b7bc-17b030d889d7R01